UNSUPERVISED ADAPTIVE FILTERING

Volume I: Blind Source Separation

ADAPTIVE AND LEARNING SYSTEMS FOR SIGNAL PROCESSING, COMMUNICATIONS, AND CONTROL

Editor: Simon Haykin

UNSUPERVISED ADAPTIVE FILTERING

Volume I: Blind Source Separation

EDITED BY

Simon Haykin
McMaster University

A WILEY-INTERSCIENCE PUBLICATION

JOHN WILEY & SONS, INC.

New York / Chichester / Weinheim / Brisbane / Singapore / Toronto

Copyright © 2000 by John Wiley & Sons, Inc. All rights reserved

Published simultaneously in Canada.

For ordering and customer service, call 1-800-CALL WILEY.

Library of Congress Cataloging-in-Publication Data:

Unsupervised adaptive filtering / Simon Haykin, ed.
    p.   cm.
  "A Wiley-Interscience publication."
  Includes bibliographical references and index.
  ISBN 0-471-29412-8 (alk. paper)
  1. Adaptive filters. 2. Adaptive signal processing. I. Haykin, Simon S., 1931–

TK7872.F5   U57   2000
621.3815′324—dc21   99-029980

Printed in the United States of America
10 9 8 7 6 5 4 3 2 1

CONTENTS

5 Blind Source Separation: Models, Concepts, Algorithms and Performance   191

Pierre Comon and Pascal Chevalier

6 Information Theory, Independent-Component Analysis, and Applications   237

Anthony J. Bell

7 Information-Theoretic Learning   265

Jose C. Principe, Dongxin Xu, and John W. Fisher III

8 Blind Separation of Delayed and Convolved Sources   **321**

Kari Torkkola

9 Blind Deconvolution of Multipath Mixtures   **377**

Russell H. Lambert and Chrysostomos L. Nikias

CONTRIBUTORS

Shun-ichi Amari
Laboratory of Information Synthesis
RIKEN Brain Science Institute
Wako-shi, Hirosawa 2-1, Saitama, Japan
e-mail: amari@brain.riken.go.jp

Anthony J. Bell
523 Utah Street
San Francisco, California
e-mail: tony@salk.edu

Jean-François Cardoso
Ecole Nationale
Supérieure de Télécommunications
Department Signal
46 rue Barrault
Paris, France
e-mail: cardoso@si.enst.fr.

Pascal Chevalier
Thomson-CSF Communications
Gennevilliers, France
e-mail: pascal.chevalier@tcc.thomson.fr

Andrzej Cichocki
Laboratory for Open Information Systems
RIKEN Brain Science Institute
Wako-shi, Hirosawa 2-1, Saitama, Japan
e-mail: cia@brain.riken.go.jp

Pierre Comon
I3S Laboratory, Algorithmes/Euclide B
Sophia-Antipolis, Biot, France
e-mail: comon@i3s.unice.fr.

Scott C. Douglas
Department of Electrical Engineering
School of Engineering and Applied Science
Southern Methodist University
Dallas, Texas, USA
e-mail; douglas@seas.smu.edu

John W. Fisher III
Artificial Intelligence Laboratory
Massachusetts Institute of Technology
Cambridge, Massachusetts, USA
e-mail: fisher@ai.mit.edu

Simon Haykin
Electrical and Computer Engineering
Communications Research Laboratory
McMaster University
Hamilton, Ontario, Canada
e-mail: haykin@mcmaster.ca

Russell H. Lambert
Pivotal Technologies
Pasadena, California, USA
e-mail: russ@pivotaltech.com

Chrysostomos L. Nikias
Department of Electrical Engineering Systems
Signal and Image Processing Institute
University of Southern California
Los Angeles, California, USA

Jose C. Principe
Computational NeuroEngineering Laboratory
University of Florida
Gainsville, Florida, USA
e-mail: principe@cnel.ufl.edu

Kari Torkkola
Human Interface Laboratory
Motorola Corporation
Tempe, Arizona, USA
e-mail: a540aa@email.mot.com

Dongxin Xu
BBN Technologies
Cambridge, Massachusetts, USA
e-mail: dxu@bbn.com

Howard H. Yang
Department of Electrical and Computer Engineering
Oregon Graduate Institute of Science and Technology
Beaverton, Oregon, USA
e-mail: hyang@cse.ogi.edu

PREFACE

In 1994 I edited a book on "blind deconvolution," which presented an account of the various algorithms that had been developed essentially for solving the blind channel-equalization problem. The material presented in that book spanned a period of over 25 years, going back to the pioneering work of Robert Lucky in 1966 on the decision-directed mode of operating the least-mean-square algorithm and that of Y. Sato in 1975 on a blind channel-equalization algorithm that bears his name. These two pioneering contributions were followed by another pioneering contribution to blind channel equalization, namely, the constant-modulus algorithm that was developed independently by Godard in 1980 and Treichler and Agee in 1983. Subsequently, it was recognized that these three blind equalization algorithms are members of the family of Bussgang algorithms

In 1994 Pierre Comon published a paper in a signal-processing journal on "independent component analysis," which was followed by Tony Bell and Terry Sejnowski's 1995 paper in a neural computation journal on the Infomax (or, more precisely, the maximum-entropy) algorithm for blind signal separation. Although, indeed, work on the blind signal-separation problem could be traced to a much earlier paper by J. Herault, C. Jutten, and B. Ans that was published in 1985, it would be fair to say that Pierre Comon's paper and that of Tony Bell and Terry Sejnowski served as catalysts for raising the profile of research interests in blind source separation to the extent that the subject has become a "hot" area with potential applications in a variety of diverse fields.

Despite the fact that blind channel equalization and blind source separation have originated in their own somewhat independent ways, they are in actual fact intimately related to each other. Indeed, they constitute the two pillars of unsupervised adaptive filtering. By bringing them together under the umbrella of this new book, organized in two volumes, not only have we provided an up-to-date treatment of blind signal-separation and blind channel-equalization algorithms and their underlying theoretical formalisms but also opened an avenue for the cross-fertilization of new ideas. Volume I of the book covers blind source-separation algorithms, and Volume II covers blind deconvolution (i.e., blind equalization) and its relationship to blind source separation.

I would like to take this opportunity to express my deep gratitude to each and every one of my coauthors for making the writing of this unique two-volume work a reality.

SIMON HAYKIN

Ancaster, Ontario, Canada
March 2000

UNSUPERVISED ADAPTIVE FILTERING

Volume I: Blind Source Separation

1

INTRODUCTION

Simon Haykin

1.1 WHY ADAPTIVE FILTERING?

In the signal-processing, communications, and control literature, the term *filter* is commonly used to refer to a device or algorithm that is applied to a set of noisy data in order to extract a prescribed quantity of interest. In much of the work done in the past, optimization of the filter has been based on an objective function or index of performance using second-order statistics. When the filter is linear (i.e., it satisfies the principle of superposition) and all the pertinent statistics are known, the solution is defined by the *Wiener filter* (Haykin 1996). When, however, the filter is required to operate in an environment of unknown statistics or a nonstationary environment, an adaptive filter provides an elegant solution to this more difficult problem. The filter starts from an arbitrary initial condition, knowing nothing about the environment, and proceeds in a step-by-step manner toward an optimum solution.

An *adaptive filter* is formally defined as *a self-designing system* that relies on a recursive algorithm for adjusting its free parameters to operate satisfactorily in an unknown environment.

There are different ways of classifying adaptive filters, depending on the feature of interest. With input–output mapping as the feature to focus on, we can classify adaptive filters into two main groups: linear and nonlinear. *Linear adaptive filters* compute an estimate of a desired response by using a linear combination of the available set of observables applied to the input. This form of input–output mapping is satisfied by having a single layer of computational (processing) units or simply a single computational unit as the output layer. On the other hand, when the input–output mapping is required to be nonlinear, we

Unsupervised Adaptive Filtering, Volume I, Edited by Simon Haykin.
ISBN 0-471-29412-8 © 2000 John Wiley & Sons, Inc.

naturally need to use a *nonlinear adaptive filter*. Typically, nonlinear adaptive filters involve the use of one or more layers of hidden computational units in addition to the output layer (Haykin 1999). As such, nonlinear adaptive filters are capable of tackling more difficult signal-processing tasks than linear adaptive filters.

1.2 SUPERVISED AND UNSUPERVISED FORMS OF ADAPTIVE FILTERING

Adaptive filters may also be classified in another way, depending on whether a desired response is available or not. Specifically, we have the following two classes:

- *Supervised Adaptive Filters.* The algorithms used to design this class of filters assume the availability of a training sequence that specifies the desired response for a certain input signal. In a popular approach to the design of supervised adaptive filters, the desired response is compared against the actual response of the filter, and the resulting error signal is used to adjust the free parameters of the filter. The process of parameter adjustments is continued in a step-by-step manner until a steady-state condition is established. In effect, the information contained in the training sequence about the environment is stored as the design values of the filter's free parameters. Thereafter, the filter is ready for testing data not seen before. The *least-mean-square (LMS) algorithm* invented by Widrow and Hoff (1960) some 40 years ago is an example of supervised adaptive filtering that is simple and yet capable of achieving high performance in a robust manner.
- *Unsupervised Adaptive Filtering.* In this second class of adaptive filters, adjustments to the free parameters of the filter are performed without the need for a desired response. For the filter to perform its function, however, its design includes a set of rules that enable the filter to compute an input-output mapping with specific desirable properties. In the signal-processing literature, unsupervised adaptive filtering is often referred to as *blind adaptation*.

In this two-volume book we study the many facets of unsupervised adaptive filters applied to two important signal-processing tasks: as discussed next, Volume I covers blind source separation, and Volume II covers blind deconvolution.

1.3 TWO IMPORTANT UNSUPERVISED SIGNAL-PROCESSING TASKS

1.3.1 Blind Source (Signal) Separation

A filtering problem humans are familiar with is the *cocktail party phenomenon*. We have a remarkable ability to focus on a speaker in the noisy environment of

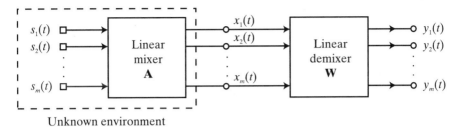

Unknown environment

Figure 1.1 Block diagram illustrating the background to the blind signal separation problem in its most basic form.

a cocktail party, despite the fact that the speech signal originating from that speaker is buried in an undifferentiated noise background due to other interfering conversations in the room. In the context of unsupervised adaptive filtering a similar filtering problem arises under the umbrella of *blind signal (source) separation*, though it must be said at the outset that the present status of this subject is rather primitive in sophistication compared to the cocktail party phenomenon. To formulate the blind signal-separation problem in its basic form, consider a set of unknown source signals $s_1(t), s_2(t), \ldots, s_m(t)$ that are mutually independent of each other. These signals are linearly mixed in an unknown environment to produce the m-by-1 observation vector (see Fig. 1.1)

$$\mathbf{x}(t) = \mathbf{A}\mathbf{s}(t) \tag{1.1}$$

where

$$\mathbf{s}(t) = [s_1(t), s_2(t), \ldots, s_m(t)]^T$$

$$\mathbf{x}(t) = [x_1(t), x_2(t), \ldots, x_m(t)]^T$$

and $\mathbf{A}$ is an unknown nonsingular *mixing matrix* of dimensions m-by-m. That is, the number of sensors where $\mathbf{x}(t)$ is observed is equal to the number of sources that produce $\mathbf{s}(t)$. Given the observation vector $\mathbf{x}(t)$, the requirement is to recover the original source signals $s_1(t), s_2(t), \ldots, s_m(t)$ in an unsupervised manner. The solution to this problem is feasible, except for an arbitrary *scaling* of each source signal and possible *permutation* of indices under certain but fairly general conditions. It other words, provided the original source signals are independent and the mixing matrix $\mathbf{A}$ is nonsingular, it is possible to find a *demixing matrix* $\mathbf{W}$ defined ideally as follows:

$$\mathbf{y} = \mathbf{W}\mathbf{x} = \mathbf{W}\mathbf{A}\mathbf{x} \rightarrow \mathbf{DPs} \tag{1.2}$$

where $\mathbf{y}$ is the output signal vector produced by the demixer, $\mathbf{D}$ is a nonsingular diagonal matrix, and $\mathbf{P}$ is a permutation matrix.

The blind signal-separation problem may be traced to a paper by Herault et al., (1985) published in French. The underlying principle involved in the solution to this problem is nowadays called *independent-components analysis* (ICA) (Comon 1994), which may be viewed as an extension of the widely known principal-components analysis (PCA). Whereas PCA imposes independence in a statistical sense only to the second order while constraining the direction of vectors to be orthogonal, ICA imposes statistical independence on all the individual components of the output vector **y**, but has no orthogonality constraint. In practice, an algorithmic implementation of ICA can only go for "as statistically independent as possible." In any event, the terms "independent components analysis" and "blind signal (source) separation" mean essentially the same thing; as such, they are used interchangeably in the literature and in this book.

1.3.2   Blind Deconvolution

Turning next to blind deconvolution, the motivation is different. To begin with, *deconvolution* is a signal-processing operation that ideally unravels the effects of convolution performed by a linear time-invariant system operating on an input signal. More specifically, in deconvolution the output signal and the system are both known, and the requirement is to reconstruct what the input signal must have been. In *blind deconvolution*, only the output signal is known (both the system and the input signal are unknown), and the requirement is to find both the input signal and the system itself (Haykin 1994). Clearly, blind deconvolution is a more difficult signal-processing task than ordinary deconvolution.

To be specific, consider an unknown linear time-invariant system $\mathcal{L}$ with input $s(t)$ assumed to consist of independent and identically distributed symbols. The only thing known about the input is its probability distribution. The requirement is to restore $x(t)$, or equivalently to identify the inverse $\mathcal{L}^{-1}$ of the system $\mathcal{L}$, given the observed signal $x(t)$ at the output of system $\mathcal{L}$, as illustrated in Fig. 1.2.

If the system $\mathcal{L}$, assumed to be a discrete-time system, is *minimum phase* (i.e., the transfer function of the system has all of its poles and zeros confined

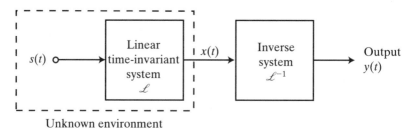

Unknown environment

Figure 1.2 Block diagram illustrating the background to the blind deconvolution problem.

to the interior of the unit circle in the z-plane), then not only is the system $\mathscr{L}$ stable but so is the inverse system $\mathscr{L}^{-1}$. In this case we can view the unknown input signal $s(t)$ as the "innovation" of the unknown system output $x(t)$, in which case the inverse system $\mathscr{L}^{-1}$ is just a *whitening filter*; with this observation, the blind deconvolution problem is solved.

In many practical situations, however, the system $\mathscr{L}$ may *not* be minimum phase. A discrete-time system is said to be *nonminimum phase* if its transfer function has any of its zeros located outside the unit circle in the z-plane; exponential stability of the system dictates that the poles must be located inside the unit circle. Practical examples of a nonminimum-phase channel include a telephone channel and a fading (multipath) radio channel. In such a situation, channel equalization [i.e., the restoration of the source signal $s(t)$], given only the channel output, is a difficult signal-processing task.

Although the blind signal separation and blind deconvolution problems are usually formulated in their own individual ways, in reality they are types of *inverse problems* with similarities and subtle differences between them. Typically, blind deconvolution involves a single source of information and a single observable (measurement). On the other hand, blind source separation involves multiple but independent sources of information and multiple observables (sensors); preferably, the number of sensors is equal to the number of sources.

1.4 THREE FUNDAMENTAL APPROACHES TO UNSUPERVISED ADAPTIVE FILTERING

Earlier we mentioned that an unsupervised adaptive filter operates by invoking a set of rules that enable the filter to compute an input–output mapping with desirable properties of interest. In this context we can identify three different approaches for the design of unsupervised adaptive filters.

(1) *Bussgang Statistics.* Consider the blind equalization of a communication channel; such a signal-processing operation is a form of blind deconvolution as illustrated in Fig. 1.2. The unsupervised adaptive filter is now configured as shown in Fig. 1.3. The filter has three constituents, a finite-duration impulse-response (FIR) filter, a zero-memory nonlinear output unit, and an algorithm for adjusting the filter's parameters in an iterative manner. The adaptive filtering algorithm is the ubiquitous LMS algorithm described simply as

$$w_k(t+1) = w_k(t) + \mu e(t)x(t-k), \qquad k = -L, \ldots, -1, 0, 1, \ldots, L \qquad (1.3)$$

where μ is the step-size (learning-rate) parameter and $e(t)$ is the error signal defined by

$$e(t) = s(t) - y(t) \qquad (1.4)$$

$$= g(y(t)) - y(t) \qquad (1.5)$$

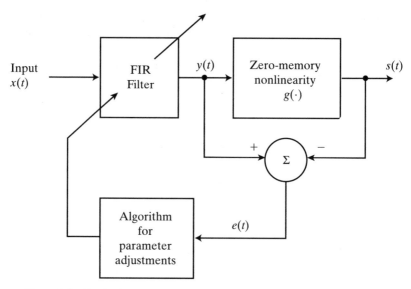

Figure 1.3 Block diagram of a blind adaptive filter of the Bussgang type.

and

$$y(t) = \sum_{k=-L}^{L} w_k(t)x(t-k) \qquad (1.6)$$

Note that although the LMS algorithm needs a desired response for its operation, the output of the nonlinearity, $s(t)$, is treated as the desired response. Provided that the filter length $(2L+1)$ is sufficiently large and the algorithm has converged, the output of the filter approximately satisfies the following condition:

$$E[y(t)y(t-k)] \simeq E[y(t)g(y(t-k))] \qquad (1.7)$$

where the function $g(\cdot)$ is the zero-memory nonlinearity. A process that satisfies this condition is called a *Bussgang process* (Bussgang 1952). In other words, a Bussgang process has the property that its autocorrelation function is equal to the cross-correlation between that process and the output of a zero-memory nonlinearity produced by that process, with both correlations being measured for the same lag. Unsupervised adaptive filters that satisfy Eq. (1.7) are referred to as *Bussgang algorithms*. The Bussgang family of unsupervised adaptive filters include the *decision-directed algorithm* (Lucky 1966), *Sato algorithm* (Sato 1975), and the *constant-modulus algorithm* (CMA) for blind equalization (Godard 1980; Treichler and Agee 1983).

(2) *Higher-Order Statistics.* The Bussgang family of algorithms uses the higher-order statistics of the observed (received) signal in an *implicit* sense. In

contrast, the second family of unsupervised adaptive-filtering algorithms uses higher-order statistics of the observed signal in an *explicit* sense. The higher-order statistics of a stationary process are described in terms of *cumulants*, and their Fourier transforms are known as *polyspectra*. Indeed, the cumulants and polyspectra may be viewed as generalizations of the autocorrelation function and power spectrum, respectively. Polyspectra provide a basis for the identification (and therefore blind equalization) of a nonminimum-phase system by virtue of their ability to preserve phase information in the observed signal (Hatzenakos and Nikias 1991).

Let $x(t), x(t + \tau_1), \ldots, x(t + \tau_{l-1})$ denote random variables obtained by observing a stationary stochastic process at times $t, t + \tau_1, \ldots, t + \tau_{l-1}$, respectively. The lth-order cumulant of the process is defined in terms of the joint moments of orders up to l. Specifically, the second-, third-, and fourth-order cumulants are defined by

$$c_2(\tau) = E[x(t)x(t + \tau)]$$

$$c_3(\tau_1, \tau_2) = E[x(t)x(t + \tau_1)x(t + \tau_2)]$$

$$c_4(\tau_1, \tau_2, \tau_3) = E[x(t)x(t + \tau_1)x(t + \tau_2)x(t + \tau_3)]$$
$$- E[x(t)x(t + \tau_1)]E[x(t + \tau_2)x(t + \tau_3)]$$
$$- E[x(t)x(t + \tau_2)]E[x(t + \tau_1)x(t + \tau_3)]$$
$$- E[x(t)x(t + \tau_3)]E[x(t + \tau_1)x(t + \tau_2)]$$

From these definitions, we note the following:

- The second-order cumulant $c_2(\tau)$ is the same as the autocorrelation function of $x(t)$.
- The third-order cumulant $c_3(\tau)$ is the same as the third-order joint moment of $x(t)$.
- The fourth-order cumulant $c_4(\tau_1, \tau_2, \tau_3)$ is different from the fourth-order joint moment of $x(t)$. The difference requires knowledge of six different values of the autocorrelation function of $x(t)$.

Note also that the lth-order cumulant does not depend on time t. For this to be valid, the process $x(t)$ has to be stationary up to order l.

The *lth-order polyspectrum* is defined as the multidimensional Fourier transform of the kth-order cumulant, as shown by

$$C_k(\omega_1, \omega_2, \ldots, \omega_{l-1}) = \sum_{\tau_1=-\infty}^{\infty} \cdots \sum_{\tau_{l-1}=-\infty}^{\infty} c_k(\tau_1, \tau_2, \ldots, \tau_{l-1})$$

$$\cdot \exp(-j(\omega_1\tau_1 + \omega_2\tau_2 + \cdots + \omega_{l-1}\tau_{l-1})) \quad (1.8)$$

For this polyspectrum to exist we require the lth-order cumulant $c_l(\tau_1, \tau_2, \ldots, \tau_{l-1})$ to be absolutely summable over all $\tau_1, \tau_2, \ldots, \tau_{l-1}$.

Shalvi and Weinstein (1990) have derived a set of criteria based on higher-order cumulants for blind deconvolution. These criteria may be viewed as "universal" in the sense that they do not impose any restrictions on the probability distribution of the source signal, except for the fact that a Gaussian distribution is not permitted. The blind deconvolution algorithms of Shalvi and Weinstein, based on third- and fourth-order cumulants, are of particular interest because of their simplicity.

(3) *Information-Theoretic Models.* The third family of unsupervised adaptive filtering algorithms exploits concepts rooted in Shannon's information theory, namely, entropy and mutual information.

Consider a source of information whose output is denoted by the vector **x**. Let $f(\mathbf{x})$ denote the probability density of **x**. The *differential entropy* of this source, signified as X, is defined by

$$h(X) = -E[\log f(\mathbf{x})]$$

$$= -\int_{-\infty}^{\infty} f(\mathbf{x}) \log f(\mathbf{x}) \, d\mathbf{x} \tag{1.9}$$

where E is the expectation operator and the integration is carried out over all the components of **x**. Bell and Sejnowski (1995) have devised a simple and yet effective algorithm for blind source separation, which is based on maximization of the differential entropy defined in a certain way. It is noteworthy that the maximum-entropy algorithm so derived for blind source separation is equivalent to a maximum-likelihood approach (Cardoso 1997).

Mutual information, by definition, involves input and output quantities. Let the vector **y** denote the output of a system produced in response to the vector **x** applied to the input of the system. The mutual information between these random vectors, denoted by $I(X; Y)$, is defined as the difference between the entropy of **x** and the conditional entropy of **x** given **y**, as shown by

$$I(X; Y) = h(X) - h(X|Y)$$

$$= -\int_{\infty}^{\infty} \int_{-\infty}^{\infty} f(\mathbf{x}, \mathbf{y}) \log\left(\frac{f(\mathbf{x}|\mathbf{y})}{f(\mathbf{x})}\right) d\mathbf{x} \, d\mathbf{y} \tag{1.10}$$

where $f(\mathbf{x})$ is the probability density function of **x**; $f(\mathbf{x}, \mathbf{y})$ is the joint-probability density function of **x** and **y**; and $f(\mathbf{x}|\mathbf{y})$ is the conditional-probability density function of **x** given **y**. In effect, the entropy $h(X)$ measures the uncertainty that we have about the system input *before* observing the system output, and the conditional entropy $h(X|Y)$ measures our uncertainty about the system input *after* observing the system output. The difference between these two entropies represents the uncertainty about the system input that is resolved by observing the system output.

The mutual information $I(X; Y)$ may also be interpreted as the Kullback-

Leibler divergence between two probability distributions, namely, the joint-probability density function $f(\mathbf{x}, \mathbf{y})$ and the product of marginal-probability density functions $f(\mathbf{x})$ and $f(\mathbf{y})$. In general, the *Kullback-Leibler divergence* between two probability density functions $f_1(\mathbf{y})$ and $f_2(\mathbf{y})$ is formally defined by

$$D_{f_1 \| f_2} = \int_{-\infty}^{\infty} f_1(\mathbf{y}) \log \left(\frac{f_1(\mathbf{y})}{f_2(\mathbf{y})} \right) d\mathbf{y} \qquad (1.11)$$

The Kullback-Leibler divergence provides a natural basis for solving the blind source-separation problem by viewing $f_1(\mathbf{y})$ as the joint-probability density function of the components constituting the vector $\mathbf{y}$ at the demixer output in Fig. 1.1 and $f_2(\mathbf{y})$ as the product of the marginal-probability density functions of those components (Amari et al. 1996; Comon 1994). The Kullback-Leibler divergence between the probability density functions $f_1(\mathbf{y})$ and $f_2(\mathbf{y})$ is also denoted by D_{12}, or $D(f_1, f_2)$.

1.5   ORGANIZATION OF VOLUME I

The main part of Volume I is organized in 8 chapters, which collectively provide a detailed exposition of unsupervised adaptive-filtering theory and algorithms for solving the important signal-processing task of blind signal separation.

Chapter 2 by Scott Douglas and Shun-ichi Amari discusses *natural-gradient adaptation* as an alternative to adaptation based on the standard-gradient descent. (The natural gradient is also referred to as the relative gradient.) The primary motivation of natural-gradient adaptation is to overcome the poor convergence properties of standard-gradient adaptation. This improvement in convergence performance is made possible by exploiting knowledge of the underlying Riemannian structure of the parameter (weight) space of the adaptive-filtering algorithm. Examples are presented to illustrate the utility and, in some cases, the simplicity of adaptation using the natural gradient.

Chapter 3 by Shun-ichi Amari, Andrzej Cichocki, and Howard Yang presents an extensive overview of the many facets of the blind signal-separation problem, based on *neural and information-theoretic approaches*. Criteria based on Shannon's mutual information and the Kullback-Leibler divergence are reviewed, with emphasis on on-line adaptation using noisy data. A novel *recurrent neural network* is described for simultaneous estimation of the unknown demixing matrix, blind signal separation, and reduction of noise in the extracted signals at the demixer output. The optimal choice of the nonlinear activation function, a fundamental component in the solution to blind signal separation, is discussed for various noise distributions.

A basic approach for the derivation of blind signal-separation algorithms is to optimize the so-called *contrast function*. Chapter 4 by Jean-François Cardoso provides a theoretical discussion of contrast functions in the framework of *in-*

formation geometry that was originally developed by Shun-ichi Amari (1985). The important issues of identifiability, consistency, and stability of the algorithms (based on the optimization of contrast functions) are discussed therein. The notion of natural gradient or relative gradient is also reviewed in this chapter.

Chapter 5 by Pierre Comon and Pascal Chevalier continues the discussion of the blind signal-separation problem by using algebraic and statistical tools in the development of algorithms. The issues of numerical complexity and performance analysis of blind signal-separation algorithms feature prominently in the chapter. The application of these techniques to *array signal processing* in radar and communications environments is given special attention; unlike standard techniques, blind signal-separation algorithms do *not* require array calibration or knowledge of the underlying probability distributions of the sources responsible for the incident electromagnetic waves. As with Chapter 2, Chapter 5 also includes an extensive listing of the literature of blind signal separation.

Chapter 6 by Anthony Bell takes a viewpoint different from that of the preceding chapters on how to solve the blind signal-separation problem. After a brief review of some basic information-theoretic ideas, the derivation of a simple and yet effective algorithm for solving the blind signal-separation problem is presented. The algorithm, known as the *information-maximization (Infomax) algorithm* or more precisely *the maximum-entropy algorithm*, was originally derived by Anthony Bell and Terry Sejnowski in 1995. Formulation of the algorithm is based on maximization of the entropy measured at the output of a zero-memory nonlinearity following the demixer. A novel application of the Infomax algorithm is that of extracting the independent basis functions of a natural image, a discussion of which is included in the chapter. Another interesting application of the Infomax algorithm to the decomposition of electroencephalographic (EEG) data and artifact removal is also discussed in this chapter.

The information-theoretic approaches discussed in Chapters 2 through 6, in one form or another, focus on the use of entropy and mutual information as originally defined by Claude Shannon in his classic paper (1948). Chapter 7 by Jose Principe, Dongxin Xu, and John Fisher III takes yet another information-theoretic approach based on the definition of *Renyi's quadratic entropy* (1970). The approach integrates this definition with the notion of nonparametric density estimation. An attractive feature of this approach is the fact that it lends itself to both unsupervised and supervised learning. Moreover, this new approach makes it possible to solve not only the blind signal-separation problem but also other "information filtering" problems such as pose estimation and classification of synthetic aperture-radar images.

Chapter 8 by Kari Torkkola takes a bold step forward in tackling the blind source-separation problem when we are confronted with practical realities, namely, *delayed* or *convolved mixtures* of independent source signals. This chapter builds on the Infomax algorithm of Chapter 6 due to the simplicity and intuitive appeal of the algorithm. The chapter places particular emphasis on practical applications in two important areas: audio and wireless radio communication.

The important practical issue of how to separate a convolved mixture of independent source signals is continued by Russell Lambert and Chrysostomos (Max) Nikias in Chapter 9. The approach taken here for solving this difficult signal-processing task is based on the *Bussgang family of cost functions*. A novel tool of general applicability, namely *FIR matrix algebra*, is introduced. Through this powerful tool, existing algorithms developed for ordinary blind channel equalization are expanded to deal with the blind deconvolution of convolved (multipath) mixtures of source signals. The chapter includes the results of real-life speech-separation experiments, confirming the power and robustness of this new approach.

REFERENCES

Amari, S., 1985, *Differential-Geometrical Methods in Statistics* (New York: Springer-Verlag).

Amari, S., A. Cichocki, and H. H. Yang, 1996, "A new learning algorithm for blind signal separation," *Advances in Neural Information Processing Systems*, vol. 8 (Cambridge, MA: MIT Press), pp. 757–763.

Bell, A. J., and T. J. Sejnowski, 1995, "An information maximization approach to blind separation and blind deconvolution," *Neural Computation*, vol. 6, pp. 1129–1159.

Bussgang, J. J., 1952, "Cross correlation functions of amplitude-distributed Gaussian signals," *Technical Rept. 216*, MIT Research Laboratory of Electronics, Cambridge, MA.

Cardoso, J. F., 1997, "Infomax and maximum likelihood for blind source separation," *IEEE Signal Processing Lett.*, vol. 4, pp. 112–114.

Comon, P., 1994, "Independent component analysis: A new concept?" *Signal Processing*, vol. 36, pp. 287–314.

Godard, D. N., 1980, "Self-recovering equalization and carrier tracking in a two-dimensional data communication system," *IEEE Trans. Communic.*, vol. COM-28, pp. 1867–1875.

Hatzenakos, D., and C. L. Nikias, 1991, "Blind equalization using a trispectrum based algorithm," *IEEE Trans. Communic.*, vol. COM-39, pp. 669–682.

Haykin, S., ed., 1994, *Blind Deconvolution* (Englewood Cliffs, NJ: Prentice-Hall).

Haykin, S., 1996, *Adaptive Filter Theory* (Englewood Cliffs, NJ: Prentice-Hall).

Haykin, S., 1999, *Neural Networks: A Comprehensive Foundation*, 2nd ed. (Englewood Cliffs, NJ: Prentice-Hall).

Herault, J., C. Jutten, and B. Ans, 1985, "Detection de grandeurs primitives dans un message composite par une architecture de calul neuromimetique un apprentissage non supervise," Proc. GRETSI, Nice, France.

Lucky, R. W., 1966, "Techniques for adaptive equalization of digital communication systems," *Bell System Tech. J.*, vol. 45, pp. 255–286.

Rényi, A., 1970, *Probability Theory* (Amsterdam: North-Holland).

Shannon, C. E., 1948, "A mathematical theory of communication," *Bell System Tech. J.*, vol. 27, pp. 379–423, 623–656.

Sato, Y., 1975, "Two extensional applications of the zero-forcing equalization method," *IEEE Trans. Communic.*, vol. COM-23, pp. 684–687.

Shalvi, O., and E. Weinstein, 1990, "New criteria for blind equalization of non-minimum phase systems (channels)," *IEEE Trans. Inform. Theory*, vol. 36, pp. 312–321.

Treichler, J. R., and B. F. Agee, 1983, "A new approach to multipath correction of constant modulus signals," *IEEE Trans. Acoust, Speech Signal Processing*, vol. ASSP-31, pp. 459–471.

Widrow, B., and M. C. Hoff, Jr., 1960, "Adaptive switching circuits," IRE WESCON Convention Record, Pt. 4, pp. 96–104.

2

NATURAL-GRADIENT ADAPTATION

Scott C. Douglas and Shun-ichi Amari

ABSTRACT

Gradient adaptation is a celebrated technique for adjusting a set of parameters to minimize a chosen cost function. While simple to understand and often easy to implement in an on-line setting, the convergence speed of gradient adaptation can be slow when the slope of the cost function varies widely for small changes in the adjusted parameters. This chapter describes an alternative technique, termed natural gradient adaptation, that in many cases can overcome the poor convergence properties of gradient adaptation, is described. The natural gradient is based on differential geometry and employs knowledge of the Riemannian structure of the parameter space to adjust the gradient search direction. Unlike Newton's method, the natural-gradient adaptation method does not assume that the cost function is locally quadratic. Moreover, for maximum-likelihood estimation tasks, natural-gradient adaptation is Fisher-efficient in its adaptation capabilities. Examples drawn both from classic estimation theory and from modern real-world applications indicate the utility and, in some cases, the simplicity of the natural-gradient adaptation method for on-line parameter estimation, filtering, and inverse modeling tasks.

2.1 INTRODUCTION

Parameter estimation is the task of determining useful numerical values for certain constants within some problem structure. The pervasiveness of this task in

Unsupervised Adaptive Filtering, Volume I, Edited by Simon Haykin.
ISBN 0-471-29412-8 © 2000 John Wiley & Sons, Inc.

the scientific, engineering, social science, and financial arenas shows how useful and important this task is. While other techniques for performing parameter estimation exist, perhaps the most popular method is *gradient descent*. Gradient descent is an iterative optimization procedure that is computationally simple to implement. Because it is an iterative technique, its ability both to converge quickly to and to identify the best parameter values for a particular problem are limited in some cases. Even so, it has found wide use in numerous areas and is extremely popular in the adaptive filtering and neural-network fields as exemplified by the well-known least-mean-square (LMS) and backpropagation algorithms (Haykin 1994; Haykin 1996; Widrow and Stearns 1985). While other optimization techniques such as genetic algorithms, simulated annealing, and fuzzy logic can be employed for parameter estimation (Jang et al. 1997), gradient descent remains the technique to which other methods are likely to be compared.

Although often not explicitly stated, the usefulness of gradient descent depends on the structure of the cost function being optimized. The behavior of gradient descent is most useful for cost functions (i) that have a single minimum and (ii) whose gradients are isotropic in magnitude with respect to any direction away from this minimum. In practice, however, the cost function being optimized is multimodal, and the gradient magnitudes are nonisotropic about any minimum. In such cases, the parameter estimates are only guaranteed to locally minimize the cost function, and convergence to any local minimum can be extremely slow. In addition, the structure of some tasks defines certain constraints that the parameters must satisfy within the adaptation process. For example, in principal component analysis the optimum parameters form a matrix whose rows are orthonormal (Diamantaras and Kung 1996). It is desirable to specify an adaptation procedure that preserves these constraints implicitly, so that periodic projection of the parameters back to the allowable parameter space is unnecessary. Such methods can be more convenient to implement in practice than methods that explicitly reduce the dimension of the parameterized space via the constraints.

This chapter outlines a useful alternative to standard-gradient adaptation. Termed *natural-gradient adaptation*, the proposed iterative procedure modifies the standard gradient search direction according to the Riemannian structure of the parameter space (Amari 1998). While not removing local minima and thus guaranteeing globally optimum parameter estimates, natural-gradient adaptation can provide isotropic convergence properties independent of the dependencies within the data being processed by the algorithm. Moreover, when used with statistically efficient cost functions, natural-gradient adaptation overcomes many of the limitations of Newton's method, which assumes that the cost function being minimized is approximately quadratic about any local minimum (Luenberger 1984). For this reason, natural-gradient adaptation is appropriate for a large class of cost functions and for nonlinear models used in parameter estimation, such as neural networks. The chief drawback of natural-gradient adaptation is the knowledge required to determine the Riemannian structure of the parameter space, although examples indicate that its form can be quite simple for certain estimation problems.

The organization of this chapter is as follows. In the next section, we review the basic concepts behind standard-gradient descent and indicate some of its performance limitations. The natural gradient is introduced in Section 2.3, where its general properties are discussed. Section 2.4 provides several examples of natural-gradient adaptation as applied to particular problems in signal processing. Conclusions are drawn in Section 2.5.

The term "Riemannian" comes from the field of differential geometry. While a full appreciation of natural-gradient adaptation is best obtained from the study of differential geometry, our discussion assumes a mathematical background only in basic calculus and linear algebra. An accessible introduction to the field of differential geometry can be found in Morgan (1993), and a more detailed treatment of the field can be found in Boothby (1986). To aid the reader, we have included in the following Notation section a list of all mathematical symbols and notation used throughout this chapter, along with the function definitions used. All algorithms are given in finite difference form in discrete time with time index k, although continuous-time expressions can also be found via limiting arguments. In addition, we follow the standard mathematical convention that partial derivatives of functions take precedence over function evaluations; thus, $\partial g(c)/\partial x$ is the partial derivative of the function $g(x)$ with respect to x evaluated at $x = c$.

2.1.1 Notation

GENERAL SIGNAL AND PARAMETER DESCRIPTIONS

Variable	Description
k	Time index
N	Number of parameters
$\mathbf{w}$ and $\mathbf{v}$	Parameter vectors
w_i and v_i	ith parameter
$\mathbf{w}_{\text{opt}}$	Optimum parameter vector
$w_{i,\text{opt}}$	ith optimum parameter
$\delta\mathbf{w}$ and $\delta\mathbf{v}$	Small change in parameter vector
δw_i and δv_i	ith element of $\delta\mathbf{w}$ and $\delta\mathbf{v}$
$\mathscr{J}(\mathbf{w})$	Cost function
$(\mathbf{w},\mathbf{v})$	Distance between $\mathbf{w}$ and $\mathbf{v}$
$\mathbf{H}(\mathbf{w})$	Hessian matrix
$h_{ij}(\mathbf{w})$	(i,j)th element of Hessian
$\mathbf{G}(\mathbf{w})$	Riemannian metric tensor
$g_{ij}(\mathbf{w})$	(i,j)th element of Riemannian metric tensor
$\mathbf{t}_{\mathbf{G}}(k)$	Gradient in the tangent space
$\lambda_{max}[\mathbf{M}]$	Maximum eigenvalue of matrix $\mathbf{M}$
$\mu(k)$	Step size
ε	Small constant

VARIABLES USED IN ILLUSTRATIVE EXAMPLES

L	Number of signal measurements
x, y, and z	Original coordinates
r, θ, and φ	Transformed coordinates
$\mathbf{T}$	Transformation matrix
t_{ij}	(i, j)th element of transformation matrix
$\mathbf{x}$, $\mathbf{y}$, and $\mathbf{s}$	Signal vectors
$f_x(\mathbf{x})$, $f_y(\mathbf{y})$, and $f_s(\mathbf{s})$	pdf's of $\mathbf{x}$, $\mathbf{y}$, and $\mathbf{s}$
$f_y(\mathbf{y}|\mathbf{w})$	Parametric model of pdf of $\mathbf{y}$
$\mathscr{J}(\mathbf{w})$	Estimate of cost function
ρ	Correlation parameter
c	Arbitrary constant

VARIABLES USED IN SIMULATION EXAMPLES

Maximum-Likelihood Estimation of a Noisy Sinusoid

ω	Angular frequency constant
$\eta(k)$	Noise signal
σ_η^2	Variance of $\eta(k)$

Single-Layer Perceptron Training

$u(k)$	Linear perceptron output signal
$g(u)$	Output nonlinearity
α and β	Output nonlinearity parameters
$\mathbf{R}_{xx}$	Autocorrelation matrix
$\bar{\mathbf{R}}(\mathbf{w})$	Perceptron matrix function
$\bar{\mu}(k)$	Modified step-size

Instantaneous Blind Source Separation

$\mathbf{s}(k)$	Source signal vector
$\mathbf{x}(k)$	Measured signal vector
$\mathbf{y}(k)$	Output signal vector
$\mathbf{A}$	Mixing matrix
$\mathbf{W}$	Demixing matrix
$\mathbf{P}$	Permutation matrix
$\mathbf{D}$	Diagonal scaling matrix
$\mathbf{I}$	Identity matrix
$\mathbf{C}$	Combined system matrix
$f_{\mathbf{w}}(\mathbf{s})$	Parametric model of $f_s(\mathbf{s})$
$f_i(y_i)$	pdf of ith output signal
$\phi(\mathbf{y})$	Output nonlinearity vector
$\phi_i(y_i)$	ith element of $\phi(\mathbf{y})$
$[\mathbf{G}(\mathbf{W})]_{ij,kl}$	(i, j, k, l)th element of the Riemannian metric tensor
$\mathbf{E}(\mathbf{y}(k))$	Multiplicative update matrix
$\zeta(k)$	Performance factor

Blind Equalization

$s(k)$	Source signal
$S(z)$	z-transform of $s(k)$
$x(k)$	Received signal
$X(z)$	z-transform of $x(k)$
$y(k)$	Output signal
$Y(z, k)$	z-transform of $y(k)$
a_p	Unknown system impulse response
$A(z)$	z-transform of a_p
$w_p(k)$	Equalizer impulse response
$W(z, k)$	z-transform of $w_p(k)$
z^{-1}	Unit delay
$u(k)$	Filtered output signal
$\mathbf{u}(k)$	Filtered output signal vector
$d(k)$	Training sequence
$\mathbf{y}_k(k)$	Output signal vector
$\mathbf{X}(k)$	Hankel input signal matrix
A_2	Magnitude parameter

Principal and Minor Subspace Analysis

n	Signal dimension
m	Subspace dimension
$\mathbf{W}$	Orthonormal subspace matrix
$\hat{\mathbf{x}}(k)$	Estimated signal vector
$\mathbf{E}_1$	Principal eigenvector matrix
$\mathbf{E}_2$	Minor eigenvector matrix
$\mathbf{Q}$	Rotation matrix
$\Delta(\mathbf{W})$	Differential update matrix

FUNCTION DEFINITIONS

Function	Description
$\mathbf{M}^*$	Element-by-element conjugate of $\mathbf{M}$
$\mathbf{M}^T$	Transpose of $\mathbf{M}$
$\mathbf{M}^{-1}$	Inverse of $\mathbf{M}$
$E\{\mathbf{v}\}$	Expectation of $\mathbf{v}$
$\partial f(\mathbf{M})/\partial \mathbf{M}$	Matrix of partial derivatives $\partial f(\mathbf{M})/\partial m_{ij}$

2.2 STANDARD-GRADIENT ADAPTATION

2.2.1 Parameters and Cost Functions

To describe standard-gradient adaptation, define a vector of N adjustable parameters as

$$\mathbf{w}(k) = [w_1(k) \ w_2(k) \cdots w_N(k)]^T \tag{2.1}$$

where $w_i(k)$ is the ith parameter value at time k. The parameters $\{w_i(k)\}$ typically correspond to terms within some useful computational structure. For example, they could correspond to impulse-response values of a linear finite-duration impulse-response (FIR) filter, or they could correspond to weights within a multilayer perceptron as in a neural network. The number of parameters N is finite, but this number could be arbitrarily large.

For any set of numerical values for $\{w_i(k)\}$, we define a scalar *cost function* $\mathcal{J}(\mathbf{w})$ satisfying the following properties (Luenberger 1984):

1. *Smoothness.* The cost function $\mathcal{J}(\mathbf{w})$ is twice differentiable with respect to any pair $\{w_i, w_j\}$ for $1 \le i \le j \le N$.
2. *Existence of Solution.* At least one parameter vector $\mathbf{w}_{\mathrm{opt}} = [w_{1,\mathrm{opt}} \cdots w_{N,\mathrm{opt}}]^T$ exists such that
   (i)

$$\frac{\partial \mathcal{J}(\mathbf{w}_{\mathrm{opt}})}{\partial w_i} = 0, \qquad 1 \le i \le N \tag{2.2}$$

   (ii) The $(N \times N)$ *Hessian matrix* $\mathbf{H}(\mathbf{w})$ with entries $h_{ij}(\mathbf{w})$ given by

$$h_{ij}(\mathbf{w}) = \frac{\partial^2 \mathcal{J}(\mathbf{w})}{\partial w_i \partial w_j} \tag{2.3}$$

   is positive definite for $\mathbf{w} = \mathbf{w}_{\mathrm{opt}}$.

The vector

$$\frac{\partial \mathcal{J}(\mathbf{w})}{\partial \mathbf{w}} = \left[\frac{\partial \mathcal{J}(\mathbf{w})}{\partial w_1} \ \frac{\partial \mathcal{J}(\mathbf{w})}{\partial w_2} \cdots \frac{\partial \mathcal{J}(\mathbf{w})}{\partial w_N} \right]^T \tag{2.4}$$

is known as the *gradient* of $\mathcal{J}(\mathbf{w})$. Therefore, the condition in Eq. (2.2) means that all of the entries of the gradient are zero at $\mathbf{w} = \mathbf{w}_{\mathrm{opt}}$. Furthermore, the condition that $\mathbf{H}(\mathbf{w}_{\mathrm{opt}})$ in Eq. (2.3) is positive definite implies that $\mathcal{J}(\mathbf{w})$ has a positive curvature at $\mathbf{w} = \mathbf{w}_{\mathrm{opt}}$ such that for any small nonzero deviation vector $\delta\mathbf{w}$ the relationship

$$\mathcal{J}(\mathbf{w}_{\mathrm{opt}} + \delta\mathbf{w}) > \mathcal{J}(\mathbf{w}_{\mathrm{opt}}) \tag{2.5}$$

holds. Thus, $\mathbf{w}_{\mathrm{opt}}$ represents a local minimum of the cost function in parameter space.

Generally, we define a cost function as being either *deterministic* or *stochastic* in the following manner. A deterministic cost function depends directly on signal measurements taken from a physical environment. An example of a deterministic cost function is the least-squares cost function for linear or non-linear regression tasks. A stochastic cost function is an approximation to a well-defined cost function whose form depends on the probability distributions of a set of signals within a physical environment. When an iterative procedure is used to minimize the stochastic cost function with respect to a set of parameters, a form of time averaging across successive signal measurements is effectively employed to approximate the ensemble-averaged value of the stochastic cost function. Stochastic algorithms include the LMS algorithm for linear adaptive filtering (Widrow and Stearns 1985) and the backpropagation algorithm for multilayer perceptron training (Haykin 1994).

2.2.2 Steepest Descent

Given a cost function $\mathscr{J}(\mathbf{w})$, we would like to obtain parameter estimates within $\mathbf{w}$ that minimize $\mathscr{J}(\mathbf{w})$. Many procedures for achieving this purpose can be derived. In this section, we outline one of the simplest and most popular methods, that of *steepest descent*. We first state the procedure before exploring some of its properties.

Steepest-Descent Method   For any cost function $\mathscr{J}(\mathbf{w})$ satisfying the criteria in Eqs. (2.2)–(2.5), the steepest-descent method is given by

$$\mathbf{w}(k+1) = \mathbf{w}(k) - \mu(k)\frac{\partial \mathscr{J}(\mathbf{w}(k))}{\partial \mathbf{w}} \tag{2.6}$$

where $\mathbf{w}(0)$ is any initial-parameter vector and $\mu(k)$, $k = \{0, 1, 2, \ldots\}$ is a positive-valued sequence of *step sizes*.

The vector $-\partial \mathscr{J}(\mathbf{w})/\partial \mathbf{w}$ in Eq. (2.6) represents the steepest-descent direction of the cost function evaluated at $\mathbf{w}$. The steepest-descent direction $\delta \mathbf{w} = -\mu \partial \mathscr{J}(\mathbf{w})/\partial \mathbf{w}$, where μ is a proportionality constant, can be obtained by minimizing $\mathscr{J}(\mathbf{w} + \delta \mathbf{w})$ with respect to the entries of $\delta \mathbf{w}$ under the condition that

$$\sqrt{\delta \mathbf{w}^T \delta \mathbf{w}} = \sqrt{\sum_{i=1}^{N} \delta w_i^2} = \varepsilon \tag{2.7}$$

where ε is a small constant. The steepest-descent method is an iterative technique in which a fraction of the gradient of the cost function with respect to each parameter is subtracted from each parameter. This process is continued indefinitely or until the value of $\mathscr{J}(\mathbf{w}(k))$ reaches a suitably small value, at which point $\mathbf{w}(k)$ is "close" to $\mathbf{w}_{\text{opt}}$.

To understand why the steepest-descent method works, consider the two conditions satisfied by the cost function $\mathscr{J}(\mathbf{w})$ in Eqs. (2.2) and (2.5), respectively. Equation (2.2) states that the gradient of the cost function vanishes at

$\mathbf{w} = \mathbf{w}_{\mathrm{opt}}$. Substituting $\mathbf{w}(k) = \mathbf{w}_{\mathrm{opt}}$ into the update in Eq. (2.6), we obtain an identity condition, indicating that this procedure does not change the value of $\mathbf{w}(k)$ if it has reached its optimum value. Furthermore, by letting $\mathbf{w}(k) = \mathbf{w}_{\mathrm{opt}} + \delta\mathbf{w}$, we have

$$w_i(k+1) = w_{i,\,\mathrm{opt}} + \delta w_i - \mu(k)\frac{\partial \mathscr{J}(\mathbf{w}_{\mathrm{opt}} + \delta\mathbf{w})}{\partial w_i} \qquad (2.8)$$

which for infintesimally small $\delta\mathbf{w}$ becomes

$$w_i(k+1) - w_{i,\,\mathrm{opt}} = \left[1 - \mu(k)\frac{\mathscr{J}(\mathbf{w}_{\mathrm{opt}} + \delta\mathbf{w}) - \mathscr{J}(\mathbf{w}_{\mathrm{opt}})}{\delta w_i^2}\right]\{w_i(k) - w_{i,\,\mathrm{opt}}\} \qquad (2.9)$$

Because $\mathscr{J}(\mathbf{w}_{\mathrm{opt}} + \delta\mathbf{w}) - \mathscr{J}(\mathbf{w}_{\mathrm{opt}}) > 0$ from Eq. (2.5), then for small positive values of $\mu(k)$ the first term within brackets on the right-hand side (RHS) of Eq. (2.9) is less than one in magnitude, such that $|w_i(k+1) - w_{i,\,\mathrm{opt}}| < |w_i(k) - w_{i,\,\mathrm{opt}}|$. The steepest-descent method therefore reduces the error in the ith component of $\mathbf{w}(k)$ at each iteration around $\mathbf{w}_{\mathrm{opt}}$. With proper selection of $\mu(k)$, the steepest-descent method adjusts $\mathbf{w}(k)$ so that

$$\lim_{k\to\infty} \mathbf{w}(k) = \mathbf{w}_{\mathrm{opt}} \qquad (2.10)$$

for values of $\mathbf{w}(0)$ that are suitably close to $\mathbf{w}_{\mathrm{opt}}$. Such an algorithm causes $\mathbf{w}(k)$ to *converge* to $\mathbf{w}_{\mathrm{opt}}$.

2.2.3 Stochastic-Gradient Descent

The cost function $\mathscr{J}(\mathbf{w})$ can be defined either in terms of signal measurements or in terms of the probability distributions corresponding to particular signal measurements. If $\mathscr{J}(\mathbf{w})$ depends on signal measurements, as is the case for deterministic criteria such as least-squares cost functions, then the gradient $\partial \mathscr{J}(\mathbf{w})/\partial\mathbf{w}$ also depends on signal measurements and thus can be evaluated once a signal set has been obtained.

  If $\mathscr{J}(\mathbf{w})$ is defined in terms of unknown probability distributions of a set of signal measurements, then one must use the signal measurements in a procedure to *estimate* the gradient. The following example considers one such case.

Example 1. *Maximum-Likelihood Estimation* (Poor 1994). Let $\mathbf{y}(m)$ be a sequence of M-dimensional vectors with a joint probability density function (pdf) $f_\mathbf{y}(\mathbf{y}|\mathbf{w}_{\mathrm{opt}})$, where $\mathbf{w}_{\mathrm{opt}}$ is a vector of unknown parameters. We wish to estimate the values of the parameters in $\mathbf{w}_{\mathrm{opt}}$ given the parametric model $f_\mathbf{y}(\mathbf{y}|\mathbf{w})$ for the pdf of $\mathbf{y}(k)$. A useful cost function for such a task is

$$\mathcal{J}_{ML}(\mathbf{w}) = -\int f_{\mathbf{y}}(\mathbf{y}|\mathbf{w}_{opt}) \log f_{\mathbf{y}}(\mathbf{y}|\mathbf{w}) \, dy \qquad (2.11)$$

$$= -E\{\log f_{\mathbf{y}}(\mathbf{y}|\mathbf{w})\} \qquad (2.12)$$

where $E\{\cdot\}$ denotes statistical expectation.

Such a cost function cannot be used directly within the steepest-descent procedure because $\mathbf{w}_{opt}$ is unknown, and thus the integral in Eq. (2.11) cannot be evaluated. However, we can use signal measurements to estimate $\mathcal{J}_{ML}(\mathbf{w})$ as

$$\hat{\mathcal{J}}_{ML}(\mathbf{w}) = -\frac{1}{L} \sum_{m=1}^{L} \log f_{\mathbf{y}}(\mathbf{y}(m)|\mathbf{w}) \qquad (2.13)$$

where L is the number of signal measurements used to estimate $\mathcal{J}_{ML}(\mathbf{w})$ for a given $\mathbf{w}$. Then, the coefficients $\mathbf{w}(k)$ are updated as

$$\mathbf{w}(k+1) = \mathbf{w}(k) - \mu(k)\frac{\partial \hat{\mathcal{J}}_{ML}(\mathbf{w}(k))}{\partial \mathbf{w}} \qquad (2.14)$$

This procedure effectively uses an *averaged gradient*

$$\frac{\partial \hat{\mathcal{J}}_{ML}(\mathbf{w}(k))}{\partial \mathbf{w}} = \frac{1}{L}\sum_{m=1}^{L} -\frac{\partial \log f_{\mathbf{y}}(\mathbf{y}(m)|\mathbf{w}(k))}{\partial \mathbf{w}} \qquad (2.15)$$

in place of $\partial \mathcal{J}_{ML}(\mathbf{w}(k))/\partial \mathbf{w}$ in the steepest-descent procedure.

More generally, if $\mathcal{J}(\mathbf{w})$ depends on the unknown probability distribution of the signal measurements, we define a cost function $\hat{\mathcal{J}}(\mathbf{w})$ based on the signal measurements such that

$$E\{\hat{\mathcal{J}}(\mathbf{w})\} = \mathcal{J}(\mathbf{w}) \qquad (2.16)$$

Then, $\mathbf{w}(k)$ is adjusted according to

$$\mathbf{w}(k+1) = \mathbf{w}(k) - \mu(k)\frac{\partial \hat{\mathcal{J}}(\mathbf{w}(k))}{\partial \mathbf{w}} \qquad (2.17)$$

Since Eq. (2.17) is based on a statistically approximate or stochastic implementation of a true steepest-descent procedure, it is often referred to as *stochastic-gradient adaptation*.

The number of measurements L used to calculate $\partial \hat{\mathcal{J}}(\mathbf{w}(k))/\partial \mathbf{w}$ in Eq. (2.17) depends on the application. If $\hat{\mathcal{J}}(\mathbf{w})$ consists of an average of L gradients for L independent measurements, then by the law of large numbers, the value of $\partial \hat{\mathcal{J}}(\mathbf{w}(k))/\partial \mathbf{w}$ tends to $\partial \mathcal{J}(\mathbf{w}(k))/\partial \mathbf{w}$ as L is increased (Cover and Thomas 1991), providing a behavior that is closer to that of a true steepest-descent procedure. For many practical situations, however, calculating this gradient estimate from a single measurement $\mathbf{y}(k)$ (such that $L = 1$) yields a simple but still useful algorithm. The LMS algorithm in adaptive filtering is perhaps the best known method of this class (Widrow and Stearns 1985). This technique has also been extensively studied in the statistics literature, where it is known as *stochastic approximation* (Kushner and Clark 1978). In general, the transient behaviors of Eqs. (2.6) and (2.17) can be shown to be similar for small step-sizes, although the steady-state characteristics of the algorithms usually differ in practice.

2.2.4 Behavior of Gradient-Descent Methods

The transient behavior of any gradient-descent method depends on the form of $\mathscr{J}(\mathbf{w})$ [or, equivalently, $\hat{\mathscr{J}}(\mathbf{w})$ for stochastic-gradient methods]. The convergence of $\mathbf{w}(k)$ to $\mathbf{w}_{\text{opt}}$ can be fast or slow, depending on the choice of $\mu(k)$ and on the gradient components $\partial \mathscr{J}(\mathbf{w}(k))/\partial w_i$. In general, it is difficult or impossible to choose $\mu(k)$ in such a way as to provide fast convergence from all possible initial coefficient values $\mathbf{w}(0)$. This difficulty stems from the fact that the gradient components $\partial \mathscr{J}(\mathbf{w}(k))/\partial w_i$ vary widely in magnitude in different directions from $\mathbf{w}_{\text{opt}}$ for typical parametrizations and cost functions. Without knowledge of where $\mathbf{w}(k)$ is with respect to $\mathbf{w}_{\text{opt}}$ in the space of possible parameters, one cannot choose $\mu(k)$ property to obtain fast convergence of each $w_i(k)$ to $w_{i,\text{opt}}$. The following example illustrates this fact.

Example 2. *Gradient Descent for Different Cost Functions.* Consider the following cost function for the two-dimensional parameter vector $\mathbf{w} = [w_1 \; w_2]^T$:

$$\mathscr{J}_E(\mathbf{w}) = \tfrac{1}{2}[(w_1 - 1)^2 + w_2^2] \tag{2.18}$$

This cost function is twice-differentiable at all points, and it has a single minimum at $\mathbf{w}_{\text{opt}} = [1 \; 0]^T$, at which point the value of the gradient vanishes. Moreover, the Hessian of the cost function, given by $\mathbf{H}(\mathbf{w}) = \mathbf{I}$ where $\mathbf{I}$ is the identity matrix, is positive definite for all $\mathbf{w}$. The negative of the gradient of this cost function for any $\mathbf{w}$ is

$$-\frac{\partial \mathscr{J}_E(\mathbf{w})}{\partial \mathbf{w}} = -\begin{bmatrix} w_1 - 1 \\ w_2 \end{bmatrix} \tag{2.19}$$

Figure 2.1*a* plots the contours of this cost function as well as the gradients for various $\mathbf{w}$. For this particular cost function, all of the gradients point toward $\mathbf{w}_{\text{opt}}$, and there is no angular variation of the gradients for different directions away from $\mathbf{w}_{\text{opt}}$. Such a surface is ideal for steepest-descent adaptation, as it is possible to obtain fast adaptation from any initial $\mathbf{w}(k)$ to $\mathbf{w}_{\text{opt}}$ with a proper choice of $\mu(k)$. Figure 2.1*b* indicates the paths taken by the gradient-

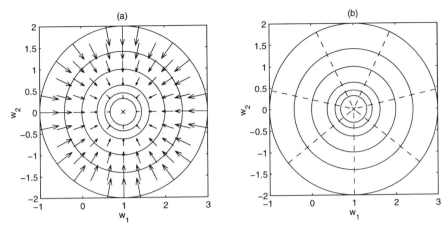

Figure 2.1 (a) Contours and gradients of the cost function $\mathcal{J}_E(\mathbf{w}(k))$. (b) Coefficient trajectories obtained from the steepest-descent method for seven different initial conditions $\mathbf{w}(0)$.

descent method for this cost function for seven different initial conditions $\mathbf{w}(0) = \mathbf{w}_{opt} + [2\cos(\pi/14 + j2\pi/7)\ 2\sin(\pi/14 + j2\pi/7)]^T$, $0 \leq j \leq 6$, where we have chosen $\mu(k) = 0.01$. In each case, the coefficients take a direct path to $\mathbf{w}_{opt}$, irrespective of the initial value of $\mathbf{w}(k)$.

We now consider the cost function given by

$$\mathcal{J}_p(\mathbf{w}) = \tfrac{1}{2}[\{w_1 - 1\}^2 + w_2^2 + 2\rho w_2\{w_1 - 1\}] \tag{2.20}$$

where $\rho = 0.8$. This cost function also has a single minimum at $\mathbf{w}_{opt} = [1\ 0]^T$, and the negative of its gradient is

$$-\frac{\partial \mathcal{J}_p(\mathbf{w})}{\partial \mathbf{w}} = -\begin{bmatrix} w_1 - 1 + \rho w_2 \\ w_2 + \rho(w_1 - 1) \end{bmatrix} \tag{2.21}$$

Figure 2.2a shows the contours as well as the gradients of this cost function. In this case, each of the gradients no longer points to $\mathbf{w}_{opt}$. In addition, the magnitudes of the gradients vary for different directions away from $\mathbf{w}_{opt}$. Figure 2.2b shows the coefficient trajectories for the steepest-descent method as applied to this cost function, where it is seen that the coefficients take a curved path from a typical initial $\mathbf{w}(k)$ toward $\mathbf{w}_{opt}$. It is difficult to choose $\mu(k)$ to reduce each component of the error vector $\mathbf{w}(k) - \mathbf{w}_{opt}$ in a rapid fashion in this example.

Consider the third cost function given by

$$\mathcal{J}_P(\mathbf{w}) = \tfrac{1}{2}[\{r\cos\theta - 1\}^2 + r^2\sin^2\theta] \tag{2.22}$$

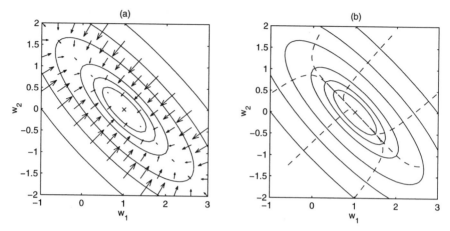

Figure 2.2 (a) Contours and gradients of the cost function $\mathscr{J}_p(\mathbf{w}(k))$. (b) Coefficient trajectories obtained from the steepest-descent method for seven different initial conditions $\mathbf{w}(0)$.

where we define $\mathbf{w} = [r \; \theta]^T$. This cost function is nothing more than $\mathscr{J}_E(\mathbf{w})$ as represented in the polar-coordinate system

$$w_1 = r\cos\theta, \qquad w_2 = r\sin\theta \tag{2.23}$$

and it also has its minimum at $[r_{\text{opt}} \; \theta_{\text{opt}}]^T = [1 \; 0]^T$. The error surface is markedly different from the previous two cases, however, as indicated in the contour plot in Fig. 2.3a. Here, we see that the cost function has multiple minima at

$$\mathbf{w}_{\text{opt}}^{(n)} = \begin{bmatrix} (-1)^n \\ \pi n \end{bmatrix} \tag{2.24}$$

for integer values of n. All of these minima are equivalent, however, in that $\mathscr{J}(\mathbf{w}_{\text{opt}}^{(n)})$ is the same for all n. Also shown in Fig. 2.3a are the gradients of the cost function for different $\mathbf{w}$, given by

$$-\frac{\partial \mathscr{J}_P(\mathbf{w})}{\partial \mathbf{w}} = -\begin{bmatrix} r - \cos\theta \\ r\sin\theta \end{bmatrix} \tag{2.25}$$

The magnitudes of the gradients vary widely across the surface, and they do not point toward any $\mathbf{w}_{\text{opt}}^{(n)}$ in general. In particular, the gradients are small along the locus of points $2|\theta| = \pi(1 - r)$ over the range $r > 0$ and $|\theta| < \pi$. Figure 2.3b shows the coefficient trajectories of the steepest-descent method as applied to

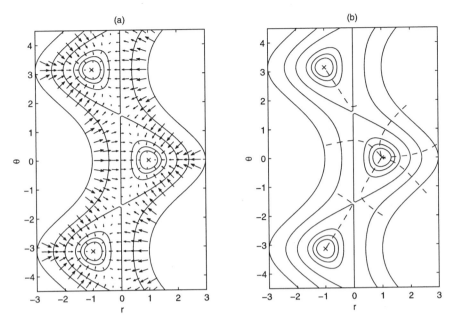

Figure 2.3 (a) Contours and gradients of the cost function $\mathcal{J}_P(\mathbf{w}(k))$. (b) Coefficient trajectories obtained from the steepest-descent method for seven different initial conditions $\mathbf{w}(0)$.

this cost function for seven different initial values $\mathbf{w}(0)$. Depending on the initial coefficient values, $\mathbf{w}(k)$ converges to different minima on the error surface. Moreover, the coefficient trajectories are far from straight paths, and it is challenging to choose $\mu(k)$ to quickly cause $\mathbf{w}(k)$ to converge to any $\mathbf{w}_{\text{opt}}^{(n)}$ in this case.

2.3 NATURAL-GRADIENT ADAPTATION

2.3.1 Euclidean and Riemannian Geometry

To understand natural-gradient learning as a useful tool for optimization requires us to reconsider the fundamental notion of *distance*. In everyday life, we possess an intuitive concept of distance. The familiar folk axiom "the shortest distance between two points is a straight line" is an embodiment of our common notion of distance. In mathematical terms, we refer to such a measure of distance as *Euclidean distance*, in deference to the the mathematician Euclid of Alexandria whose treatise entitled *The Elements* forms the foundation of plane geometry. In Euclidean terms, the distance between two points $\mathbf{v}$ and $\mathbf{v} + \delta\mathbf{v}$ in the N-dimensional space spanned by $\mathbf{v} = [v_1 \ v_2 \cdots v_N]^T$ is

$$d_E(\mathbf{v}, \mathbf{v} + \delta\mathbf{v}) = \|\delta\mathbf{v}\|_2 \qquad (2.26)$$

$$= \sqrt{\sum_{i=1}^{N} \delta v_i^2} = \sqrt{\delta\mathbf{v}^T \delta\mathbf{v}} \qquad (2.27)$$

where $\delta\mathbf{v} = [\delta v_1 \; \delta v_2 \cdots \delta v_N]^T$ and $\|\mathbf{v}\|_2$ denotes the L_2 or *Euclidean norm* of $\mathbf{v}$.

In reality, however, "straight line" or Euclidean distance represents an approximation to a more complex notion of distance. For example, most geographical maps depict the world as a flat two-dimensional plane instead of the curved surface that it is. While it shouldn't come as a surprise, we find it unusual that the shortest physical route between two locations is generally represented as an arc on such a map, which actually is along the geodesic defined by the global curvature of the earth's surface. In effect, the map's representation of the world as flat is a local or tangential approximation that is not accurate for travel over large distances, such as across or between continents.

Although man-made planar maps represent an artificial distortion of distance, non-Euclidean measures of distance regularly occur in the physical world. In his famous Theory of Relativity, Einstein predicted that mass distorts physical space, such that the path of light would bend when passing by a large heavenly mass such as star. This prediction was verified in 1919 when photographs of the sun during a solar eclipse showed that the visual positions of stars were indeed apparently altered by the sun's mass (see Fig. 2.4). The path taken by the stars' light is the shortest in terms of gravitational space, although it is clearly "bent" in Euclidean terms. That Euclidean geometry was found not to adequately describe fundamental physical laws was astounding to the scientific community, and it led to a renewed focus on the mathematics of curved space,

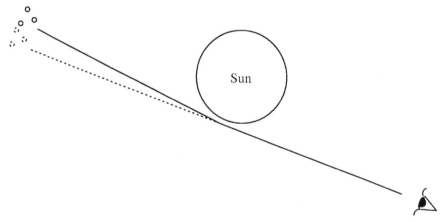

Figure 2.4 Einstein's Theory of Relativity correctly predicts the curvature of a star's light caused by the sun's mass.

a field known as *differential* or *Riemannian geometry*. It is Riemannian geometry upon which natural-gradient adaptation is based (Amari 1998).

In Riemannian geometry, distance is not measured according to the Euclidean norm in Eq. (2.27). Rather, for two vectors $\mathbf{w}$ and $\mathbf{w} + \delta\mathbf{w}$, where the elements of $\delta\mathbf{w}$ are of small magnitude, we define the distance metric $d_{\mathbf{w}}(\cdot, \cdot)$ at $\mathbf{w}$ as

$$d_{\mathbf{w}}(\mathbf{w}, \mathbf{w} + \delta\mathbf{w}) = \sqrt{\sum_{i=1}^{N}\sum_{j=1}^{N} \delta w_i \delta w_j g_{ij}(\mathbf{w})} \tag{2.28}$$

$$= \sqrt{\delta\mathbf{w}^T \mathbf{G}(\mathbf{w}) \delta\mathbf{w}} \tag{2.29}$$

where $\mathbf{G}(\mathbf{w})$, the *Riemannian metric tensor*, is an $(N \times N)$ positive-definite matrix whose (i, j)th entry is $g_{ij}(\mathbf{w})$. The Riemannian metric tensor characterizes the intrinsic curvature of a particular manifold in N-dimensional space. In the case of the Euclidean coordinate system in which variations in each component of $\mathbf{v}$ are orthogonal to each other, $\mathbf{G}(\mathbf{v}) = \mathbf{I}$ is the identity matrix, such that Eq. (2.28) reduces to Eq. (2.27).

When the nature of the manifold can be described in terms of a transformation of Euclidean orthogonal space with coordinate vector $\mathbf{v}$ to $\mathbf{w}$, then one can determine the form of $\mathbf{G}(\mathbf{w})$ through the relationship

$$d_E^2(\mathbf{v}, \mathbf{v} + \delta\mathbf{v}) = d_{\mathbf{w}}^2(\mathbf{w}, \mathbf{w} + \delta\mathbf{w}) \tag{2.30}$$

where $\delta\mathbf{v}$ is small and $\mathbf{w} + \delta\mathbf{w}$ is the transformed value of $\mathbf{v} + \delta\mathbf{v}$. Examples below illustrate this calculation.

Example 3. *The Riemannian Metric Tensor in Polar Coordinate Space.* Define $\mathbf{v} = [x \ y]^T$ for notational simplicity, such that the pair $\{x, y\}$ define a point in two-dimensional Euclidean space. We can represent this point using polar coordinates via the relationship

$$x = r\cos\theta, \qquad y = r\sin\theta \tag{2.31}$$

where $\mathbf{w} = [r \ \theta]^T$ represents the same point in polar space. The inverse transformation is

$$r = \sqrt{x^2 + y^2}, \qquad \theta = \arctan(y/x) \tag{2.32}$$

where the arctan $(\cdot)$ function ranges from $-\pi$ to π, depending on the signs of x and y.

The distance between $\mathbf{v}$ and $\mathbf{v} + \delta\mathbf{v}$ in Euclidean terms is

$$d_E(\mathbf{v}, \mathbf{v} + \delta\mathbf{v}) = \|\delta\mathbf{v}\|_2^2 = \sqrt{\delta x^2 + \delta y^2} \tag{2.33}$$

The Riemannian metric tensor $\mathbf{G}(\mathbf{w})$ for polar space is defined such that Eq. (2.30) holds. Noting the relationship in Eq. (2.31), we have

$$\mathbf{v} + \delta\mathbf{v} = \begin{bmatrix} (r + \delta r)\cos(\theta + \delta\theta) \\ (r + \delta r)\sin(\theta + \delta\theta) \end{bmatrix} \tag{2.34}$$

$$= \begin{bmatrix} r\cos\theta + \delta r\cos\theta - \delta\theta r\sin\theta \\ r\sin\theta + \delta r\sin\theta + \delta\theta r\cos\theta \end{bmatrix} \tag{2.35}$$

where we have neglected higher-order terms of the form $\delta r\delta\theta^i$ and $\delta\theta^{i+1}$ for $i \geq 1$. Subtracting $\mathbf{v}$ from both sides of Eq. (2.35) gives

$$\delta\mathbf{v} = \begin{bmatrix} \delta r\cos\theta - \delta\theta r\sin\theta \\ \delta r\sin\theta + \delta\theta r\cos\theta \end{bmatrix} \tag{2.36}$$

Therefore, Eq. (2.33) becomes

$$d_E^2(\mathbf{v}, \mathbf{v} + \delta\mathbf{v}) = \delta r^2 + r^2\delta\theta^2 \tag{2.37}$$

$$= \delta\mathbf{w}^T\mathbf{G}(\mathbf{w})\delta\mathbf{w} \tag{2.38}$$

where the Riemannian metric tensor for $\mathbf{w}$ is found to be

$$\mathbf{G}(\mathbf{w}) = \begin{bmatrix} 1 & 0 \\ 0 & r^2 \end{bmatrix} \tag{2.39}$$

Example 4. *The Riemannian Metric Tensor in Spherical-Coordinate Space.* A similar calculation for the three-dimensional spherical-coordinate transformation can be performed, in which $\mathbf{v} = [x\ y\ z]^T$, $\mathbf{w} = [r\ \theta\ \varphi]^T$, and

$$\mathbf{v} = \begin{bmatrix} r\cos\theta \\ r\sin\theta\cos\varphi \\ r\sin\theta\sin\varphi \end{bmatrix} \tag{2.40}$$

In this case, it is found that

$$\|\delta\mathbf{v}\|_2^2 = \delta r^2 + r^2\delta\theta^2 + r^2\sin^2\theta\delta\varphi^2 \tag{2.41}$$

such that

$$\mathbf{G}(\mathbf{w}) = \begin{bmatrix} 1 & 0 & 0 \\ 0 & r^2 & 0 \\ 0 & 0 & r^2\sin^2\theta \end{bmatrix} \tag{2.42}$$

Example 5. *The Riemannian Metric Tensor in a Linearly Transformed Euclidean Space.* Although the previous examples indicate that $\mathbf{G}(\mathbf{w})$ depends on the value of $\mathbf{w}$ in general, it can be constant valued in certain cases. The most celebrated example is when $\mathbf{w}$ is obtained from $\mathbf{v}$ in standard Euclidean space via the linear transformation

$$\mathbf{w} = \mathbf{T}\mathbf{v} \tag{2.43}$$

where $\mathbf{T}$ is an $(N \times N)$ transformation matrix with entries t_{ij}. Then it is straightforward to show that

$$\|\delta\mathbf{v}\|_2^2 = \delta\mathbf{w}^T\mathbf{T}^T\mathbf{T}\delta\mathbf{w} \tag{2.44}$$

such that

$$\mathbf{G}(\mathbf{w}) = \mathbf{T}^T\mathbf{T} \tag{2.45}$$

independent of the value of $\mathbf{w}$. Note that, unlike the previous cases, $\mathbf{G}(\mathbf{w})$ in Eq. (2.45) is not diagonal, as is the case in general.

2.3.2 Natural-Gradient Descent

As discussed previously, gradient adaptation involves iteratively calculating the gradient of a cost-function surface and then changing the current parameter estimates according to this gradient estimate. In Euclidean coordinates, the gradient is defined in Eq. (2.4). It should come as no surprise, therefore, that cost functions that emulate the Euclidean distance measure in Eq. (2.27) are well matched to gradient-descent adaptation. In fact, if

$$\mathcal{J}(\mathbf{w}) = c\|\mathbf{w} - \mathbf{w}_{\text{opt}}\|_2^2 \tag{2.46}$$

where c is an arbitrary constant, then

$$-\frac{\partial\mathcal{J}(\mathbf{w}(k))}{\partial\mathbf{w}} = -2c(\mathbf{w}(k) - \mathbf{w}_{\text{opt}}) \tag{2.47}$$

such that a step-size value of $\mu = 1/(2c)$ allows the steepest-descent procedure in Eq. (2.6) to converge to $\mathbf{w}_{\text{opt}}$ in one step, irrespective of $\mathbf{w}(k)$! Such is the case for $\mathcal{J}_E(\mathbf{w})$ in Eq. (2.18) as indicated by Figure 2.1a. This result is obtained because the negative gradient of this particular cost function at any $\mathbf{w}$ points toward the minimum of the error surface.

In practice, however, the cost function to be minimized is not of the form in Eq. (2.46), and thus a gradient in the Euclidean space of the parameters is not the best for searching this cost function. Moreover, the underlying space of parameters is not Euclidean, but is curved and distorted, that is, Riemannian.

In this case, the negative of the standard gradient $-\partial \mathscr{J}(\mathbf{w})/\partial\mathbf{w}$ does not represent the steepest-descent direction of the cost function $\mathscr{J}(\mathbf{w})$ in the parameter space. The steepest-descent direction of $\mathscr{J}(\mathbf{w})$ at $\mathbf{w}$ is instead defined by the direction vector $\delta\mathbf{w}$ that minimizes $\mathscr{J}(\mathbf{w}+\delta\mathbf{w})$ under the constraint

$$d_{\mathbf{w}}(\mathbf{w}, \mathbf{w}+\delta\mathbf{w}) = \sqrt{\delta\mathbf{w}^T \mathbf{G}(\mathbf{w})\delta\mathbf{w}} = \varepsilon \qquad (2.48)$$

when ε is small. Simple variational calculus yields

$$\delta\mathbf{w} = -\mu\mathbf{G}^{-1}(\mathbf{w})\frac{\partial\mathscr{J}(\mathbf{w})}{\partial\mathbf{w}} \qquad (2.49)$$

where μ is a positive constant. This result motivates an alternative iterative procedure for minimizing $\mathscr{J}(\mathbf{w})$, a procedure that we refer to as *natural-gradient adaptation*. We begin by stating the natural-gradient algorithm for parameter estimation.

Natural-Gradient Adaptation (Amari 1998) For any suitably, smooth gradient-searchable cost function $\mathscr{J}(\mathbf{w}(k))$, natural-gradient adaptation is defined as

$$\mathbf{w}(k+1) = \mathbf{w}(k) - \mu(k)\mathbf{G}^{-1}(\mathbf{w}(k))\frac{\partial\mathscr{J}(\mathbf{w}(k))}{\partial\mathbf{w}} \qquad (2.50)$$

where $\mathbf{G}(\mathbf{w})$ is the Riemannian metric tensor for the parameter vector $\mathbf{w}$, as defined by the manifold of parameters. The form of $\mathscr{J}(\mathbf{w}(k))$ can be defined explicitly as in the Euclidean case, by measurements as in deterministic cost functions, or by instantaneous approximations as in stochastic gradient descent.

The natural-gradient search method applies our notion of curved parameter space to the gradient-adaptation task. For parameter space that is curved, the standard-gradient direction is no longer appropriate for use within a gradient search method. The natural-gradient direction, on the other hand, uses the knowledge of the Riemannian distance structure of the parameter space to alter the gradient direction, thus providing fast and accurate adaptation behavior.

As is shown in Section 2.4, the Riemannian distance structure is naturally defined from the characteristics of the parameter space in most practical applications. To introduce these concepts in a more simplified manner, we first consider parameter spaces that are obtained from transformations of the orthogonal Euclidean space where $\mathscr{J}_E(\mathbf{w})$ in Eq. (2.46) is most appropriate. In such situations, $\mathscr{J}(\mathbf{w}+\delta\mathbf{w})$ is identical to the squared value of the Riemannian distance measure $d_{\mathbf{w}}(\mathbf{w}, \mathbf{w}+\delta\mathbf{w})$ in Eq. (2.28), and thus $\mathbf{G}(\mathbf{w})$ can be determined from the relationship in Eq. (2.30). In what follows, we use this correspondence to determine the form of Eq. (2.50) for the cost functions in Eqs. (2.20) and (2.22) in the linearly transformed Euclidean and polar-coordinate parameter spaces, respectively.

Example 6. *Natural-Gradient Adaptation in Linearly Transformed Euclidean Space.* The cost function in Eq. (2.20) can be written as

$$\mathcal{J}_p(\mathbf{w}) = \{\mathbf{w} - \mathbf{w}_{\text{opt}}\}^T \begin{bmatrix} 1 & \rho \\ \rho & 1 \end{bmatrix} \{\mathbf{w} - \mathbf{w}_{\text{opt}}\} \qquad (2.51)$$

$$= \|\mathbf{T}\{\mathbf{w} - \mathbf{w}_{\text{opt}}\}\|_2^2 \qquad (2.52)$$

where $\mathbf{T}$ is any (2×2) matrix satisfying the condition

$$\mathbf{T}^T\mathbf{T} = \begin{bmatrix} 1 & \rho \\ \rho & 1 \end{bmatrix} \qquad (2.53)$$

Observe that $\mathcal{J}_p(\mathbf{w})$ is equivalent to the squared Euclidean distance between $\mathbf{Tw}$ and $\mathbf{Tw}_{\text{opt}}$, and thus the form of $\mathbf{G}(\mathbf{w})$ is

$$\mathbf{G}(\mathbf{w}) = \begin{bmatrix} 1 & \rho \\ \rho & 1 \end{bmatrix} \qquad (2.54)$$

from Eq. (2.45). Hence, the natural-gradient coefficient update is given by

$$\mathbf{w}(k + 1) = \mathbf{w}(k) - \mu(k) \begin{bmatrix} 1 & \rho \\ \rho & 1 \end{bmatrix}^{-1} \begin{bmatrix} w_1(k) - 1 + \rho w_2(k) \\ w_2(k) + \rho(w_1(k) - 1) \end{bmatrix} \qquad (2.55)$$

$$= \mathbf{w}(k) - \mu(k) \begin{bmatrix} w_1(k) - 1 \\ w_2(k) \end{bmatrix}. \qquad (2.56)$$

Comparing Eqs. (2.56) and (2.21), we see that they are identical. Therefore, natural-gradient adaptation provides behavior that is identical to standard steepest-descent adaptation using the squared Euclidean distance metric $\mathcal{J}_E(\mathbf{w})$ in this case. Since standard-gradient adaptation is ideal for searching the cost function $\mathcal{J}_E(\mathbf{w})$, the natural-gradient method provides ideal adaptation performance in this situation.

Example 7. *Natural-Gradient Adaptation in Polar-Coordinate Space.* We now consider the cost function in Eq. (2.22). Noting that Eqs. (2.23) and (2.31) define the same rectangular-to-polar coordinate transformation, we have immediately from our previous polar-coordinate example that

$$\mathbf{G}(\mathbf{w}(k)) = \begin{bmatrix} 1 & 0 \\ 0 & r^2(k) \end{bmatrix} \qquad (2.57)$$

and thus the natural-gradient algorithm in this case is

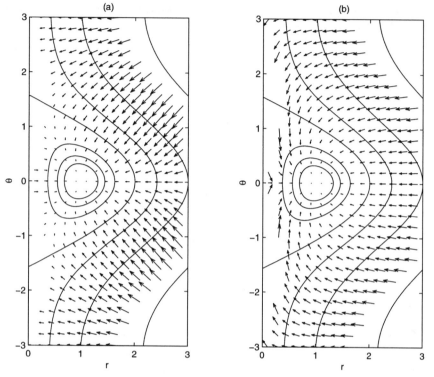

Figure 2.5 (*a*) Standard gradients of the cost function $\mathcal{J}_P(\mathbf{w}(k))$. (*b*) Natural gradients of the cost function $\mathcal{J}_P(\mathbf{w}(k))$.

$$\mathbf{w}(k+1) = \mathbf{w}(k) - \mu(k)\begin{bmatrix} 1 & 0 \\ 0 & r^2(k) \end{bmatrix}^{-1}\begin{bmatrix} r(k) - \cos\theta(k) \\ r(k)\sin\theta(k) \end{bmatrix} \tag{2.58}$$

$$= \mathbf{w}(k) - \mu(k)\begin{bmatrix} r(k) - \cos\theta(k) \\ \dfrac{\sin\theta(k)}{r(k)} \end{bmatrix} \tag{2.59}$$

Figure 2.5*b* shows the natural-gradient flows as superimposed onto the error surface $\mathcal{J}_P(\mathbf{w})$ in this example. Shown for comparison are the standard-gradient flows as determined from Eq. (2.25) in Figure 2.5*a*. In this case, the natural-gradient flow lines are of a more equal magnitude across the error surface as compared to the standard-gradient flows, and more importantly, the magnitudes of the natural gradients no longer vary to the same degree with angle away from $\mathbf{w}_{\mathrm{opt}}$ as do those of the standard-gradients.

Using the relationship in Eq. (2.36), we can translate the natural gradient search direction $\delta\mathbf{w} = -\mu\mathbf{G}^{-1}(\mathbf{w})\partial\mathcal{J}_P(\mathbf{w})/\partial\mathbf{w}$ for small μ into rectangular coordinates $\mathbf{v} = [x\ y]^T$. We can rewrite Eq. (2.36) as

$$\delta \mathbf{v} = \begin{bmatrix} \cos\theta & -r\sin\theta \\ \sin\theta & r\cos\theta \end{bmatrix} \delta \mathbf{w} \tag{2.60}$$

$$= -\mu \begin{bmatrix} \cos\theta & -r\sin\theta \\ \sin\theta & r\cos\theta \end{bmatrix} \begin{bmatrix} r - \cos\theta \\ \dfrac{\sin\theta}{r} \end{bmatrix} \tag{2.61}$$

$$= -\mu \begin{bmatrix} r\cos\theta - 1 \\ r\sin\theta \end{bmatrix} \tag{2.62}$$

Noting the relationship in Eq. (2.31), we have

$$\delta \mathbf{v} = -\mu \begin{bmatrix} x - 1 \\ y \end{bmatrix} \tag{2.63}$$

which is identical to the steepest-descent direction for $\mathcal{J}_E(\mathbf{w})$ in Eq. (2.19). Thus, for small step sizes, the behavior of natural-gradient adaptation in polar coordinates is identical to that of standard-gradient adaptation in rectangular coordinates, as in the previous case. To verify this statement, Figure 2.6 shows the adaptive trajectories for the standard-gradient and natural-gradient methods, respectively, as applied in $\{r, \theta\}$-space for seven different initial conditions $\mathbf{w}(0)$ with $\mu(k) = 0.01$. Here, we have mapped the coefficient trajectories back to the $\{x, y\}$-coordinates to allow direct comparison with Figure 2.1b. Note the similarity of the natural-gradient trajectories with those in Figure 2.1b, which are in sharp contrast with the circuitous paths taken by the standard-gradient method in this Euclidean distance measure.

2.3.3 Relationships to Other Methods

Relationship to Newton's Method We now compare natural-gradient adaptation with a similar search technique known as *Newton's method*. Newton's method employs a coefficient update given by (Luenberger 1984)

$$\mathbf{w}(k+1) = \mathbf{w}(k) - \mu(k)\mathbf{H}^{-1}(\mathbf{w}(k))\frac{\partial \mathcal{J}(\mathbf{w}(k))}{\partial \mathbf{w}} \tag{2.64}$$

where $\mathbf{H}(\mathbf{w})$ is the Hessian of the cost function $\mathcal{J}(\mathbf{w})$ as defined in Eq. (2.3).

Comparing Eqs. (2.50) and (2.64), we see that natural-gradient adaptation is identical to Newton's method when $\mathbf{H}(\mathbf{w}(k)) = \mathbf{G}(\mathbf{w}(k))$. Such will be the case if $\mathcal{J}(\mathbf{w})$ is quadratic in its parameters, as is $\mathcal{J}_p(\mathbf{w})$ in Eq. (2.20). In more general contexts, natural-gradient adaptation is different from Newton's method. As an example, the Hessian for the cost function in Eq. (2.22) is

$$\mathbf{H}(\mathbf{w}) = \begin{bmatrix} 1 & \sin\theta \\ \sin\theta & r\cos\theta \end{bmatrix} \tag{2.65}$$

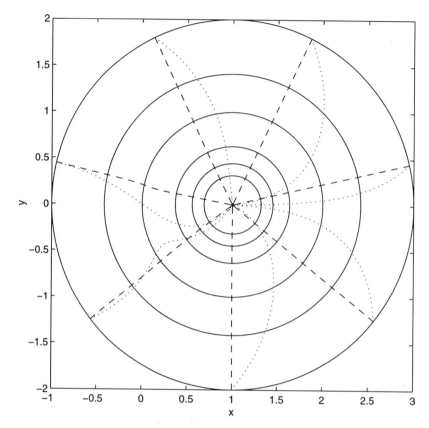

Figure 2.6 Coefficient trajectories for the standard-gradient (dotted) and natural-gradient (dashed) algorithms for iteratively minimizing $\mathscr{J}_P(\mathbf{w})$, as expressed in Euclidean coordinates.

such that the Newton update direction in this case is

$$
-\mathbf{H}^{-1}(\mathbf{w}(k))\frac{\partial \mathscr{J}(\mathbf{w}(k))}{\partial \mathbf{w}}
$$

$$
= -\frac{1}{r(k)\cos\theta(k) - \sin^2\theta(k)}
\begin{bmatrix}
r^2(k)\cos\theta(k) - r(k) \\
\sin\theta(k)\cos\theta(k)
\end{bmatrix}
\quad (2.66)
$$

Figure 2.7 plots the Newton update directions for various points on the surface $\mathscr{J}_P(\mathbf{w})$. As can be seen, they clearly differ from those of the natural gradient in Figure 2.5b.

Although providing efficient adaptive performance locally about a cost-function minimum, one feature of Newton's method makes its use undesirable

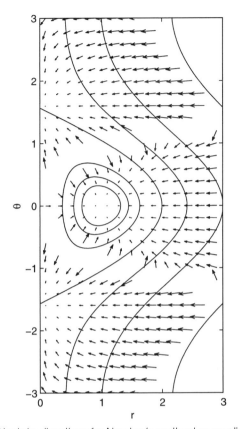

Figure 2.7   Update directions for Newton's method as applied to $\mathscr{J}_P(\mathbf{w})$.

in a general context: *Newton's method can converge to spurious local minima.* This result is due to the fact that the Hessian $\mathbf{H}(\mathbf{w})$ is not guaranteed to be positive definite for all $\mathbf{w}$, and thus saddle points in $\mathscr{J}(\mathbf{w})$ can become attractive solutions for the Newton update. Such is the case in Figure 2.7, where it is seen that the saddle points at $[0\ (\pi/2 + \pi n)]^T$ become local minima when Newton's method is used. Of course, one can alter the matrix $\mathbf{H}(\mathbf{w})$ used in Newton's method to force its positive definiteness (by regularization, for example), but such methods destroy the useful local convergence properties of the technique. As the Riemannian metric tensor $\mathbf{G}(\mathbf{w})$ is always positive definite (Boothby 1986), no spurious minima are created when using natural-gradient adaptation.

Relationship to a True-Gradient Method   Natural-gradient adaptation can be closely related to a true-gradient procedure on a Riemannian manifold (Edelman et al. 1998; Smith 1993). A true gradient procedure is defined as follows.

- Calculate the vector

$$\mathbf{t_G}(k) = -\mu(k)\mathbf{G}^{-1}(\mathbf{w}(k))\frac{\partial \mathscr{J}(\mathbf{w}(k))}{\partial \mathbf{w}} \qquad (2.67)$$

which corresponds to the scaled component of the gradient of $\mathscr{J}(\mathbf{w})$ in the *tangent space* of the Riemannian manifold at $\mathbf{w}(k)$.

- Move a distance equal to $\|\mathbf{t_G}(k)\|$ along a geodesic within the manifold in the direction of $\mathbf{t_G}(k)$.

A *geodesic* is a curve within the parameter space that connects two arbitrary points by an arc of shortest length.

Note that $\mathbf{t_G}(k)$ is identical to the update term on the RHS of Eq. (2.50). Thus, true-gradient adaptation differs from natural-gradient adaptation only in the way the updates are represented in discrete iterative form. Natural-gradient adaptation assumes that $\mu(k)$ is small relative to the components in $\partial \mathscr{J}(\mathbf{w}(k))/\partial \mathbf{w}$ to approximate true geodesic motion within the parameter manifold. In fact, the differential form of Eq. (2.50) given by

$$\frac{d\mathbf{w}(t)}{dt} = -\mu(t)\mathbf{G}^{-1}(\mathbf{w}(t))\frac{\partial \mathscr{J}(\mathbf{w}(t))}{\partial \mathbf{w}} \qquad (2.68)$$

guarantees geodesic motion within the parameter space as measured by the Riemannian metric. As an example, deviations from true geodesic trajectories can be observed in the polar coordinate example of Figure 2.6, in which slight deviations of the natural gradient coefficient trajectories from straight-line paths can be detected through careful study.

2.3.4 Properties of Natural-Gradient Adaptation

In this subsection, we summarize several properties about the natural-gradient adaptive algorithm in Eq. (2.50). Details concerning these properties can be found in the references provided.

- *Necessary conditions on $\mu(k)$ for local convergence of the natural-gradient algorithm in (2.50) about any minimum $\mathbf{w}_{opt}$ satisfying Eqs. (2.2)–(2.5) are*

$$0 < \mu(k) < \frac{2}{\lambda_{max}[\mathbf{G}^{-1}(\mathbf{w}_{opt})\mathbf{H}(\mathbf{w}_{opt})]} \qquad (2.69)$$

where $\lambda_{max}[\mathbf{G}^{-1}(\mathbf{w}_{opt})\mathbf{H}(\mathbf{w}_{opt})]$ is the maximum eigenvalue of the matrix $\mathbf{G}^{-1}(\mathbf{w}_{opt})\mathbf{H}(\mathbf{w}_{opt})$. This result is proven in the Appendix.

- *Natural-Gradient Adaptation Involves Nonlinear Parameter Updates in General.* Thus, characterizing the transient performance characteristics of natural-gradient adaptation is a challenging task. Analysis of natural-gradient algorithms begins with a clear specification of the parameter estimation problem and the form of the natural-gradient algorithm, from which it is sometimes possible to determine the behavior of the parameter updates using standard averaging or ordinary differential equation (ODE) tools. The behaviors of several natural-gradient algorithms have already been studied, among them algorithms for blind source separation (Amari and Cardoso 1997; Amari et al. 1996, 1997a; 1999; Amari and Cichocki 1998; Bell and Sejnowski 1995; Cardoso, 1997; 1998a, 1998b; Cardoso 1994; Cardoso and Laheld 1996; Cichocki et al. 1994), spatial and temporal decorrelation (Douglas and Cichocki 1997; Silva and Almeida 1991), blind equalization (Douglas et al. 1996, 1999), multichannel blind deconvolution (Amari et al. 1999, 1997a, 1997b), and multilayer perceptron training (Amari 1985; Yang and Amari 1998).

The nonlinear form of the natural-gradient algorithm implies that sufficient bounds on the step-size $\mu(k)$ to guarantee stability and convergence of the algorithm's parameters are difficult to obtain. Experience with other existing parameter-estimation algorithms often provides insight into the selection of $\mu(k)$ for the natural-gradient algorithm in specific cases (Douglas and Cichocki 1996, 1997, 1999). In some problem contexts, a step-size sequence can be calculated to guarantee parameter convergence if sufficient information about the problem structure is available (Amari et al. 1997a).

• *For Maximum-Likelihood Estimation, Natural-Gradient Adaptation Is Fisher-Efficient.* The Riemannian metric tensor in the space of the parameters within a statistical model is the Fisher information matrix (Amari 1985; Campbell 1985)

$$\mathbf{G}(\mathbf{w}) = E\left\{ \frac{\partial \log f_{\mathbf{y}}(\mathbf{y}|\mathbf{w})}{\partial \mathbf{w}} \frac{\partial \log f_{\mathbf{y}}(\mathbf{y}|\mathbf{w})}{\partial \mathbf{w}^T} \right\} \qquad (2.70)$$

Employing the natural-gradient algorithm for the negative log-likelihood cost function in Eq. (2.11) in this situation with a step size of $\mu(k) = 1/k$, one obtains parameter estimates whose errors $\tilde{\mathbf{w}}(k) = \mathbf{w}(k) - \mathbf{w}_{\text{opt}}$ have an asymptotic covariance matrix given by (Amari 1998)

$$E\{\tilde{\mathbf{w}}(k)\tilde{\mathbf{w}}^T(k)\} = \frac{1}{k}\mathbf{G}^{-1}(\mathbf{w}_{\text{opt}}) + \mathbf{O}\left(\frac{1}{k^2}\right) \qquad (2.71)$$

which, for large k, approaches the well-known Cramèr-Rao lower bound for unbiased parameter estimates (Rao 1945). In practical terms, this result means that natural-gradient adaptation provides fast adaptation performance in such problems.

• *Natural-Gradient Adaptation Requires Extensive Knowledge of the Structure of the Parameter-Estimation Problem.* Although the natural gradient is local in nature and only depends on the parameter values $\mathbf{w}(k)$, determining $\mathbf{G}(\mathbf{w}(k))$ for use within the procedure usually requires precise knowledge of the problem structure. For example, if signal measurements are corrupted by additive noise terms, then the pdf's of these noise terms must be completely known to determine the form of $\mathbf{G}(\mathbf{w})$ in most problems. However, the amount of information needed to form $\mathbf{G}(\mathbf{w})$ varies from problem to problem, and there exist several practical cases where this information is easily obtained from the problem's inherent structure (Amari et al. 1996; Cardoso and Laheld 1996; Douglas et al. 1996). Examples provided in the next section illustrate these issues.

· *Natural-Gradient Adaptation can be Simpler to Implement than Standard-Gradient Adaptation.* As is shown in the next section, there exist problems for which the update in Eq. (2.50) is simpler to compute than that in Eq. (2.6). Such problems generally involve the estimation of a linear system (e.g., a gain matrix or one or more transfer functions) that is related to the inverse of the unknown signal model. In some cases, the matrix $\mathbf{G}(\mathbf{w})$ is a function only of the system's parameters, and these parameters can combine with the components of $\partial \mathscr{J}(\mathbf{w})/\partial \mathbf{w}$ to simplify the algorithm's structure.

2.4 SIMULATION EXAMPLES

In this section, we apply the natural-gradient adaptation method to several signal-processing tasks. Although not inclusive of all possible uses of the technique, the examples provided here illustrate the forms of the natural-gradient algorithms in specific contexts, and we compare the behaviors of the standard- and natural-gradient algorithms via computer simulations.

2.4.1 Maximum-Likelihood Estimation of a Noisy Sinusoid

Let $y(k)$ be a discrete time series given by

$$y(k) = r_{\text{opt}} \cos(\omega k + \theta_{\text{opt}}) + \eta(k) \tag{2.72}$$

where ω is a known angular frequency constant; r_{opt} and θ_{opt} are unknown amplitude and phase parameters, respectively; and $\eta(k)$ is an independent and identically distributed (i.i.d.) zero-mean random process with pdf $f_\eta(\eta)$. The conditional pdf of $y(k)$ given $\mathbf{w}_{\text{opt}} = [r_{\text{opt}} \ \theta_{\text{opt}}]^T$ is

$$f_y(y(k)|\mathbf{w}_{\text{opt}}) = f_\eta(y(k) - r_{\text{opt}} \cos(\omega k + \theta_{\text{opt}})) \tag{2.73}$$

We wish to estimate the values of r_{opt} and θ_{opt} from knowledge of $f_y(y(k)|\mathbf{w})$.

The structure of this problem fits that of maximum-likelihood estimation, and thus we employ the cost function

$$\mathscr{I}_{ML}(\mathbf{w}(k)) = -E\{\log f_\eta(y(k) - r(k)\cos(\omega k + \theta(k)))\} \qquad (2.74)$$

To further simplify the problem structure, assume that $\eta(k)$ is Gaussian distributed, such that

$$f_y(y(k)|\mathbf{w}) = \frac{1}{\sqrt{2\pi\sigma_\eta^2}} \exp\left(-\frac{1}{2\sigma_\eta^2}|y(k) - r\cos(\omega k + \theta)|^2\right) \qquad (2.75)$$

where σ_η^2 is the variance of $\eta(k)$. Then, Eq. (2.74) becomes

$$\mathscr{I}_{ML}(\mathbf{w}) = \frac{1}{2}\left[\log(2\pi\sigma_\eta^2) + \frac{1}{\sigma_\eta^2}E\{|y(k) - r\cos(\omega k + \theta)|^2\}\right] \qquad (2.76)$$

Finally, to allow the use of signal measurements in place of statistical expectation, we employ the stochastic-gradient algorithm in Eq. (2.17), where in our case $\hat{\mathscr{I}}_{ML}(\mathbf{w}(k))$ is

$$\hat{\mathscr{I}}_{ML}(\mathbf{w}(k)) = \frac{1}{2}\left[\log(2\pi\sigma_\eta^2) + \frac{1}{\sigma_\eta^2}|y(k) - r(k)\cos(\omega k + \theta(k))|^2\right] \qquad (2.77)$$

The standard-gradient descent algorithm is therefore

$$r(k+1) = r(k) + \mu(k)e(k)\cos(\omega k + \theta(k)) \qquad (2.78)$$
$$\theta(k+1) = \theta(k) - \mu(k)e(k)r(k)\sin(\omega k + \theta(k)) \qquad (2.79)$$

where we have absorbed $1/\sigma_\eta^2$ into $\mu(k)$ and

$$e(k) = y(k) - r(k)\cos(\omega k + \theta(k)) \qquad (2.80)$$

is the instantaneous error between the actual and estimated sinusoidal output signals.

For the model in Eq. (2.72), we can write $y(k)$ as

$$y(k) = w_{1,\text{opt}}\cos(\omega k) - w_{2,\text{opt}}\sin(\omega k) \qquad (2.81)$$

where the relationship between $[w_{1,\text{opt}}\ w_{2,\text{opt}}]^T$ and $[r_{\text{opt}}\ \theta_{\text{opt}}]^T$ is given by Eq. (2.23). Therefore, it can be shown that

$$\mathscr{I}_{ML}(\mathbf{w}) = \frac{1}{2}\left[\log(2\pi\sigma_\eta^2) + \frac{1}{2\sigma_\eta^2}\{|r\cos\theta - w_{1,\text{opt}}|^2 + |r\sin\theta - w_{2,\text{opt}}|^2\}\right] \qquad (2.82)$$

where we have taken both statistical and time averages of the quantities depending on $y(k)$ in Eq. (2.76). Since the negative log-likelihood cost function for this problem is the squared Euclidean metric in $\mathbf{v} = [r\cos\theta \; r\sin\theta]^T$, $\mathbf{G}(\mathbf{w})$ is given by Eq. (2.39), and thus the natural gradient algorithm is

$$r(k+1) = r(k) + \mu(k)e(k)\cos(\omega k + \theta(k)) \tag{2.83}$$

$$\theta(k+1) = \theta(k) - \mu(k)e(k)\frac{\sin(\omega k + \theta(k))}{r(k)} \tag{2.84}$$

Note that the standard-gradient and natural-gradient updates differ only in the way $\theta(k)$ is adjusted.

We have simulated the behaviors of Eqs. (2.78)–(2.79) and of Eqs. (2.83)–(2.84) in this situation. Here, we have chosen $\omega = 0.5$ and $r_{opt} = 1$, and we have selected $r(0)$, $\theta(0)$, and θ_{opt} as independent uniformly distributed random variables in the range $0 \le r(0) \le 2$, $-\pi < \theta(0) \le \pi$, and $-\pi < \theta_{opt} < \pi$ for each simulation run. For each algorithm, we compute the ensemble average of the noiseless squared error signal $|r_{opt}\cos(\omega k + \theta_{opt}) - r(k)\cos(\omega k + \theta(k))|^2$ over 1000 simulation runs. The step sizes for the standard gradient and natural-gradient algorithms were both chosen as $\mu(k) = 0.2$, such that the average error for both algorithms is the same in steady state.

Figure 2.8 shows the average behaviors of the standard gradient and natural-gradient algorithms as applied to the stochastic cost function in Eq. (2.77). As can be seen, the natural-gradient algorithm's squared output error converges to approximately 10^{-5} in about 70 iterations, whereas the standard-gradient algorithm requires about 140 iterations to achieve the same performance level. Both algorithms perform identically in steady state. Similar relative performance behaviors were observed for different choices of ω and μ. The natural-gradient algorithm provides inherently faster convergence to a low steady-state error as compared to standard-gradient descent in this situation.

2.4.2 Single-Layer Perceptron Training

We now consider the problem of estimating the parameters of a single-layer perceptron in a system identification task (Haykin 1994). In our model, the output $y(k)$ of the perceptron is computed as

$$\hat{y}(k) = g(u(k)) \tag{2.85}$$

$$u(k) = \sum_{i=1}^{N} w_i(k)x_i(k) = \mathbf{w}^T(k)\mathbf{x}(k) \tag{2.86}$$

where $\mathbf{x}(k) = [x_1(k) \; x_2(k) \cdots x_N(k)]^T$ is an input signal vector whose elements are zero-mean jointly Gaussian distributed with a pdf of

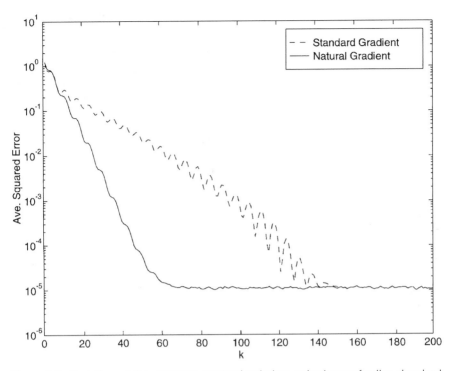

Figure 2.8 Evolution of the average squared noiseless output error for the standard-and natural-gradient algorithms in the sinusoidal estimation task.

$$f_{\mathbf{x}}(\mathbf{x}) = \frac{1}{(2\pi)^{N/2}(\det \mathbf{R})_{\mathbf{xx}}^{1/2}} \exp\left(-\frac{1}{2}\mathbf{x}^T \mathbf{R}_{\mathbf{xx}}^{-1}\mathbf{x}\right) \qquad (2.87)$$

and

$$\mathbf{R}_{\mathbf{xx}} = E\{\mathbf{x}(k)\mathbf{x}^T(k)\} \qquad (2.88)$$

is the *autocorrelation matrix* of the input signal vector.

Several choices for the output nonlinearity $g(u)$ are possible. In order to make the evaluation of the natural gradient tractable, we assign

$$g(u) = \mathrm{erf}\left(\sqrt{\frac{\alpha}{4}}u\right) \qquad (2.89)$$

$$= \sqrt{\frac{\alpha}{\pi}} \int_0^u \exp\left(-\frac{\alpha t^2}{4}\right) dt \qquad (2.90)$$

where α is a positive parameter. A plot of $\mathrm{erf}(\sqrt{\alpha/4}u)$ for different values of α is

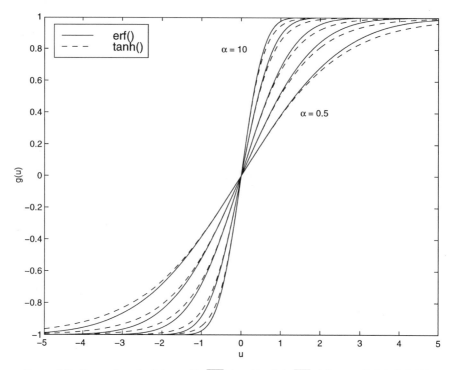

Figure 2.9 Examples of $g(u) = \text{erf}(\sqrt{\alpha/4}u)$ and $\tanh(\sqrt{\alpha/\pi}u)$ for $\alpha = \{0.5, 1, 2, 5, 10\}$.

provided in Figure 2.9, along with the often-chosen $\tanh(\beta u)$ nonlinearity for $\beta = \sqrt{\alpha/\pi}$.

We assume that $y(k)$ is generated according to the model

$$y(k) = g(\mathbf{w}_{\text{opt}}^T(k)\mathbf{x}(k)) + \eta(k) \tag{2.91}$$

where $\eta(k)$ is a sequence of uncorrelated zero-mean Gaussian random variables. With this choice, a suitable cost function to minimize is the mean-squared-error (MSE) cost given by

$$\mathscr{J}_{MSE}(\mathbf{w}(k)) = \tfrac{1}{2} E\{|y(k) - g(\mathbf{w}^T(k)\mathbf{x}(k))|^2\} \tag{2.92}$$

or its instantaneous equivalent

$$\hat{\mathscr{J}}_{MSE}(\mathbf{w}(k)) = \tfrac{1}{2}|y(k) - g(\mathbf{w}^T(k)\mathbf{x}(k))|^2 \tag{2.93}$$

The standard steepest-descent procedure for this task is (Haykin 1994)

$$\mathbf{w}(k+1) = \mathbf{w}(k) + \mu(k)[y(k) - g(u(k))]g'(u(k))\mathbf{x}(k) \qquad (2.94)$$

$$g'(u) = \sqrt{\frac{\alpha}{\pi}} e^{-\alpha u^2/4}, \qquad (2.95)$$

where $u(k)$ is as defined in Eq. (2.86).

From Amari (1998), the Riemannian metric tensor for this task is equal to the Fisher information matrix as given by

$$\mathbf{G}(\mathbf{w}) = \frac{1}{\sigma_\eta^2} E\{g'^2(\mathbf{w}^T\mathbf{x})\mathbf{x}\mathbf{x}^T\} \qquad (2.96)$$

Moreover, using the results of (Bershad 1988), we can evaluate the expectation on the RHS of Eq. (2.96) as

$$E\{g'^2(\mathbf{w}^T\mathbf{x})\mathbf{x}\mathbf{x}^T\} = \frac{\alpha}{\pi} \frac{\det \overline{\mathbf{R}}(\mathbf{w})}{\det \mathbf{R}_{xx}} \overline{\mathbf{R}}(\mathbf{w}) \qquad (2.97)$$

where $\overline{\mathbf{R}}(\mathbf{w})$ is defined as

$$\overline{\mathbf{R}}(\mathbf{w}) = [\mathbf{R}_{xx}^{-1} + \alpha\mathbf{w}\mathbf{w}^T]^{-1} = \mathbf{R}_{xx} - \frac{\mathbf{R}_{xx}\mathbf{w}\mathbf{w}^T\mathbf{R}_{xx}}{\mathbf{w}^T\mathbf{R}_{xx}\mathbf{w} + \alpha^{-1}} \qquad (2.98)$$

Therefore, the natural gradient algorithm in this case is

$$\mathbf{w}(k+1) = \mathbf{w}(k) + \bar{\mu}(k)[y(k) - g(u(k))]g'(u(k))[\mathbf{R}_{xx}^{-1}\mathbf{x}(k) + \alpha u(k)\mathbf{w}(k)] \qquad (2.99)$$

$$\bar{\mu}(k) = \mu(k)\frac{\pi}{\alpha}\frac{\det \mathbf{R}_{xx}}{\det \overline{\mathbf{R}}(\mathbf{w}(k))} \qquad (2.100)$$

In this algorithm, the update direction is proportional to the sum of the two vectors $\mathbf{R}_{xx}^{-1}\mathbf{x}(k)$ and $u(k)\mathbf{w}(k)$. The former vector is the same one used in Newton's method for cost functions that are quadratic in the parameter estimates, and the latter vector represents an anti-Hebbian term based on the perceptron's linear output $u(k)$. The nonlinearity parameter α controls the relative contributions of these two terms within the update.

Figure 2.10 illustrates the behaviors of the stochastic forms of the standard- and natural-gradient algorithms in a parameter-estimation task. In this case, we have generated sequences of $N = 4$-element independent jointly Gaussian random vectors $\mathbf{x}(k)$ with autocorrelation matrix

$$\mathbf{R}_{xx} = \begin{bmatrix} 1.0 & 1.1 & 0.8 & 0.3 \\ 1.1 & 2.0 & 0.9 & 0.5 \\ 0.8 & 0.9 & 0.7 & 0.3 \\ 0.3 & 0.5 & 0.3 & 0.3 \end{bmatrix} \qquad (2.101)$$

and we have chosen $\mathbf{w}_{\mathrm{opt}} = [2\ 1\ -0.5\ 0.2]^T$ to generate $y(k)$ in Eq. (2.91),

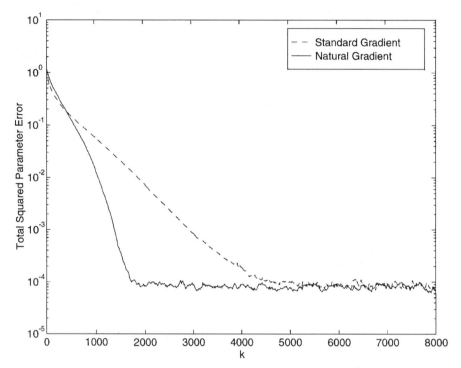

Figure 2.10 Evolution of the averaged parameter error $\|\mathbf{w}(k) - \mathbf{w}_{\text{opt}}\|_2^2$ for the standard-gradient and natural-gradient algorithms in the single-layer perceptron training example.

where $\sigma_\eta^2 = 0.0001$ and $\alpha = 10$. For each algorithm, $\mathbf{w}(0)$ was generated by adding an i.i.d. uniformly distributed-$[-1, 1]$ random variable to each element of $\mathbf{w}_{\text{opt}}$. Ensemble averages of the parameter error $\|\mathbf{w}(k) - \mathbf{w}_{\text{opt}}\|_2^2$ from 100 different simulation runs have been computed for each algorithm. The behaviors of both Eq. (2.94) and Eqs. (2.99)–(2.100) are shown, where we have chosen $\mu(k)$ equal to 0.5 and 0.001 for Eq. (2.94) and Eqs. (2.99)–(2.100), respectively, to allow each algorithm to have the same average parameter error in steady state. The standard-gradient and natural-gradient algorithms converge to an average error level of 8.5×10^{-5} in approximately 2000 and 5000 iterations, respectively. Interestingly, a modified algorithm employing a search direction of $\mathbf{R}_{\mathbf{xx}}^{-1}\mathbf{x}(k)$ in place of $\mathbf{x}(k)$ in Eq. (2.94) performed poorly with respect to both the standard-gradient and natural-gradient algorithms in this situation. Thus, the anti-Hebbian term is necessary to improve the system's adaptation performance in this perceptron training example.

2.4.3 Instantaneous Blind Source Separation

We now consider the problem of instantaneous blind source separation, also known as independent-component analysis (Amari et al. 1996; Bell and

Sejnowski 1995; Cardoso and Laheld 1996; Oja and Karhunen 1995). As will be seen, natural-gradient adaptation is ideally suited to this task. In fact, it was the blind source-separation problem for which similar techniques were first successfully applied (Cardoso et al. 1994; Cichocki et al. 1994), although their relationship within the framework of natural gradient adaptation has been identified more recently (Amari et al. 1996).

Let $\mathbf{s}(k) = [s_1(k)\ s_2(k) \cdots s_n(k)]^T$ be a sequence of n-dimensional vectors whose elements are statistically independent of one another. We measure an n-dimensional vector $\mathbf{x}(k) = [x_1(k)\ x_2(k) \cdots x_n(k)]^T$ given by

$$\mathbf{x}(k) = \mathbf{A}\mathbf{s}(k) \tag{2.102}$$

where $\mathbf{A}$ is an unknown $(n \times n)$-dimensional mixing matrix. In blind source separation, the measurements in $\mathbf{x}(k)$ and any approximate knowledge of the source signal distributions are used to calculate a set of n output signals in the vector $\mathbf{y}(k) = [y_1(k)\ y_2(k) \cdots y_n(k)]^T$ so that a scaled version of each of the signals in $\mathbf{s}(k)$ is contained in $\mathbf{y}(k)$. Because of the linear nature of Eq. (2.102), a linear system of the form

$$y(k) = \mathbf{W}(k)\mathbf{x}(k) = \mathbf{W}(k)\mathbf{A}\mathbf{s}(k) \tag{2.103}$$

where $\mathbf{W}(k)$ is an $(n \times n)$-dimensional matrix of adjustable coefficients, is sufficient to solve this task. The goal is to adjust $\mathbf{W}(k)$ such that

$$\lim_{k \to \infty} \mathbf{W}(k)\mathbf{A} = \mathbf{P}\mathbf{D} \tag{2.104}$$

where $\mathbf{P}$ and $\mathbf{D}$ are permutation and nonsingular diagonal scaling matrices, respectively.

A useful cost function for the blind source-separation task is the Kullback-Leibler divergence given by (Amari et al. 1996; Cardoso and Laheld 1996; Cover and Thomas 1991)

$$\mathscr{J}_{KL}(\mathbf{W}) = \int f_s(\mathbf{s}) \log\left(\frac{f_s(\mathbf{s})}{f_{\mathbf{W}}(\mathbf{s})}\right) ds \tag{2.105}$$

where $f_s(\mathbf{s})$ is the joint pdf of the source signal vector $\mathbf{s}(k)$ and $f_{\mathbf{W}}(\mathbf{s})$ is a parametric model of the joint pdf of $\mathbf{s}(k)$. For the parametric model given by

$$f_{\mathbf{W}}(\mathbf{s}(k)) = |\det(\mathbf{W}\mathbf{A})| \prod_{i=1}^{n} f_i(y_i(k)) \tag{2.106}$$

$$= |\det \mathbf{A}|\,|\det \mathbf{W}| \prod_{i=1}^{n} f_i(y_i(k)) \tag{2.107}$$

where $f_i(s)$ is the pdf of the source signal extracted at the ith output of the system, the Kullback-Leibler divergence measures the degree of statistical independence between the output signals. By minimizing this criterion with respect to **W**, one can obtain a solution to the blind source-separation task even when $f_i(y_i)$ is not an exact match to any of the marginal distributions of $s_j(k)$ within **s**(k), so long as $f_i(y_i)$ is "close enough" to the true marginal distribution of the ith extracted source signal (Amari et al. 1997a; Cardoso and Laheld 1996). Such a cost function can be shown to be equivalent (Cardoso 1997) both to maximizing the information transfer from inputs to outputs (Bell and Sejnowski 1995), and minimizing the mutual information between the system's outputs (Amari et al. 1996).

To obtain a stochastic-gradient algorithm that approximately minimizes $\mathcal{J}_{KL}(\mathbf{W})$ in an iterative fashion, we use the equivalent instantaneous cost function

$$\hat{\mathcal{J}}_{KL}(\mathbf{W}) = -\log\left[|\det \mathbf{W}| \prod_{i=1}^{n} f_i(y_i(k))\right] \tag{2.108}$$

$$= -\log|\det \mathbf{W}| - \sum_{i=1}^{n} \log f_i(y_i(k)) \tag{2.109}$$

Taking derivatives of $\hat{\mathcal{J}}_{KL}(\mathbf{W})$ with respect to each element of **W** yields

$$-\frac{\partial \hat{\mathcal{J}}_{KL}(\mathbf{W})}{\partial \mathbf{W}} = \mathbf{W}^{-T} - \boldsymbol{\phi}(\mathbf{y}(k))\mathbf{x}^T(k) \tag{2.110}$$

where we have defined

$$\boldsymbol{\phi}(\mathbf{y}) = \left[-\frac{\partial \log f_1(y_1)}{\partial y_1} - \frac{\partial \log f_1(y_2)}{\partial y_2} \cdots - \frac{\partial \log f_n(y_n)}{\partial y_n}\right]^T \tag{2.111}$$

Thus, the stochastic-gradient-descent algorithm for this cost function is (Amari et al. 1996; Bell and Sejnowski 1995)

$$\mathbf{W}(k+1) = \mathbf{W}(k) - \mu(k)\frac{\partial \hat{\mathcal{J}}_{KL}(\mathbf{W}(k))}{\partial \mathbf{W}} \tag{2.112}$$

$$= \mathbf{W}(k) + \mu(k)[\mathbf{W}^{-T}(k) - \boldsymbol{\phi}(\mathbf{y}(k))\mathbf{x}^T(k)] \tag{2.113}$$

The algorithm in Eq. (2.113) requires the inverse of the estimated demixing matrix $\mathbf{W}(k)$, which is computationally intensive to compute. Moreover, the algorithm is known to have a poor convergence speed (Bell and Sejnowski). Therefore, we desire a modified algorithm that is both simpler to implement and has an enhanced performance. The natural-gradient algorithm provides such an update in this case. In (Amari 1998), it is shown that the (i, j, k, l)th entry of the the Riemannian metric tensor in the space of separating matrices $\mathbf{W}$ with the output probability model in Eq. (2.107) is

$$[\mathbf{G}(\mathbf{W})]_{ij,kl} = \sum_{m=1}^{n} \delta_{ij} [\mathbf{W}^{-1}]_{jm} [\mathbf{W}^{-1}]_{lm} \tag{2.114}$$

where δ_{ij} is the Kronnecker delta function, and the natural-gradient algorithm for this task is

$$\mathbf{W}(k+1) = \mathbf{W}(k) - \mu(k) \frac{\partial \hat{\mathscr{J}}_{KL}(\mathbf{W}(k))}{\partial \mathbf{W}} \mathbf{W}^T(k) \mathbf{W}(k) \tag{2.115}$$

Combining Eqs. (2.110) and (2.115), we obtain

$$\mathbf{W}(k+1) = \mathbf{W}(k) + \mu(k) [\mathbf{I} - \phi(\mathbf{y}(k)) \mathbf{y}^T(k)] \mathbf{W}(k) \tag{2.116}$$

This simple update no longer requires $\mathbf{W}^{-T}(k)$ to implement. Its performance is also much improved over the standard stochastic-gradient algorithm. The asymptotic performance and stability behaviors of this natural-gradient algorithm have been extensively studied; for details, the reader is referred to Amari and Cardoso (1997), Amari and Cichocki (1998), Cardoso (1998b), and Cardoso and Laheld (1996).

This algorithm can also be derived by determining a multiplicative update for $\mathbf{W}(k)$ of the form

$$\mathbf{W}(k+1) = [\mathbf{I} + \mu(k) \mathbf{E}(\mathbf{y}(k))] \mathbf{W}(k) \tag{2.117}$$

where the matrix $\mathbf{E}(\mathbf{y}(k))$ depends on the output signals alone. Such a methodology is considered in Cardoso and Laheld (1996), in which the algorithm is termed the *relative gradient*. In blind source-separation tasks, the natural and relative gradients are equivalent (Cardoso 1998a), as they are both based on an invariance property that can be imposed on the space of demixing matrices $\mathbf{W}(k)$ due to its Lie group structure (Amari 1998). The algorithms so obtained from this invariance property are *equivariant* with respect to all possible non-singular mixing matrices $\mathbf{A}$, so that they provide equivalent performance independent of the nature of $\mathbf{A}$. This remarkable property can be readily seen by considering the combined system matrix

$$\mathbf{C}(k) = \mathbf{W}(k) \mathbf{A} \tag{2.118}$$

The update in Eq. (2.116) can be written in terms of $\mathbf{C}(k)$ as

$$\mathbf{C}(k+1) = \mathbf{C}(k) + \mu(k) [\mathbf{I} - \phi(\mathbf{C}(k) \mathbf{s}(k)) \mathbf{s}^T(k) \mathbf{C}^T(k)] \mathbf{C}(k) \tag{2.119}$$

which does not depend explicitly on $\mathbf{A}$.

We have simulated the standard-gradient and natural-gradient algorithms in Eqs. (2.113) and (2.116), respectively, for a particular source-separation task. Three independent random sources—one uniform-$[-1, 1]$-distributed and two binary-$\{\pm1\}$-distributed—have been generated to produce the source vector sequence $\mathbf{s}(k)$, and Eq. (2.102) is used to create $\mathbf{x}(k)$ where

$$\mathbf{A} = \begin{bmatrix} 0.4 & 1.0 & 0.7 \\ 0.6 & 0.5 & 0.5 \\ 0.3 & 0.7 & 0.2 \end{bmatrix} \qquad (2.120)$$

The condition number of $\mathbf{A}E\{\mathbf{s}(k)\mathbf{s}^T(k)\}\mathbf{A}^T$ is 115.2 in this case. For the two algorithms, we have chosen $\phi_i(y) = y^3$ for each element of $\boldsymbol{\phi}(\mathbf{y})$, as this choice enables the algorithms to separate sources with a negative kurtosis (Amari et al. 1997a; Cardoso and Laheld 1996). One hundred trials have been run, in which $\mathbf{W}(0)$ was a different random orthogonal matrix such that $\mathbf{W}(0)\mathbf{W}^T(0) = 0.25\mathbf{I}$, and the ensemble-averaged value of the performance factor

$$\zeta(k) = \frac{1}{m} \sum_{i=1}^{m} \left(\sum_{j=1}^{m} \frac{c_{ij}^2(k)}{\max_{1 \le l_i \le m} c_{il_i}^2(k)} - 1 \right), \qquad l_i \ne l_j \quad \text{for} \quad i \ne j \qquad (2.121)$$

has been computed for each algorithm, where $c_{ij}(k)$ is the (i, j)th entry of $\mathbf{C}(k)$. The value of $\zeta(k)$ measures the total source signal crosstalk in the output signals $\{y_i(k)\}$.

Figure 2.11 shows the evolution of $\zeta(k)$ for both the standard- and natural-gradient algorithms in this situation, where the step-size values of 0.016 and 0.02, respectively, yield nearly the same value of $\zeta(k) \approx 0.0074$ for both algorithms in steady state. As can be seen, the standard-gradient algorithm converges after approximately 6500 iterations on average, whereas the natural-gradient converges in approximately 600 iterations on average for these step-size choices.

When the linear vector function $\boldsymbol{\phi}(\mathbf{y}) = \mathbf{y}$ is used in Eq. (2.116), one obtains an algorithm that both normalizes and decorrelates the components of the output signal vector, such that $E\{\mathbf{y}(k)\mathbf{y}^T(k)\} \to \mathbf{I}$ (Silva and Almeida 1991). Such an algorithm is obtained from $\hat{\mathcal{J}}_{KL}(\mathbf{w})$ when each $f_i(y)$ is a Gaussian pdf with zero mean and unit variance. A convergence analysis and study of the properties of this and other multichannel decorrelating algorithms is provided in Douglas and Cichocki (1997).

2.4.4 Blind Equalization

In digital communications, a digital message represented by the signal $s(k)$ is encoded, modulated, and transmitted through a physical medium or channel before arriving at a desired receiver. Once received, the task of demodulating

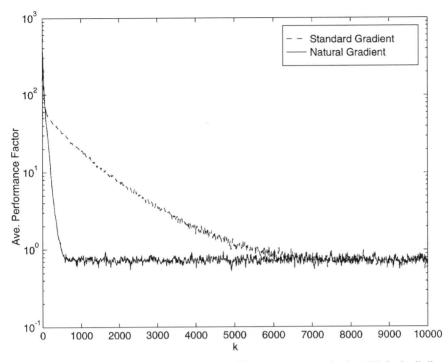

Figure 2.11 Evolution of the average value of the performance factor $\zeta(k)$ for both the standard- and natural-gradient algorithms in the blind source-separation example.

and decoding the digital message usually requires some knowledge of the channel in order to estimate the original message with the highest accuracy. In many cases, the effects of the modulation–transmission–demodulation process can be accurately represented by the linear discrete-time filter (Proakis and Salehi 1994)

$$x(k) = \sum_{p=-\infty}^{\infty} a_p s(k - p) \tag{2.122}$$

where $x(k)$ is the baseband-sampled received signal and a_p is the impulse response of the channel model. The simplest method by which to "unravel" the effects of the channel is *equalization*, in which $x(k)$ is processed by a second discrete-time filter as

$$y(k) = \sum_{p=-\infty}^{\infty} w_p(k) x(k - p) \tag{2.123}$$

where $w_p(k)$ is the impulse response of the equalizer at time k. The task of the equalizer is to adjust each $w_p(k)$ so that

$$y(k) \approx cs(k - \Delta) \qquad (2.124)$$

where c and Δ are an arbitrary constant and integer, respectively. In practice, the equalizer in Eq. (2.123) is unrealistic, as it involves an infinite number of noncausal computations, although we shall assume this equalizer model for purposes of the algorithm description in what follows.

When communications occur only between a single transmitter and receiver, the transmitter and receiver can establish a protocol by which the coefficients $w_p(k)$ of the equalizer can be trained. This protocol usually employs a training sequence $d(k)$ that is known to both the transmitter and receiver. In broadcast applications, however, it is difficult for the transmitter to simultaneously establish and maintain such protocols with multiple receivers. In such cases, equalization must be performed without a training sequence, a technique known as *blind equalization* (Benveniste and Goursat 1984).

Several algorithms for blind equalization have been developed and their properties described. A large class of these algorithms known as *Bussgang techniques* are defined by the update (Bellini 1994)

$$w_p(k + 1) = w_p(k) - \mu(k) \frac{\partial \hat{\mathscr{J}}_B(y(k))}{\partial w_p} \qquad (2.125)$$

$$= w_p(k) + \mu(k)g(y(k))x^*(k - p) \qquad (2.126)$$

where $*$ denotes complex-conjugate and $g(y) = -\partial \hat{\mathscr{J}}_B(y)/\partial y$ is the derivative of a cost function $\hat{\mathscr{J}}(y)$ that depends on the equalizer's output signal. These algorithms are simply stochastic-gradient adaptive procedures on the instantaneous cost function $\hat{\mathscr{J}}_B(y(k))$. As an example, consider the instantaneous *constant-modulus* cost function (Godard 1980; Treichler and Agee 1983)

$$\hat{\mathscr{J}}_{CMA}(y(k)) = \tfrac{1}{4} |A_2^2 - |y(k)|^2|^2 \qquad (2.127)$$

where A_2 is a magnitude parameter. Then the constant-modulus algorithm is given by Eq. (2.126) with

$$g(y(k)) = [A_2^2 - |y(k)|^2]y(k) \qquad (2.128)$$

We now describe the natural-gradient algorithm for the blind equalization task. We can write the coefficient updates for the infinite-length equalizer in Eq. (2.123) in z-domain form as

$$W(z, k + 1) = W(z, k) + \mu(k)g(y(k))z^{-k}X^*(z^{-1}) \qquad (2.129)$$

where

$$W(z,k) = \sum_{p=-\infty}^{\infty} w_p(k)z^{-p} \quad \text{and} \quad X(z) = \sum_{p=-\infty}^{\infty} x(p)z^{-p} \quad (2.130)$$

We can also express the z-transform of the received signal $x(k)$ as

$$X(z) = A(z)S(z) \quad (2.131)$$

where $A(z)$ and $S(z)$ are the z-transforms of a_p and $s(p)$, respectively. Thus, the equalizer output signal can be expressed as

$$Y(z,k) = W(z,k)A(z)S(z) \quad (2.132)$$

Comparing Eqs. (2.132) and (2.103), we see that they are similar in form. In fact, the derivation of the natural gradient for blind deconvolution and equalization tasks specifies an invariance within the space of filters that is similar in spirit to that used to derive Eq. (2.116) for instantaneous blind source separation. The details of the derivation are provided in (Amari et al. 1999b), where it is shown that the natural gradient algorithm for single-channel blind deconvolution is given by

$$w_p(k+1) = w_p(k) + \mu(k)\left[\frac{\partial \hat{\mathcal{J}}_B(y(k))}{w_p}\right]W^*(z^{-1},k)W(z,k) \quad (2.133)$$

where the filter $W^*(z^{-1},k)W(z,k)$ acts on the gradient terms $\partial \hat{\mathcal{J}}_B(y(k))/\partial w_p(k)$ in a particular way. In z-domain form, the updates are

$$W(z,k+1) = W(z,k) + \mu(k)g(y(k))z^{-k}X^*(z^{-1})W^*(z^{-1},k)W(z,k) \quad (2.134)$$

$$= [1 + \mu(k)g(y(k))z^{-k}Y^*(z^{-1},k)]W(z,k) \quad (2.135)$$

which is somewhat similar to Eq. (2.117) in form.

In practice, a finite-length $(L+1)$-coefficient equalizer of the form

$$y(k) = \sum_{p=0}^{L} w_p(k)x(k-p) = \mathbf{w}^T(k)\mathbf{x}(k) \quad (2.136)$$

is used, where

$$\mathbf{w}(k) = [w_0(k) \; w_1(k) \cdots w_L(k)]^T \quad (2.137)$$

$$\mathbf{x}(k) = [x(k) \; x(k-1) \cdots x(k-L)]^T \quad (2.138)$$

Then, the coefficient updates of the standard-gradient algorithm for blind equalization can be represented in vector form as

$$\mathbf{w}(k+1) = \mathbf{w}(k) + \mu(k)g(y(k))\mathbf{x}^*(k) \qquad (2.139)$$

Unfortunately, when the finite-length equalizer in Eq. (2.136) is employed, the group structure of the parameter space that allows the natural gradient to be specified is lost. Even so, we can give approximate implementations of the natural-gradient algorithm assuming that the length L of the equalizer is sufficient to accurately model the inverse of the channel $H(z)$ up to some delay. One such implementation is

$$\mathbf{w}(k+1) = \mathbf{w}(k) + \mu(k)[\mathbf{w}^T(k)\mathbf{g}(\mathbf{y}_k(k))]\mathbf{y}_k^*(k) \qquad (2.140)$$

$$\mathbf{y}_k(k) = \mathbf{X}(k)\mathbf{w}(k) \qquad (2.141)$$

where the $(i+1)$th element of $\mathbf{g}(\mathbf{y}_k(k))$ is $g(\mathbf{w}^T(k)\mathbf{x}(k-i))$ and $\mathbf{X}(k)$ is an $(L+1) \times (L+1)$ Hankel matrix whose $(i+1, j+1)$th element is $x(k-i-j)$.

The algorithm in Eqs. (2.136) and (2.139) uses approximately $2L$ multiply/accumulates (MACs) at each sample time. In contrast, the natural-gradient update in Eqs. (2.140)–(2.141) requires $O(L^2)$ MACs per sample time, which is an order of magnitude more operations than Eqs. (2.136) and (2.139). It is possible to simplify this update further by assuming that $\mu(k)$ is small such that $\mathbf{w}(k) \approx \mathbf{w}(k-1) \approx \cdots \approx \mathbf{w}(k-L)$, and thus

$$\mathbf{y}_k(k) \approx \mathbf{y}(k) = [y(k)\ y(k-1)\cdots y(k-L)]^T \qquad (2.142)$$

Using Eq. (2.142), a simplified version of Eqs. (2.140)–(2.141) is (Douglas et al. 1999)

$$\mathbf{w}(k+1) = \mathbf{w}(k) + \mu(k)[\mathbf{w}^T(k)\mathbf{g}(\mathbf{y}(k))]\mathbf{y}^*(k) \qquad (2.143)$$

where the $(i+1)$th element of $\mathbf{g}(\mathbf{y}(k))$ is $f(y(k-i))$. This simplified update uses about $3L$ MACs per sample time. An alternative implementation with similar complexity and performance is (Douglas et al. 1996, 1999)

$$\mathbf{w}(k+1) = \mathbf{w}(k) + \mu(k)g(y(k-L))\mathbf{u}^*(k) \qquad (2.144)$$

where $\mathbf{u}(k) = [u(k)\ u(k-1)\cdots u(k-L)]^T$ and

$$u(k) = \sum_{p=0}^{L} w_{L-p}^*(k)y(k-p) \qquad (2.145)$$

The algorithms in Eqs. (2.143) and (2.144) are expressed in block diagram form

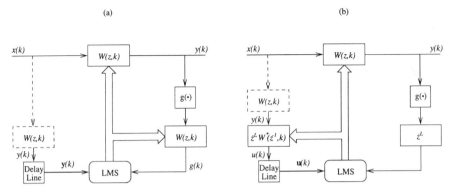

Figure 2.12 Block diagrams of the approximate natural-gradient algorithms for blind equalization. (a) Equation (2.143). (b) Equation (2.144).

in Figure 2.12a and b, respectively. From this figure, it is seen that these algorithms effectively use the equalizer as a prewhitening filter within the coefficient updates. A discussion of the relationships of these algorithms with similar prewhitening algorithms and an exploration of their performance using both trained and blind adaptation criteria can be found in Douglas et al. (1999).

To illustrate the improvements that natural-gradient adaptation provide in this situation, we plot the gradients of the constant-modulus error surface for a particular two-coefficient equalization example (Ding and Johnson 1993). In this example, the impulse response of the channel is

$$a_p = 0.7^p u(p) \tag{2.146}$$

where $u(p)$ is the unit step function, and $s(k)$ is assumed to be an i.i.d. binary-$\{\pm 1\}$ signal. Such a channel can be exactly equalized using the causal finite-length equalizer in Eq. (2.136) for $L = 1$ when $\mathbf{w}(k) = \mathbf{w}_{opt} = [1 \ -0.7]^T$. Moreover, $E\{\mathscr{J}_{CMA}(\mathbf{w})\}$ has a local minimum at $\mathbf{w}_{opt}$ for the chosen $s(k)$ when $A_2 = 1$. Shown in Figure 2.13a are the contours of $E\{\mathscr{J}_{CMA}(\mathbf{w})\}$ about $\mathbf{w}_{opt}$ in this case.

Figure 2.13b and c show the directions and magnitude contours of the expected value of the standard gradient update in Eq. (2.139) for this two-coefficient example. In this case, the gradient magnitudes vary widely around $\mathbf{w}_{opt}$, causing slow convergence of $\mathbf{w}(k)$ for this algorithm, and the gradient directions do not point toward $\mathbf{w}_{opt}$ in general. Figure 2.13d and e show the directions and magnitude contours of the expected value of the natural-gradient update. Here, we see the advantages provided by natural-gradient adaptation. The natural gradients have similar magnitudes in different directions about $\mathbf{w}_{opt}$, and their orientations cause $\mathbf{w}(k)$ to take a more direct path to $\mathbf{w}_{opt}$. In fact, the observed performance of this algorithm locally about $\mathbf{w}_{opt}$ is similar to that obtained when the received signals are prewhitened prior to the application of the equalizer.

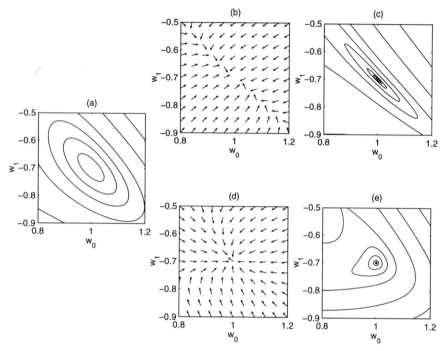

Figure 2.13 (a) Contours of the cost function $E\{\hat{\mathscr{J}}_{CMA}(\mathbf{w})\}$. (b) Directions of the standard gradients of $E\{\hat{\mathscr{J}}_{CMA}(\mathbf{w})\}$. (c) Contours of the magnitudes of the standard gradients of $E\{\hat{\mathscr{J}}_{CMA}(\mathbf{w})\}$. (d) Directions of the natural gradients of $E\{\hat{\mathscr{J}}_{CMA}(\mathbf{w})\}$. (e) Contours of the magnitudes of the natural gradients of $E\{\hat{\mathscr{J}}_{CMA}(\mathbf{w})\}$.

The performance behaviors of the algorithms in Eqs. (2.143) and (2.144) for complex-valued constant-modulus signals are considered in Douglas et al. (1999). In addition, the extension of Eq. (2.133) to the multichannel deconvolution task can be found in Amari et al. (1997b, 1997c, 1999b), which also presents the extension of the instantaneous blind source-separation algorithm in Eq. (2.116) to the multichannel deconvolution task.

2.4.5 Principal and Minor Subspace Analysis

In principal subspace analysis (PSA), one represents a set of n-dimensional vectors $\mathbf{x}(k)$ using another set of m-dimensional vectors $\mathbf{y}(k)$, $m < n$, defined as

$$\mathbf{y}(k) = \mathbf{W}(k)\mathbf{x}(k) \qquad (2.147)$$

where $\mathbf{W}(k) = [\mathbf{w}_1(k) \cdots \mathbf{w}_m(k)]^T$ is an adjustible $(m \times n)$-dimensional orthonormal matrix satisfying

$$\mathbf{W}(k)\mathbf{W}^T(k) = \mathbf{I} \qquad (2.148)$$

Defining the estimated signal vector $\hat{\mathbf{x}}(k)$ as

$$\hat{\mathbf{x}}(k) = \mathbf{W}^T(k)\mathbf{y}(k) = \mathbf{W}^T(k)\mathbf{W}(k)\mathbf{x}(k) \tag{2.149}$$

the goal of PSA is to iteratively adjust $\mathbf{W}(k)$ to minimize the MSE difference between $\hat{\mathbf{x}}(k)$ and $\mathbf{x}(k)$, defined as $E\{\hat{\mathscr{J}}(\mathbf{W})\}$, where

$$\hat{\mathscr{J}}_{\mathrm{MSE}}(\mathbf{W}) = \tfrac{1}{2}\|\mathbf{x}(k) - \hat{\mathbf{x}}(k)\|^2 \tag{2.150}$$

under the constraint in Eq. (2.148). It can be shown that if $\mathbf{x}(k)$ is wide-sense stationary, the value of $\mathbf{W}(k)$ that both minimizes $E\{\hat{\mathscr{J}}(\mathbf{W})\}$ and satisfies Eq. (2.148) is of the form

$$\mathbf{W}_{\mathrm{opt}} = \mathbf{Q}\mathbf{E}_1^T \tag{2.151}$$

where $\mathbf{E}_1 = [\pm\mathbf{e}_1 \cdots \pm\mathbf{e}_m]$ contains the m eigenvectors associated with the m largest eigenvalues of the autocorrelation matrix $\mathbf{R}_{\mathbf{xx}}$ defined in Eq. (2.88) and $\mathbf{Q}$ is any $(m \times m)$-dimensional rotation matrix satisfying $\mathbf{Q}\mathbf{Q}^T = \mathbf{Q}^T\mathbf{Q} = \mathbf{I}$. The specific solution given by

$$\mathbf{W}_{\mathrm{opt}} = \mathbf{E}_1^T \tag{2.152}$$

provides principal component analysis (PCA) of the vector sequence $\mathbf{x}(k)$. Solutions to PSA and PCA are useful for many signal-processing tasks, including data compression, signal analysis, and subspace tracking (Diamantaras and Kung 1996).

Iterative gradient-descent algorithms that solve either the PSA or PCA tasks must minimize the expected value of $\hat{\mathscr{J}}(\mathbf{W})$ over either an example within the matrix subspace defined by Eq. (2.148) or the space of matrices satisfying Eq. (2.148). Perhaps the conceptually most straightforward technique for the PSA task calculates a new matrix $\tilde{\mathbf{W}}(k+1)$ as

$$\tilde{\mathbf{W}}(k+1) = \mathbf{W}(k) - \mu(k)\frac{\partial\hat{\mathscr{J}}(\mathbf{W}(k))}{\partial\mathbf{W}} \tag{2.153}$$

$$= \mathbf{W}(k) + \mu(k)\mathbf{y}(k)\mathbf{x}^T(k) \tag{2.154}$$

where the constraint in Eq. (2.148) has been used to simplify Eq. (2.150) in going from Eq. (2.153) to Eq. (2.154). Then, the rows of $\tilde{\mathbf{W}}(k+1)$ are orthonormalized via a Gram-Schmidt procedure to extract a matrix $\mathbf{W}(k+1)$ spanning the same space as $\tilde{\mathbf{W}}(k+1)$ and satisfying $\mathbf{W}(k+1)\mathbf{W}^T(k+1) = \mathbf{I}$. If the orthonormalization procedure for the rows of $\tilde{\mathbf{W}}(k)$ is consistently ordered, then such an algorithm iteratively solves the PCA task.

While the preceding technique functions properly, the Gram-Schmidt procedure is computationally intensive, requiring $O(m^2n)$ operations to compute.

A more elegant approach adjusts $\mathbf{W}(k)$ within either the matrix subspace or the space of matrices satisfying the constraint in Eq. (2.148), respectively. The manifold of subspaces satisfying this constraint is called the *Grassman manifold*, and the manifold of matrices satisfying this constraint is called the *Stiefel manifold*. The former manifold is appropriate for homogeneous cost functions in which $E\{\hat{\mathscr{J}}(\mathbf{W})\} = E\{\hat{\mathscr{J}}(\mathbf{QW})\}$ for any nonidentity rotation matrix $\mathbf{Q}$, whereas the latter manifold is appropriate for inhomogeneous cost functions in which $E\{\hat{\mathscr{J}}(\mathbf{W})\} \neq E\{\hat{\mathscr{J}}(\mathbf{QW})\}$. To adjust $\mathbf{W}(k)$ in either of these manifolds, one must calculate the components of the gradient elements that lie within either of these manifolds and adjust $\mathbf{W}(k)$ according to these components. The components of the gradient elements for an example matrix within the Grassman manifold for any instantaneous cost function $\hat{\mathscr{J}}(\mathbf{W})$ are (Edelman et al. 1998)

$$\Delta(\mathbf{W}) = \mathbf{WW}^T \frac{\partial \hat{\mathscr{J}}(\mathbf{W})}{\partial \mathbf{W}} - \frac{\partial \hat{\mathscr{J}}(\mathbf{W})}{\partial \mathbf{W}} \mathbf{W}^T \mathbf{W} \tag{2.155}$$

The corresponding components of the gradient elements within the Stiefel manifold are (Chen et al. 1998; Edelman et al. 1998)

$$\Delta(\mathbf{W}) = \mathbf{WW}^T \frac{\partial \hat{\mathscr{J}}(\mathbf{W})}{\partial \mathbf{W}} - \mathbf{W} \left[\frac{\partial \hat{\mathscr{J}}(\mathbf{W})}{\partial \mathbf{W}} \right]^T \mathbf{W} \tag{2.156}$$

Then, the natural-gradient algorithms for adjusting $\mathbf{W}$ in continuous-time form are (Chen et al. 1998)

$$\frac{d\mathbf{W}}{dt} = -\mu \Delta(\mathbf{W}) \tag{2.157}$$

In the case of PSA, the instantaneous MSE cost function $\hat{\mathscr{J}}_{\mathrm{MSE}}(\mathbf{W})$ in Eq. (2.150) is homogeneous, and thus the Grassman manifold is the appropriate choice for the constraint on $\mathbf{W}(k)$. In this case, Eq. (2.155) evaluates to

$$\Delta(\mathbf{W}(k)) = \mathbf{W}(k)\mathbf{W}^T(k)\mathbf{y}(k)\mathbf{x}^T(k) - \mathbf{y}(k)\mathbf{y}^T(k)\mathbf{W}(k) \tag{2.158}$$

Approximating the continuous-time differential update in Eq. (2.157) with a discrete-time finite difference update yields the algorithm

$$\mathbf{W}(k+1) = \mathbf{W}(k) + \mu(k)[\mathbf{W}(k)\mathbf{W}^T(k)\mathbf{y}(k)\mathbf{x}^T(k) - \mathbf{y}(k)\mathbf{y}^T(k)\mathbf{W}(k)] \tag{2.159}$$

This update suffers from a marginal instability problem, however, that causes the span of $\mathbf{W}(k)$ to slowly deviate from the Grassman manifold due to numerical error accumulation. To maintain stable behavior, the principal subspace rule given by (Oja and Karhunen 1985)

$$\mathbf{W}(k+1) = \mathbf{W}(k) + \mu(k)[\mathbf{y}(k)\mathbf{x}^T(k) - \mathbf{y}(k)\mathbf{y}^T(k)\mathbf{W}(k)] \qquad (2.160)$$

can be used instead. This update approximately maintains the constraint in Eq. (2.148) over time.

Alternatively, if $E\{\hat{\mathscr{J}}_{\mathrm{MSE}}(\mathbf{W})\}$ in Eq. (2.150) is maximized under the constraint in Eq. (2.148), minor subspace analysis (MSA) is performed. MSA is useful for several signal-processing tasks, including bias removal in parameter estimation (Douglas and Rupp 1996; Reddy et al. 1982; Regalia 1994). A stabilized algorithm for MSA that is similar in form and spirit to Eq. (2.160) is (Douglas et al. 1998b)

$$\mathbf{W}(k+1) = \mathbf{W}(k) - \mu(k)[\mathbf{W}(k)\mathbf{W}^T(k)\mathbf{W}(k)\mathbf{W}^T(k)\mathbf{y}(k)\mathbf{x}^T(k)$$
$$- \mathbf{y}(k)\mathbf{y}^T(k)\mathbf{W}(k)] \qquad (2.161)$$

Details regarding the stability behavior of this and related algorithms can be found in Douglas et al. (1998a, 1998b, 1999).

2.5 CONCLUSIONS

This chapter has outlined a novel optimization method known as natural-gradient adaptation. The technique overcomes one of the chief drawbacks of standard-gradient adaptation by providing a more uniform convergence performance about a local minimum of the cost function, as shown in various examples involving maximum-likelihood estimation, perceptron training, blind source separation, and blind equalization. In addition, its form can be simple to implement in some cases. The main disadvantage of natural-gradient adaptation is the knowledge of the parameter space or cost-function error surface that is required to determine the final form of the algorithm in any particular case, although in some cases its exact form can be readily implemented without specific knowledge of the signal statistics.

APPENDIX

In this Appendix we show that the bounds on $\mu(k)$ in Eq. (2.69) are necessary to guarantee convergence of Eq. (2.50). Suppose that the difference between $\mathbf{w}(k)$ and $\mathbf{w}_{\mathrm{opt}}$ is small such that

$$\frac{\partial \mathscr{J}(\mathbf{w}(k))}{\partial \mathbf{w}} = \frac{\partial \mathscr{J}(\mathbf{w}_{\text{opt}})}{\partial \mathbf{w}} + \mathbf{A}(\mathbf{w}_{\text{opt}})\{\mathbf{w}(k) - \mathbf{w}_{\text{opt}}\}$$

$$+ \mathbf{o}[(w_i(k) - w_{i,\text{opt}})(w_j(k) - w_{j,\text{opt}})] \qquad (2.162)$$

$$\approx \mathbf{A}(\mathbf{w}_{\text{opt}})\{\mathbf{w}(k) - \mathbf{w}_{\text{opt}}\} \qquad (2.163)$$

where the Hessian $\mathbf{H}(\mathbf{w})$ is defined in Eq. (2.3). Subtracting $\mathbf{w}_{\text{opt}}$ from both sides of Eq. (2.50) gives

$$\mathbf{w}(k+1) - \mathbf{w}_{\text{opt}} = \mathbf{w}(k) - \mathbf{w}_{\text{opt}} - \mu(k)\mathbf{G}^{-1}(\mathbf{w}(k))\frac{\partial \mathscr{J}(\mathbf{w}(k))}{\partial \mathbf{w}} \qquad (2.164)$$

$$= [\mathbf{I} - \mu(k)\mathbf{G}^{-1}(\mathbf{w}_{\text{opt}})\mathbf{H}(\mathbf{w}_{\text{opt}})]\{\mathbf{w}(k) - \mathbf{w}_{\text{opt}}\} \qquad (2.165)$$

where we have substituted Eq. (2.163) into Eq. (2.164) and neglected terms in $O[(w_i(k) - w_{i,\text{opt}})(w_j(k) - w_{j,\text{opt}})]$ to obtain Eq. (2.165).

For $\mathbf{w}_{\text{opt}}$ that minimize $\mathscr{J}(\mathbf{w})$, all of the eigenvalues of $\mathbf{H}(\mathbf{w}_{\text{opt}})$ are positive, and the eigenvalues of $\mathbf{G}(\mathbf{w})$ are all positive by construction. Noting the exponential form of Eq. (2.165), the natural-gradient algorithm is locally convergent if all of the eigenvalues of $[\mathbf{I} - \mu(k)\mathbf{G}^{-1}(\mathbf{w}_{\text{opt}})\mathbf{H}(\mathbf{w}_{\text{opt}})]$ are less than one in magnitude, which leads to Eq. (2.69).

REFERENCES

Amari, S., 1985, *Differential-Geometrical Methods in Statistics* (New York: Springer-Verlag).

Amari, S., Feb. 1998, "Natural gradient works efficiently in learning," *Neural Computation*, vol. 10, pp. 251–276.

Amari, S., Apr. 1999, "Superefficiency in blind source separation," *IEEE Trans. Signal Processing*, vol. 47, pp. 936–944.

Amari, S., and J.-F. Cardoso, Nov. 1997, "Blind source separation—semi-parametric statistical approach," *IEEE Trans. Signal Processing*, vol. 45, pp. 2692–2700.

Amari, S., and A. Cichocki, Oct. 1998, "Adaptive blind signal processing—neural network approaches," *Proc. IEEE*, vol. 86, pp. 2026–2048.

Amari, S., A. Cichocki, and H. H. Yang, 1996, "A new learning algorithm for blind signal separation," *Adv. Neural Inform. Processing Sys. 8* (Cambridge, MA: MIT Press), pp. 757–763.

Amari, S., T.-P. Chen, and A. Cichocki, Nov. 1997a, "Stability analysis of learning algorithms for blind source separation," *Neural Networks*, vol. 10, pp. 1345–1351.

Amari, S., S. C. Douglas, A. Cichocki, and H. H. Yang, Apr. 1997b, "Multichannel blind deconvolution and equalization using the natural gradient," *Proc. IEEE Workshop Signal Processing Adv. Wireless Communic.*, Paris, France, pp. 101–104.

Amari, S., S. C. Douglas, A. Cichocki, and H. H. Yang, July 1997c, "Novel on-line adaptive learning algorithms for blind deconvolution using the natural gradient approach," *Proc. 11th IFAC Symp. Syst. Ident.*, Kitakyushu, Japan, vol. 3, pp. 1057–1062.

Amari, S., T.-P. Chen, and A. Cichocki, 1999a, "Nonholonomic orthogonal learning algorithm for blind source separation," *Neural Computation* (in press).

Amari, S., S. C. Douglas, and A. Cichocki, Jan. 1999b, "Multichannel blind deconvolution and source separation with the natural gradient," *Tech. Rept. No. EE-99-002*, Dept. of Electrical Engineering, Southern Methodist Univ., Dallas, TX.

Bell, A. J., and T. J. Sejnowski, Nov. 1995, "An information maximization approach to blind separation and blind deconvolution," *Neural Computation*, vol. 7, pp. 1129–1159.

Bellini, S., 1994, "Bussgang techniques for blind deconvolution and equalization," in S. Haykin, ed., *Blind Deconvolution* (Englewood Cliffs, NJ: Prentice-Hall) pp. 8–59.

Benveniste, A., and M. Goursat, Aug. 1984, "Blind equalizers," *IEEE Trans. Communic.*, vol. 32, pp. 871–883.

Bershad, N. J., Apr. 1988, "On error saturation nonlinearities in LMS adaptation," *IEEE Trans. Acoust., Speech, Signal Processing*, vol. 36, pp. 440–452.

Boothby, W. M., 1986, *An Introduction to Differential Manifolds and Riemannian Geometry*, 2nd ed. (Orlando, FL: Academic Press).

Campbell, L. L., June 1985, "The relation between information theory and the differential-geometric approach to statistics," *Inform. Sci.*, vol. 35, pp. 199–210.

Cardoso, J.-F., Apr. 1997, "Infomax and maximum likelihood in source separation," *IEEE Signal Processing Lett.*, vol. 4, pp. 112–114.

Cardoso, J.-F., Sept. 1998a, "Learning in manifolds: the case of source separation," *Proc. IEEE Workshop Statistical Signal Array Processing*, Portland, OR, pp. 136–139.

Cardoso, J.-F., Oct. 1998b, "Blind signal separation: Statistical principles," *Proc. IEEE*, vol. 86, pp. 2009–2025.

Cardoso, J.-F., and B. Laheld, Dec. 1996, "Equivariant adaptive source separation," *IEEE Trans. Signal Processing*, vol. 44, pp. 3017–3030.

Cardoso, J. F., A. Belouchrani, and B. Laheld, Apr. 1994, "A new composite criterion for adaptive and iterative blind source separation," *IEEE Int. Conf. Acoust., Speech, Signal Processing*, Adelaide, Australia, vol. IV, pp. 273–276.

Chen, T.-P., S. Amari, and Q. Lin, Apr. 1998, "A unified algorithm for principal and minor components extraction," *Neural Networks*, vol. 11, pp. 385–390.

Cichocki, A., R. Unbehauen, and E. Rummert, Aug. 1994, "Robust learning algorithm for blind separation of signals," *Electron. Lett.*, vol. 30, pp. 1386–1387.

Cover, T. M., and J. A. Thomas, 1991, *Elements of Information Theory* (New York: Wiley).

Diamantaras, K. I., and S.-Y. Kung, 1996, *Principal Component Neural Networks: Theory and Applications* (New York: Wiley).

Ding, Z., and C. R., Johnson, Jr., May 1993, "On the nonvanishing stability of undesirable equilibria for FIR Godard blind equalizers," *IEEE Trans. Signal Processing*, vol. 41, pp. 1940–1944.

Douglas, S. C., and A. Cichocki, Nov. 1997, "Neural networks for blind decorrelation of signals," *IEEE Trans. Signal Processing*, vol. 45, pp. 2829–2842.

Douglas, S. C., and M. Rupp, Nov. 1996, "On bias removal and unit norm constraints in stochastic gradient adaptive IIR filters," *Proc. 30th Asilomar Conf. Signals, Syst., Comput.*, Pacific Grove, CA, vol. 2, pp. 1093–1097.

Douglas, S. C., S. Amari, and S.-Y. Kung, Sept. 1998a, "Gradient adaptation under unit-norm constraints," *Proc. IEEE Workshop Statistical Signal Array Processing*, Portland, OR, pp. 144–147.

Douglas, S. C., S. Amari, and S.-Y. Kung, Feb. 1999, "Gradient adaptation with unit-norm constraints," *Tech. Rept. No. EE-99-003*, Dept. of Electrical Engineering, Southern Methodist Univ., Dallas, TX.

Douglas, S. C., S.-Y. Kung, and S. Amari, Dec. 1998b, "A self-stabilized minor subspace rule," *IEEE Signal Processing Lett.*, vol. 5, pp. 328–330.

Douglas, S. C., A. Cichocki, and S. Amari, Nov. 7, 1996, "Fast-convergence filtered regressor algorithms for blind equalisation," *Electron. Lett.*, vol. 32, pp. 2114–2115.

Douglas, S. C., A. Cichocki, and S. Amari, Apr. 1999, "Self-whitening algorithms for adaptive equalization and deconvolution," *IEEE Trans. Signal Processing*, vol. 47, pp. 1161–1165.

Edelman, A., T. Arias, and S. T. Smith, 1999, "The geometry of algorithms with orthogonality constraints," *SIAM J. Matrix Anal. Appl.*, vol. 20, pp. 303–353, 1998.

Godard, D. N., Nov. 1980, "Self-recovering equalization and carrier tracking in two-dimensional data communication systems," *IEEE Trans. Communic.*, vol. 28, pp. 1867–1875.

Haykin, S., 1994, *Neural Networks: A Comprehensive Foundation* (New York: Macmillan).

Haykin, S., 1996, *Adaptive Filter Theory*, 3rd ed. (Saddle River, NJ: Prentice-Hall).

Jang, J.-S. R., C.-T. Sun, and E. Mizutani, 1997, *Neuro-Fuzzy and Soft Computing* (Upper Saddle River, NJ: Prentice-Hall).

Kushner, H. J., and D. S. Clark, 1978, *Stochastic Approximation Methods for Constrained and Unconstrained Systems* (New York: Springer-Verlag).

Luenberger, D. G., 1984, *Linear and Nonlinear Programming*, 2nd ed. (Reading, MA: Addison-Wesley).

Morgan, F., 1993, *Riemannian Geometry: A Beginner's Guide* (Boston, MA: Jones and Bartlett).

Oja, E., and J. Karhunen, Feb. 1985, "On stochastic approximation of the eigenvectors and eigenvalues of the expectation of a random matrix," *J. Math Anal. Appl.*, vol. 106, pp. 69–84.

Oja, E., and J. Karhunen, 1995, "Signal separation by nonlinear Hebbian learning," in *Computational Intelligence—A Dynamic System Perspective*, ed. M. Palaniswami et al. (New York: IEEE Press), pp. 83–97.

Poor, H. V., 1994, *Introduction to Signal Detection and Estimation*, 2nd ed. (New York: Springer-Verlag).

Proakis, J. G., and M. Salehi, 1994, *Communication Systems Engineering* (Englewood Cliffs, NJ: Prentice-Hall).

Rao, C. R., 1945, "Information and accuracy attainable in the estimation of statistical parameters," *Bull. Calcutta Math. Soc.*, vol. 37, pp. 81–91.

Rattray, M., D. Saad, and S. Amari, Dec. 14, 1998, "Natural gradient descent for on-line learning," *Phys. Rev. Lett.*, vol. 81, no. 24, pp. 5461–5464.

Reddy, V. U., B. Egardt, and T. Kailath, June 1982, "Least squares type algorithm for adaptive implementation of Pisarenko's harmonic retrieval method," *IEEE Trans. Acoust., Speech, Signal Processing*, vol. 30, pp. 399–405.

Regalia, P. A., June 1994, "An unbiased equation error identifier and reduced-order approximations," *IEEE Trans. Signal Processing*, vol. 42, pp. 1297–1412.

Silva, F. M., and L. B. Almeida, June 1991, "A distributed solution for data ortho-normalization," *Proc. Int. Conf. Artificial Neural Networks*, Espoo, Finland, vol. 1, pp. 943–948.

Smith, S. T., 1993, Geometric optimization methods for adaptive filtering, Ph.D. thesis, Harvard Univ., Cambridge, MA.

Treichler, J. R., and B. G. Agee, Apr. 1983, "A new approach to multipath correction of constant modulus signals," *IEEE Trans. Acoust., Speech, Signal Processing*, vol. 31, pp. 349–372.

Widrow, B., and S. D. Stearns, 1985, *Adaptive Signal Processing* (Englewood Cliffs, NJ: Prentice-Hall).

Yang, H. H., and S. Amari, Oct. 1997, "Adaptive on-line learning algorithms for blind separation: maximum entropy and minimum mutual information," *Neural Computation*, vol. 9, pp. 1457–1482.

Yang, H. H., and S. Amari, Nov. 1998, "Complexity issues in natural gradient descent method for training multilayer perceptrons," *Neural Computation*, vol. 10, pp. 2137–2157.

3

BLIND SIGNAL SEPARATION AND EXTRACTION: NEURAL AND INFORMATION-THEORETIC APPROACHES

Shun-ichi Amari, Andrzej Cichocki, and Howard H. Yang

ABSTRACT

Fundamental neural and information-theoretic approaches together with learning algorithms are presented for the problem of adaptive blind signal processing, especially instantaneous blind separation, blind extraction (of single source or specified number of sources), and multichannel blind deconvolution/ equalization of independent source signals. Recent developments of adaptive learning algorithms based on the natural-gradient approach in the general linear, orthogonal and Stiefel manifolds are discussed. Mutual information, Kullback-Leibler divergence, and several promising schemes are proposed and reviewed in the chapter, especially for noisy data. Emphasis is given to an information-theoretical unifying approach, adaptive filtering models, and associated on-line adaptive nonlinear learning algorithms. In this chapter, the focus is mainly on approaches to blind separation/extraction of sources when the measured signals are contaminated by additive noise. Existing adaptive algorithms with equivariant properties are extended in order to considerably reduce the bias in the demixing-matrix estimation caused by measurement noise. Moreover, a novel recurrent dynamic neural network for simultaneous

Unsupervised Adaptive Filtering, Volume I, Edited by Simon Haykin.
ISBN 0-471-29412-8 © 2000 John Wiley & Sons, Inc.

estimation of the unknown mixing matrix, blind source separation, and reduction of noise in the extracted output signals is discussed. The optimal choice of nonlinear activation functions for various noise distributions, for example, Gaussian, Laplacian, impulsive, and uniformly distributed noise assuming a generalized-Gaussian-distributed noise model, is discussed. Computer simulations have confirmed the usefulness and excellent performance of the developed algorithms. Some research results presented in this chapter are new and published for the first time.

3.1 INTRODUCTION AND PROBLEM FORMULATION

3.1.1 Formulation of General Blind Signal Processing Problems

The problems of independent-component analysis (ICA) and/or blind separation and multichannel deconvolution of source signals have received wide attention in various fields such as biomedical signal analysis and processing (EEG, MEG, ECG), geophysical data processing, data mining, speech and image recognition and enhancement, and wireless communications (Amari et al. 1997a,b; Cichocki and Unbehauen 1994; Haykin 1994; Lee et al. 1997; Makeig et al. 1996).

The most general blind signal-processing problem can be formulated as follows. We observe records of signals from a multiple-input multiple-output (MIMO) nonlinear dynamic system. The objective is to find an inverse system, termed a *reconstruction system*, if it exists and is stable, in order to estimate the primary input signals. This estimation is performed on the basis of the output signals as well as *a priori* knowledge about the source signals and the system. Preferably, it is required that the inverse system should be adaptive, so that it has some tracking capability in nonstationary environments (see Fig. 3.1).

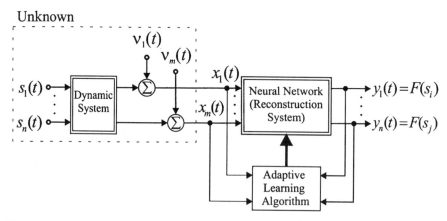

Figure 3.1 Block diagram illustrating blind signal processing or blind identification problem.

Alternatively, instead of estimating the source signals directly, it is sometimes more convenient to identify an unknown system first (e.g., when the inverse system does not exist), and then estimate source signals implicitly by applying a suitable optimization procedure (Hua 1996; Lewicki and Sejnowski 1998, Ljung and Söderstrom 1983; Tong et al. 1994, Tugnait 1996). In many cases, source signals are simultaneously linearly filtered and mutually mixed. It is desired to process the observations such that the original source signals are extracted by neural networks [see Amari (1967, 1977, 1985, 1987, 1991, 1996, 1998, 1999a, 1999b), Amari and Cardoso (1997), Amari et al. (1995, 1997a,b,c, 1998, 1999), Amari and Kawanabe (1997a, 1997b), Cichocki and Unbehauen (1994).] These problems are known as an ICA, blind source separation (BSS), blind extraction of sources (BES), or multichannel blind deconvolution (MBD) (Haykin 1994; Jutten and Herault 1991; Tong et al. 1991, 1993). They can be formulated as the problems of separating or estimating the waveforms of the original sources from an array of sensors or transducers without knowing the characteristics of the transmission channels. For some models, however, there is no guarantee that the estimated or extracted signals can be of exactly the same waveforms as the source signals, and so the requirements are sometimes relaxed to the extent that the extracted waveforms are distorted (filtered) versions of the source signals (Chan et al. 1996; Weinstein et al. 1993) (see Fig. 3.1).

Recently, a number of efficient adaptive on-line learning algorithms have been developed for ICA, BSS, BSE, and MBD [6]–[146]. Although the underlying principles and approaches are different, many of the techniques have very similar forms. Most of these algorithms assume that any measurement noise within the mixed signals can be neglected. However, in real-world applications most measured signals are contaminated by additive noise $(v_1, \ldots, v_m)$. Thus, the problem arises, of efficiently reducing the influence of noise on the performance of algorithms for ICA, and in particular, methods are desired to reduce noise in the stochastically independent extracted components. This chapter also addresses this difficult and challenging problem (Cichocki et al. 1996d, 1998b, 1998c; Hyvarinen 1998; Moulines 1997).

3.1.2 Mathematical Formulation of the Basic Problems

The mixing and filtering processes of the unknown input sources $s_j(t)$ $(j = 1, 2, \ldots, n)$ may have different mathematical or physical models, depending on specific applications. In this section, we focus mainly on the simplest cases when m mixed signals $x_i(t)$ are a linear combination of the n (typically $m \geq n$) unknown statistically independent, zero-mean source signals $s_j(t)$, and are noise-contaminated (see Fig. 3.2). They are written as

$$x_i(t) = \sum_{j=1}^{n} a_{ij} s_j(t) + v_i(t) \qquad i = 1, 2, \ldots, m \qquad (3.1)$$

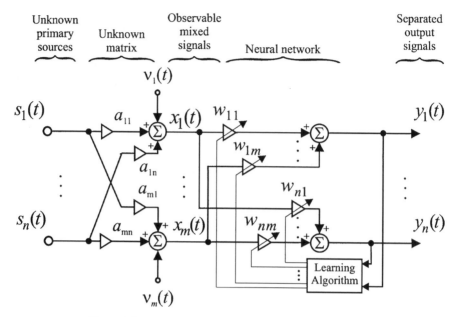

Figure 3.2   Block diagram illustrating basic BSS problem.

or in the matrix notation

$$\mathbf{x}(t) = \mathbf{A}\mathbf{s}(t) + \mathbf{v}(t) \tag{3.2}$$

where $\mathbf{x}(t) = [x_1(t) \cdots x_m(t)]^T$ is a noisy sensor vector; $\mathbf{s}(t) = [s_1(t) \cdots s_n(t)]^T$ is a source signal vector; $\mathbf{v}(t) = [v_1(t) \cdots v_m(t)]^T$ is the noise vector; and $\mathbf{A}$ is an unknown full-rank $m \times n$ mixing matrix. In this section, it is assumed that the number of source signals n is, in general, unknown unless stated otherwise. It is assumed that only the sensor vector $\mathbf{x}(t)$ is available, and it is necessary to design a feedforward or recurrent neural network and an associated adaptive learning algorithm that enables estimation of sources and/or identification of the mixing matrix $\mathbf{A}$ with good tracking abilities.

The preceding problems are often referred to as noisy BSS or ICA: the BSS of a noisy random vector $\mathbf{x} = [x_1 \cdots x_m]^T$ is obtained by finding an $n \times m$, full-rank, linear transformation (unmixing) matrix $\mathbf{W}$ such that the output signal vector $\mathbf{y} = [y_1 \cdots y_n]^T$, defined as $\mathbf{y} = \mathbf{W}\mathbf{x}$, contains components that are as independent as possible, as measured by an information-theoretic cost function such as the Kullback-Leibler divergence. In other words, it is required to adapt the synaptic weights w_{ij} of the $n \times m$ matrix $\mathbf{W}$ of the linear system $\mathbf{y}(t) = \mathbf{W}(t)\mathbf{x}(t)$ (often referred to as a *single-layer feedforward neural network*) to combine the observations $x_i(t)$ to generate estimates of the source signals

$$\hat{s}_j(t) = y_j(t) = \sum_{i=1}^{m} w_{ji} x_i(t) \tag{3.3}$$

The optimal weights correspond to the statistical independence of the output signals $y_j(t)$ (see Fig. 3.2).

Remark. In general, when the number of source signals is unknown and may change dynamically in time we assume that the separating matrix $\mathbf{W}$ is a square matrix, that is, the number of outputs is equal to the number of sensors (Cichocki et al. 1997e, 1998a).

There are several definitions of ICA. In this chapter, we employ the three distinct definitions given below.

Definition 1. The ICA of a noisy random vector $\mathbf{x}(t) = [x_1(t) \cdots x_m(t)]^T$ is obtained by finding an $n \times m$ (with $m \geq n$), full-rank, separating matrix $\mathbf{W}$ such that the output signal vector $\mathbf{y}(t) = [y_1(t) \cdots y_n(t)]^T$, defined as

$$\mathbf{y}(t) = \mathbf{W}\mathbf{x}(t) \tag{3.4}$$

contains the source components that are as independent as possible, evaluated by an information-theoretic cost function such as the minimum Kullback-Leibler divergence.

Definition 2. For a random noisy vector $\mathbf{x}(t)$, defined as

$$\mathbf{x}(t) = \mathbf{A}\mathbf{s}(t) + \mathbf{v}(t) \tag{3.5}$$

where $\mathbf{A}$ is an $(m \times n)$ mixing matrix, $\mathbf{s}(t) = [s_1(t) \cdots s_n(t)]^T$ is a source vector of stochastically independent signals, and $\mathbf{v}(t)$ is a vector of uncorrelated noise terms, ICA is obtained by estimating both the mixing matrix $\mathbf{A}$ and the additive noise $\mathbf{v}(t)$.

Definition 3. Estimate simultaneously all the source signals and their numbers and/or identify a mixing matrix $\hat{\mathbf{A}}$ or its pseudoinverse demixing matrix $\mathbf{W} = \hat{\mathbf{A}}^+$, assuming only the stochastic independence of the primary sources, the linear independence of columns of $\mathbf{A}$, and $m \leq n$.

We can define BES similarly.

Definition 4. Estimate one source or a selected number of the sources sequentially one by one or estimate a specified group of sources and identify the corresponding vector(s) $\hat{\mathbf{a}}_j$ of a mixing matrix $\hat{\mathbf{A}}$ and their pseudoinverses $\mathbf{w}_j$, which are rows of the demixing matrix $\mathbf{W} = \hat{\mathbf{A}}^+$, assuming only the stochastic independence of its primary sources and the linear independence of columns of $\mathbf{A}$.

Remark. It is worth emphasizing that in the literature the terms BSS, BES, and ICA are often confused, although they refer to the same model, usually with the same assumptions, pursue the same tasks, and are solved with the same algorithms.

Because the estimation of a separating (demixing) matrix $\mathbf{W}$ and a mixing matrix $\hat{\mathbf{A}}$ in the presence of noise is rather difficult, the majority of past research has been devoted to the noiseless case where $v(t) = \mathbf{0}$. The objective of this chapter is to develop novel approaches and learning algorithms that are more robust with respect to noise than existing techniques or that can reduce the noise in the estimated output vector $\mathbf{y}(t)$. In this chapter, we assume that the source signals and additive noise components are statistically independent.

In general, the problem of noise cancellation is difficult and is even impossible to treat, because we have $(m + n)$ unknown source signals (n sources and m noise signals). In many practical situations, however, we can measure or model the environmental noise. Such noise is termed referenced noise. The vector of reference noise signals is denoted by v_R (see Fig. 3.3). For example, in the acoustic "cocktail party" problem we can measure or record the environmental noise by using an isolated microphone. In a similar way, noise in biomedical applications, such as EEG or ECG, can be measured by auxiliary sensors (or electrodes) placed appropriately. The noise $v_R(t)$ may influence each sensor in some unknown manner due to environmental effects. Hence, effects such as delays, reverberations, echos, and nonlinear distortions may occur. It may be assumed that the reference noise is processed by some unknown dynamic

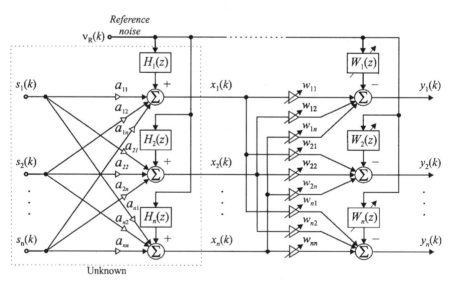

Figure 3.3   Block diagrams illustrating blind separation of sources with convolutive reference noise (blocks $H_i(z)$ and $W_i(z)$ mean transfer functions of FIR filters) (Cichocki et al. 1996d; Karhunen et al. 1997; Kasprzak et al. 1997).

system before reaching the sensors. In a simple case, a convolutive model of noise is assumed where the reference noise is processed by some finite-duration impulse-response (FIR) filters (see Fig. 3.3). In this case, two learning processes are performed simultaneously. An unsupervised learning procedure performs blind separation and a supervised learning algorithm performs noise reduction (Cichocki et al. 1996d). This approach has been successfully applied to the elimination of colored noise under the assumption that the reference noise is available (Cichocki et al. 1996d; Karhunen et al. 1997).

While several recently developed algorithms have shown promise for solving practical tasks, they could fail to separate on-line (nonstationary) signal mixtures containing both sub- and super-Gaussian distributed source signals, especially when the number of sources is unknown and changes dynamically over time. The problem of on-line estimation of sources in the case where the number of sources is unknown is relevant in many practical applications, such as analysis of EEG signals and the cocktail party problem where the number of source signals usually changes over time. In this chapter we also propose a solution to this problem under certain assumptions regarding mixing and demixing models and statistical independence of sources.

A more general problem is a multichannel blind deconvolution and equalization task, which can be considered as a natural extension or generalization of the instantaneous blind separation problem. In the multidimensional blind deconvolution problem an m-dimensional vector of received discrete-time signals $\mathbf{x}(k) = [x_1(k) \cdots x_m(k)]^T$ at time k is assumed to be produced from an n-dimensional vector of source signals $\mathbf{s}(k) = [s_1(k) \cdots s_n(k)]^T$, $m \geq n$, by using the mixture model (Amari et al. 1997a,b, 1998; Cichocki et al. 1997a; Cichocki 1997; Ding et al. 1994; Douglas 1996; Inouye 1997; Thi and Jutten 1995; Yelin and Weinstein 1996)

$$\mathbf{x}(k) = \sum_{p=-\infty}^{\infty} \mathbf{H}_p \mathbf{s}(k-p) = \mathbf{H}_p * \mathbf{s}(k) = \mathbf{H}(z)\mathbf{s}(k) \tag{3.6}$$

where $\mathbf{H}_p$ is an $(m \times n)$ matrix of mixing coefficients at lag p,

$$\mathbf{H}(z) = \sum_{p=-\infty}^{\infty} \mathbf{H}_p z^{-p} \tag{3.7}$$

is a matrix transfer function, and z^{-1} is the delay operator defined by $z^{-p}[s_i(k)] = s_i(k-p)$. The goal is to calculate possibly scaled and delayed estimates of the source signals from the received signals by using approximate knowledge of the source signal distributions and statistics. Typically, each source signal $s_i(k)$ is an independent and identically distributed (i.i.d.) sequence that is independent of all the other source sequences. In order to recover the source signals, we can use the neural-network models depicted in Fig. 3.4a and

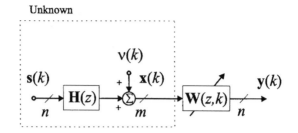

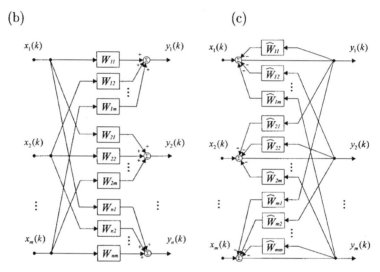

Figure 3.4 (*a*) Functional block diagram illustrating the multichannel blind deconvolution problem. (*b*) Architecture of feedforward neural network (each synaptic weight $W_{ij}(z, k)$ is an adaptive FIR filter described by Eq. (3.8)). (*c*) Architecture of simple recurrent neural network.

b but the synaptic weights should be generalized to dynamic filters (e.g., FIR or infinite-duration impulse-response (IIR) filters). In this chapter, we consider one example of this generalization: a multichannel blind deconvolution where each synaptic weight (Amari et al. 1997a; Cichocki et al. 1997a; Inouye 1997; Johnson 1991).

$$W_{ij}(z, k) = \sum_{p=0}^{L} w_{ijp}(k)z^{-p} \tag{3.8}$$

is described by an FIR adaptive filter at discrete time k (Inouye 1997, Thi and Jutten 1995). In other words, we consider a feedforward model that estimates the source signals directly using a truncated version of a doubly infinite multichannel equalizer of the form (Inouye 1997) (see Fig. 3.4a and b)

$$\mathbf{y}(k) = \sum_{p=-\infty}^{\infty} \mathbf{W}_p(k)\mathbf{x}(k-p) = \mathbf{W}(z,k)[\mathbf{x}(k)]$$

$$= \mathbf{W}_p(k) * \mathbf{x}(k), \tag{3.9}$$

where $\mathbf{y}(k) = [y_1(k) \cdots y_n(k)]^T$ is an n-dimensional vector of outputs and $\mathbf{W}(k) = \{\mathbf{W}_p(k), -\infty \le p \le \infty\}$ is a sequence of $n \times m$ coefficient matrices used at time k, and the transfer function is

$$\mathbf{W}(z,k) = \sum_{p=-\infty}^{\infty} \mathbf{W}_p(k)z^{-p} \tag{3.10}$$

Then the goal of adaptive blind deconvolution or equalization is to adjust $\mathbf{W}(z,k)$ such that

$$\lim_{k \to \infty} \mathbf{G}(z,k) = \mathbf{W}(z,k)\mathbf{H}(z) = \mathbf{P}_0\mathbf{D}(z) \tag{3.11}$$

where $\mathbf{P}_0$ is an $n \times n$ permutation matrix; $\mathbf{D}(z)$ is an $n \times n$ diagonal matrix whose (i,i)th entry is $c_i z^{-\Delta_i}$; c_i is a nonzero scalar weighting; and Δ_i is an integer delay. We assume that both $\mathbf{H}(z)$ and $\mathbf{W}(z,k)$ are stable with nonzero eigenvalues on the unit circle $|z| = 1$. In addition, derivatives of quantities with respect to $\mathbf{W}(z,k)$ can be understood as a series of matrices indexed by the lag p of $\mathbf{W}_p(k)$ in Eq. (3.4) (Amari et al. 1997b; Inouye 1997).

Figure 3.4c shows an alternative recurrent neural-network model with the same number of inputs and outputs $(n = m)$. In these models the synaptic weights W_{ij} or $\hat{W}_{ij}$ may be generalized to real- or complex-valued dynamic FIR filters, as illustrated in Fig. 3.5a. A further variation is to consider the constrained IIR, such as the gamma or Laguerre filters (see Fig. 3.5b,c) or other structures, such as state-space models that might have some useful properties (Cichocki and Zhang 1998; Cichocki et al. 1998d; Zhang and Cichocki 1998b, 1998c).

3.2 PREPROCESSING: WHITENING, PCA, AND PRELIMINARY NOISE REDUCTION

In some applications it is beneficial to apply some kind of preprocessing, for example, to eliminate redundancy (in the case of more sensors than sources), to reduce additive noise, or to improve convergence properties of adaptive algorithms by decorrelating mixing signals. In this section we discuss briefly two closely related preprocessing techniques: prewhitening and principal-component analysis (PCA) for real-valued signals.

(a)

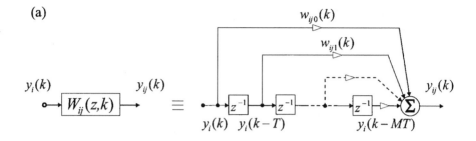

(b)

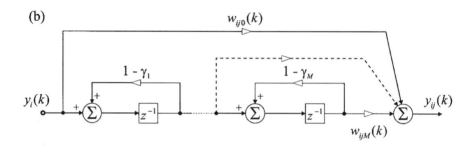

(c)

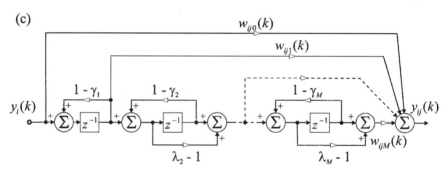

Figure 3.5 Exemplary models of synaptic weights for the recurrent neural network shown in Fig. 3.4c: (a) basic FIR filter model; (b) gamma filter model; (c) Laguerre filter model.

3.2.1 Simple On-Line Adaptive Learning Algorithms for Prewhitening

Some adaptive algorithms for blind separation require prewhitening of mixed (sensor) signals. A random zero-mean vector $\mathbf{x}_1$ is said to be *white* if its auto-correlation matrix is the identity matrix, that is, $\mathbf{R}_{\mathbf{x}_1\mathbf{x}_1} = E\{\mathbf{x}_1\mathbf{x}_1^T\} = \mathbf{I}_n$ or $E\{x_{1i}x_{1j}\} = \delta_{ij}$. In whitening, the data vectors $\mathbf{x}(k)$ are preprocessed by using a whitening transformation (Fig. 3.6):

$$\mathbf{x}_1(k) = \mathbf{Q}(k)\mathbf{x}(k) \tag{3.12}$$

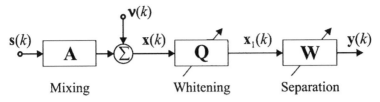

Figure 3.6 Block diagram illustrating sequential prewhitening and BSS.

Here $\mathbf{x}_1(t)$ denotes the whitened vector, and $\mathbf{Q}(k)$ is an $n \times m$ whitening matrix. When $m > n$, and n is known in advance, $\mathbf{Q}(k)$ simultaneously reduces the dimension of the data vectors from m to n. In whitening, the matrix $\mathbf{Q}(k)$ is chosen so that the autocorrelation matrix $E\{\mathbf{x}_1(k)\mathbf{x}_1(k)^T\}$ becomes the identity matrix $\mathbf{I}_n$. Thus the components of the whitened vectors $\mathbf{x}_1(k)$ are mutually uncorrelated and they have unit variance. Note that the prewhitening matrix $\mathbf{Q}$ is not unique. When $\mathbf{Q}$ is a prewhitening matrix, $\mathbf{WQ}$ is also prewhitening for arbitrary $n \times n$ orthogonal matrix $\mathbf{W}$.

Uncorrelatedness is a necessary condition (but not sufficient) for the stronger statistical independence condition. After prewhitening, the blind separation task usually becomes somewhat easier, because the subsequent separating matrix $\mathbf{W}$ can be constrained to be an orthogonal matrix for real-valued signals and a unitary matrix for complex-valued signals and weights, that is, $\mathbf{WW}^T = \mathbf{I}_n$, (see Fig. 3.6), since

$$\mathbf{R}_{yy} = E\{\mathbf{yy}^T\} = E\{\mathbf{Wx}_1\mathbf{x}_1^T\mathbf{W}^T\} = \mathbf{WW}^T = \mathbf{I}_n \qquad (3.13)$$

under the condition that the output signals are normalized to unit variance.

There exist several solutions for whitening the input data. Simple adaptive on-line learning rules for prewhitening have the following forms (Cichocki et al. 1997b; Douglas and Cichocki 1997; Karhunen 1996; Silva and Almeida 1991):

1. Robust learning rule (Cichocki and Unbehauen 1994; Douglas and Cichocki 1997; Silva and Almeida 1991)

$$\mathbf{Q}(k+1) = \mathbf{Q}(k) + \eta(k)[\mathbf{I} - \mathbf{x}_1(k)\mathbf{x}_1^T(k)]\mathbf{Q}(k) \qquad (3.14)$$

2. Local learning rule (Cichocki and Unbehauen 1994; Douglas and Cichocki 1997):

$$q_{ij}(k+1) = q_{ij}(k) + \eta(k)[\delta_{ij} - \tilde{x}_{1i}(k)\tilde{x}_{1j}(k)] \qquad (3.15)$$

or in the matrix form

$$\mathbf{Q}(k+1) = \mathbf{Q}(k) + \eta(k)[\mathbf{I} - \mathbf{x}_1(k)\mathbf{x}_1^T(k)] \qquad (3.16)$$

The first algorithm, Eq. (3.14), is a robust one with *equivariant property* (Cardoso and Laheld 1996) as the global system, described by the combined mixing and decorrelation process

$$\mathbf{B}(k+1) \overset{df}{=} \mathbf{Q}(k+1)\mathbf{A} = \mathbf{B}(k) + \eta(t)[\mathbf{I} - \mathbf{B}(k)\mathbf{s}(k)\mathbf{s}^T(k)\mathbf{B}^T(k)]\mathbf{B}(k) \quad (3.17)$$

is completely independent from the mixing matrix $\mathbf{A}$. The second algorithm, Eq. (3.15), is a simple local algorithm and can be applied only to square matrices with $m = n$. The algorithms that iteratively apply either the rule embodied in Eq. (3.14) or Eq. (3.16) achieve the equilibrium point when the autocorrelation matrix of the output signal satisfies the condition:

$$\mathbf{R}_{\mathbf{x}_1\mathbf{x}_1} = E\{\mathbf{x}_1\mathbf{x}_1^T\} = E\{\mathbf{Q}\mathbf{x}\mathbf{x}^T\mathbf{Q}^T\} = \mathbf{Q}E\{\mathbf{x}\mathbf{x}^T\}\mathbf{Q}^T = \mathbf{Q}\mathbf{R}_{\mathbf{x}\mathbf{x}}\mathbf{Q}^T = \mathbf{I}_n \quad (3.18)$$

When we restrict $\mathbf{Q}$ to be a nonsingular and symmetrical matrix, we get the equilibrium point defined by

$$\mathbf{Q}_* = \hat{\mathbf{R}}_{xx}^{-1/2} = \langle \mathbf{x}\mathbf{x}^T \rangle^{-1/2} \quad (3.19)$$

It is interesting to note that for the special case when a mixing matrix $\mathbf{A}$ is a nonsingular and symmetrical, prewhitening the algorithms in Eqs. (3.14) and (3.16) can perform blind signal separation successfully, since $\mathbf{R}_{xx} = \mathbf{A}\mathbf{A} = \mathbf{A}^2$ (under the standard assumption that $\mathbf{R}_{ss} = \mathbf{I}_n$), and hence the equilibrium point satisfies $\mathbf{Q}_* = \mathbf{A}^{-1}$ (Cichocki et al. 1997b).

It should be noted that when $\mathbf{x}(k)$ is noisy such that $\mathbf{x}(k) = \hat{\mathbf{x}}(k) + \mathbf{v}(k)$ where $\hat{\mathbf{x}}(k)$ and $\hat{\mathbf{y}}(k) = \mathbf{Q}(k)\hat{\mathbf{x}}(k)$ are the noiseless estimates of the input and output vectors, respectively, it is easy to show that the additive noise $\mathbf{v}(k)$ within $\mathbf{x}(k)$ introduces a bias in the estimated decorrelation matrix $\mathbf{Q}$ (Cichocki et al. 1998a,b; Douglas et al. 1998). The autocorrelation matrix of the output is defined by

$$\mathbf{R}_{\mathbf{x}_1\mathbf{x}_1} = E\{\mathbf{x}_1(k)\mathbf{x}_1^T(k)\} = \mathbf{Q}\mathbf{R}_{\hat{x}\hat{x}}\mathbf{Q}^T + \mathbf{Q}\mathbf{R}_{vv}\mathbf{Q}^T \quad (3.20)$$

where $\mathbf{R}_{\hat{x}\hat{x}} = E\{\hat{\mathbf{x}}(k)\hat{\mathbf{x}}^T(k)\}$ and $\mathbf{R}_{vv} = E\{\mathbf{v}(t)\mathbf{v}^T(t)\}$

Assuming that the autocorrelation matrix of the noise is known (e.g., $\mathbf{R}_{vv} = \sigma_v^2\mathbf{I}$) or can be estimated, a modified learning algorithm employing bias removal is given by (Douglas et al. 1998)

$$\Delta\mathbf{Q}(k) = \eta(k)[\mathbf{I} - E\{\mathbf{x}_1(k)\mathbf{x}_1^T(k)\} + \mathbf{Q}(k)\mathbf{R}_{vv}\mathbf{Q}^T(k)]\mathbf{Q}(k) \quad (3.21)$$

The stochastic gradient version of this algorithm for $\mathbf{R}_{vv} = \sigma_v^2\mathbf{I}$ is [cf. Eq. (3.14)]

$$\Delta\mathbf{Q}(k) = \eta(k)[\mathbf{I} - \mathbf{x}_1(k)\mathbf{x}_1^T(k) + \sigma_v^2\mathbf{Q}(k)\mathbf{Q}^T(k)]\mathbf{Q}(k) \quad (3.22)$$

Derivation of Whitening Learning Rule   The learning rule, Eq. (3.14), can be derived by minimizing the following objective function (Cichocki and Unbehauen 1996):

$$l_Q(\mathbf{Q}) = \frac{1}{4}\|\mathbf{R}_{\mathbf{x}_1\mathbf{x}_1} - \mathbf{I}\|_F \tag{3.23}$$

where $\|\mathbf{X}\|_F$ denotes the Frobenius norm of matrix $\mathbf{X}$ and $\mathbf{R}_{\mathbf{x}_1\mathbf{x}_1} = E\{\mathbf{x}_1\mathbf{x}_1^T\}$. Alternatively, we can use

$$\tilde{l}_Q(\mathbf{Q}) = \frac{1}{2}[\|\mathbf{x}_1\|^2 - \log\det(\mathbf{Q}\mathbf{Q}^T)] \tag{3.24}$$

where the first term ensures minimization of the energy of output signals and the second term guarantees that the prewhitening matrix $\mathbf{Q}$ is of full rank. It is interesting to note that

$$\begin{aligned}\mathbf{R}_{\mathbf{x}_1\mathbf{x}_1} &= E\{\mathbf{Q}(k)\mathbf{x}(k)\mathbf{x}^T(k)\mathbf{Q}^T(k)\} \\ &= \mathbf{Q}(k)\mathbf{R}_{xx}\mathbf{Q}^T(k) = \mathbf{Q}(k)\mathbf{A}\mathbf{R}_{ss}\mathbf{A}^T\mathbf{Q}^T(k)\end{aligned} \tag{3.25}$$

Hence, without loss of generality, assuming that $\mathbf{R}_{ss} = E\{\mathbf{s}(k)\mathbf{s}^T(k)|\} = \mathbf{I}_n$, we have

$$\mathbf{R}_{\mathbf{x}_1\mathbf{x}_1} = \mathbf{B}(k)\mathbf{B}^T(k) \tag{3.26}$$

where matrix $\mathbf{B}(k) = \mathbf{Q}(k)\mathbf{A}$ describes the global system of mixing and whitening operations. Multiplying Eq. (3.14) by the mixing matrix $\mathbf{A}$ from the right-hand side, we get

$$\mathbf{Q}(k+1)\mathbf{A} \overset{df}{=} \mathbf{B}(k+1) = \mathbf{B}(k) + \eta(k)[\mathbf{I} - \mathbf{B}(k)\mathbf{s}(k)\mathbf{s}^T(k)\mathbf{B}^T(k)]\mathbf{B}(k) \tag{3.27}$$

Taking the expectation of both sides of Eq. (3.27) and assuming, without loss of generality, that $\mathbf{R}_{ss} = \mathbf{I}$, it is evident that the learning algorithm, employing the learning rule in Eq. (3.14), achieves the equilibrium condition when the matrix $\mathbf{B}(k)$ becomes orthogonal, that is, $\mathbf{B}^{-1} = \mathbf{B}^T$.

3.2.2 Robust PCA and Preliminary Noise Reduction

PCA is a standard technique for the computation of eigenvectors and eigenvalues of an estimated autocorrelation matrix $\hat{\mathbf{R}}_{xx} = \langle \mathbf{x}\mathbf{x}^T \rangle = (1/M)\sum_{k=1}^{M}\mathbf{x}(k)\mathbf{x}^T(k) = \mathbf{V}\mathbf{\Lambda}\mathbf{V}^T \in R^{m\times m}$. PCA enables mixed signals to decompose into two subspaces: the signal subspace corresponding to the principal components associated with the largest eigenvalues $\lambda_1, \lambda_2, \ldots, \lambda_n$ $(m > n)$ and the noise subspace corresponding to the minor components associated with the smallest eigenvalues $\lambda_{n+1}, \ldots, \lambda_m$, where $\lambda_1 \geq \lambda_2 \geq \cdots \geq \lambda_m \geq 0$ is assumed. The standard numerical procedure is as follows. First, estimate the $m \times m$ autocorrelation matrix $\mathbf{R}_{xx}$ of the zero-mean mixed-signal vector $\mathbf{x}(k)$. Then compute the m eigenvalues $\lambda_1 \geq \lambda_2 \geq \cdots \geq \lambda_m \geq 0$ and the corresponding eigenvectors $\mathbf{v}_1, \mathbf{v}_2, \ldots, \mathbf{v}_m$. These eigenvalues and eigenvectors are obtained by standard PCA. The subspace spanned by the n eigenvectors $\mathbf{v}_i$ corresponding to n largest eigenvalues can be considered an approximation of the noiseless signal subspace. The data vector $\mathbf{x}(k)$ is projected on the estimated signal subspace and then rescaled such that each component has unit variance. This procedure is written as

$$\mathbf{x}_1(k) = \mathbf{Q}\mathbf{x}(k) = \mathbf{\Lambda}^{-1/2}\mathbf{V}^T\mathbf{x}(k) \tag{3.28}$$

where $\mathbf{\Lambda} = \mathrm{diag}\{\lambda_1, \lambda_2, \ldots, \lambda_n\}$ is the diagonal matrix containing the n largest eigenvalues and $\mathbf{V} = [\mathbf{v}_1, \mathbf{v}_2, \ldots, \mathbf{v}_n]$ is the matrix of the associated eigenvectors.

One important advantage of this approach is that it not only enables reduction in the noise level, but also allows us to estimate the number of sources. A problem arising from this approach, however, is how to correctly set or estimate the threshold that divides eigenvalues into the two subsets, especially when the noise is large [i.e., the signal-to-noise ratio (SNR) is low].

Fast Adaptive Algorithm for Robust Principal-Component Extraction
Instead of using a classic numerical approach for the problem, a neural-network approach with an on-line adaptive algorithm can be used (Amari 1997; Chen et al. 1998; Cichocki and Unbehauen 1993, 1994b; Cichocki et al. 1996a; Karhunen and Joutsensalo 1993; Karhunen et al. 1995; Oja and Karhunen 1995; Oja 1997). In what follows we propose a robust PCA for extracting principal components (PCs), corresponding to real-valued zero-mean signals without estimating a large-dimension autocorrelation matrix (Cichocki and Unbehauen 1993; Cichocki et al. 1996a). We extract the principal components sequentially as long as the eigenvalues λ_i are larger than some user-specified small threshold. We assume that minor components for $i > n$ correspond to additive noise. Let us assume that we want to extract PCs sequentially by employing the concept of the self-supervising principle (replicator) and cascade (hierarchical) neural network architecture (Cichocki and Unbehauen 1993; Cichocki et al. 1996a).

To be specific, consider a single linear neuron (see Fig. 3.7a)

$$y_1(k) = \mathbf{v}_1^T\mathbf{x} = \sum_{p=1}^{n} v_{1p}x_p(k) \tag{3.29}$$

which is to extract the first principal component, with $\lambda_1 = E\{y_1^2\}$. The vector $\mathbf{v}_1$ is optimally determined in such a way that the reconstruction vector $\hat{\mathbf{x}} = \mathbf{v}_1 y_1$ reproduces the input training vector $\mathbf{x}$ as faithfully as possible, according to a suitable optimization criterion.

(a)

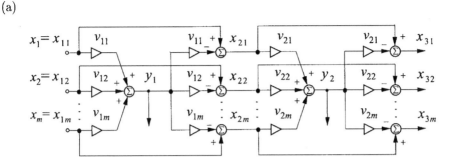

(b)

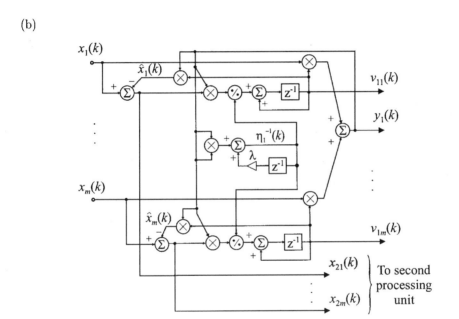

Figure 3.7 Block diagrams illustrating (*a*) sequential extraction of principal components; (*b*) on-line on-chip implementation of fast RLS learning algorithm.

In general, the objective function may be expressed as (Cichocki and Unbehauen 1993)

$$l_{1\rho}(\mathbf{v}_1) \triangleq \rho(\mathbf{e}_1) = \sum_{p=1}^{n} \rho(e_{1p}) \qquad (3.30)$$

with

$$\mathbf{e}_1 \triangleq \mathbf{x}_2 = \mathbf{x}_1 - \mathbf{v}_1 y_1, \qquad y_1 \triangleq \mathbf{v}_1^T \mathbf{x}_1, \qquad \mathbf{x}_1(t) \triangleq \mathbf{x}(k)$$

Table 3.1   Typical Robust Objective Functions $\rho(e)$ and Corresponding Influence Functions $\Psi(e) = d\rho(e)/de$

Name	Loss Function $\rho(e)$	Influence Functions $\Psi(e)$
Logistic	$\rho_L = \beta \log(\cosh(e/\beta))$	$\Psi_L = \tanh(e/\beta)$
Huber	$\rho_H = \begin{cases} e^2/2, & \text{for } \lvert e \rvert \le \beta; \\ \beta \lvert e \rvert - \beta^2/2, & \text{otherwise} \end{cases}$	$\Psi_H = \begin{cases} e, & \text{for } \lvert e \rvert \le \beta; \\ \beta\, \text{sign}(e), & \text{otherwise} \end{cases}$
L_p	$\rho_{Lp} = \dfrac{1}{r}\lvert e \rvert^r$	$\Psi_{Lp} = \lvert e \rvert^{r-1}\, \text{sign}(e)$
Cauchy	$\rho_C = \dfrac{\sigma^2}{2}\log[1+(e/\sigma)^2]$	$\Psi_C = \dfrac{e}{1+(e/\sigma)^2}$
Geman, McCulre	$\rho_G = \dfrac{1}{2}\dfrac{e^2}{1+e^2}$	$\Psi_G = \dfrac{e}{(1+e^2)^2}$
Welsh	$\rho_W = \dfrac{\sigma^2}{2}[1-\exp(-(e/\sigma)^2)]$	$\Psi_W = e\exp(-(e/\sigma)^2)$
Fair	$\rho_F = \sigma^2\left[\dfrac{\lvert e \rvert}{\sigma} - \log(1+\lvert e \rvert/\sigma)\right]$	$\Psi_F = \dfrac{e}{1+\lvert e \rvert/\sigma}$
$L_1 - L_2$	$\rho_{L12} = 2(\sqrt{1+e^2/2}-1)$	$\Psi_{L12} = \dfrac{e}{\sqrt{1+e^2/2}}$
Talvar	$\rho_{Ta} = \begin{cases} e^2/2, & \text{for } \lvert e \rvert \le \beta; \\ \beta^2/2, & \text{otherwise} \end{cases}$	$\Psi_{Ta} = \begin{cases} e, & \text{for } \lvert e \rvert \le \beta; \\ 0, & \text{otherwise} \end{cases}$
Hampel	$\rho_{Ha} = \begin{cases} \dfrac{\beta^2}{\pi}(1-\cos(\pi e/\beta)), & \text{for } \lvert e \rvert \le \beta; \\ 2\beta/\pi, & \text{otherwise} \end{cases}$	$\Psi_H = \begin{cases} \sin(\pi e/\beta) \\ 0 \end{cases}$

where $\rho(e)$ is a typically convex function, for example, $\rho_2(e) = \frac{1}{2}e^2$, $\rho_1(e) = \lvert e \rvert$, or $\rho_L(e) = \beta \log\cosh(e/\beta)$, where $\beta > 0$ is a problem-dependent parameter called *the cutoff parameter* (see Table 3.1).

Formulation of the objective function $l_{i\rho}(v_i)$ is a key step in our approach because it enables us to transform the minimization problem into a set of differential or difference equations, which determines an adaptive learning algorithm. Minimization of Eq. (3.30) according to the standard gradient descent approach leads, after some mathematical manipulations, to the following learning algorithm (Cichocki and Unbehauen 1993; Cichocki et al. 1996a):

$$v_1(k+1) = v_1(k) + \eta_1(k)y_1(k)\Psi_1[e_1(k)] \tag{3.31}$$

where

$$\Psi_1(\mathbf{e}_1) = [\Psi_1(e_{11}), \Psi_1(e_{12}), \ldots, \Psi_1(e_{1n})]^T$$

$$\Psi_1(e_{1p}) = \frac{\partial \rho(e_{1p})}{\partial e_{1p}}$$

for example, $\Psi_1(e_{1p}) = \tanh(e_{1p}/\beta)$ for $\rho_L = \beta \log(\cosh(e_{1p}/\beta))$. It should be noted that the preceding algorithm for the special cases of the functions $\rho_2(e) = \frac{1}{2}e^2$, yielding $\Psi_1(\mathbf{e}_1) = \mathbf{e}_1$, simplifies to the well-known Oja's learning rule (Oja and Karhunen 1995; Oja 1997). The preceding learning algorithms can be easily extended for higher principal components using the self-supervising principle and cascade hierarchical neural network shown in Fig. 3.7a. In other words, the learning algorithm for the extraction of the second principal component corresponding to the second largest eigenvalue $\lambda_2 = E[y_2^2]$ is performed in the same way as for the first principal component. However, we do not carry out the extraction process directly from the input data $\mathbf{x}_1(k) = \mathbf{x}(k)$, but from the error $\mathbf{e}_1(k) \triangleq \mathbf{x}_2(k) = \mathbf{x}_1(k) - \hat{\mathbf{x}}_1(k) = \mathbf{x}_1(k) - \mathbf{v}_1 y_1(k)$ and $y_2(k) \triangleq \mathbf{v}_2^T \mathbf{e}_1(k)$ (not $y_2(k) = \mathbf{v}_2^T \mathbf{x}(k)$) as is usually assumed. It can be shown that the learning algorithm for the ith principal component can be written in a general form as follows (Cichocki and Unbehauen 1993; Cichocki et al. 1996a)

$$\mathbf{v}_i(k+1) = \mathbf{v}_i(k) + \eta_i(k) y_i(k) \Psi_i[\mathbf{x}_{i+1}(k)] \tag{3.32}$$

where

$$\mathbf{e}_i = \mathbf{x}_{i+1} \triangleq \mathbf{x}_i - \mathbf{v}_i y_i, \qquad y_i \triangleq \mathbf{v}_i^T \mathbf{x}_i, \qquad \mathbf{x}_1(k) \triangleq \mathbf{x}(k)$$

$$\Psi_i(\mathbf{e}_i) = [\Psi_i(e_{i1}), \Psi_i(e_{i2}), \ldots, \Psi_i(e_{in})]^T$$

$$\Psi_i(e_{ip}) = \frac{\partial \rho(e_{ip})}{\partial e_{ip}}$$

for example, $\Psi_i(e_{ip}) = \tanh(e_{ip}/\beta)$ for $\rho_L = \beta \log(\cosh(e_{ip}/\beta))$. The optimal choice of the activation function $\Psi_i(\mathbf{e}_i)$ depends on the distribution of signal (see Table 3.1). For Gaussian noise the linear function $\Psi_i(\mathbf{e}_i) = \mathbf{e}_i$ is optimal. In the special case of L_2-norm: $\rho(e) = \frac{1}{2}e^2$ we have $\Psi(e_i) = e_i$, and so the algorithm simplifies to the form

$$\mathbf{v}_i(k+1) = \mathbf{v}_i(k) + \eta_i(k) y_i(k) [\mathbf{x}_i(k) - y_i(k) \mathbf{v}_{i*}] \tag{3.33}$$

After applying the preceding learning procedure, the extracted output signals $y_i(k)$ will be decorrelated with decreasing values of variances $\lambda_i = E[y_i^2]$, $(i = 1, 2, \ldots, n)$. In the accelerated version of the preceding algorithm, the key role is played by the learning rate $\eta_i(k) \geq 0$ and forgetting factor λ. If the learning rate is too large, the algorithm is unstable. Otherwise, if it is fixed or an exponentially decreasing parameter, the convergence speed of the algorithm may be very slow (Amari, 1997).

In order to increase the convergence speed, we can employ recursive least-squares (RLS) or Kalman filtering for optimal updating of the learning rate η_i (Cichocki and Unbehauen 1993) (see Fig. 3.7b):

$$y_i(k) = \mathbf{v}_i^T(k)\mathbf{x}_i(k) \tag{3.34}$$

$$\eta_i^{-1}(k) = \lambda \eta_i^{-1}(k-1) + |y_i(k)|^2 \tag{3.35}$$

$$\mathbf{v}_i(k+1) = \mathbf{v}_i(k) + \frac{y_i(k)}{\eta_i^{-1}(k)}[\mathbf{x}_i(k) - y_i(k)\mathbf{v}_i(k)] \tag{3.36}$$

$$\mathbf{x}_{i+1}(k) = \mathbf{x}_i(k) - y_i(k)\mathbf{v}_{i*} \tag{3.37}$$

$$\mathbf{x}_1(k) = \mathbf{x}(k) \tag{3.38}$$

where $\eta_i^{-1}(0) = \sigma_{xi}^{-2}$ and λ $(0 < \lambda \leq 1)$ is the forgetting factor.

3.3   FUNDAMENTALS OF ADAPTIVE LEARNING ALGORITHMS FOR BSS

The present section is devoted to the analysis of learning algorithms for a typical but simple instantaneous blind source-separation problem. Here, we assume that the number of source signals is known and is equal to the number of sensors (with $m = n$), so that both $\mathbf{A}$ and $\mathbf{W}$ are nonsingular $n \times n$ matrices. The source signals $s_i(t)$ are assumed to be mutually independent and have zero mean $(E\{s_i(t)\} = 0)$. We also assume that additive noise terms $v(t)$ are negligible or reduced to be at negligible levels by the preprocessing stated in the previous section. The present section gives a prototype of mathematical analysis. We relax most of these constraints in later sections.

In order to obtain a good estimate $\mathbf{y} = \mathbf{W}\mathbf{x}$ of the source signals $\mathbf{s}$, we introduce an objective or contrast function $l(\mathbf{y}, \mathbf{W})$ in terms of the estimated $\mathbf{y}$ and $\mathbf{W}$. Its expectation with respect to $\mathbf{y}$,

$$L(\mathbf{W}) = E\{l(\mathbf{y}, \mathbf{W})\} \tag{3.39}$$

should be a function of $\mathbf{W}$ that represents the performance of demixing by $\mathbf{W}$. In other words, $L(\mathbf{W})$ should be minimized when the components of $\mathbf{y}$ become independent, that is, when $\mathbf{W}$ is a rescaled permutation of $\mathbf{A}^{-1}$. We use the Kullback-Leibler divergence for this purpose. Let $f_y(\mathbf{y}; \mathbf{W})$ be the probability density function of the random variable $\mathbf{y} = \mathbf{W}\mathbf{x} = \mathbf{W}\mathbf{A}\mathbf{s}$. Let $q(\mathbf{y})$ denote another probability density function of $\mathbf{y}$, from which all the y_i are statistically independent. In this case, $q(\mathbf{y})$ can be decomposed in a product form as

$$q(\mathbf{y}) = \prod_{i=1}^{n} q_i(y_i) \tag{3.40}$$

This independent distribution is called the *reference function* and is arbitrary for the moment.

We use the Kullback-Leibler divergence between the distribution $f_y(\mathbf{y}; \mathbf{W})$ of $\mathbf{y}$ obtained by $\mathbf{W}$ and the reference distribution $q(\mathbf{y})$,

$$D_{fq}(\mathbf{W}) = D[f_y(\mathbf{y}; \mathbf{W}) \| q(\mathbf{y})]$$

$$= \int f_y(\mathbf{y}; \mathbf{W}) \log \frac{f_y(\mathbf{y}; \mathbf{W})}{q(\mathbf{y})} \, d\mathbf{y} \qquad (3.41)$$

The Kullback-Leibler divergence is a natural measure of deviation for two probability distributions (see Appendix 3.1). Hence, $D_{fq}(\mathbf{W})$ shows how far the distribution $f_y(\mathbf{y}; \mathbf{W})$ is from the reference distribution.

It is easy to show that (see Appendix 3.1)

$$D_{fq}(\mathbf{W}) = -H(\mathbf{W}) - E\{\log q(\mathbf{y})\}$$

$$= -H(\mathbf{W}) - \sum_{i=1}^{n} E\{\log q_i(y_i)\} \qquad (3.42)$$

where $H(\mathbf{W})$ is the entropy of $f_y(\mathbf{y}; \mathbf{W})$. It is also easy to show that for

$$l_q(\mathbf{y}, \mathbf{W}) = -\log |\det(\mathbf{W})| - \sum_{i=1}^{n} \log q_i(y_i) \qquad (3.43)$$

its expectation $L_q(\mathbf{W})$ coincides with $D_{fq}(\mathbf{W})$, except for a constant term not depending on $\mathbf{W}$. Hence, minimization of $D_{fq}(\mathbf{W})$ is equivalent to that of $L(\mathbf{W})$.

It is interesting to note that most learning algorithms proposed so far from heuristic considerations can be understood in terms of the preceding contrast function. It is remarkable that the entropy maximization (Bell and Sejnowski 1995; Yang and Amari 1997), independent components analysis (ICA) (Amari et al. 1996; Comon 1994), nonlinear PCA (Oja 1997; Oja et al. 1995), and the maximum-likelihood approach (Cardoso 1997; Moulines et al. 1997; Pham et al. 1992; Yang and Amari 1997) are formulated in this framework, where the only differences arise in the choices of the reference function $q(\mathbf{y})$. If we choose the true distribution of the sources as q, we have the maximum-likelihood approach. Note that the true distribution is unknown in general, so that we need to estimate $q(\mathbf{y})$. If we choose the marginalized independent distribution of $f_y(\mathbf{y}; \mathbf{W})$, this leads to ICA (Amari et al. 1996; Comon 1994). The entropy maximization uses nonlinear transformations $z_i = g_i(y_i)$ and tries to maximize the joint entropy of $\mathbf{z}$. This is easily shown to be equivalent to choosing (Bell and Sejnowski 1995)

$$q_i(y_i) = \frac{d}{dy_i} g_i(y_i) \qquad (3.44)$$

A fundamental problem of how to define the contrast function $L_q(\mathbf{W})$ or $D_{fq}(\mathbf{W})$ depends on $q(\mathbf{y})$. More precisely, for a given $q(\mathbf{y})$, it is a fundamental problem to see if the true separating solution $\mathbf{W}$ is a local minimum of $D_{fq}(\mathbf{W})$. Information geometry is useful for elucidating such fundamental problems, but we do not enter into it here (Amari 1985, 1998). We will show that whichever q is chosen, the true separating solution is a critical point of $D_{fq}(\mathbf{W})$. However, this does not imply that the true solution is a local minimum. Next we show that, for a wide range of $q(\mathbf{y})$, the true solution is a local minimum. It depends on the statistics of the true distribution and functions q_i whether the true solution is a local minimum or not. In this section we explicitly give the necessary and sufficient condition for the true solution to be a local minimum. This is related to the stability of the learning algorithm derived therefrom. We also show how to adaptively transform unstable learning algorithms into stable ones. In the rest of this section, the subscript q in $l_q(\mathbf{y}, \mathbf{W})$ is omitted. Since our target is to minimize of the expectation of Eq. (3.42), a simple idea is to use the ordinary stochastic descent on-line learning algorithm given by

$$\Delta \mathbf{W}(k) = \mathbf{W}(k+1) - \mathbf{W}(k) = -\eta(k) \frac{\partial l[\mathbf{y}(k), \mathbf{W}]}{\partial \mathbf{W}} \tag{3.45}$$

where $\eta(k)$ is a learning rate depending on k, and $\partial l / \partial \mathbf{W}$ is the $n \times n$ gradient matrix whose entries are $\partial l / \partial w_{ij}$. We can calculate the gradient matrix $\partial l / \partial \mathbf{W}$ by differentiating by component. Appendix 3.2 gives a simple differential matrix calculus to obtain $\partial l / \partial \mathbf{W}$. The result is

$$\Delta \mathbf{W}(k) = \eta(k)[\mathbf{W}^{-T}(k) - \boldsymbol{\varphi}[\mathbf{y}(k)]\mathbf{x}^T(k)] \tag{3.46}$$

where $\mathbf{W}^{-T}$ is the transpose of the inverse of $\mathbf{W}$, and $\boldsymbol{\varphi}(\mathbf{y})$ is the column vector whose ith component is (see Table 3.2)

$$\varphi_i(y_i) = -\frac{d}{dy_i} \log q_i(y_i) \tag{3.47}$$

The gradient $-\partial l / \partial \mathbf{W}$ represents the steepest decreasing direction of function l when the parameter space is Euclidean. In the present case, the parameter space consists of all the nonsingular $n \times n$ matrices $\mathbf{W}$. This is a multiplicative group where the identity matrix $\mathbf{I}_n$ is the unit. Moreover, it is a manifold so that it forms a Lie group. Amari et al. (1996, 1998) exploited this fact to introduce a natural Riemannian metric to the space of $\mathbf{W}$. They showed that the true steepest-descent direction in the Riemannian space of parameters $\mathbf{W}$ is not $\partial l / \partial \mathbf{W}$ but

$$-\frac{\partial l(\mathbf{y}; \mathbf{W})}{\partial \mathbf{W}} \mathbf{W}^T \mathbf{W} = [\mathbf{I} - \boldsymbol{\varphi}(\mathbf{y})\mathbf{y}^T]\mathbf{W} \tag{3.48}$$

Table 3.2 Typical Probability Density Functions $q(y)$ and Corresponding Optimal Activation Functions $\varphi(y) = -d\log q(y)/dy$

Name	Density Function $q(y)$	Activation Function $\varphi(y)$
Gaussian	$\dfrac{1}{\sqrt{2\pi}\sigma}\exp(-\lvert y\rvert^2/2\sigma^2)$	y
Laplace	$\dfrac{1}{2\sigma}\exp(-\lvert y\rvert/\sigma)$	$\mathrm{sign}(y)$
Cauchy	$\dfrac{1}{\pi\sigma}\dfrac{1}{1+(y/\sigma)^2}$	$\dfrac{2y}{\sigma^2+y^2}$
Hyperbolic cosine	$\dfrac{1}{\pi\cosh(\gamma y)}$	$\tanh(\gamma y)$
Unimodal	$\dfrac{\exp(-2\gamma y)}{(1+\exp(-2\gamma y))^2}$	$\tanh(\gamma y)$
Triangular	$\dfrac{1}{a}(1-\lvert y\rvert/a)$ $\lvert y\rvert < a$	$\dfrac{\mathrm{sign}(y)}{a(1-\lvert y\rvert/a)}$
Generalized Gaussian	$\dfrac{r}{2a\Gamma(1/r)}\exp(-(\lvert y\rvert/a)^r)$ $r\geq 1$	$\lvert y\rvert^{r-1}\mathrm{sign}(y)$
Robust Generalized Gaussian	$\dfrac{r}{2a\Gamma(1/r)}\exp(-\lvert\rho(y)/a\rvert^r)$ $\rho(y)$—robust function (see Table 3.1)	$\lvert\rho(y)\rvert^{r-1}\dfrac{\partial\rho}{\partial y}$

Hence, the learning algorithm takes the form (Amari et al. 1995, 1996, 1997c)

$$\Delta\mathbf{W}(k) = -\eta(k)\frac{\partial l(\mathbf{y};\mathbf{W})}{\partial\mathbf{W}}\mathbf{W}^T\mathbf{W} = \eta(k)[\mathbf{I} - \varphi[\mathbf{y}(k)]\mathbf{y}^T(k)]\mathbf{W}(k) \qquad (3.49)$$

This type of learning rule was first introduced by Cichocki et al. (1994a,b; Cichocki and Unbehauen 1996). This is similar to the relative gradient introduced independently by Cardoso and Laheld (1996). The equivariant property (Cardoso and Laheld 1996) holds in this learning rule, because the underlying Riemannian metric is based on the Lie group structure. Amari (1998) showed that the natural gradient learning has a number of desirable properties including the Fisher efficiency.

Remark. We can easily derive similar algorithms for a fully recurrent neural network described by $\mathbf{y}(k) = \mathbf{x}(k) - \hat{\mathbf{W}}\mathbf{y}(k)$ (see Fig. 3.4c). Assuming that the demixing matrix $\mathbf{W}$ [giving $\mathbf{y}(k) = \mathbf{W}\mathbf{x}(k)$] is nonsingular, we have simple relations: $\hat{\mathbf{W}} = \mathbf{W}^{-1} - \mathbf{I}$ and

$$\frac{d\mathbf{W}}{dt}\mathbf{W}^{-1} = -\mathbf{W}\frac{d\mathbf{W}^{-1}}{dt}$$

Hence, we obtain a learning rule first developed by Cichocki and Unbehauen (1994a)

$$\Delta\hat{\mathbf{W}}(k) = -\eta(k)[\hat{\mathbf{W}}(k) + \mathbf{I}][\mathbf{I} - \boldsymbol{\varphi}[\mathbf{y}(k)]\mathbf{y}^T(k)] \tag{3.50}$$

or equivalently

$$\Delta\hat{\mathbf{A}}(k) = -\eta(k)\hat{\mathbf{A}}(k)[\mathbf{I} - \boldsymbol{\varphi}[\mathbf{y}(k)]\mathbf{y}^T(k)] \tag{3.51}$$

where $\hat{\mathbf{A}}(k) = \mathbf{W}^{-1}(k) = \hat{\mathbf{W}}(k) + \mathbf{I}$ is an estimating matrix of the unknown mixing matrix $\mathbf{A}$ (up to permutations and scale factors).

Before stating variations and modifications of the basic learning algorithm, Eq. (3.49), we show its stability. The learning equation (3.49) is a stochastic difference equation depending on random inputs $\mathbf{y}(k)$. To analyze its behavior, we consider the ensemble average of the equation and approximate it by the differential equation with continuous time t,

$$\frac{d\mathbf{W}(t)}{dt} = \eta(t)E\{\mathbf{I} - \boldsymbol{\varphi}[\mathbf{y}(t)]\mathbf{y}^T(t)\}\mathbf{W}(t) \tag{3.52}$$

We thus find that the true separating solution $\mathbf{W}$ with which y_i and y_j are independent is an equilibrium solution of the averaged equation, because, when y_i and y_j are independent, the off-diagonal term of the equilibrium is

$$E\{\varphi_i(y_i)y_j\} = 0, \qquad i \neq j \tag{3.53}$$

This condition is satisfied when y_i and y_j are independent. The diagonal term is

$$E\{\varphi_i(y_i)y_i\} = 1 \tag{3.54}$$

which determines the scaling of the recovered signals.

The stability of the equilibrium is analyzed by the variational equation

$$\frac{d}{dt}\delta\mathbf{W}(t) = \eta(t)\frac{\partial E\{\mathbf{I} - \boldsymbol{\varphi}[\mathbf{y}(t)]\mathbf{y}^T(t)\}}{\partial\mathbf{W}}\delta\mathbf{W}(t) \tag{3.55}$$

which shows the dynamic behavior of small perturbations $\delta\mathbf{W}(t)$ in the neighborhood of the true solution $\mathbf{W}$. In order to show the stability of the algorithm, we need to check the eigenvalues of the expectation of the extended Hessian

$$-E\left\{\frac{\partial l(\mathbf{y}, \mathbf{W})}{\partial \mathbf{W} \partial \mathbf{W}^T} \mathbf{W}^T \mathbf{W}\right\}$$

at the equilibrium point $\mathbf{W}_*$. When all the real parts of the eigenvalues of the preceding quantity are negative, the solution is stable. We present the analysis in terms of differential calculus in Appendix 3.2. The result is summarized as follows.

The true solution is the stable equilibrium of the on-line learning algorithm when $q(\mathbf{y})$ or the derived $\varphi(\mathbf{y})$ satisfies the following conditions (Amari et al. 1977c)

$$m_i + 1 > 0 \tag{3.56}$$

$$\kappa_i > 0 \tag{3.57}$$

$$\sigma_i^2 \sigma_j^2 \kappa_i \kappa_j > 0 \tag{3.58}$$

where $m_i = E\{y_i^2 \varphi_i'(y_i)\}$, $\kappa_i = E\{\varphi_i'(y_i)\}$, and $\sigma_i^2 = E\{y_i^2\}$.

Now we may state modifications of the basic learning equation, Eq. (3.49). Amari and Cardoso (1997) studied this problem from the point of view of semi-parametric statistical models of information geometry (Amari and Kawanabe 1997a,b). They treat a general form of learning equations,

$$\Delta \mathbf{W}(k) = \eta(k) \mathbf{F}[\mathbf{y}(k), \mathbf{W}(k)] \mathbf{W}(k) \tag{3.59}$$

for arbitrary smooth function $\mathbf{F}$. The results shows that (1) the diagonal entries f_{ii} of $\mathbf{F}$ can be arbitrary, and (2) general admissible (efficient) forms of $\mathbf{F}$ are spanned by $\varphi(\mathbf{y})\mathbf{y}^T$. This implies that another form $\mathbf{y}\varphi(\mathbf{y}^T)$ is also a good candidate of the learning equation, because it is a linear combination of the former. They have shown that

$$\mathbf{F}[\mathbf{y}, \mathbf{W}] = \alpha \varphi(\mathbf{y})\mathbf{y}^T + \beta \mathbf{y}\varphi(\mathbf{y}^T) \tag{3.60}$$

is the general form of the admissible learning function, where the diagonal entries of $\mathbf{F}$ are arbitrarily assigned nonnegative values. Amari et al. (1999) demonstrated that, when the magnitudes of source signals rapidly change over time or when some of them become zero for a while, the learning algorithm with $\mathbf{F}$ such that its diagonal elements are put equal to 0 has a good performance. In other words,

$$\Delta \mathbf{W}(k) = \eta(k)[\mathbf{\Lambda}(k) - \varphi[\mathbf{y}(k)]\mathbf{y}^T(k)]\mathbf{W}(k) \tag{3.61}$$

where $\mathbf{\Lambda}(k)$ is the diagonal matrix with the diagonal entries $\lambda_{ii} = \varphi(y_i)y_i$, works very well in such cases. This introduces nonholonomic constraints in the learning dynamics, and the behavior of the nonholonomic learning algorithm is analyzed in (Amari et al. 1999).

Another modification is to use the learning algorithm of the form (Cichocki et al. 1994a, 1997e; Cichocki and Unbehauen 1996)

$$\Delta\mathbf{W}(k) = \eta(k)[\mathbf{\Lambda}(k) - \mathbf{y}(k)\boldsymbol{\psi}[\mathbf{y}^T(k)]]\mathbf{W}(k) \tag{3.62}$$

This is the filtering gradient (Attick and Redlich 1993; Cichocki et al. 1997e), which is the dual to the natural-gradient descent (see Appendix 3.3),

$$\Delta\mathbf{W}(k) = -\eta(k)\mathbf{W}(k)\left[\frac{\partial D_{fq}(\mathbf{W})}{\partial\mathbf{W}}\right]^T\mathbf{W}(k) \tag{3.63}$$

Here, the nonlinearity $\psi_i(y_i)$ is the inverse of (dual to) the function $\varphi_i(y_i) = -\{\log q(y_i)\}'$.

More generally, Amari et al. (Amari 1998; Amari et al. 1997b,c) have proposed the following learning rule

$$\Delta\mathbf{W}(k) = \eta(k)[\mathbf{\Lambda}(k) - (\alpha\boldsymbol{\varphi}[\mathbf{y}(k)]\mathbf{y}^T(k) + \beta\mathbf{y}(k)\boldsymbol{\varphi}[\mathbf{y}^T(k)])]\mathbf{W}(k) \tag{3.64}$$

where $\mathbf{\Lambda}$ is a diagonal matrix configured to eliminate the diagonal entries in the bracket of the right-hand side, and α and β are adequate parameters to be adaptively determined. This guarantees the best performance, that is, the most efficient on-line estimation, provided constants α and β are adequately determined. See Amari et al. (1997b) for details of α and β.

We have so far excluded from our considerations the learning rules of the type developed first by Cichocki et al. (1994a,b, 1996b, 1997b,e; Cichocki and Unbehauen 1996)

$$\Delta\mathbf{W}(k) = \eta(k)[\mathbf{\Lambda}(k) - \boldsymbol{\varphi}[\mathbf{y}(k)]\boldsymbol{\psi}[\mathbf{y}^T(k)]]\mathbf{W}(k) \tag{3.65}$$

where $\mathbf{\Lambda}$ is an arbitrary positive-definite diagonal matrix and $\boldsymbol{\psi}(\mathbf{y})$ may be a nonlinear function other than $\boldsymbol{\varphi}(\mathbf{y})$. Here, $\boldsymbol{\varphi}$ or $\boldsymbol{\psi}$ should be chosen such that either

$$E\{\boldsymbol{\varphi}(\mathbf{y})\} = 0$$

or

$$E\{\boldsymbol{\psi}(\mathbf{y})\} = 0$$

is satisfied. This learning rule is not admissible from the point of view of efficiency (Amari and Cardoso 1997) in the noiseless case. However, when additive noise exists, it can give a robust performance (Cichocki et al. 1994a, 1997b; Cichocki and Unbehauen 1996) (see Section 3.7).

We finally show the case where prewhitening has taken place. In this case, the demixing matrix $\mathbf{W}$ is restricted to be an orthogonal matrix, because we have

$$E\{\mathbf{ss}^T\} = E\{\mathbf{xx}^T\} = \mathbf{I}_n \tag{3.66}$$

where $\mathbf{x}$ is the prewhitened signal. It should be remarked that $d\mathbf{X} = d\mathbf{W}\mathbf{W}^{-1}$ is a skew symmetric matrix because of the orthogonality constraint. Hence, the natural gradient in the case of orthogonal matrices is given by $-[\boldsymbol{\varphi}(\mathbf{y})\mathbf{y}^T - \mathbf{y}\boldsymbol{\varphi}(\mathbf{y})^T]\mathbf{W}$, where the term inside brackets is an antisymmetric matrix. Hence, the corresponding natural-gradient learning algorithm is easily derived (see Section 3.4),

$$\Delta\mathbf{W}(k) = \eta(k)[\boldsymbol{\varphi}[\mathbf{y}(k)]\mathbf{y}^T(k) - \mathbf{y}(k)\boldsymbol{\varphi}[\mathbf{y}^T(k)]]\mathbf{W}(k) \tag{6.67}$$

This type of equation is used by Cardoso and Laheld in (1996). The stability of the preceding learning rule is also very simple and elegant, and is given by

$$\kappa_i\sigma_j^2 + \kappa_j\sigma_i^2 > 2 \tag{6.68}$$

where $\kappa_i = E\{\varphi_i'(y_i)\}$ and $\sigma_i^2 = E\{y_i^2\}$.

3.4 BLIND EXTRACTION OF SOURCE SIGNALS

In many applications, a large number of sensors (microphones or transducers) are available, but only a very few source signals are a subject of interest. For example, in EEG or MEG, we observe typically more than 64 sensor signals and only a few source signals are interesting; the rest can be considered interfering noise. As another example, in the cocktail party problem it is usually essential to extract voices of specific persons rather than separate all source signals available from a large array of microphones. For such applications, it is essential to develop robust and effective learning algorithms that enable us to extract only a small number of source signals that are potentially interesting and contain useful information. This problem is the subject of this section.

Before we begin to explain the derivation of learning algorithms, we should emphasize some advantages of this approach. The blind signal-extraction approach may have several advantages over simultaneous blind separation (Cichocki et al. 1996f, 1997c; Hyvärinen and Oja 1997; Hyvärinen 1998):

1. Signals can be extracted in a specified *order* according to the stochastic features of the source signals, for example, in the order of absolute values of generalized normalized kurtosis. The blind extraction can be considered as a generalization of PCA, where decorrelated output signals are extracted according to the decreasing order of their variances.

2. Only "interesting" signals need to be extracted. For example, if the source signals are mixed with a large number of Gaussian noise terms, we should extract only those signals with some desired stochastic properties.

3. The developed learning algorithms for BES are purely local and biologically plausible. In fact, the learning algorithms derived below can be considered as extensions or modifications of the Hebbian/anti-Hebbian learning rule. Typically, they are simpler than those of instantaneous blind source separation.

In summary, blind signal extraction is a useful approach when it is desired that several source signals with specific stochastic properties be extracted from a large number of mixtures. Extraction of a single source is closely related to the problem of blind deconvolution (Haykin 1994; Inouye and Sato 1997; Inouye 1997; Shynk and Gooch 1996).

3.4.1   Sequential Blind Extraction of Sources

Our objective in this subsection is to extract the source signals sequentially one by one, rather than to separate all of them simultaneously. This procedure is called the *sequential blind signal extraction* in contrast with the instantaneous BSS.

Sequential BES can be performed by using a cascade neural network similar to the one used for the extraction of principal components (compare Fig. 3.7*a* and Fig. 3.8*a*). However, in contrast with PCA, the optimization criteria for BES are different. For the principal-component extraction, we have applied the optimization criterion that ensures the best possible reconstruction of vector $\mathbf{x}(k)$ after its compression using a single processing unit. In order to extract independent source signals we use a completely different criterion: maximization of the absolute value of generalized normalized kurtosis, which is a measure of the distance of the extracted source signal from Gaussianity. Several approaches have been recently developed for blind signal extraction and blind deconvolution (Cichocki et al. 1996f, 1997c; Delfosse and Loubaton 1995; Hyvärinen and Oja 1997; Inouye and Sato 1997; Inouye 1997; Shynk and Gooch 1996). A single processing unit (artificial neuron) is used in the first step to extract one source signal with specified stochastic properties. In the next step, a deflation technique is used in order to eliminate the already extracted signals from the mixtures.

Extraction Procedure for Single Source   Let us consider a single processing unit described by (see Fig. 3.8*a*)

$$y_1(k) = \mathbf{w}_1^T(k)\mathbf{x}(k) = \sum_{j=1}^m w_{1j}(k)x_j(k) \qquad (3.69)$$

The unit successfully extracts a source signal, say the jth signal, if $\mathbf{w}_1(\infty) = \mathbf{w}_{1*},$

(a)

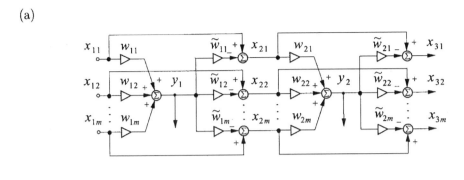

(b)

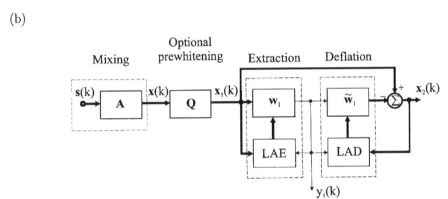

Figure 3.8 Block diagrams illustrating: (a) sequential blind extraction of sources and independent components; (b) implementation of extraction and deflation principles. LAE and LAD mean learning algorithm for extraction and deflation, respectively.

which satisfies the relation $\mathbf{w}_{1*}^T \mathbf{A} = \mathbf{e}_j^T$, where $\mathbf{e}_j$ denotes the jth column of a $n \times n$ nonsingular diagonal matrix.

As a contrast function we can employ (Cichocki et al. 1997c)

$$l_\kappa(\mathbf{w}_1) = -\frac{1}{4}|\kappa_4(y_1)| \tag{3.70}$$

where $\kappa_4(y_1)$ is the normalized kurtosis defined by

$$\kappa_4(y_1) = \frac{E\{|y_1|^4\}}{E^2\{|y_1|^2\}} - 3 \tag{3.71}$$

It should be noted that the absolute value of the kurtosis can be taken as a measure of Gaussianity of the extracted signal y_1.

It is easy to show that for the prewhitened sensor signals $\mathbf{x}$, the normalized kurtosis satisfies the following relations:

$$\kappa_4(y_1) = \kappa_4(\mathbf{w}_1^T \mathbf{A s}) = \kappa_4(\mathbf{e}^T \mathbf{s}) = \sum_{i=1}^{m} \frac{e_i^4}{\|\mathbf{e}\|^4} \kappa_4(s_i) = \sum_{i=1}^{m} \bar{e}_i^4 \kappa_4(s_i) \qquad (3.72)$$

where $\mathbf{e}^T = \mathbf{w}_1^T \mathbf{A}$ is a $1 \times m$ row vector; e_i are its elements; and $\bar{\mathbf{e}} = \mathbf{e}/\|\mathbf{e}\|$ such that $\|\bar{\mathbf{e}}\| = 1$. Thus our normalized kurtosis satisfies the lemma given by Delfosse and Loubaton (1995), so that the contrast function does not contain any spurious local minima. In other words, all the minima correspond to the source signals. In order to derive a global convergent learning rule, we apply the standard stochastic-gradient descent technique. Minimization of the contrast function, Eq. (3.70), leads to a simple local learning rule (Cichocki et al. 1996f, 1997c)

$$\mathbf{w}_1(k+1) = \mathbf{w}_1(k) - \eta_1(k) \frac{\partial l_\kappa}{\partial \mathbf{w}_1}$$
$$= \mathbf{w}_1(k) - \eta_1(k)\varphi_1(y_1(k))\mathbf{x}(k) \qquad (3.73)$$

where $\mathbf{x}$ is the vector of sensor signals, and the nonlinear activation function is evaluated adaptively as

$$\varphi_1(y_1) = \text{sign}[\kappa_4(y_1)]\left[y_1 - \frac{m_2(y_1)}{m_4(y_1)} y_1^3\right] \frac{m_4(y_1)}{m_2^3(y_1)} \qquad (3.74)$$

where $m_k(y_i) = E[(y_i)^k]$ is the kth moment of y_i. The higher-order moments m_2 and m_4 and the sign of the kurtosis $\kappa_4(y_1)$ can be estimated on-line by using the following averaging formula

$$m_p(k+1) = (1 - \eta)m_p(k) + \eta|y_1(k)|^p \qquad (3.75)$$

with $\eta > 0$ and $m_p(k) \cong E[|y_1(k)|^p]$, $p = 2, 4$. This contrast function is similar to that already proposed by Shalvi and Weinstein (1990) and in an extended form by Inouye (Inouye and Sato 1997; Inouye 1997) and Comon (1996) for the blind deconvolution problem (Haykin 1994). However, instead of the standard kurtosis we employ a normalized one, which eliminates any constraint on the magnitudes of the output signals and therefore considerably simplifies the optimization procedure.

It should be noted that the term $E\{|y(k)|^4\}/E^3\{|y(k)|^2\}$ is always positive, so it can be absorbed by the learning rate $\eta(k)$. Moreover, communication signals are usually sub-Gaussian, so that they have, in many applications, negative kurtosis. In such cases the nonlinear function can be simplified as

$$\varphi_1[y_1] = y_1 - \alpha_1 y_1^3 \qquad (3.76)$$

where $\alpha_1 = E\{|y(k)|^2\}/E\{|y(k)|^4\}$ and $\kappa_4\{y_1(k)\} < 0$.

The preceding nonlinear activation function can be generalized by minimizing the following contrast function

$$J[\mathbf{w}_1(k)] = l_{pq}[\mathbf{w}_1(k)] = -\frac{1}{p}|\kappa_{p,q}\{y_1(k)\}|$$ (3.77)

where $\kappa_{p,q}\{y_1(k)\}$ is the generalized normalized kurtoses or Gray's variable norm (Lambert 1996), defined as

$$\kappa_{p,q}\{y_1(k)\} = \frac{E\{|y_1(k)|^p\}}{E^q\{|y_1(k)|^{p/q}\}} - c_{pq}$$ (3.78)

where c_{pg} is a positive constant such that for the Gaussian distribution, $\kappa_{p,q} = 0$. For this contrast function, $\varphi_1[y_1(k)]$ has the general form (assuming for simplicity that all the source signals are sub-Gaussian with negative kurtoses)

$$\varphi_1[y_1] = \text{sign}(y_1)\left(|y_1|^{q-1} - \frac{E\{|y_1|^q\}}{E\{|y_1|^p\}}|y_1|^{p-1}\right)$$ (3.79)

In the special case, where $p = 1$ and $q = 2$, for sub-Gaussian signals, we obtain the modified Sato algorithm with nonlinearity (Papadias and Paulraj 1997; Sato 1975, 1994; Treichler and Larimore 1985)

$$\varphi_1[y_1] = y_1 - \frac{E\{|y_1|^2\}}{E\{|y_1|\}}\text{sign}(y_1)$$ (3.80)

Alternatively, for the sub-Gaussian signals, choosing $q = 2$ produces a modified generalized Godard or constant modulus algorithm (CMA), Eq. (3.73), with the adaptive nonlinear activation function (Cichocki et al. 1996f, 1997c; Douglas et al. 1997a; Hyvarinen and Oja 1997; Lambert 1996; Papadias and Paulraj 1997; Treichler and Larimore 1985):

$$\varphi_1[y_1] = y_1 - \frac{E\{|y_1|^2\}}{E\{|y_1|^p\}}|y_1|^{p-1}\text{sign}[y_1]$$ (3.81)

where $p = 3, 4, \ldots$.

It should be noted that the nonlinear function $\varphi_1(y_1) = y_1(|y_1|^{q-2} - \alpha_1|y_1|^{p-2})$ can be fixed if the statistics of the source signal are known, by taking

$$\alpha_1 = \frac{E\{|\hat{s}_1(k)|^q\}}{E\{|\hat{s}_1(k)|^p\}} = \text{const.}$$ (3.82)

In general, however, when the statistics of the source signals are not known or cannot be estimated, the parameter $\alpha(k)$ is not fixed but can be adapted during the learning process, depending on the higher-order moments of the absolute values of the estimated output signal $y(k)$. In this case, the higher-order moments of the form $m_r(k) = E\{|y(k)|^r\}$ appearing in the nonlinearities can be estimated on-line by using the moving-average (MA) procedure

$$m_r(k+1) = (1 - \eta)m_r(k) + \eta|y(k)|^r \tag{3.83}$$

where η is a small positive constant.

It is interesting to note that the preceding algorithms can be easily extended for complex-valued signals by noting that in such a case $y_1(k) = \mathbf{x}_k^H \mathbf{w}_1(k)$ (where superscript H means the complex-conjugate transpose of Hermitian operation) and by replacing $\text{sign}[y_1(k)]$ by $y_1(k)/|y_1(k)|$. For example, CMA for $p = 4$ and $q = 2$, for complex-valued coefficients, and signals, takes the form:

$$\mathbf{w}_1(k+1) = \mathbf{w}_1(k) \pm \eta(k) \left[y_1(k) - \frac{E\{|y_1(k)|^2\}}{E\{|y_1(k)|^4\}} y_1(k)|y_1(k)|^2 \right] \mathbf{x}_k^* \tag{3.84}$$

where $\mathbf{x}_k^*$ is the vector of complex-conjugate sensor signals and the plus sign is for sub-Gaussian and the minus sign for super-Gaussian signals.

The preceding algorithms can be considered as the blind least-mean-square (BLMS) algorithm with a "blind" error signal equal to

$$y_1(k) - \frac{E\{|y_1(k)|^2\}}{E\{|y_1(k)|^4\}} y_1(k)|y_1(k)|^2$$

So many powerful and efficient techniques developed for standard least mean square (LMS) (Haykin 1996) also can be applied to these algorithms.

It should be noted that in general the activation function is not fixed but rather adaptive during the learning process. This observation was first made by S. Haykin (1992). Adapting the nonlinearities are important in the multi-channel case, as the signals may consist of mixtures of several sources with different distributions. The optimal choice for the values of p and q depends on the statistics of the input signals, implying a trade-off between the tracking ability and the estimation accuracy of the algorithm (Lambert 1996). Such methods also have a natural extension to the multichannel blind deconvolution problems (Choi and Cichocki 1998; Papadias and Paulraj 1997). This is discussed later. Figure 3.8a and b illustrate the extraction and deflation process that is explained in the next subsection. The presented neural network employs two different types of processing units, one for blind extraction and the other for deflation, alternately connected with each other in a cascade fashion. Namely, the kth extraction processing unit extracts a source signal from its inputs that are linear mixtures of the remaining source signals yet to be extracted. The kth deflation processing unit then deflates (removes) the newly extracted source signal from the mixtures and feeds the resulting outputs to the next $(k + 1)$th extraction processing unit. Note that a successfully extracted signal may have a reversed sign with any nonzero magnitude scale, compared to the original source signal, that is, sign and/or scale indeterminacy arises.

Remark. It should be noted that the preceding method does not need prewhitening. For ill-conditioned problems (when a mixing matrix is ill-conditioned and/or source signals have different powers), however, we can apply preprocessing (prewhitening) in the form of $\mathbf{x}_1 = \mathbf{Q}\mathbf{x}$, where the decorrelation matrix $\mathbf{Q} \in \mathbf{R}^{n \times m}$ ensures that the autocorrelation matrix $\mathbf{R}_{\mathbf{x}_1 \mathbf{x}_1} = E\{\mathbf{x}_1 \mathbf{x}_1^T\} = \mathbf{I_n}$. Prewhitening can simultaneously reduce the dimensional redundancy of the signals from m to n if we select $\mathbf{Q} \in \mathbf{R}^{n \times m}$. It should be noted that after the decorrelation process, the new unknown mixing matrix defined as $\mathbf{B} = \mathbf{Q}\mathbf{A}$, is an orthogonal matrix satisfying the relation $\mathbf{B}^T \mathbf{B} = \mathbf{I}_n$, that is, $\mathbf{b}_i^T \mathbf{b}_j = \delta_{ij}$, where $\mathbf{b}_i$ is the ith vector of the mixing matrix $\mathbf{B}$.

Deflation Procedure We now describe the deflation process. After the successful extraction of the first source signal $y_1(k) \sim s_i(k)(i \in \overline{1,n})$, we can apply the deflation procedure, which removes the previously extracted signals from the mixtures. This procedure may be recursively applied to extract sequentially the rest of the estimated source signals. This means that we require an on-line linear transformation given by (see Fig. 3.8a,b) (Cichocki et al. 1996f, 1997c)

$$\mathbf{x}_{j+1}(k) = \mathbf{x}_j(k) - \tilde{\mathbf{w}}_j(k)y_j(k) \qquad j = 1, 2, \ldots \tag{3.85}$$

which ensures minimization of the generalized cost (loss) function

$$l_j(\tilde{\mathbf{w}}_j) = \rho(\mathbf{x}_{j+1}) = \sum_{p+1}^{m} \rho(x_{j+1,p}) \tag{3.86}$$

where $\rho(x_{j+1,p})$ is a objective function (see Table 3.1) and $y_j = \mathbf{w}_j^T \mathbf{x}_j$, is the jth extracted signal subject to

$$\mathbf{w}_j(k + 1) = \mathbf{w}_j(k) - \eta_j(k)\varphi_j[y_j(k)]\mathbf{x}_j(k) \tag{3.87}$$

$$\varphi_j(y_j) = \text{sign}[\kappa_4(y_j)] \left[y_j - \frac{m_2(y_j)}{m_4(y_j)} y_j^3 \right] \frac{m_4(y_j)}{m_2^3(y_j)} \tag{3.88}$$

Note that $\tilde{\mathbf{w}}_j$ is different from $\mathbf{w}_j$. Minimization of the objective function, Eq. (3.86), leads to the simple local-type LMS learning rule

$$\tilde{\mathbf{w}}_j(k + 1) = \tilde{\eta}_j(k)y_j(k)\mathbf{g}[\mathbf{x}_{j+1}(k)] \qquad j = 1, 2, \ldots \tag{3.89}$$

where $\tilde{\mathbf{w}}_j$ is an estimate of $\hat{\mathbf{a}}_j$ and $\mathbf{g}[\mathbf{x}_{j+1}] = [g(x_{j+1,1}) \cdots g(x_{j+1,m})]^T = \partial l_j / \partial \mathbf{x}_{j+1}$, with $g(x_{j+1,p}) = \partial p(x_{j+1,p})/\partial x_{j+1,p}$; typically $p(x_{j+1,p}) = x_{j+1,p}^2/2$, hence $g(x_{j+1,p}) = x_{j+1,p}$ (see also Table 3.1). The procedure can be continued until all the estimated source signals are recovered, that is, until the amplitude of each signal x_{j+1} will be below a preassigned threshold. This procedure means that it is not required to know the number of source signals in advance, but it is assumed that the number is constant and that the mixing system is stationary (e.g., the sources do not "change places" during the convergence of the algorithm). In Appendix 3.5 we prove that this deflation optimization procedure has no spurious (undesired) minima and after the convergence the algorithm, Eq. (3.89), estimates one column of the mixing matrix $\mathbf{A}$ with scaling and permutation indeterminacy.

In order to extract signals in a specified order, for example, in the order of the decreasing absolute values of the normalized kurtosis, we can apply global optimization to achieve a global minimum. In order to avoid local minima, an auxiliary Gaussian noise v can be applied to the nonlinear functions $f(y)$, artificially, that is, $f(\tilde{y}_i) = f(y_i + v_i)$, where v is a Gaussian noise whose variance is decreasing over time. In this manner, it is possible to avoid local minima in most practical cases (Cichocki et al. 1997a,c).

3.4.2 Blind Extraction of Group of Sources

Consider the case where we do not want to extract one single source at a time, but wish to extract simultaneously a specified number of sources, say p, where $1 \leq p \leq n$, with $m \geq n$ (the number of sensors m is greater or equal to the number of sources n and the number of sources is generally unknown) (Amari 1999a). Let us also assume that the sensor signals are prewhitened (sphered), for example, by using the PCA technique described previously. Then, the transformed sensor signals satisfy the condition

$$E\{\mathbf{x}_1 \mathbf{x}_1^T\} = \mathbf{I}_n \tag{3.90}$$

where $\mathbf{x}_1 = \mathbf{Q}\mathbf{x}$ and new global $n \times n$ mixing matrix $\mathbf{B} = \mathbf{Q}\mathbf{A}$ is orthogonal, that is, $\mathbf{B}\mathbf{B}^T = \mathbf{B}^T\mathbf{B} = \mathbf{I}_n$. Hence the ideal $n \times n$ separating Matrix is $\mathbf{W}_* = \mathbf{B}^{-1} = \mathbf{B}^T$ for $p = n$.

In order to solve this problem we can formulate the appropriate contrast function expressed by the Kullback-Leibler divergence:

$$D_{fq} = D(f(\mathbf{y}, \mathbf{W}) \| q(\mathbf{y})) = \int f(\mathbf{y}, \mathbf{W}) \log \frac{f(\mathbf{y}, \mathbf{W})}{q(\mathbf{y})} \, d\mathbf{y} \tag{3.91}$$

where $q(\mathbf{y}) = \prod_{i=1}^{p} q_i(y_i)$ represents an adequate independent probability distribution of output signals. Hence the contrast function takes the form

$$l(\mathbf{y}, \mathbf{W}) = -\sum_{i=1}^{p} \log q_i(y_i) \tag{3.92}$$

subject to constraints $\mathbf{WW}^T = \mathbf{I}_p$ or equivalently $\mathbf{w}_i\mathbf{w}_j^T = \delta_{ij}$, where $\mathbf{W}$ is a $p \times n$ demixing matrix, with $p \leq n$, and $\mathbf{w}_i$ is the ith row vector of matrix $\mathbf{W}$.

These constraints follow from the simple fact that the mixing matrix $\mathbf{B} = \mathbf{QA}$ is square orthogonal matrix and the separating matrix $\mathbf{W}$ should satisfy the following relationship after successful extraction of p sources (ignoring scaling and permutation for simplicity):

$$\mathbf{W}_*\mathbf{B} = [\mathbf{I}_p, \mathbf{0}_{n-p}] \tag{3.93}$$

We say that the matrix $\mathbf{W}$ satisfying this condition forms a Stiefel manifold since its rows are mutually orthogonal ($\mathbf{w}_i\mathbf{w}_j = \delta_{ij}$). In order to satisfy the constraints during the learning process, we employ the following natural gradient formula (Amari 1999a):

$$\Delta\mathbf{W}(k) = \mathbf{W}(k+1) - \mathbf{W}(k) = -\eta(k)\left[\frac{\partial l(\mathbf{y}, \mathbf{W})}{\partial\mathbf{W}} - \mathbf{W}\left(\frac{\partial l(\mathbf{y}, \mathbf{W})}{\partial\mathbf{W}}\right)^T\mathbf{W}\right] \tag{3.94}$$

It can be shown that the separating matrix $\mathbf{W}$ satisfies the relation $\mathbf{W}(k)\mathbf{W}^T(k) = \mathbf{I}_p$ in each iteration step under the condition that $\mathbf{W}(0)\mathbf{W}^T(0) = \mathbf{I}_p$.

Applying the natural gradient formula, Eq. (3.94), we obtain a learning rule:

$$\mathbf{W}(k+1) = \mathbf{W}(k) + \eta(k)[\boldsymbol{\varphi}[\mathbf{y}(k)]\mathbf{x}_1^T(k) - \mathbf{y}(k)\boldsymbol{\varphi}[\mathbf{y}^T(k)]\mathbf{W}(k)] \tag{3.95}$$

with initial conditions $\mathbf{W}(0)$ satisfying the condition $\mathbf{W}(0)\mathbf{W}^T(0) = \mathbf{I}_p$. It is very interesting that this generalized formula for $p = 1$ (extraction of only one single source) is reduced to the well-known nonlinear principal-component analysis proposed by Oja and Karhunen in 1996 as (Karhunen 1996; Karhunen et al. 1995; Oja and Karhunen 1995)

$$\Delta\mathbf{w}_1(k) = \eta(k)\varphi[y_1(k)][\mathbf{x}_1^T(k) - y_1(k)\mathbf{w}_1(k)] \tag{3.96}$$

where $\mathbf{w}_1 = [w_{11} \cdots w_{1n}]$ is a row vector and $y_1(k) = \mathbf{w}_1(k)\mathbf{x}_1(k)$. Assuming that $\mathbf{v}_1 = \mathbf{w}_1^T$, we obtain an algorithm similar to that derived in Section 3.2:

$$\Delta\mathbf{v}_1(k) = \eta(k)\varphi[y_1(k)][\mathbf{x}_1(k) - y_1(k)\mathbf{v}_1(k)] \tag{3.97}$$

On the other hand, for $p = n$, our learning rule simplifies to the well-known algorithm proposed by Cardoso and Laheld in 1996 (Cardoso and Laheld 1996):

$$\mathbf{W}(k+1) = \mathbf{W}(k) + \eta(k)[\boldsymbol{\varphi}[\mathbf{y}(k)]\mathbf{y}^T(k) - \mathbf{y}(k)\boldsymbol{\varphi}[\mathbf{y}^T(k)]]\mathbf{W}(k) \tag{3.98}$$

since $\mathbf{x}_1 = \mathbf{W}^{-1}\mathbf{y} = \mathbf{W}^T\mathbf{y}$. The preceding algorithm can be extended by simultaneously performing prewhitening and blind separation (Cardoso and Laheld

1996):

$$\Delta \mathbf{W}(k) = \eta(k)[\mathbf{I}_n - \mathbf{y}(k)\mathbf{y}^T(k) + \boldsymbol{\varphi}[\mathbf{y}(k)]\mathbf{y}^T(k) - \mathbf{y}(k)\boldsymbol{\varphi}[\mathbf{y}^T(k)]]\mathbf{W}(k) \quad (3.99)$$

It can be shown that the separating matrix $\mathbf{W}(k)$ approximately satisfies [for the sufficiently small value of learning rate $\eta(k)$] the relation $\mathbf{W}(k)\mathbf{W}^T(k) = \mathbf{I}_p$ in each iteration step k under the condition that $\mathbf{W}(0)\mathbf{W}^T(0) = \mathbf{I}_p$, since the preceding learning rule satisfies the relation

$$\Delta \mathbf{W}(k)\mathbf{W}^T(k) + \mathbf{W}(k)\Delta \mathbf{W}^T(k) = \mathbf{0} \quad (3.100)$$

3.5   NATURAL-GRADIENT LEARNING ALGORITHM FOR BLIND ESTIMATION OF THE MIXING MATRIX: UNDERCOMPLETE BASE IN ICA

Now let us consider the more challenging problem of estimating the mixing matrix when the number of sources is larger than or equal to the number of sensors (with $m \leq n$). Without loss of generality, let us assume that the sensor signals are prewhitened, that is,

$$E\{\mathbf{x}_1 \mathbf{x}_1^T\} = \mathbf{I}_m \quad (3.101)$$

where $\mathbf{x}_1 = \mathbf{Q}\mathbf{x} = \mathbf{Q}\mathbf{A}\mathbf{s} = \mathbf{B}\mathbf{s}$; $\mathbf{Q}$ is the prewhitening matrix; and we assume that the source signals have a unit variance, that is,

$$E\{\mathbf{s}\mathbf{s}^T\} = \mathbf{I}_n \quad (3.102)$$

Taking these assumptions into account, we have

$$E\{\mathbf{x}_1 \mathbf{x}_1^T\} = E\{\mathbf{B}\mathbf{s}\mathbf{s}^T\mathbf{B}^T\} = \mathbf{B}\mathbf{B}^T = \mathbf{I}_m \quad (3.103)$$

It should be noted that the constraint $\mathbf{B}\mathbf{B}^T = \mathbf{I}_m$ implies that m rows of $\mathbf{B} = \mathbf{Q}\mathbf{A}$ are mutually orthogonal. Moreover, when $\Delta \mathbf{B}$ is a small modification of $\mathbf{B}$, it satisfies the relation

$$\Delta \mathbf{B}\mathbf{B}^T + \mathbf{B}\Delta \mathbf{B}^T = \mathbf{0} \quad (3.104)$$

This set of such matrices forms a Stiefel manifold, and it reduces to the orthogonal group when $m = n$ (Amari 1999a; Edeleman et al. 1999).

To solve the problem we introduce the following contrast function:

$$l(\mathbf{y}, \hat{\mathbf{B}}) = -\sum_{i=1}^{m} \log q_i(y_i) - \frac{1}{2}\log(\det(\hat{\mathbf{B}}\hat{\mathbf{B}})^T) \quad (3.105)$$

where $\mathbf{y} = \hat{\mathbf{B}}^T \mathbf{x}_1$ or a simpler one

$$l(\mathbf{y}, \hat{\mathbf{B}}) = -\sum_{i=1}^{m} \log q_i(y_i) \qquad (3.106)$$

subject to the constraint $\hat{\mathbf{B}}\hat{\mathbf{B}}^T = \mathbf{I}_m$.

This constraint can be automatically satisfied by applying the following natural gradient formula (Amari 1999a):

$$\Delta\hat{\mathbf{B}}(k) = -\eta(k)\left[\frac{\partial l(\mathbf{y}, \hat{\mathbf{B}})}{\partial \hat{\mathbf{B}}} - \hat{\mathbf{B}}\left(\frac{\partial l(\mathbf{y}, \hat{\mathbf{B}})}{\partial \hat{\mathbf{B}}}\right)^T \hat{\mathbf{B}}\right] \qquad (3.107)$$

Applying this gradient formula, we obtain the learning algorithm for estimating the mixing matrix

$$\Delta\hat{\mathbf{B}}(k) = -\eta(k)\hat{\mathbf{B}}(k)[\varphi[\mathbf{y}(k)]\mathbf{y}^T(k)\hat{\mathbf{B}}(k)\hat{\mathbf{B}}^T(k) - \mathbf{y}(k)\varphi[\mathbf{y}^T(k)]] \qquad (3.108)$$

$$= -\eta(k)[\hat{\mathbf{B}}(k)\varphi[\mathbf{y}(k)]\mathbf{y}^T(k) - \mathbf{x}_1(k)\varphi[\mathbf{y}^T(k)]] \qquad (3.109)$$

with an initial matrix $\mathbf{B}(0)$ satisfying $\hat{\mathbf{B}}(0)\hat{\mathbf{B}}(0)^T = \mathbf{I}_m$.

It is interesting to note that, in the special case when $m = n$, the estimating matrix satisfies the relation $\hat{\mathbf{B}}\hat{\mathbf{B}}^T = \hat{\mathbf{B}}^T\hat{\mathbf{B}} = \mathbf{I}_n$ and the learning algorithm just given is simplified as

$$\Delta\hat{\mathbf{B}}(k) = -\eta(k)\hat{\mathbf{B}}(k)[\varphi[\mathbf{y}(k)]\mathbf{y}^T(k) - \mathbf{y}(k)\varphi[\mathbf{y}^T(k)]] \qquad (3.110)$$

Moreover, the learning algorithm, Eq. (3.98), derived in the previous section also can be obtained from the algorithm in Eq. (3.110) and vice versa. Hence, after simple calculations, we obtain the following relationships between learning rules for the estimation of the mixing and separating matrices:

$$\Delta\mathbf{W} = -\mathbf{W}\Delta\hat{\mathbf{B}}\mathbf{W} \qquad (3.111)$$

and

$$\Delta\hat{\mathbf{B}} = -\hat{\mathbf{B}}\Delta\mathbf{W}\hat{\mathbf{B}} \qquad (3.112)$$

under the assumption that $\Delta\hat{\mathbf{B}}$ and $\Delta\mathbf{W}$ are small changes.

Remark. It should be noted that estimation of the mixing matrix $\mathbf{A}$ can be considered a preliminary step in the estimation of source signals. This is particularly useful in the case when the number of sensors is less than the number of sources (Chen et al. 1996; Lewicki and Sejnowski 1998). In the special case when the source signals are spiky and "sparse" in the sense that they fluctuate mostly around zero and only occasionally have nonzero values, the problem of estimating unknown source signals can be converted to the extended linear programming problem: find the optimal sequence of estimated source signals $\hat{s}_i(k)$ that minimize the cost function:

$$\sum_k \sum_{i=0}^n |\hat{s}_i(k)| \tag{3.113}$$

subject to the constraints

$$\hat{\mathbf{B}}\hat{\mathbf{s}}(k) = \mathbf{x}(k) \qquad \forall k$$

3.6 GENERALIZED DISTRIBUTION MODELS FOR ICA: PRACTICAL IMPLEMENTATION OF THE ALGORITHMS

The performance of a learning algorithm depends on the shape of the activation functions. Optimal selection of nonlinearities depends on the probability density function (pdf) of the source signals. Various generalized models can be used, such as, generalized Gaussian, generalized Cauchy, mixture of Gaussian distributions, and the Pearson model.

Let us assume that the source signals have generalized Gaussian distributions of the form (Cichocki et al. 1997e, 1998a,c):

$$q_i(y_i) = \frac{r_i}{2\sigma_i \Gamma(1/r_i)} \exp\left(-\frac{1}{r_i}\left|\frac{y_i}{\sigma_i}\right|^{r_i}\right) \tag{3.114}$$

where $r_i > 0$ is a variable parameter; $\Gamma(r) = \int_0^\infty y^{r-1} \exp(-y)\, dy$ is the gamma function; and $\sigma_i^r = E\{|y_i|^r\}$ is a generalized measure of variance known as the dispersion of the distribution. The parameter r_i can change from zero, through 1 (Laplace distribution) and $r_i = 2$ (standard Gaussian distribution), to r_i going to infinity (for uniform distribution) (see Fig. 3.9a,b).

The optimal normalized nonlinear activation functions can be expressed in such cases as:

$$\varphi_i(y_i) = -\frac{d\log(q_i(y_i))}{dy_i} = |y_i|^{r_i-1}\operatorname{sign}(y_i), \qquad r_i \geq 1 \tag{3.115}$$

Taking into account that $\operatorname{sign}(y) = y/|y|$, we obtain (see Fig. 3.9b)

$$\varphi_i(y_i) = \frac{y_i}{|y_i|^{2-r_i}} \tag{3.116}$$

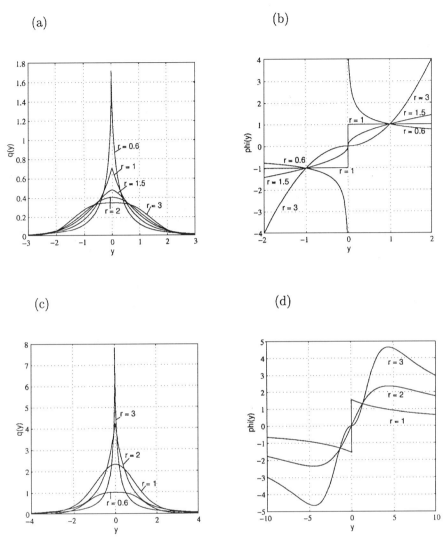

Figure 3.9 Plot of generalized Gaussian and Cauchy pdf models for various values of parameter r and corresponding nonlinear activation functions: (a) Generalized Gaussian distribution ($\sigma^2 = 1$); (b) corresponding optimal activation functions; (c) generalized Cauchy distribution ($v = 1, \sigma^2 = 1$); (d) corresponding optimal activation functions.

In the special case of spiky or a impulsive signals, the parameters r_i can take the value between zero and one. In that case, we can use the slightly modified activation functions

$$\varphi_i(y_i) = \frac{y_i}{[|y_i|^{2-r_i} + \varepsilon]} \qquad 0 < r_i < 1 \qquad (3.117)$$

where ε is a very small positive parameter (typically 10^{-4}) that avoids a singularity of the function when $y_i = 0$. Alternatively, we can take the moving average of the instantaneous values of the nonlinear function as

$$\varphi_i(y_i) = \frac{y_i}{\langle|y_i|^{2-r_i}\rangle} = \frac{y_i(k)}{\hat{\sigma}_i^{(2-r_i)}(k)}, \qquad 0 \le r_i < \infty \tag{3.118}$$

with estimation of $\hat{m}_{2-r_i} = \hat{\sigma}_i^{(2-r_i)}$ by the moving average as

$$\hat{\sigma}_i^{(2-r_i)}(k+1) = (1-\eta)\hat{\sigma}_i^{(2-r_i)}(k) + \eta|y_i(k)|^{2-r_i} \tag{3.119}$$

Such an activation function can be considered as a "linear" time-variable function modulated in time by the fluctuating estimated moment, $m_{2-r_i} = \hat{\sigma}_i^{(2-r_i)}$.

The parameters r_i can be fixed if some *a priori* knowledge about the statistics of source signals is available or obtained through the learning. The stochastic-gradient-based rule for adjusting parameter $r_i(k)$ takes the form

$$\Delta r_i(k) = -\eta_i(k)\frac{\partial l_q}{\partial r_i} = \eta_i(k)\frac{\partial \log(q_i(y_i))}{\partial r_i} \tag{3.120}$$

$$\cong \eta_i(k)\frac{0.1r_i(k) + |y_i(k)|^{r_i(k)}(1 - \log(|y_i(k)|^{r_i(k)}))}{r_i^2(k)} \tag{3.121}$$

It is interesting to note that in the special case of spiky signals for $r_i = 0$ we obtain as the optimal function the "linear" time-variable function proposed by Matsuoka et al. (1995) as

$$\varphi_i(y_i) = \frac{y_i}{\langle|y_i|^2\rangle} = \frac{y_i(k)}{\hat{\sigma}_i^2(k)} \tag{3.122}$$

Analogously for $r_i = 1$ (Laplace distribution), we obtain

$$\varphi_i(y_i) = \frac{y_i}{\langle|y_i|\rangle} = \frac{y_i(k)}{\hat{\sigma}_i(k)} \tag{3.123}$$

and for large $r_i \gg 1$, say $r_i = 10$ (approximately uniform distribution)

$$\varphi_i(y_i) = y_i(k)\langle|y_i(k)|^8\rangle = y_i(k)\hat{\sigma}_i^8(k) \tag{3.124}$$

Remark. It should be noted that the Matsuoka activation function satisfies the conditions

$$\langle\varphi_i(y_i)y_i\rangle = 1 \qquad \forall i \tag{3.125}$$

independently of the nonzero variance of the output signals. Hence the normalization of scales of the output signals in the learning algorithm is automatically taken [cf. nonholonomic constraints (Amari et al. 1999)].

Remark. The Matsuoka activation function does not satisfy the stability conditions (Amari 1997), since in this case we have (see Appendix 3.3)

$$\gamma_{ij} = \sigma_i^2 \sigma_j^2 E\{\varphi_i'(y_i)\} E\{\varphi_j'(y_j)\} = 1 \tag{3.126}$$

Therefore in the basic learning algorithm $\Delta \mathbf{W}(k) = \eta \mathbf{F}(\mathbf{y})\mathbf{W}(k)$, we should apply entries $f_{ij} = -y_i \varphi_j(y_j)$ of matrix $\mathbf{F}(\mathbf{y})$ instead of $f_{ij} = -\varphi_i(y_i)y_j$.

More generally, for mixture of sub- and super-Gaussian signals distorted by impulsive noise (with large outliers), using this model and Amari's natural-gradient approach, we can theorize using a learning algorithm of the form

$$\Delta \mathbf{W}(k) = \eta \mathbf{F}(\mathbf{y})\mathbf{W}(k) \tag{3.127}$$

with diagonal matrix $\Lambda(k) = \text{diag}\{\varphi(\mathbf{y}(k))\psi(\mathbf{y}^T(k))\}$ and the elements of matrix $\mathbf{F}(\mathbf{y})$ defined by

$$f_{ij} = \begin{cases} -\varphi_i(y_i)y_j, & \text{for } \gamma_{ij} < 1 \\ -y_i \varphi_j(y_j), & \text{otherwise} \end{cases} \tag{3.128}$$

where $\gamma_{ij} = \sigma_i^2 \sigma_j^2 E\{\varphi_i'(y_i)\} E\{\varphi_j'(y_j)\}$, and $\varphi_i(y_i) = y_i/(|y_i|^{2-r_i} + \varepsilon)$, with r_i between zero and one (see Appendix 3.3).

Alternatively, after some simplifications we can use the learning rule, Eq. (3.65), with diagonal matrix $\Lambda(k) = \text{diag}\{\varphi(\mathbf{y}(k))\psi(\mathbf{y}^T(k))\}$ and with robust adaptive time-variable activation functions with respect to outliers (Cichocki et al. 1997f, 1998a)

$$\varphi_i(y_i) = \begin{cases} \dfrac{y_i}{\hat{\sigma}_i^{2-r_i}} & \text{for } \kappa_4(y_i) > \delta \\ y_i & \text{otherwise} \end{cases} \tag{3.129}$$

$$\psi_i(y_i) = \begin{cases} y_i & \text{for } \kappa_4(y_i) > -\delta \\ \dfrac{y_i}{\hat{\sigma}_i^{2-r_i}} & \text{otherwise} \end{cases} \tag{3.130}$$

where $\kappa_4(y_i) = E\{y_i^4\}/E^2\{y_i^2\} - 3$ is the normalized value of kurtosis, and $\delta \geq 0$ is a small threshold. The value of the kurtosis can be evaluated on-line using

$$E\{y_i^q(k+1)\} = (1-\eta)E\{y^q(k)\} + \eta|y_i(k)|^q, \qquad (q = 2, 4)$$

The learning algorithms—Eqs. (3.65), (3.129), (3.130)—monitor and estimate the statistics of each output signal, depending on the sign or value of its normalized kurtosis (which is a measure of distance from the Gaussianity). It then automatically selects (or switches) suitable nonlinear activation functions, such that successful (stable) separation of all the non-Gaussian source signals is possible. In this approach, the activation functions are adaptive time-varying nonlinearities.

Similar methods can be applied for other parameterized distributions. For example, for the generalized Cauchy distribution defined in terms of three parameters $r_i > 0$, $v_i > 0$, and σ_i^2 (see Fig. 3.9c and d),

$$q_i(y_i) = \frac{B(r_i, v_i)}{\left(1 + \frac{1}{v_i}\left[\frac{y_i}{A(r_i)}\right]^{r_i}\right)^{v_i + 1/r_i}} \tag{3.131}$$

with $A(r_i) = [\sigma_i^2 \Gamma(1/r_i)/\Gamma(3/r_i)]^{1/2}$ and $B(r_i, v_i) = r_i v_i^{-1/r_i} \Gamma(r_i + 1/r_i)/2A(r_i)\Gamma(v_i)\Gamma(1/r_i)$, we have the following activation function (see Fig. 3.9d)

$$\varphi_i(y_i) = \frac{(v_i r_i + 1)}{(v_i|A(r_i)|^{r_i} + |y_i|^{r_i})}|y_i|^{r_i - 1}\,\text{sign}(y_i) \tag{3.132}$$

Similarly, for the generalized Rayleigh distribution, one obtains $\varphi_i(y_i) = |y_i|^{r_i - 2} y_i$ for complex-valued signals and coefficients.

Summarizing, in this subsection we derived and analyzed unsupervised adaptive on-line and microbatch algorithms for BSS, especially when the source signals are nonstationary and they have spiky (impulsive) behavior, and the algorithms are applicable to mixtures of an unknown number of independent source signals with unknown statistics (Cichocki et al. 1997f, 1998a). Nonlinear activation functions are rigorously derived by assuming that the sources are modeled by generalized Gaussian distributions. As a special case, we derived and justified the time-variable "linear" activation function

$$\varphi(y_i) = y_i(k)/\langle y_i^2 \rangle = y_i(k)/\hat{\sigma}_i^2$$

proposed by Matsuoka et al. (1995) for the blind separation of nonstationary signals. Applying the MA estimation to the output variance $\hat{\sigma}_i^2 = \langle y_i^2 \rangle$, we can use the same concept even for stationary signals, since the estimation of variance continuously fluctuates in time. Extensive computer simulations have confirmed that the proposed algorithms are able to separate spiky and nonstationary sources (like biomedical signals, especially MEG signals) (Amari et al. 1999; Cichocki et al. 1997f, 1998a,e; Jahn et al. 1998).

3.7 BIAS REMOVAL FOR BSS ADAPTIVE ALGORITHMS

The learning algorithms, Eqs. (3.49)–(3.65) possess excellent performances for separating noiseless signal mixtures; however, their performance deteriorates with noisy measurements due to undesirable coefficient biases and the existence of noise in the separated signals.

In order to estimate the coefficient biases, we use the Taylor series expansions of the nonlinearities $\varphi_i(y_i)$ and $\psi_j(y_j)$ about the estimated noiseless values $\hat{y}_i$. The generalized covariance matrix $R_{\varphi\psi}$ can be approximately evaluated as (Douglas et al. 1998; Cichocki et al. 1998b,c,e)

$$\mathbf{R}_{\varphi\psi} = E\{\boldsymbol{\varphi}[\mathbf{y}(k)]\boldsymbol{\psi}[\mathbf{y}^T(k)]\}$$
$$= E\{\boldsymbol{\varphi}|\hat{\mathbf{y}}(k)]\boldsymbol{\psi}[\hat{\mathbf{y}}^T(k)]\} + \mathbf{k}_\varphi\mathbf{W}\mathbf{R}_{vv}\mathbf{W}^T\mathbf{k}_\psi \qquad (3.133)$$

where $\mathbf{k}_\varphi$ and $\mathbf{k}_\psi$ are diagonal matrices with entries $k_{\varphi i} = E\{d\varphi_i(y_i(t))/dy_i\}$ and $k_{\psi i} = E\{d\psi_i(y_i(t))/dy_i\}$, respectively. Thus a modified adaptive learning algorithm with reduced bias has the form

$$\Delta\mathbf{W}(k) = \eta(k)[\mathbf{I} - \boldsymbol{\varphi}[\mathbf{y}(k)]\boldsymbol{\psi}[\mathbf{y}^T(k)] + \mathbf{k}_\varphi\mathbf{W}(k)\mathbf{R}_{vv}\mathbf{W}^T(k)\mathbf{k}_\psi]\mathbf{W}(k)$$
$$= \eta(k)[\mathbf{I} - \boldsymbol{\varphi}[\mathbf{y}(k)]\boldsymbol{\psi}[\mathbf{y}^T(k)] + \mathbf{C} \circ \mathbf{W}(k)\mathbf{R}_{vv}\mathbf{W}^T(k)]\mathbf{W}(k) \qquad (3.134)$$

where $\mathbf{C} = [c_{ij}]$ is an $n \times n$ scaling matrix with entries $c_{ij} = k_{\varphi i}k_{\psi j}$ and $\circ$ means the Hadamard product. In the special case when all of the source distributions are identical, $\varphi_i(y_i) = \varphi(y_i) \; \forall i$, $\psi_i(y_i) = \psi(y_i) \; \forall i$, and $\mathbf{R}_{vv} = \sigma_v^2\mathbf{I}$, the bias-correction term is simplified to $\mathbf{B}_c = \sigma_v^2 k_\varphi k_\psi \mathbf{W}\mathbf{W}^T$. It is interesting to note that we can almost always select nonlinearities such that the global scaling coefficient $c = k_\varphi k_\psi$ can be close to zero for a wide class of signals. For example, when $\varphi(y_i) = |y_i|^p \text{sign}(y_i)$ and $\psi(y_i) = \tanh(\gamma y_i)$ are chosen, or when $\varphi(y_i) = \tanh(\gamma y_i)$ and $\psi(y_i) = |y_i|^p \text{sign}(y_i)$ are chosen, the scaling coefficient is equal to $c = k_\varphi k_\psi = p\gamma E\{|y_i|^{p-1}(t)\}[1 - E\{\tanh^2(\gamma y_i(t))\}]$ for $p \geq 1$, which is smaller over the range $|y_i| \leq 1$ than the case when $\psi(y_i) = y_i$ is chosen. Moreover, we can optimally design the parameters p and γ so that, within a specified range of y_i, the absolute value of the scaling coefficient $c = k_\varphi k_\psi$ is minimal.

Another possible solution to mitigate the bias is to employ nonlinearities of the form $\tilde{\varphi}(y_i) = \varphi(y_i) - \alpha_i y_i$ and $\psi(y_i) = y_i$, with $\alpha_i \geq 0$. The motivation behind the use of linear terms $-\alpha_i y_i$ is to reduce the values of the scaling coefficients as $c_{ij} = k_{\varphi i} - \alpha_i$ as well as to reduce the influence of large outliers. Alternatively, we can use the generalized Fahlman functions given by $\tanh(\beta_i y_i) - \alpha_i y_i$ for either $\varphi_i(y_i)$ or $\psi_i(y_i)$ (Cichocki et al. 1998b).

One disadvantage of these proposed techniques for bias removal is that any equivariant properties for the resulting algorithm are lost when a bias-compensating term is added, and thus the algorithm may perform poorly or even fail to separate sources if the mixing matrix is ill conditioned. For this reason, it is necessary to design nonlinearities that correspond as closely as possible to those produced from the true pdf's of the source signals while also maximally reducing the bias caused by noise.

3.8 RECURRENT NEURAL NETWORK APPROACH FOR NOISE CANCELLATION

3.8.1 Basic Concept and Algorithm Derivation

Assume that we have successfully estimated an unbiased estimate of the separating matrix $\mathbf{W}$ via one of the previously described approaches. Then, we can estimate a mixing matrix $\hat{\mathbf{A}} = \mathbf{W}^+ = \mathbf{APD}$, where $\mathbf{W}^+$ is the pseudoinverse of $\mathbf{W}$, $\mathbf{P}$ is any $n \times n$ permutation matrix, and $\mathbf{D}$ is an $n \times n$ nonsingular diagonal scaling matrix. We now propose approaches for canceling the effects of noise in the estimated source signals.

In order to develop a viable neural-network approach for noise cancellation, we define the error (noise) vector (Cichocki et al. 1998b,c)

$$\mathbf{e}(t) = \mathbf{x}(t) - \hat{\mathbf{A}}\hat{\mathbf{y}}(t) \qquad (3.135)$$

where $\mathbf{e}(t) = [e_1(t) \cdots e_m(t)]^T$ and $\hat{\mathbf{y}}(t)$ is an estimate of the source $\mathbf{s}(t)$ to be defined. To compute $\hat{\mathbf{y}}(t)$, consider the minimum entropy (ME) cost function

$$J(\hat{\mathbf{y}}, \hat{\mathbf{A}}) = E\{\rho(\mathbf{e}(t))\} = -\sum_{i=1}^{m} E\{\log[f_i(e_i(t))]\} \qquad (3.136)$$

where $f_i(e_i)$ is the true pdf of the additive noise $v_i(t)$. It should be noted that we have assumed that the noise sources are i.i.d.; thus, stochastic-gradient descent of the ME function yields stochastic independence of the error components as well as the minimization of their magnitude in an optimal way. The resulting system of differential equations is

$$\frac{d\hat{\mathbf{y}}(t)}{dt} = \boldsymbol{\mu}(t)\hat{\mathbf{A}}^T\boldsymbol{\Psi}[\mathbf{e}(t)] \qquad (3.137)$$

where $\boldsymbol{\mu}(t)$ is a symmetrical positive-definite matrix of learning rates and $\boldsymbol{\Psi}[\mathbf{e}(t)] = [\Psi_1[e_1(t)] \cdots \Psi_m[e_m(t)]]^T$, with nonlinearities $\Psi_i(e_i) = -\partial \log f_i(e_i)/\partial e_i$ (Cichocki and Unbehauen 1994).

A block diagram illustrating the implementation of the preceding algorithm is shown in Fig. 3.10, where "learning algorithm" denotes an appropriate bias removal learning rule, Eq. (3.134). In the proposed algorithm, the optimal choice of nonlinearities $\Psi_i(e_i)$ depends on the noise distributions. Table 3.2 shows typical pdf's and corresponding optimal nonlinearities. Assume for

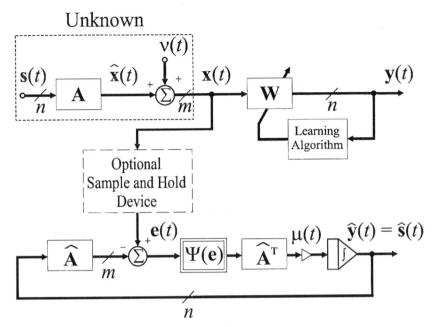

Figure 3.10 Neural network architecture for estimating the separating matrix and efficient noise reduction.

example that all of the noise signals have generalized Gaussian distributions of the form (Cichocki et al. 1997e, 1998a)

$$f_i(e_i) = c_1 \exp(c_2 |e_i|^{r_i}) \tag{3.138}$$

where $r_i > 0$ is a variable parameter and c_1, c_2 are suitable normalization constants. Note that a unit value of r_i yields a Laplacian distribution, a value of $r_i = 2$ yields the standard Gaussian distribution, and $r_i \to \infty$ yields a uniform distribution. In general, we can select any value of $r_i \geq 1$, in which case the locally optimal nonlinear activation functions are of the form

$$\Psi_i(e_i) = -\frac{\partial \log(f_i(e_i))}{\partial e_i} = |e_i|^{r_i - 1} \operatorname{sign}(e_i), \qquad r_i \geq 1 \tag{3.139}$$

For impulsive (spiky) noise with a high value of kurtosis, the optimal parameter r_i typically takes a value between zero and one. In such cases, we can use the modified activation functions $\Psi_i(e_i) = e_i / [|e_i|^{2-r_i} + \varepsilon]$, where ε is a small positive constant, to avoid the singularity of the function at $e_i = 0$. Moreover, when we don't have exact *a priori* knowledge about the noise distributions, we can adapt the value of $r_i(t)$ for each error signal $e_i(t)$ according to its estimated distance from Gaussianity. A simple gradient-based rule for adjusting each parameter $r_i(k)$ is

$$\Delta r_i(k) = -\eta_i(k)\frac{\partial J}{\partial r_i} = \eta_i(k)\frac{\partial \log(f_i(e_i))}{\partial r_i} \tag{3.140}$$

$$\cong \eta_i(k)|e_i(k)|^{r_i(k)}\log(|e_i(k)|) \tag{3.141}$$

Similar methods can be applied for other parameterized noise distributions. For example, when $f_i(e_i)$ is generalized for the generalized Rayleigh distribution, we obtain $\Psi_i(e_i) = |e_i|^{r_i-2}e_i$ for complex-valued signals and coefficients.

It should be noted that the continuous-time algorithm in Eq. (3.137) can be easily converted to a discrete-time algorithm as

$$\hat{\mathbf{y}}(k+1) = \hat{\mathbf{y}}(k) + \eta(k)\hat{\mathbf{A}}^T\mathbf{\Psi}[\mathbf{x}(k) - \hat{\mathbf{A}}\hat{\mathbf{y}}(k)] \tag{3.142}$$

The proposed system depicted in Fig. 3.10 can be considered as a form of nonlinear filtering (postprocessing) that effectively reduces the additive noise components in the estimated source signals. In the next subsection, we propose a more efficient architecture that estimates the mixing matrix $\mathbf{A}$ while simultaneously reducing the amount of noise in the separated sources.

3.8.2 Simultaneous Estimation of a Mixing Matrix and Noise Reduction

Let us consider on-line estimation of the mixing matrix $\mathbf{A}$ rather than estimation of the separating matrix $\mathbf{W}$. It is easy to derive such a learning algorithm taking into account that $\hat{\mathbf{A}} = \mathbf{W}^+$, and for $m \geq n$, from $\mathbf{W}\hat{\mathbf{A}} = \mathbf{W}\mathbf{W}^+ = \mathbf{I}_n$ we have the simple relation (Cichocki and Unbehauen 1996; Cichocki et al. 1998b)

$$\frac{d\mathbf{W}}{dt}\hat{\mathbf{A}} + \mathbf{W}\frac{d\hat{\mathbf{A}}}{dt} = 0 \tag{3.143}$$

Hence, by using Eq. (3.61), we obtain the learning algorithm:

$$\frac{d\hat{\mathbf{A}}(t)}{dt} = -\hat{\mathbf{A}}\frac{d\mathbf{W}}{dt}\hat{\mathbf{A}} = -\mu_1(t)\hat{\mathbf{A}}(t)[\mathbf{\Lambda}(t) - \boldsymbol{\varphi}[\mathbf{y}(t)]\mathbf{y}^T(t)] \tag{3.144}$$

We can replace the output vector $\mathbf{y}(t)$ by an improved estimate $\hat{\mathbf{y}}(t)$ to derive a novel learning algorithm as

$$\frac{d\hat{\mathbf{A}}(t)}{dt} = -\mu_1(t)\hat{\mathbf{A}}(t)[\mathbf{\Lambda}(t) - \boldsymbol{\varphi}[\hat{\mathbf{y}}(t)]\hat{\mathbf{y}}^T(t)] \tag{3.145}$$

and

$$\frac{d\hat{\mathbf{y}}(t)}{dt} = \mu(t)\hat{\mathbf{A}}^T(t)\mathbf{\Psi}[\mathbf{e}(t)] \tag{3.146}$$

or in discrete-time,

$$\Delta\hat{\mathbf{A}}(k) = \hat{\mathbf{A}}(k+1) - \hat{\mathbf{A}}(k)$$
$$= -\eta_1(k)\hat{\mathbf{A}}(k)[\mathbf{\Lambda}(t) - \boldsymbol{\varphi}[\hat{\mathbf{y}}(k)]\hat{\mathbf{y}}^T(k)] \tag{3.147}$$

and

$$\hat{\mathbf{y}}(k+1) = \hat{\mathbf{y}}(k) + \eta(k)\hat{\mathbf{A}}^T(k)\mathbf{\Psi}[\mathbf{e}(k)] \tag{3.148}$$

where $\mathbf{e}(k) = \mathbf{x}(k) - \hat{\mathbf{A}}(k)\hat{\mathbf{s}}(k)$ and $\mathbf{x}(k) = \hat{\mathbf{x}}(k) + \mathbf{v}(k)$. A functional block diagram illustrating this algorithm's implementation is shown in Fig. 3.11.

Regularization For some systems, the mixing matrix $\mathbf{A}$ may have highly ill-conditioned coefficients, and in order to estimate reliable and stable solutions, it is necessary to apply optional prewhitening (e.g., PCA) and/or regularization methods.

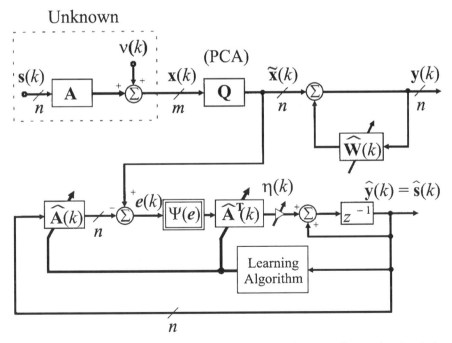

Figure 3.11 Architecture of the Amari-Hopfield kind of recurrent neural network for simultaneous noise reduction and mixing-matrix identification: conceptual discrete-time model.

For ill-conditioned cases, we can use the contrast function with a regularization term

$$l_p(\hat{\mathbf{y}}, \hat{\mathbf{A}}) = \rho(\mathbf{x} - \hat{\mathbf{A}}\hat{\mathbf{y}}) + \frac{\alpha}{p}\|\mathbf{L}\hat{\mathbf{y}}\|_p^p \qquad (3.149)$$

where $\alpha > 0$ is a regularization constant chosen to control the size of the solution, and $\mathbf{L}$ is a regularization matrix that defines a (semi)norm of the solution. Typically, the matrix $\mathbf{L}$ is equal to the identity matrix $\mathbf{I}_n$, and sometimes it represents the first- or second-order operator. If $\mathbf{L}$ is the identity matrix, $\rho(\mathbf{e}) = \frac{1}{2}\|\mathbf{e}\|_2^2$, and $p = 2$, then the problem reduces to the least-squares problem with the standard Tikhonov regularization term (Cichocki and Unbehauen 1994b):

$$l_2(\hat{\mathbf{y}}, \hat{\mathbf{A}}) = \frac{1}{2}\|\mathbf{x} - \hat{\mathbf{A}}\hat{\mathbf{y}})\|_2^2 + \frac{\alpha}{2}\|\hat{\mathbf{y}}\|_2^2 \qquad (3.150)$$

Minimization of the contrast function, Eq. (3.149), leads to the learning rule

$$\frac{d\hat{\mathbf{y}}(t)}{dt} = \mu(t)[\hat{\mathbf{A}}^T(t)\mathbf{\Psi}[\mathbf{x}(t) - \hat{\mathbf{A}}(t)\hat{\mathbf{y}}(t)] - \alpha\hat{\boldsymbol{\varphi}}[\hat{\mathbf{y}}(t)] \qquad (3.151)$$

where $\hat{\boldsymbol{\varphi}}(\hat{\mathbf{y}}) = [\hat{\varphi}_1[\hat{y}_1(t)] \cdots \hat{\varphi}_n[\hat{y}_n(t)]^T$ with nonlinearities $\hat{\varphi}_i(\hat{y}_i) = |\hat{y}_i|^{p-1}\operatorname{sign}(\hat{y}_i)$. By combining with the learning equation, Eq. (3.145), we obtain an algorithm for simultaneous estimation of the mixing matrix and source signals for ill-conditioned problems. Figure 3.12 illustrates the detailed architecture of the recurrent neural network according to Eq. (3.151). It can be proved that the preceding learning algorithm is stable if nonlinear activation functions $\psi_i(y_i)$ are monotonic increasing odd functions (see, e.g., Tables 3.1 and 3.2).

3.9 EXTENSION OF BSS LEARNING ALGORITHMS FOR MULTICHANNEL BLIND DECONVOLUTION

The previously derived learning algorithms for BSS are very effective (Amari et al. 1996, 1997c, 1999; Bell and Sejnowski 1995; Cardoso and Laheld 1996; Cichocki et al. 1994a, 1997e, 1998a; Cichocki and Unbehauen 1996). However, a key question arises as to how to extend or generalize these (and other and future algorithms for BSS/BES) for more real-world problems: robust multichannel MBD. In other words, the question arises: Is it possible to establish a direct relationship between BSS and MBD and convert a learning algorithm for BSS directly to an "equivalent" algorithm for MBD? The problem is the objective of this section. We show that the learning algorithms for BSS can be generalized or extended to the multichannel MBD (Amari et al. 1997a,b; Douglas et al. 1996). As a tool we use relationships between convolution and cross-correlation and abstract algebra with their nice properties (Douglas and Haykin 1997; Lambert 1996; Sabala et al. 1998).

Notational conventions used in this section are as follows (Sabala et al. 1998). The discrete-time signals and parameters are assumed to be real-valued.

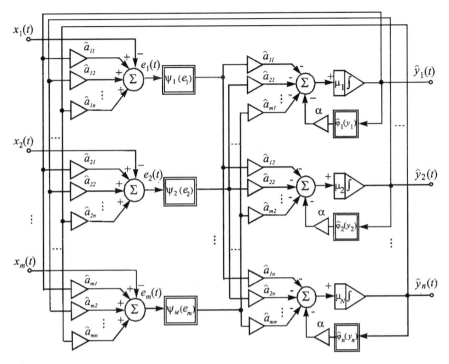

Figure 3.12 Detailed architecture of the Amari-Hopfield continuous-time (analog) model of a recurrent neural network with regularization.

An infinite time series (in both sides of time) is denoted as $\underline{x} = x(k) = \{\ldots, x(k-1), x(k), x(k+1), \ldots\}$. A reversed-order time series is defined as $\underline{x}(-k) = \{\ldots, x(k+1), x(k), x(k-1), \ldots\}$. A multivariable time series, that is, a vector time series, is denoted $\underline{\mathbf{x}}(k) = \{\ldots, \mathbf{x}(k-1), \mathbf{x}(k), \mathbf{x}(k+1), \ldots\}$, where $\mathbf{x}(k) = [x_1(k) \cdots x_m(k)]^T$. Analogously, an $m \times n$ time-domain impulse response is denoted as $\underline{\underline{\mathbf{H}}} = \underline{\underline{\mathbf{H}}}_p = \{\mathbf{H}_p\} = \{\ldots, \mathbf{H}_{p-1}, \mathbf{H}_p, \mathbf{H}_{p+1}, \ldots\}$ and its reverse order is $\underline{\underline{\mathbf{H}}}_{(-p)} = \{\mathbf{H}_{-p}\} = \{\ldots, \mathbf{H}_{p+1}, \mathbf{H}_p, \mathbf{H}_{p-1}, \ldots\}$. Equivalently by using the z-transform transfer function, we can write $\mathbf{H}(z) = \sum_{p=-\infty}^{\infty} \mathbf{H}_p z^{-p}$, where $\mathbf{H}_p$ is a matrix of mixing coefficients at lag p and $\mathbf{H}(z^{-1}) = \sum_{p=-\infty}^{\infty} \mathbf{H}_p z^p$.

We aim to recover the original signal $\underline{s}(k)$ by applying stable separating/deconvolving filters with an impulse response $\underline{\underline{\mathbf{W}}} = \{\mathbf{W}_p\} = \{\ldots, \mathbf{W}_{p-1}, \mathbf{W}_p, \mathbf{W}_{p+1}, \ldots\}$. The recovered signals can be described as $\underline{y}(k) = \underline{\underline{\mathbf{W}}} * \underline{\mathbf{x}}(k)$, where the kth element of the resulting $\underline{y}(k)$ series is computed as $\mathbf{y}(k) = \sum_{p=-\infty}^{\infty} \mathbf{W}_p \mathbf{x}(k-p)$ $(k = -\infty, \ldots, \infty)$. In the scalar form the ith output can be expressed as $y_i(k) = \sum_{j=1}^{m} \sum_{p=-\infty}^{\infty} w_{ijp} x_j(k-p)$.

A global system that describes the convolution-deconvolution process with input $\mathbf{s}(k)$ and output $\mathbf{y}(k)$ is

$$\underline{y}(k) = \underline{\underline{\mathbf{G}}} * \underline{s}(k) = \underline{\underline{\mathbf{W}}} * \underline{\underline{\mathbf{H}}} * \underline{s}(k) \qquad (3.152)$$

Notice that $\underline{\underline{\mathbf{H}}} = \{\mathbf{H}_p\}$, $\underline{\underline{\mathbf{W}}} = \{\mathbf{W}_p\}$, and $\underline{\underline{\mathbf{G}}} = \{\mathbf{G}_p\}$ are double-infinite non-causal linear filters, which, in implementation applications, are replaced by their truncated FIR versions. An equivalent description in the z-transform domain is as follows

$$\mathbf{Y}(z) = \mathbf{W}(z)\mathbf{X}(z) = \mathbf{W}(z)\mathbf{H}(z)\mathbf{S}(z) = \mathbf{G}(z)\mathbf{S}(z) \qquad (3.153)$$

where $\mathbf{Y}(z) = \sum_{k=-\infty}^{\infty} \mathbf{y}(k)z^{-k}$, $\mathbf{W}(z) = \sum_{p=-\infty}^{\infty} \mathbf{W}_p z^{-p}$ are the z-transforms of the time-domain infinite series $\underline{\mathbf{y}}(k)$ and $\underline{\underline{\mathbf{W}}}(k)$, respectively. $\mathbf{H}(z), \mathbf{G}(z), \mathbf{X}(z)$, $\mathbf{S}(z)$ are defined accordingly.

If $\underline{\underline{\mathbf{H}}} = \{\dots, \mathbf{0}, \mathbf{H}_0, \mathbf{0}, \dots\}$ and $\underline{\underline{\mathbf{W}}} = \{\dots, \mathbf{0}, \mathbf{W}_0, \mathbf{0}, \dots\}$, we obtain a much simpler task of multichannel BSS for which the derivation of learning algorithms is relatively easily.

3.9.1 Algebraic Equivalence of Various Approaches

Different ways of describing physical phenomena of signal propagation [IIR filter, FIR filter, discrete Fourier transform (DFT) z-transform, wavelets, other transforms] result in different but equivalent mathematical models. In blind deconvolution, the key operation is linear convolution (FIR and/or IIR filtering) of time series. The usual notations in the discrete domain involve the z-transform, z^{-1}-delay operator, DFT, or convolution (others are also possible). Such transformations are used to create suitable notations for the available data set and algebraic operations (Sabala et al. 1998).

The basic data set in our case consists of infinite series of elements such as w_p, $x(k)$, and $y(k)$, that is, $\underline{\mathbf{w}}_p = \{\dots, w_{p-1}, w_p, w_{p+1}, \dots\}$, $\underline{\mathbf{x}}(k) = \{\dots, x(k-1),$ $x(k), x(k+1), \dots\}$, and $\underline{\mathbf{y}}(k) = \{\dots, y(k-1), y(k), y(k+1), \dots\}$, where p and k are indices. In the case of time-domain filtering, $\underline{\mathbf{x}}(k)$ and $\underline{\mathbf{y}}(k)$ are time series of samples and k is a discrete-time index. The data can be rearranged to other forms that best suit solving a specific problem, provided that the new data set with the new operators defined fall in one of several algebraic categories. In the case of the z-transform, the data set is formed in polynomials with polynomial multiplication and addition. In the case of DFT, the data set is transformed into a complex number series (the frequency domain) with point-by-point multiplication and the addition of complex series.

The case of when time series is used in its time-domain format $\underline{\mathbf{y}}(k)$ and a convolution is used as the multiplicative operator is described in the next subsection.

3.9.2 Convolution as Multiplicative Operator

Let us consider the inner product for a complete signal space (Hilbert space). For a finite-energy signal space, the inner product is (for example, for time series: $\underline{\mathbf{x}}(k)$ and $\underline{\mathbf{y}}(k)$)

$$(\underline{\mathbf{x}}, \underline{\mathbf{y}}) = \sum_{k=-\infty}^{\infty} x(k) y(k) \qquad (3.154)$$

For a finite-power signal space, the inner product is

$$(\underline{\mathbf{x}}, \underline{\mathbf{y}}) = \lim_{M \to \infty} \frac{1}{2M+1} \sum_{k=-M}^{M} x(k) y(k) \qquad (3.155)$$

The complex conjugate of $y(k)$ would be used instead for complex-valued signals in Eqs. (3.154), (3.155).

Let the symbols $\underline{\mathbf{0}}$ (zero element) and $\underline{\mathbf{1}}$ (one element) denote the time series $\{\ldots, 0, 0, 0, \ldots\}$, $\{\ldots, 0, 1, 0, \ldots\}$, respectively, with 1 for $k = 0$. Let symbols $+$ (additive operation) denote point-by-point series addition. Let $*$ (multiplicative operation) denote linear convolution. The convolution of two time series, $\underline{\mathbf{x}}(k)$ and $\underline{\mathbf{y}}(k)$, results in a series $\underline{\mathbf{u}}(k) = \underline{\mathbf{x}}(k) * \underline{\mathbf{y}}(k)$, where the kth entry $(k = -\infty, \ldots, \infty)$ of the series $\underline{\mathbf{u}}(k)$ is $u(k) = \sum_{p=-\infty}^{\infty} x(p) y(k-p) = \sum_{p=-\infty}^{\infty} x(k-p) y(p)$. For finite-power signals, it may be infinite, which is why a more informative quantity is the average $\langle \underline{\mathbf{x}}(k) * \underline{\mathbf{y}}(k) \rangle$ defined by analogy to Eq. (3.155), with the kth entry equal to

$$\langle \underline{\mathbf{x}}(k) * \underline{\mathbf{y}}(k) \rangle = \lim_{M \to \infty} \frac{1}{2M+1} \sum_{p=-M}^{M} x(p) y(k-p) \qquad (3.156)$$

A set of time series with the zero element, one element, additive operation, and convolution operation is a general algebra that fulfills the standard conditions of a commutative ring. By adding some assumptions it becomes a field. We define symbol $/$ (division operator) by $\underline{\mathbf{x}}/\underline{\mathbf{y}} = \underline{\mathbf{x}}(k) * \underline{\mathbf{y}}^{-1}(k), \underline{\mathbf{y}} \neq \underline{\mathbf{0}}$. Since $\underline{\mathbf{y}}(k) * \underline{\mathbf{y}}^{-1}(k) = \underline{\mathbf{1}}$, a problem may arise concerning the existence of any bounded inverse element $\underline{\mathbf{y}}^{-1}$. The problem is easily solved by a suitable choice of the origin $k = 0$ of the series $\underline{\mathbf{y}}$.

All the rules of differential and integral calculus hold since these are linear operations. For example, expressions for the differentiation of filtering $y(k) = \underline{\mathbf{w}} * \underline{\mathbf{x}}(k) = \sum_{p=-\infty}^{\infty} w_p x(k-p)$ $k = -\infty, \ldots, \infty$ is as follows:

$$\frac{\partial}{\partial \underline{\mathbf{w}}} [y(k)] = \frac{\partial}{\partial \underline{\mathbf{w}}} [\underline{\mathbf{w}} * \underline{\mathbf{x}}(k)] = \underline{\mathbf{x}}(-k) \qquad (3.157)$$

The reverse order of the resulting time series is a consequence of the multiplicative operator definition.

The time series field reduces to the real-number field for the special case of the time series $\underline{\mathbf{y}}(k) = \{\ldots, 0, y, 0, \ldots\}$ with only one nonzero entry for $k = 0$.

This is the notation of a real number y in terms of time series, which is analogous to writing a real number y by using the complex notation $y + j0$.

The probability density function of a series at the point y may be described in this way $\{\ldots, 0, f_y(y), 0, \ldots\}$. When one of the operands of a multiplicative operation is of this form, the symbol $*$ will be omitted.

The single-channel approach just outlined easily extends to a multichannel one. The resulting algebra is a noncommutative ring with division, exactly as the standard matrix algebra of the instantaneous BSS case. It is easily seen that all the algebraic properties of the instantaneous and dispersive mixing models are equivalent (Lambert 1996; Sabala 1998; Sabala et al. 1998).

3.9.3 Natural-Gradient Learning Rules for Multichannel Blind Deconvolution

Natural-gradient algorithm, Eq. (3.48) (Amari 1998; Amari et al. 1997a,b, 1998), can be generalized for MBD as (see also Table 3.3)

$$\Delta \underline{\underline{W}}(n) = \underline{\underline{W}}(n+1) - \underline{\underline{W}}(n) = -\eta \frac{\partial D_{fq}}{\partial \underline{\underline{W}}} * \underline{\underline{W}}^T_{(-p)}(n) * \underline{\underline{W}}(n) \qquad (3.158)$$

where D_{fq} is the Kullback-Leibler objective function, and $\mathbf{W}(n)$ is the estimate at time n of the inverse system for recovering the source signals. Alternatively, we can employ the natural gradient by using the z-transform as

$$\Delta \mathbf{W}(z, n) = -\eta(n) \frac{\partial D_{fq}}{\partial \mathbf{W}(z)} \mathbf{W}^T(z^{-1}, n) \mathbf{W}(z, n) \qquad (3.159)$$

The dual learning rule, Eq. (3.63) can be generalized as follows

$$\Delta \underline{\underline{W}}(n) = -\eta(n) \underline{\underline{W}}(n) * \left[\frac{\partial D_{fq}}{\partial \underline{\underline{W}}} \right]^T * \underline{\underline{W}}(n) \qquad (3.160)$$

or using the z-transform

$$\Delta \mathbf{W}(z, n) = -\eta(n) \mathbf{W}(z, n) * \left[\frac{\partial D_{fq}}{\partial \mathbf{W}(z)} \right]^T * \mathbf{W}(z, n) \qquad (3.161)$$

3.9.4 Multichannel Blind Deconvolution Algorithms

Since the dispersive case is of interest here, all the symbols follow from the properties of time series with the convolution operator $*$. To avoid confusion, we note that division is defined as the inverse operation of convolution and time series are of infinite lengths. We consider here the MIMO n-by-m case. We derive a family of algorithms by minimization of the Kullback-Leibler divergence between the actual distribution $f_y(\mathbf{y})$ and the factorizable model distribution $q_s(\mathbf{y})$, which leads to the objective function

$$l(\underline{\mathbf{W}}) = -\frac{1}{2}\log|\det(\underline{\mathbf{W}}_{(-p)} * \underline{\mathbf{W}}^T)| - \sum_{i=1}^{m}\log q_{s_i}(y_i) \qquad (3.162)$$

Applying the natural-gradient rule of the form Eq. (3.158), we obtain the learning algorithm first derived rigorously by Amari et al. (1997a,b)

$$\underline{\mathbf{u}}(k) = \underline{\mathbf{W}}_{(-p)}^T(n) * \mathbf{y}(k) \qquad (3.163)$$

$$\underline{\mathbf{C}}_{\varphi,u}(n) = \langle \varphi[\underline{\mathbf{y}}(k)] * \underline{\mathbf{u}}^T(-k)\rangle \qquad (3.164)$$

$$\underline{\mathbf{W}}(n+1) = \underline{\mathbf{W}}(n) + \eta(n)(\underline{\mathbf{W}}(n) - \underline{\mathbf{C}}_{fu}(n)) \qquad (3.165)$$

with nonlinearities $\varphi_i(y_i) = -q'_{s_i}(y_i)/q_{s_i}(y_i)$, where samples of $\underline{\mathbf{u}}(k)$ can be computed as

$$\mathbf{u}(k) = \sum_{p=-\infty}^{\infty} \mathbf{W}_p^T \mathbf{y}(k+p), \qquad k = -\infty, \ldots, \infty \qquad (3.166)$$

The averaging operator $\langle \cdot \rangle$ is used here in the same sense as in Eq. (3.156). Equation (3.164) defines the multichannel cross-correlation function. The dual gradient rule, Eq. (3.160), leads to a new learning algorithm (see also Table 3.3).

$$\underline{\mathbf{v}}(k) = \underline{\mathbf{W}}_{(-p)}^T(n) * \psi[\underline{\mathbf{y}}(k)] \qquad (3.167)$$

$$\underline{\mathbf{C}}_{\psi,v}(n) = \langle \underline{\mathbf{y}}(k) * \underline{\mathbf{v}}^T(-k)\rangle \qquad (3.168)$$

$$\underline{\mathbf{W}}(n+1) = \underline{\mathbf{W}}(n) + \eta(n)\underline{\mathbf{W}}(n) - \underline{\mathbf{C}}_{fu}(n)) \qquad (3.169)$$

where samples of $\underline{\mathbf{v}}(k)$ can be computed as

$$\mathbf{v}(k) = \sum_{p=-\infty}^{\infty} \mathbf{W}_p^T \psi[\mathbf{y}(k+p)], \qquad (k = -\infty, \ldots, \infty) \qquad (3.170)$$

Detailed learning rules for calculation of the equalizer output, filtered output, cross-correlation matrix, and separating/deconvolving matrices $\mathbf{W}_p$ are given in the subsection implementation.

Nonlinearities $\psi_i(y_i)$ are now inverse (dual) to the function $\varphi_i(y_i) = -q'_{s_i}(y_i)/q_{s_i}(y_i)$. For example, instead of $\varphi(y_i) = y_i^{1/3}$, we use the cubic function $\psi_i(y_i) = y_i^3$, or instead $\varphi_i(y_i) = \tanh(y_i)$, we use the inverse function $\psi_i(y_i) = \text{artanh}(y_i) = \frac{1}{2}\log((1+y_i)/(1-y_i))$.

Table 3.3 Relationships Between Instantaneous Blind Source Separation and Multichannel Blind Deconvolution for Complex-Valued Signals and Parameters

Blind Source Separation	Multichannel Blind Deconvolution— Time Domain	Multichannel Blind Deconvolution —z-Transform Domain
	Mixing–Unmixing Model	
$\mathbf{x}(k) = \mathbf{A}\mathbf{s}(k)$	$\underline{\mathbf{x}}(k) = \underline{\mathbf{H}} * \underline{\mathbf{s}}(k)$	$\mathbf{X}(z) = \mathbf{H}(z)\mathbf{S}(z)$
$\mathbf{y}(k) = \mathbf{W}(n)\mathbf{x}(k)$	$\underline{\mathbf{y}}(k) = \underline{\mathbf{W}}(n) * \underline{\mathbf{x}}(k)$	$\mathbf{Y}(z) = \mathbf{W}(z)\mathbf{X}(z)$
$x_i(k) = \sum_{j=1}^{n} a_{ij} s_j(k)$	$\underline{\mathbf{x}}_i(k) = \sum_{j=1}^{n} \underline{\mathbf{h}}_{ij} * \mathbf{s}_j(k)$	$X_i(z) = \sum_{j=1}^{n} H_{ij}(z)[S_j(z)]$
$y_i(k) = \sum_{j=1}^{n} w_{ij}(n) x_j(k)$	$\underline{\mathbf{y}}_i(k) = \sum_{j=1}^{n} \underline{\mathbf{w}}_{ij}(n) * \underline{\mathbf{x}}_j(k)$	$Y_i(z) = \sum_{j=1}^{n} W_{ij}(z)[X_j(z)]$
	Contrast Functions: $\phi(\mathbf{y}, \mathbf{W})$ or $\phi(\mathbf{y}, \underline{\mathbf{W}})$ or $\phi(\mathbf{W}(z))$	
$-\log|\det(\mathbf{W})| - \sum_{i=1}^{n} \log(q_i(y_i))$	$-\log|\det(\underline{\mathbf{W}})| - \sum_{i=1}^{n} \log(q_i(y_i))$	$-\frac{1}{2\pi j}\oint \log|\det \mathbf{W}(z)| z^{-1}\,dz$ $-\sum_{i=1}^{n} \log(q(y_i))$

Natural Gradient Rules: $\Delta\mathbf{W}(n)$ or $\Delta\underline{\underline{\mathbf{W}}}(n)$ or $\Delta\mathbf{W}(z)$

$$-\eta\,\frac{\partial\phi}{\partial\mathbf{W}}\,\mathbf{W}^H(n)\mathbf{W}(n)$$

$$-\eta\mathbf{W}(n)\left[\frac{\partial\phi}{\partial\mathbf{W}}\right]^H\mathbf{W}(n)$$

$$-\eta\,\frac{\partial\phi}{\partial\underline{\underline{\mathbf{W}}}}*\underline{\underline{\mathbf{W}}}^H_{(-p)}(n)*\underline{\underline{\mathbf{W}}}(n)$$

$$-\eta\underline{\underline{\mathbf{W}}}(n)*\left[\frac{\partial\phi}{\partial\underline{\underline{\mathbf{W}}}}\right]^H*\underline{\underline{\mathbf{W}}}(n)$$

$$-\eta\,\frac{\partial\phi}{\partial\mathbf{W}(z)}\,\mathbf{W}^H(z^{-1})\mathbf{W}(z)$$

$$-\eta\mathbf{W}(z)\left[\frac{\partial\phi}{\partial\mathbf{W}(z)}\right]^H\mathbf{W}(z)$$

Batch Learning Algorithms: $\Delta\mathbf{W}(n)$ or $\Delta\underline{\underline{\mathbf{W}}}(n)$ or $\Delta\mathbf{W}(z,n)$

$$\eta[\mathbf{W}(n)-\langle\varphi[\mathbf{y}(k)]\mathbf{u}^H(k)\rangle]$$
where
$$\mathbf{u}(k)=\mathbf{W}^H(n)\mathbf{y}(k)$$

$$\eta[\mathbf{W}(n)-\langle\mathbf{y}(k)\mathbf{v}^H(k)\rangle]$$
where
$$\mathbf{v}(k)=\mathbf{W}^{-H}(n)\psi[\mathbf{y}(k)]$$

$$\eta[\underline{\underline{\mathbf{W}}}(n)-\langle\varphi[\underline{\mathbf{y}}(k)]*\underline{\underline{\mathbf{u}}}^H(-k)\rangle]$$
where
$$\underline{\underline{\mathbf{u}}}(k)=\underline{\underline{\mathbf{W}}}^H_{(-p)}(n)*\underline{\mathbf{y}}(k)$$

$$\eta[\underline{\underline{\mathbf{W}}}(n)-\langle\underline{\mathbf{y}}(k)*\underline{\underline{\mathbf{v}}}^H(-k)\rangle]$$
where
$$\underline{\mathbf{v}}(k)=\underline{\underline{\mathbf{W}}}^H_{(-p)}(n)*\psi[\underline{\mathbf{y}}(k)]$$

$$\eta[\mathbf{W}(z,n)-\langle Z\{\varphi[\mathbf{y}(k)]\}\mathbf{U}^H(z^{-1})\rangle]$$
where
$$\mathbf{U}(z)=\mathbf{W}^H(z^{-1},n)[\mathbf{Y}(z)]$$

$$\eta[\mathbf{W}(z,n)-\langle\mathbf{Y}(z)\mathbf{V}^H(z^{-1})\rangle]$$
where
$$\mathbf{V}(z)=\mathbf{W}^H(z^{-1},n)Z\{\psi[\mathbf{y}(k)]\}$$

Inevitable ambiguities in blind deconvolution are those of permutation, scaling, and delay. In blind separation, only those of permutation and scaling are encountered. Since, from an algebraic point of view, there is a precise equivalence between instantaneous and dispersive cases, the ambiguities should be also equivalent (Sabala et al. 1998).

The delay arises in consequence of ordering k regarded as time instants and causality of the convolving filter introduced from the practical implementation point of view. However, from the algebraic point of view, there is no restriction to incorporating the delay with the convolving filter, regarding its output as noncausal, and thus making it delayed.

Analogies and relationships between instantaneous BSS and MBD (for $m = n$ and complex-valued signals and parameters) are collected in Table 3.3. All the operations act on the algebraic fields defined for different data sets.

3.9.5 Implementation of Algorithms for MBD

Batch Update Rules The general formulas Eqs. (3.163)–(3.166) and (3.167)–(3.170) can be implemented as a generalized batch learning rule. A batch of training data $x(1), \ldots, x(M)$ is used, with zero values assumed for the range $k < 1$ and $k > M$. A biased covariance estimator is used.

For each $k = 1, \ldots, M$:

$$\mathbf{y}(k) = \sum_{p=-L}^{L} \mathbf{W}_p(n)\mathbf{x}(k - p) \tag{3.171}$$

$$\mathbf{v}(k) = \sum_{p=-L}^{L} \mathbf{W}_p^T(n)\boldsymbol{\psi}[\mathbf{y}(k + p)] \tag{3.172}$$

For each $p \in -L, \ldots, L$:

$$\hat{\mathbf{C}}_{\varphi v, p}(n) = \frac{1}{M - p} \sum_{k=p+1}^{M} \varphi[\mathbf{y}(k)]\mathbf{v}^T(k - p) \tag{3.173}$$

$$\mathbf{W}_p(n + 1) = \mathbf{W}_p(n) + \eta(n)(\mathbf{W}_p(n) - \hat{\mathbf{C}}_{\varphi v, p}(n)) \tag{3.174}$$

On-line Update Rule Since the filtering operation leading to $\mathbf{u}(k)$ is non-causal, the most recent time instant of $\mathbf{u}(k)$ that can be calculated by using samples of $\mathbf{x}(k)$ and $\mathbf{y}(k)$ up to time instant k is $\mathbf{u}(k - L)$. The covariance estimator is replaced by the instantaneous value:

$$\mathbf{y}(n) = \mathbf{y}(k) = \sum_{p=0}^{L} \mathbf{W}_p\mathbf{x}(k - p) \tag{3.175}$$

$$\mathbf{u}(n) = \mathbf{u}(k - L) = \sum_{p=0}^{L} \mathbf{W}_p^T\boldsymbol{\psi}[\mathbf{y}(k - L + p)] \tag{3.176}$$

For each $p = 0, \ldots, L$:

$$\hat{\mathbf{C}}_{\varphi u, p}(n) = \boldsymbol{\varphi}(\mathbf{y}(k - L))\mathbf{u}^T(k - L - p) \tag{3.177}$$

$$\mathbf{W}_p(n + 1) = \mathbf{W}_p(n) + \eta(n)(\mathbf{W}_p(n) - \hat{\mathbf{C}}_{\varphi u, p}(n)) \tag{3.178}$$

Equations (3.175)–(3.178) describe the $(n + 1)$th step of the on-line update rule by using samples up to the kth sample.

Block On-line Update Rule Let us assume that the data arrive continuously and are gathered into blocks: $\{\ldots, [x((n - 1)N + 1), \ldots, x(nN)], [x(nN + 1), \ldots, x((n + 1)N)], \ldots\}$. Implementation may be done in the time domain, or similarly to Amari et al. (1998), by using a fast Fourier transform (FFT) in the frequency domain.

For each $k = (n - 1)N + 1, \ldots, nN$:

$$\mathbf{y}(n) = \mathbf{y}(k) = \sum_{p=0}^{L} \mathbf{W}_p \mathbf{x}(k - p) \tag{3.179}$$

$$\mathbf{v}(n) = \mathbf{v}(k - L) = \sum_{p=0}^{L} \mathbf{W}_p^T \boldsymbol{\psi}[\mathbf{y}(k - L + p)] \tag{3.180}$$

For each $p = 0, \ldots, L$:

$$\hat{\mathbf{C}}_{\varphi u, p}(n) = \frac{1}{N} \sum_{k=(n-1)N+1}^{nN} \boldsymbol{\varphi}[\mathbf{y}(k - L)]\mathbf{u}^T(k - L - p) \tag{3.181}$$

$$\mathbf{W}_p(n + 1) = \mathbf{W}_p(n) + \eta(n)[\mathbf{W}_p(n) - \hat{\mathbf{C}}_{\varphi u, p}(n)] \tag{3.182}$$

Summarizing, relationships and equivalences between instantaneous BSS and MBD are shown in terms of algebraic properties in this section. Analysis has been done by using the notion of abstract algebra of formal series. The basic input data set consists of series of time-ordered sampled data that can be transformed into another set more suitable for a given problem. Algebraic operations are suitably modified to preserve algebraic field properties. The algorithms can be implemented by using on-line data (sampled signals, convolution operator, z-transform) or preprocessed data (DFT). Practical implementations of the algorithms have been discussed in terms of instantaneous mixing and linear convolution. Using the established analogies and relationships, the learning rules, can be almost automatically generated (see Table 3.3).

3.10 CONCLUSIONS AND DISCUSSION

In this chapter, we have presented neural-network models and a family of associated adaptive learning algorithms for independent components analysis

(ICA), that is, blind identification and blind signal separation of signals from linear instantaneous mixtures. Several robust algorithms dealing with additive noise in the mixtures have been discussed. Emphasis was given to the information-theoretic approach, the neural network, or adaptive multichannel filtering models and associated on-line nonlinear adaptive learning algorithms that have some biological resemblance or plausibility and are insensitive to additive noise. Moreover, on-line adaptive algorithms, which enable us to not only estimate waveforms of nonstationary source signals but also to estimate on-line the number of sources that have been presented. The developed approaches have been extended for the more difficult, real-world problems of blind deconvolution/equalization of independent source signals. Due to the wide interest in this fascinating area of research, we expect, in the near future, further developments of computationally efficient separation, deconvolution, self-adaptive, or self-organized equalization systems with robust on-line algorithms for many real-world applications, such as wireless communication, "cocktail party" problem, speech and image recognition, intelligent analysis of medical signals and images, and feature extraction.

A number of open problems not fully addressed in this chapter still exist concerning blind signal separation. Here we formulate 10 such basic problems (Amari and Cichocki 1998):

1. What are the necessary and sufficient conditions for successful blind separation and/or deconvolution problems? What are the practical limitations? How do we extend or modify the existing criteria, especially when the usual mutual statistical independence conditions and i.i.d. are not fully satisfied or when the data are partially correlated?

2. How do we effectively use some *a priori* available information about the linear and nonlinear dynamic system in order to successfully separate or extract the source signals?

3. How do we extract source signals with specific stochastic properties or features from the large number of source mixtures.

4. What methods can be effective in the case when there are more source signals than sensors (e.g., for EEG signals)? What *a priori* information is sufficient and/or necessary in such a case?

5. How can the number of the sources be reliably estimated in the presence of colored (non-Gaussian) and/or impulsive noises when the reference noise is not available? See (Cichocki et al. 1996d, 1998a,c).

6. What are the theoretical and practical limitations on blind source separation–deconvolution when the mixtures or channels are nonlinear? It is recognized that there nonlinearities will exist for which no solution exists, but it is desirable to quantify these limitations, and further, to establish the types of problems that are solvable effectively. Developments of the inverse models of nonlinear dynamic systems are necessary (Cichocki et al. 1998d; Yang et al. 1998).

7. How do we implement adaptive stable recurrent neural networks, especially IIR filters (e.g., gamma, Laguerre filters) and/or state-space models instead of standard FIR channels (Zhang and Cichocki 1998b, 1998c)?

8. How do we develop the inverse models and adaptive algorithms for time-varying nonstationary systems: channels with varying characteristics and/or number of sources, an unknown number of sources moving in space and varying in time (e.g., acoustic or speech signals with quiet periods)?

9. How do time- and frequency-domain algorithms, especially for ill-conditioned problems compare with respect to complexity, robustness, and biological plausibility (Amari et al. 1998; Lambert 1996; Lee et al. 1997)?

10. How do we develop computationally efficient separation, deconvolution, equalization self-adaptive systems with algorithms for real-world applications? For example, in the problem of reverberation and echo or long-channel delays, what methods can be developed to perform fast (in real time) blind source separation and deconvolution/equalization?

APPENDIX 3.1: KULLBACK-LEIBLER DIVERGENCE (RELATIVE ENTROPY) AS A MEASURE OF STOCHASTIC INDEPENDENCE

The Kullback-Leibler divergence (relative entropy) between two pdf's $f_y(\mathbf{y})$ and $q(\mathbf{y})$ on $\mathbf{R}^n$ is defined as (Amari et al. 1997c; Haykin 1998)

$$D_{fq} = D(f_y(\mathbf{y})\|q(\mathbf{y})) = \int_{-\infty}^{\infty} f_y(\mathbf{y}) \log \frac{f_y(\mathbf{y})}{q(\mathbf{y})} \, d\mathbf{y} \qquad (3.183)$$

whenever the integral exists. The Kullback-Leibler divergence always takes nonnegative values, achieving zero if and only if $f_y(\mathbf{y})$ and $q(\mathbf{y})$ are the same distribution. It is invariant with respect to the invertible (monotonic) nonlinear transformation of variables, including amplitude rescaling and permutation, in which the variables y_i are rearranged. For the independent-components analysis problem, we assume that $q(\mathbf{y})$ is the product of the distribution of independent variables y_i. It can be the product of the marginal pdf's of $\mathbf{y}$, in particular,

$$q(\mathbf{y}) = \tilde{f}(\mathbf{y}) = \prod_{i=1}^{n} f_i(y_i) \qquad (3.184)$$

where $f_i(y_i)$ are the marginal probability density functions of y_i $(i = 1, 2, \ldots, n)$. The marginal pdf is defined as

$$f_i(y_i) = \int_{-\infty}^{\infty} f_y(\mathbf{y}^i) \, dy^i \qquad (3.185)$$

where the integration is taken over $\check{\mathbf{y}}^i = [y_1 \cdots y_{i-1} \; y_{i+1} \cdots y_n]^T$, which is left after removing variable y_i. The natural measure of independence can be formulated as

$$D_{fq} = E\{l(\mathbf{W}, \mathbf{y})\} = \int_{-\infty}^{\infty} f_y(\mathbf{y}) \log \frac{f_y(\mathbf{y})}{\prod\limits_{i=1}^{n} f_i(y_i)} \, d\mathbf{y} \qquad (3.186)$$

The Kullback-Leibler divergence can be expressed in terms of mutual information as

$$D_{fq} = -H(\mathbf{y}) - \sum_{i=1}^{n} \int_{-\infty}^{\infty} f_y(\mathbf{y}) \log f_i(y_i) \, d\mathbf{y} \qquad (3.187)$$

where the differential entropy of output signals $\mathbf{y} = \mathbf{Wx}$ is defined as

$$H(\mathbf{y}) = -\int_{-\infty}^{\infty} f_y(\mathbf{y}) \log f_y(\mathbf{y}) \, d\mathbf{y} \qquad (3.188)$$

Taking into account that $d\mathbf{y} = d\check{\mathbf{y}}^i \, dy_i$ the second terms in Eq. (3.187) can be expressed by the marginal entropies as

$$\begin{aligned} \int_{-\infty}^{\infty} f_y(\mathbf{y}) \log f_i(y_i) \, d\mathbf{y} &= \int_{-\infty}^{\infty} \log f_i(y_i) \int_{-\infty}^{\infty} f_y(\mathbf{y}) \, d\check{\mathbf{y}}^i \, dy_i \\ &= \int_{-\infty}^{\infty} f_i(y_i) \log f_i(y_i) \, dy_i = E\{\log(f_i(y_i))\} \\ &= -H_i(y_i) \end{aligned} \qquad (3.189)$$

Hence, the Kullback-Leibler divergence can be expressed by the differential $H(\mathbf{y})$ and the marginal entropies $H_i(y_i)$ as

$$D_{fq} = E\{l(\mathbf{W}, \mathbf{y})\} = -H(\mathbf{y}) + \sum_{i=1}^{n} H_i(y_i) \qquad (3.190)$$

Assuming $\mathbf{y} = \mathbf{Wx}$, the differential entropy can be expressed as

$$H(\mathbf{y}) = H(\mathbf{x}) + \log|\det(\mathbf{W})| \qquad (3.191)$$

where $H(\mathbf{x}) = -\int_{-\infty}^{\infty} f_x(\mathbf{x}) \log f_x(\mathbf{x}) \, d\mathbf{x}$ is independent of matrix $\mathbf{W}$. Hence we obtain a simple contrast function

$$D_{fq} = -H(\mathbf{x}) - \log|\det(\mathbf{W})| - \sum_{i=1}^{n} E\{\log(f_i(y_i))\} \qquad (3.192)$$

The standard gradient of D_{fq} can be expressed as

$$\Delta D_{fq} = \frac{\partial D_{fq}}{\partial \mathbf{W}} = -\mathbf{W}^{-T} + \langle \boldsymbol{\varphi}(\mathbf{y})\mathbf{x}^T \rangle \tag{3.193}$$

where $\boldsymbol{\varphi}(\mathbf{y}) = [\varphi_1(y_1) \cdots \varphi_n(y_n)]^T$ with elements defined

$$\varphi_i(y_i) = -\frac{d \log f_i(y_i)}{dy_i} = -\frac{df_i(y_i)/dy_i}{f_i(y_i)} = -\frac{f_i'(y_i)}{f_i(y_i)} \tag{3.194}$$

Applying the natural gradient rules, Eqs. (3.48) and (3.63), we obtain the batch algorithms

$$\Delta \mathbf{W} = -\eta \frac{\partial D_{fq}}{\partial \mathbf{W}} \mathbf{W}^T\mathbf{W} = \eta[\mathbf{I} - \langle \boldsymbol{\varphi}(\mathbf{y})\mathbf{y}^T \rangle]\mathbf{W} \tag{3.195}$$

or

$$\Delta \mathbf{W} = -\eta \mathbf{W}\left[\frac{\partial D_{fq}}{\partial \mathbf{W}}\right]^T \mathbf{W} = \eta[\mathbf{I} - \langle \mathbf{y}\boldsymbol{\psi}(\mathbf{y}^T) \rangle]\mathbf{W} \tag{3.196}$$

which can be easily converted to on-line algorithms omitting the average (expectation) operation.

APPENDIX 3.2: DERIVATION OF BASIC LEARNING RULE (3.49) AND STABILITY CONDITIONS FOR BSS

In order to calculate the gradient of $l(\mathbf{W}, \mathbf{y})$ expressed by Eq. (3.192), we use the total differential dl of l when $\mathbf{W}$ is changed from $\mathbf{W}$ to $\mathbf{W} + d\mathbf{W}$ (Amari et al. 1997c, 1999). In the component form,

$$dl = l(\mathbf{y}, \mathbf{W} + d\mathbf{W}) - l(\mathbf{y}, \mathbf{W}) = \sum_{i,j} \frac{\partial l}{\partial w_{ij}} dw_{ij} \tag{3.197}$$

where the coefficients $\partial l/\partial w_{ij}$ of dw_{ij} represent the gradient of l. Simple algebraic and differential calculus yields

$$dl = -\text{tr}(d\mathbf{W}\mathbf{W}^{-1}) + \boldsymbol{\varphi}(\mathbf{y}^T)\,d\mathbf{y} \tag{3.198}$$

where tr is the trace of a matrix and $\boldsymbol{\varphi}(\mathbf{y})$ is a column vector whose components are $\varphi_i(y_i) = -f_i'(y_i)/f_i(y_i)$.
From $\mathbf{y} = \mathbf{W}\mathbf{x}$, we have

$$d\mathbf{y} = d\mathbf{W}\mathbf{x} = d\mathbf{W}\mathbf{W}^{-1}\mathbf{y} \tag{3.199}$$

Hence, we put

$$dX = dWW^{-1} \tag{3.200}$$

whose components dx_{ij} are linear combinations of dw_{ij}. The differentials $\{dx_{ij}\}$ form a basis of the tangent space of nonsingular matrices W, since they are linear combinations of the basis $\{dw_{ij}\}$. It should be noted that $dX = dWW^{-1}$ is a nonintegrable differential form so that we do not have a matrix function $X(W)$, which gives Eq. (3.200). Nevertheless, the nonholonomic basis dX has a definite geometrical meaning and is very useful. It is effective to analyze the differential in terms of dX, since the natural Riemannian gradient (Amari et al. 1996; Amari and Cardoso 1997; Amari 1998, 1999a,b) is automatically implemented by it, and the equivariant properties investigated in Cardoso and Laheld (1996) automatically hold in this basis. It is easy to rewrite the results in terms of dW by using Eq. (3.200).

The gradient dl in Eq. (3.198) is expressed by the differential form

$$dl = -\text{tr}(dX) + \boldsymbol{\varphi}(\mathbf{y}^T)\,dX\mathbf{y} \tag{3.201}$$

This leads to the stochastic-gradient learning algorithm:

$$\Delta \mathbf{X}(k) = \mathbf{X}(k+1) - \mathbf{X}(k) = -\eta(k)\frac{dl}{d\mathbf{X}} = \eta(k)[\mathbf{I} - \boldsymbol{\varphi}(\mathbf{y}(k))\mathbf{y}^T(k)] \tag{3.202}$$

in terms of $\Delta \mathbf{X}(k) = \Delta \mathbf{W}(k)\mathbf{W}^{-1}(k)$, or

$$\mathbf{W}(k+1) = \mathbf{W}(k) + \eta(k)[\mathbf{I} - \boldsymbol{\varphi}(\mathbf{y}(k))\mathbf{y}^T(k)]\mathbf{W}(k) \tag{3.203}$$

in terms of $\mathbf{W}(k)$.

3.2.1 Stability of the Basic Learning Rule (3.49)

We consider the expected version of the learning rule

$$\frac{d\mathbf{W}}{dt} = \mu(t)E\{\mathbf{I} - \boldsymbol{\varphi}(\mathbf{y}(t))\mathbf{y}^T(t)\}\mathbf{W}(t) \tag{3.204}$$

where we use the continuous-time version of the algorithm for simplicity's sake. By linearizing it at the equilibrium point, we have the variational equation

$$\frac{\delta\mathbf{W}(t)}{dt} = \mu(t)\frac{\partial(E\{\mathbf{I} - \boldsymbol{\varphi}(\mathbf{y}(t))\mathbf{y}^T(t)\}\mathbf{W})}{\partial\mathbf{W}}\delta\mathbf{W} \tag{3.205}$$

where $(\partial/\partial\mathbf{W})\delta\mathbf{W}$ implies $\Sigma(\partial/\partial w_{ij})\delta w_{ij}$ in the component form. This shows that, if and only if all the eigenvalues of the operator $\mathbf{K}(\mathbf{W}, \mathbf{y}) = \partial^2 l(\mathbf{W}, \mathbf{y})/$

$\partial \mathbf{W} \partial \mathbf{W} = \partial (E\{\mathbf{I} - \boldsymbol{\varphi}(\mathbf{y})\mathbf{y}^T\}\mathbf{W})/\partial \mathbf{W}$ have negative real parts, the equilibrium is asymptotically stable. Therefore, we need to evaluate all the eigenvalues of the operator. This can be done in terms of $d\mathbf{X}$, as follows. Since $\mathbf{I} - \boldsymbol{\varphi}(\mathbf{y})\mathbf{y}^T$ is derived from the gradient dl, we need to calculate its Hessian d^2l

$$d^2l = \sum \frac{\partial l(\mathbf{y}, \mathbf{W})}{\partial w_{ij} \partial w_{kl}} dw_{ij}\, dw_{kl}$$

in terms of $d\mathbf{X}$.

The equilibrium is stable if and only if the expectation of the preceding quadratic form is positive definite. We calculate the second total differential, which is the quadratic form of the Hessian of l, as (Amari et al. 1997c)

$$d^2l = \mathbf{y}^T \, d\mathbf{X}^T \boldsymbol{\varphi}'(\mathbf{y})\, d\mathbf{y} + \boldsymbol{\varphi}(\mathbf{y}^T)\, d\mathbf{X}\, d\mathbf{y}$$
$$= \mathbf{y}^T \, d\mathbf{X}^T \boldsymbol{\varphi}'(\mathbf{y})\, d\mathbf{X}\mathbf{y} + \boldsymbol{\varphi}(\mathbf{y}^T)\, d\mathbf{X}\, d\mathbf{X}\mathbf{y} \qquad (3.206)$$

where $\boldsymbol{\varphi}'(\mathbf{y})$ is the diagonal matrix whose diagonal elements are $\varphi'(y_i)$. The expectation of the first term is

$$E\{\mathbf{y}^T \, d\mathbf{X}^T \boldsymbol{\varphi}'(\mathbf{y})\, d\mathbf{X}\mathbf{y}\} = \sum E\{y_i\, dx_{ji}\varphi_j'(y_j)\, dx_{jk}\, y_k\}$$
$$= \sum_{j \neq i} E\{(y_i)^2\} E\{\varphi_j'(y_j)\}(dx_{ji})^2$$
$$+ \sum_i E\{(y_i)^2 \varphi_i'(y_i)\}(dx_{ii})^2$$
$$= \sum_{j \neq i} \sigma_i^2 \kappa_j (dx_{ji})^2 + \sum_i m_i (dx_{ii})^2$$

Here, the expectation is taken at $\mathbf{W} = \mathbf{A}^{-1}$ where the y_i are independent. Similarly,

$$E\{\boldsymbol{\varphi}(\mathbf{y})^T \, d\mathbf{X}\, d\mathbf{X}\mathbf{y}\} = \sum E\{\varphi_i(y_i)\, dx_{ij}\, dx_{jk}\, y_k\}$$
$$= \sum E\{y_i\varphi_i(y_i)\} dx_{ij}\, dx_{ji} = \sum_{i,j} dx_{ij}\, dx_{ji} \qquad (3.207)$$

because of $E\{y_i\varphi_i(y_i)\} = 1$ (the normalization condition). Hence,

$$E\{d^2l\} = \sum_{j \neq i} \{\sigma_i^2 \kappa_j (dx_{ji})^2 + dx_{ij}\, dx_{ji}\} + \sum_i (m_i + 1)(dx_{ii})^2 \qquad (3.208)$$

For a pair $(i, j), i \neq j$, the summand in the first term is rewritten as

$$k_{ij} = \sigma_i^2 \kappa_j (dx_{ji})^2 + \sigma_j^2 \kappa_i (dx_{ij})^2 + 2\, dx_{ij}\, dx_{ji} \qquad (3.209)$$

This k_{ij} $(i \neq j)$ is the quadratic form in (dx_{ij}, dx_{ji}), and

$$E\{d^2 l\} = \sum_{i \neq j} k_{ij} + \sum (m_i + 1)(dx_{ii})^2 \qquad (3.210)$$

The matrix (k_{ij}) is positive definite if and only if the following stability conditions hold:

$$m_i + 1 > 0 \qquad (3.211)$$

$$\kappa_i > 0 \qquad (3.212)$$

$$\sigma_i^2 \sigma_j^2 \kappa_i \kappa_j > 1, \qquad \text{for all} \quad 1 \leq i < j \leq m \qquad (3.213)$$

where $m_i = E\{y_i^2 \varphi_i'(y_i)\}$; $\kappa_i = E\{\varphi_i'(y_i)\}$; $\sigma_i^2 = E\{|y_i|^2\}$; y_i is the source signal extracted at the ith output; and $\varphi_i'(y)$ denotes the derivative of the activation function $\varphi_i(y)$ with respect to y.

APPENDIX 3.3: DERIVATION OF LEARNING RULE (3.63) AND STABILITY CONDITIONS FOR BSS

In order to derive the learning algorithm (3.63), we follow the notation of Appendix 3.2 (Amari et al. 1999; Attick and Redlich 1993; Cichocki et al. 1997e)

$$d\mathbf{X} \equiv d\mathbf{W}\mathbf{W}^{-1} \qquad (3.214)$$

and

$$dl(\mathbf{y}, \mathbf{W}) = -\text{tr}(d\mathbf{X}) + \psi(\mathbf{y}^T) d\mathbf{X}\mathbf{y} \qquad (3.215)$$

The standard stochastic-gradient method for $\mathbf{X}$ leads to the natural-gradient learning algorithm for updating $\mathbf{W}$. Let us apply a different update rule as follows

$$\Delta \mathbf{X} = -\eta \left(\frac{\partial l}{\partial \mathbf{X}} \right)^T = \eta[\mathbf{I} - \mathbf{y}\psi(\mathbf{y}^T)] \qquad (3.216)$$

On the other hand, we have

$$\frac{\partial l(\mathbf{y}, \mathbf{W})}{\partial \mathbf{X}} = \frac{\partial l(\mathbf{y}, \mathbf{W})}{\partial \mathbf{W}} \mathbf{W}^T \qquad (3.217)$$

Hence

$$\Delta \mathbf{W} = -\eta \mathbf{W} \left[\frac{\partial l(\mathbf{y}, \mathbf{W})}{\partial \mathbf{W}} \right]^{T} \mathbf{W} \qquad (3.218)$$

or explicitly

$$\Delta \mathbf{W}(k) = \eta(k)[\mathbf{I} - \mathbf{y}(k)\boldsymbol{\psi}[\mathbf{y}^{T}(k)]]\mathbf{W}(k) \qquad (3.219)$$

It should be noted that in general the update rule does not lead to a gradient-descent algorithm. However, we will show now that if the function $\psi_i(y_i)$ satisfies the following conditions (Zhang and Cichocki 1998a, 1998c)

$$E\{\psi_i(y_i)y_i\} = \langle \psi_i(y_i)y_i \rangle > 0, \qquad \text{for} \quad i = 1, \dots, n \qquad (3.220)$$

the learning algorithm, Eq. (3.219) is a descent learning algorithm with the contrast function $l(\mathbf{y}, \mathbf{W})$, which means that the algorithm is convergent to one of the local minima of the contrast function. It should be noted that all the odd functions $\psi_i(y_i)$ satisfy these conditions.

From Eq. (3.215) we have

$$\frac{\partial l(\mathbf{y}, \mathbf{W})}{\partial x_{ij}} = -\delta_{ij} + y_i \psi_j(y_j) \qquad (3.221)$$

We consider the decrease of the objective function during one step of learning

$$\begin{aligned}
\Delta l(\mathbf{y}, \mathbf{W}) &= l(\mathbf{y}, \mathbf{W} + \Delta \mathbf{W}) - l(\mathbf{y}, \mathbf{W}) \\
&= \left\langle \frac{\partial l(\mathbf{y}, \mathbf{W})}{\partial \mathbf{X}}, \Delta \mathbf{X}^{T} \right\rangle + O(|\Delta \mathbf{X}|^2) \\
&= -\eta \sum_{ij} \frac{\partial l}{\partial x_{ij}} \frac{\partial l}{\partial x_{ji}} + O(|\Delta \mathbf{X}|^2) \\
&= -\eta \left(\sum_{i=1}^{n}(1 - y_i\psi_i(y_i))^2 + \sum_{i \neq j} y_i\psi_j(y_j)y_j\psi_i(y_i) \right) + O(|\Delta \mathbf{X}|^2)
\end{aligned}$$
$$(3.222)$$

If $\langle y_i\psi_i(y_i) \rangle > 0$, then

$$\left\langle \frac{\partial l}{\partial \mathbf{X}}, \frac{\partial l}{\partial \mathbf{X}} \right\rangle = \text{tr}\left(\frac{\partial l}{\partial \mathbf{X}}, \frac{\partial l}{\partial \mathbf{X}^{T}} \right) > 0$$

This means that if the learning rate η is small enough, then the contrast function is decreasing during the learning process until the system achieves a minimum during the learning procedure.

3.3.1 Stability Conditions

In order to analyze the stability condition of a separating solution, we take a variation with respect to $\mathbf{W}$ of the continuous-time learning algorithm as

$$\frac{d\delta\mathbf{W}}{dt} = -\eta\delta E\{\mathbf{y}(t)\boldsymbol{\psi}[\mathbf{y}^T(t)]\}\mathbf{W}$$

$$= -\eta E\{\delta\mathbf{y}\boldsymbol{\psi}(\mathbf{y}^T) + \mathbf{y}\boldsymbol{\psi}'(\mathbf{y}^T)\delta\mathbf{y}^T)\}\mathbf{W} \qquad (3.223)$$

Substituting $d\mathbf{W} = d\mathbf{X}\mathbf{W}$ and $d\mathbf{y} = d\mathbf{X}\mathbf{y}$, we obtain

$$\frac{d\delta\mathbf{X}}{dt} = -\eta E\{d\mathbf{X}\mathbf{y}\boldsymbol{\psi}(\mathbf{y}^T) + \mathbf{y}\boldsymbol{\psi}'(\mathbf{y}^T)(\mathbf{y}^T\,d\mathbf{X}^T)\}$$

$$= -\eta(d\mathbf{X} + E\{\mathbf{y}\boldsymbol{\psi}'(\mathbf{y}^T)(\mathbf{y}^T\,d\mathbf{X}^T)\}) \qquad (3.224)$$

Since the output signals y_i are mutually independent, we have at the equilibrium point

$$E\{\psi_i(y_i)y_iy_j\} = 0 \qquad \text{for} \quad i \neq j \qquad (3.225)$$

The components of Eq. (3.224) can written as

$$\frac{d\delta x_{ii}}{dt} = -\eta(1 + E\{\psi_i'(y_i)y_i^2\})\delta x_{ii} \qquad \text{for} \quad i = 1,\ldots,n \qquad (3.226)$$

and for $i \neq j$

$$\frac{d\delta x_{ij}}{dt} = -\eta(\delta x_{ij} + E\{y_i^2\}E\{\psi_j'(y_j)\}\delta x_{ji}) \qquad (3.227)$$

$$\frac{d\delta x_{ji}}{dt} = -\eta(\delta x_{ji} + E\{y_j^2\}E\{\psi_i'(y_i)\}\delta x_{ji}) \qquad (3.228)$$

It is easy now to write the stability conditions in an explicit form:

$$1 + E\{y_i^2\psi_i'(y_i)\} > 0 \qquad \text{for} \quad i = 1,\ldots,n \qquad (3.229)$$

and

$$\gamma_{ij} = E\{y_i^2\}E\{y_j^2\}E\{\psi_i'(y_i)\}E\{\psi_j'(y_j)\} < 1 \qquad \text{for} \quad i \neq j \qquad (3.230)$$

From these stability conditions we see that one important advantage of the learning rule, Eq. (3.63), is that this algorithm is still stable in contrast to the algorithm in Eq. (3.49), even when some of the source signals become silent (decay to zero).

It is easily observed that the preceding stability conditions for the learning algorithm in Eq. (3.63) are mutually complementary to the ones for Amari's natural-gradient learning algorithm in Eq. (3.49). The observation motivates us to propose the universal convergence learning algorithm that means that the separating solutions are always a locally stable equilibrium point of the learning algorithm.

APPENDIX 3.4: UNIVERSAL ADAPTIVE LEARNING ALGORITHM

Let us consider the learning algorithm proposed by Amari et al. (1997c)

$$\Delta \mathbf{W} = \eta \mathbf{F}(\mathbf{y}, \mathbf{W}) \mathbf{W} \tag{3.231}$$

with entries of matrix $\mathbf{F}$ defined as

$$f_{ii}(\mathbf{y}, \mathbf{W}) = \lambda_i - \varphi_i(y_i) \qquad i = 1, \ldots, n \tag{3.232}$$

and

$$f_{ij}(\mathbf{y}, \mathbf{W}) = \begin{cases} -\varphi_i(y_i) y_j & \text{if } \gamma_{ij} > 1 \\ -y_i \varphi_j(y_j) & \text{if } \gamma_{ij} \leq 1 \end{cases} \tag{3.233}$$

where λ_i are given positive constants and $\gamma_{ij} = \sigma_i^2 \sigma_j^2 \kappa_i \kappa_j$. The separating solution satisfies the system of nonlinear algebraic equations

$$E\{\mathbf{F}(\mathbf{y}, \mathbf{W})\} = \mathbf{0} \tag{3.234}$$

In order to establish the stability conditions, we write the preceding learning algorithm in the continuous-time form as

$$\frac{d\mathbf{W}}{dt} = \eta E\{\mathbf{F}(\mathbf{y}, \mathbf{W})\}\mathbf{W} \tag{3.235}$$

Taking the variation of $\mathbf{W}$ at an equilibrium point, we have

$$\frac{d\delta\mathbf{W}}{dt} = \eta E\{\delta\mathbf{F}(\mathbf{y}, \mathbf{W})\}\mathbf{W} \tag{3.236}$$

Hence for $\gamma_{ij} > 1, i \neq j$, we have

$$\frac{d\delta x_{ij}}{dt} = -(E\{\varphi_i'(y_i)\}\sigma_j^2 \delta x_{ij} + \delta x_{ji}) \tag{3.237}$$

and

$$\frac{d\delta x_{ji}}{dt} = -(\delta x_{ij} + E\{\varphi_i'(y_i)\}\sigma_j^2\delta x_{ij}) \tag{3.238}$$

Similarly, if $\gamma_{ij} < 1$, then the variations δx_{ij} and δx_{ji} satisfy the conditions derived in Appendix 3.3. These results can be summarized in the form of the following Theorem (Amari et al. 1997c):

Theorem. Suppose that $\langle \varphi_i(y_i)y_i \rangle > 0, i = 1, \ldots, n$, and $\gamma_{ij} = \sigma_i^2\sigma_j^2\kappa_i\kappa_j \neq 1$, for $i \neq j$, $m_i + 1 = E\{y_i\varphi_i'(y_i)\} > 0$, $\kappa_i = E\{\varphi_i'(y_i)\} > 0$, then the separating solution by employing the learning algorithm in Eq. (3.231) is stable.

The form of the nonlinear functions $\mathbf{F}$ depends on parameters $\gamma_{ij} = \sigma_i^2\sigma_j^2\kappa_i\kappa_j$, which cannot be explicitly determined, but we can evaluate parameters γ_{ij} dynamically during the learning process.

The learning algorithm in Eq. (3.231) possesses two important properties, just as the natural-gradient learning algorithm in Eq. (3.49) does. One is the equivariant property; the other is the nonsingularity of the learning matrix $\mathbf{W}(t)$, which can be observed from the following derivation (Yang and Amari 1997; Zhang and Cichocki 1998a). We define

$$\langle \mathbf{X}, \mathbf{Y} \rangle = \mathrm{tr}(\mathbf{X}^T\mathbf{Y}) \tag{3.239}$$

and calculate

$$\frac{d\det(\mathbf{W}(t))}{dt} = \left\langle \frac{\partial \det(\mathbf{W})}{\partial \mathbf{W}}, \frac{d\mathbf{W}}{dt} \right\rangle \tag{3.240}$$

$$= \langle \det(\mathbf{W})\mathbf{W}^{-T}, \eta\mathbf{F}(\mathbf{y}, \mathbf{W})\mathbf{W} \rangle \tag{3.241}$$

$$= \eta\,\mathrm{tr}(\mathbf{F}(\mathbf{y}, \mathbf{W})\det(\mathbf{W})) \tag{3.242}$$

$$= \eta\left(\sum_{i=1}^{n}(\lambda_i - y_i\psi_i(y_i))\right)\det(\mathbf{W}) \tag{3.243}$$

Then the determinant $\det(\mathbf{W}(t))$ is expressed by

$$\det(\mathbf{W}(t)) = \det(\mathbf{W}(0))\exp\left(\eta\int_0^t\sum_{i=1}^{n}(\lambda_i - \psi_i(y_i)y_i)\,d\tau\right) \tag{3.244}$$

This means that if $\mathbf{W}(0)$ is nonsingular, then $\mathbf{W}(t)$ keeps the nonsingularity invariantly (Yang and Amari 1997).

APPENDIX 3.5: GENERAL ANALYSIS OF EXTRACTION AND DEFLATION CRITERIA

For simplicity of our considerations and without contrast of generality, we assume that the source signals are ordered in such a way that the first extracted source is s_1, the next is s_2, and so on. Let us define the diagonal matrix $D_k = \text{diag}[d_{k1}, d_{k2}, \ldots, d_{km}]$, with $d_{kj} = 0$ for $j \leq k$ and $d_{kj} = 1$ for $j > k$ (Thawonmas et al. 1998). Now let S denote the set of all the elements in the source vector $\mathbf{s}$, that is, $S = \{s_1, s_2, \ldots, s_m\}$. If we assume that an index i exists such that $\beta_i \kappa_4(s_i) < 0$, where $\beta_i = \text{sign}\,\kappa_4(s_i)$, $s_i \in S - \{s_1, s_2, \ldots, s_{k-1}\}$, then we have the following theorem for the jth extraction unit and deflation processing unit (Thawonmas et al. 1998).

Theorem. The output signal y_k converges to $\pm cs_{j^*}$ if and only if $\beta \kappa_4(s_{j^*}) < 0$, where $s_{j^*} \in S - \{s_1, s_2, \ldots, s_{j-1}\}$, and vector $\mathbf{x}_{j+1}$ of the deflated mixture converges to $\mathbf{AD}_j\mathbf{s}$.

Proof. We prove this theorem by induction. It should be noted that, according to the theorem, at the first extraction processing unit the output signal y_1 converges to $\pm cs_1$, with $\beta \kappa(s_1) < 0$. The cost function in Eq. (3.86) for $\rho(x) = \frac{1}{2}x^2$ can be expressed as

$$\tilde{\mathscr{J}}_1(\tilde{\mathbf{w}}_1) = \sum_{i=1}^{n}(a_{i1} - \tilde{w}_{1i})^2 E(s_1^2) + \sum_{i=1}^{n}\sum_{j=2}^{m} a_{ij}^2 E(s_j^2)$$

or, assuming that source signals have unit variance, as

$$\tilde{\mathscr{J}}_1(\tilde{\mathbf{w}}_1) = \sum_{i=1}^{n}(a_{i1} - \tilde{w}_{1i})^2 + \sum_{i=1}^{n}\sum_{j=2}^{m} a_{ij}^2 \tag{3.245}$$

From Eqs. (3.86) and (3.245) we have $(d\tilde{\mathscr{J}}_1(\tilde{\mathbf{w}}_1))/dt \leq 0$ and $(d\tilde{\mathscr{J}}_1(\tilde{\mathbf{w}}_1))/dt = 0$ only for $(d\tilde{\mathbf{w}}_1)/dt = 0$ at $\tilde{\mathbf{w}}_1 = \tilde{\mathbf{w}}_{1^*} = \mathbf{a}_1 = (a_{11}, a_{21}, \ldots, a_{n1})^T$. Hence, $\tilde{\mathscr{J}}_1(\tilde{\mathbf{w}}_1)$ is a Lyapunov function, and thus $\tilde{\mathbf{w}}_1$ globally converges to $\tilde{\mathbf{w}}_{1^*}$. After convergence, vector $\mathbf{x}_2$ includes no elements of s_1. In other words, vector $\mathbf{x}_2$ converges to $\mathbf{A}[0, s_2, s_3, \ldots, s_m]^T = \mathbf{AD}_1\mathbf{s}$.

Next, for the proof at the kth extraction and deflation processing units, we assume that the output signal y_{k-1} has converged to $\pm cs_{(k-1)^*}$ with $\beta \kappa_4(s_{(k-1)^*}) < 0$, where $s_{(k-1)^*} \in S - \{s_1, \ldots, s_{k-2}\}$, and that vector $\mathbf{x}_k$ has converged to $\mathbf{AD}_{k-1}\mathbf{s}$. Following the form of vector $\mathbf{x}_1 = \mathbf{As} = \mathbf{A}[s_1, s_2, \ldots, s_m]^T$ in the theorem, vector $\mathbf{x}_k$ can be expressed as $\mathbf{A}[0, 0, \ldots, 0, s_k, s_{k+1}, \ldots, s_m]^T$. Replacing the subscript 1 with k in this theorem yields that the output signal y_k converges to $\pm cs_{k^*}$ with $\beta \kappa_4(s_{k^*}) < 0$, where $s_{k^*} \in \{0, 0, \ldots, 0, s_k, s_{k+1}, \ldots, s_m\}$. Because $\kappa_4(0) = 0$, the condition $\beta \kappa_4(s_{k^*}) < 0$ does not hold for the zero elements in $\{0, 0, \ldots, 0, s_k, s_{k+1}, \ldots, s_m\}$. It therefore can be stated explicitly that the output signal y_k converges to $\pm c s_{k^*}$ with $\beta \kappa_4(s_{k^*}) < 0$, where $s_{k^*} = \{s_k, s_{k+1}, \ldots, s_m\} = S - \{s_1, s_2, \ldots, s_{k-1}\}$.

It can also be shown that $\tilde{\mathscr{J}}_k(\tilde{\mathbf{w}}_k)$ is a Lyapunov function whose $\tilde{\mathbf{w}}_k$ globally converges to $\tilde{\mathbf{w}}_{k^*} = [a_{1k^*}, a_{2k^*}, \ldots, a_{nk^*}]^T$. Because vector $\mathbf{x}_k$ includes no elements of $s_1, s_2, \ldots, s_{k-1}$ and the subscript k^* becomes k after index permutation, vector $\mathbf{x}_{k+1}(t)$, after convergence, includes no elements of $s_1, s_2, \ldots, s_k$. This means that vector $\mathbf{x}_{k+1}(t)$ converges to $\mathbf{A}[0, 0, \ldots, 0, s_{k+1}, s_{k+2}, \ldots, s_m]^T = \mathbf{A}\mathbf{D}_k\mathbf{s}$. This theorem then holds by induction.

REFERENCES

Amari, S., Mar. 1967, "A theory of adaptive pattern classifiers," *IEEE Trans. on Electronic Computers*, vol. EC-16, pp. 299–307.

Amari, S., 1977, "Neural theory of association and concept-formation," *Biological Cybernetics*, vol. 26, pp. 175–185.

Amari, S., 1985, *Differential-Geometrical Methods of Statistics, Springer Lecture Notes in Statistics*, vol. 28 (Berlin: Springer).

Amari, S., 1987, "Differential geometry of a parametric family of invertible linear systems—Riemannian metric, dual affine connections and divergence," *Mathematical Systems Theory*, vol. 20, pp. 53–82.

Amari, S., 1991, "Mathematical theory of neural learning," *New Generation of Computing*, vol. 8, pp. 135–143.

Amari, S., Jan. 1998, "Natural gradient works efficiently in learning," *Neural Computation*, vol. 10, pp. 251–276.

Amari, S., 1999a, "Natural gradient for over- and under-complete ICA," *Neural Computation*, vol. 11, no. 8. (in press).

Amari, S., 1999b, "Super-efficiency in blind source separation," *IEEE Trans. Signal Processing*, vol. 47, pp. 936–944.

Amari, S., and J. F. Cardoso, Nov. 1997, "Blind source separation—Semi-parametric statistical approach," *IEEE Trans. Signal Processing*, vol. 45, no. 11, pp. 2692–2700.

Amari, S., and A. Cichocki, Oct. 1998, "Adaptive blind signal processing—neural network approaches," *Proc. IEEE*, vol. 86, no. 10, pp. 2026–2048.

Amari, S., and M. Kawanabe, 1997a, "Information geometry of estimating functions in semi-parametric statistical models," *Bernoulli*, vol. 3, no. 1, pp. 29–54.

Amari, S., and M. Kawanabe, 1997b "Estimating functions in semi-parametric statistical models," Selected Proceedings of the Symposium on Estimating Functions, IMS Monograph Series, vol. 32, eds. I. V. Basawa et al., pp. 65–81.

Amari, S., A. Cichocki, H. H. Yang, 1995, "Recurrent neural networks for blind separation of sources," in *Proc. Int. Symp. Nonlinear Theory and its Applications*, NOLTA-95, Las Vegas, Dec. 1995, pp. 37–42.

Amari, S., A. Cichocki, and H. H. Yang, 1996, "A new learning algorithm for blind signal separation," *Adv. in Neural Inform. Processing Systems*, vol. 8, *(NIPS-1995)* (Boston, MA: MIT Press), pp. 752–763.

Amari, S., S. C. Douglas, A. Cichocki, H. H. Yang, 1997a, "Multichannel blind deconvolution and equalization using the natural gradient," *Proc. of IEEE International Workshop on Wireless Communication*, Paris, April 1997, pp. 101–104.

Amari, S., S. Douglas, A. Cichocki, and H. H. Yang, 1997b, "Novel on-line algorithms for blind deconvolution using natural gradient approach," *Proc. of 11th IFAC Symposium on System Identification, SYSID-97* Kitakyushu City, Japan, July 1997, pp. 1057–1062.

Amari, S., T.-P. Chen, and A. Cichocki, 1997c, "Stability analysis of adaptive blind source separation," *Neural Networks*, vol. 10, no. 8, pp. 1345–1351.

Amari, S., S. C. Douglas, and A. Cichocki, 1998, "Multichannel blind deconvolution and source separation using the natural gradient," submitted to *IEEE Trans. Signal Processing*.

Amari, S., T.-P. Chen, and A. Cichocki, in press, "Non-holonomic constraints in learning algorithms for blind source separation," *Neural Computation*.

Attick, J. J., and A. N. Redlich, 1993, "Convergent algorithm for sensory receptive fields development," *Neural Computation*, vol. 5, pp. 45–60.

Bell, A. J., and T. J. Sejnowski, 1995, "An information-maximization approach to blind separation and blind deconvolution," *Neural Computation*, vol. 7, pp. 1129–1159.

Belouchrani, A., A. Cichocki, and K. Abed Meraim, 1996, "A blind identification and separation technique via multi-layer neural networks," in *Progress in Neural Information Processing—ICONIP'96, Proceedings*, (Hong Kong Sept. 1996), vol. 2, Springer-Verlag Singapore Ltd., pp. 1195–1200.

Benveniste, A., M. Goursat, and G. Ruget, June 1980, "Robust identification of a nonminimum phase system: Blind adjustment of a linear equalizer in data communications," *IEEE Trans. Automat. Contr.*, vol. AC-25, no. 3, pp. 385–399.

Cao, X.-R., and R.-W. Liu, March 1996, "General approach to blind source separation," *IEEE Trans. on Signal Processing*, vol. 44, pp. 562–571.

Cardoso, J.-F., April 1997, "Infomax and maximum likelihood for blind source separation," *IEEE Signal Processing Lett.*, vol. 4, pp. 109–111.

Cardoso, J.-F., and B. Laheld, Dec. 1996, "Equivariant adaptive source separation," *IEEE Trans. Signal Processing*, vol. SP-43, pp. 3017–3029.

Capdevielle, V., C. Serviere, and J. Lacoume, May 1995, "Blind separation of wideband sources in the frequency domain," *Proc. ICASSP*, pp. 2080–2083.

Chan, D. B., P. J. W. Rayner, and S. J. Godsill, May 1996, "Multi-channel signal separation," *Proc. ICASSP*, pp. 649–652.

Chen, S., D. L. Donoho, and M. A. Suanders, 1996, Atomic decomposition by basis pursuit. Technical Report, Dept. Statistics, Stanford University, Stanford, CA.

Chen, T. P., S. Amari, and Q. Lin, 1998, "A unified algorithm for principal and minor components extraction," *Neural Networks*, vol. 11, pp. 385–390.

Choi, S., and A. Cichocki, 1998, "Cascade neural networks for multichannel blind deconvolution," *Electronics Lett.*, vol. 34, no. 12, pp. 1186–1187.

Choi, S., R.-W. Liu, and A. Cichocki, Jan. 1998, "A spurious equilibria-free learning algorithm for the blind separation of non-zero skewness signals," *Neural Processing Lett.*, vol. 7, no. 2, pp. 61–68.

Cichocki, A., "Blind separation and extraction of source signals—recent results and open problems," *Proc. of the 4th Ann. Conf. of the Institute of Systems, Control and Information Engineers, ISCIIE*, Osaka, May 21–23 1997, pp. 43–48.

Cichocki, A., 1998, "Blind identification and separation of noisy source signals—neural networks approaches," *ISCIE Journal, Japan*, 1998, vol. 42, no. 2, pp. 63–73.

Cichocki, A., and L. Moszczynski, 1992, "A new learning algorithm for blind separation of sources," *Electronics Lett.*, vol. 28, no. 21, pp. 1986–1987.

Cichocki, A., and R. Unbehauen, 1993, "Robust estimation of principal components in real time," *Electronics Lett.*, vol. 29, no. 21, pp. 1869–1870.

Cichocki, A., and R. Unbehauen, 1994, *Neural Networks for Optimization and Signal Processing*, (Chichester: John Wiley) chap. 8, pp. 416–471.

Cichocki, A., and R. Unbehauen, Nov. 1996, "Robust neural networks with on-line learning for blind identification and blind separation of sources," *IEEE Trans. Circuits and Systems I: Fundamentals Theory and Applications*, vol. 43, no. 11, pp. 894–906.

Cichocki, A., and L. Zhang, 1998, "Two-stage blind deconvolution using state-space models," *Proc. of the Fifth Int. Conf. on Neural Information Processing, ICONIP'98*, Kitakyushu, Japan, Oct. 1998, pp. 729–732.

Cichocki, A., R. Unbehauen, and E. Rummert, August 1994a, "Robust learning algorithm for blind separation of signals," *Electronics Lett.*, vol. 30, no. 17, pp. 1386–1387.

Cichocki, A., R. Unbehauen, L. Moszczyński, and E. Rummert, 1994b, "A new on-line adaptive learning algorithm for blind separation of source signals," in *Proc. of ISANN-94*, Taiwan, Dec. 1994, pp. 406–411.

Cichocki, A., R. Swiniarski, and R. Bogner, 1996a, "Hierarchical neural network for robust PCA of complex-valued signals," in *Proc. of World Congress on Neural Networks, WCNN'96—1996*, Int. Neural Network Society Annual Meeting in San Diego, USA, Sept. 1996, pp. 818–821.

Cichocki, A., S. Amari, M. Adachi, and W. Kasprzak, 1996b, "Self-adaptive neural networks for blind separation of sources," *Proc. of 1996 IEEE International Symposium on Circuit and Systems*, vol. 2, May 1996, pp. 157–160.

Cichocki, A., S. Amari, and J. Cao, 1996c, "Blind separation of delayed and convolved signals with self-adaptive learning rate," *1996 IEEE International Symposium on Nonlinear Theory and its Applications*, NOLTA-96, Kochi Japan, Oct. 7–9, 1996, pp. 229–232.

Cichocki, A., W. Kasprzak, and S. Amari, 1996d, "Adaptive approach to blind source separation with cancellation of additive and convolutional noise," ICSP'96, *3rd Int. Conf. on Signal Processing, Proceedings*, IEEE Press/PHEI Beijing, vol. I, Sept. 1996, pp. 412–415.

Cichocki, A., W. Kasprzak, and S. Amari, 1996e, "Neural network approach to blind separation and enhancement of images," *Signal Processing VIII. Theories and Applications*, Vol. I, (Trieste, Italy: EURASIP/LINT Publ.) Sept. 1996, pp. 579–582.

Cichocki, A., S. Amari, and R. Thawonmas, Oct. 1996f, "Blind signal extraction using self adaptive non-linear Hebbian learning rule," *NOLTA'96*, pp. 377–380.

Cichocki, A., S. Amari, and J. Cao, Sept. 1997a, "Neural network models for blind separation of time delayed and convolved signals," *Japanese IEICE Transaction on Fundamentals*, vol. E-82-A, no. 9, pp. 1057–1062.

Cichocki, A., R. E. Bogner, L. Moszczynski, and K. Pope, 1997b, "Modified Herault-Jutten algorithms for blind separation of sources," *Digital Signal Processing*, vol. 7, no. 2, pp. 80–93.

Cichocki, A., R. Thawonmas, and S. Amari, Jan. 1997c, "Sequential blind signal extraction in order specified by stochastic properties," *Electronics Lett.*, vol. 33, no. 1, pp. 64–65.

Cichocki, A., B. Orsier, A. D. Back, and S. Amari, 1997d, "On-line adaptive algorithms in nonstationary environments using a modified conjugate gradient approach," in *Proc. IEEE Workshop on Neural Networks for Signal Processing*, N.Y. Press, Sept. 1997, pp. 316–325.

Cichocki, A., I. Sabala, S. Choi, B. Orsier, and R. Szupiluk, 1997e, "Self adaptive independent component analysis for sub-Gaussian and super-Gaussian mixtures with unknown number of sources and additive noise," *Proceedings of 1997 Int. Symp. on Nonlinear Theory and its Applications (NOLTA-97)*, vol. 2, Hawaii, U.S.A., Dec. 1997, pp. 731–734.

Cichocki, A., J. Cao, and I. Sabala, 1997f, "On-line adaptive algorithms for blind equalization of multi-channel systems," *Progress in Connectionist-Based Information Systems*, eds. N. Kasabov, Springer, ICONIP-97, New Zealand, Nov. 1997, pp. 649–652.

Cichocki, A., I. Sabala, and S. Amari, 1998a, "Intelligent neural networks for blind signal separation with unknown number of sources," *Proc. of Conf. Engineering of Intelligent Systems*, ESI-98, Tenerife, Feb. 11–13, 1998, pp. 148–154.

Cichocki, A., S. Douglas, S. Amari, and P. Mierzejewski, 1998b, "Independent component analysis for noisy data," *Proc. of Int. Workshop on Independence and Artificial Neural Networks*, Tenerife, Feb. 9–10, 1998, pp. 52–58.

Cichocki, A., S. Douglas, and S. Amari, 1998c, "Robust techniques for independent component analysis (ICA) with noisy data," *Neurocomputing*, vol. 22, pp. 113–129.

Cichocki, A., L. Zhang, and S. Amari, 1998d, "Semi-blind and state-space approaches to nonlinear dynamic independent component analysis," *Proc. of the 1998 Int. Symposium on Nonlinear Theory and Its Applications, NOLTA'98*, Crans-Montana, Switzerland, Sept. 1998, pp. 291–294.

Cichocki, A., J. Cao, S. Amari, N. Murata, T. Taketa, and H. Endo, 1998e, Enhancement and blind identification of magnetoencephalographic signals using independent component analysis, in *Proc. 11th Int. Conf. on Biomagentism*, BIOMAG-98, Sendai, Japan (Aug. 1998) in *Recent Advances in Biomagnetism*, Tohoku University Press, Seudai, Japan 1999 T. Yashimoto et al. eds., pp. 169–172.

Comon, P., 1994, "Independent component analysis: a new concept?" *Signal Processing*, vol. 36, pp. 287–314.

Comon, P., July 1996, "Contrast functions for blind deconvolution," *IEEE Signal Processing Lett.*, vol. SPL-3, no. 7, pp. 209–211.

Delfosse, N., and P. Loubaton, 1995, "Adaptive blind separation of independent sources: A deflation approach," *Signal Processing*, vol. 45, pp. 59–83.

Deville, Y., 1997, "Analysis of the convergence properties of a self-normalized source separation neural network," *Proc. of IEEE Workshop on Signal Processing Advances in Wireless Communications*, SPAWC-97, Paris, April 1997, pp. 69–72.

Ding, Z., C. R. Johnson, Jr., and R. A. Kennedy, 1994, "Global convergence issues with linear blind adaptive equalizers," in *Blind Deconvolution*, S. Haykin, ed. (Englewood Cliffs, NJ: Prentice-Hall), pp. 60–120.

Douglas, S. C., and A. Cichocki, Nov. 1997, "Neural networks for blind decorrelation of signals," *IEEE Trans. Signal Processing*, vol. 45, no. 11, pp. 2829–2842.

Douglas, S. C., and S. Haykin, 1997, "On the Relationship Between Blind Deconvolution and Blind Source Separation," *31st Asilomar Conf. on Signals, Systems, and Computers*, Pacific Grove, CA, Nov. 1997.

Douglas, S. C., A. Cichocki, and S. Amari, Dec. 1996, "Fast-convergence filtered regressor algorithms for blind equalization," *Electronics Lett.*, vol. 32, no. 23, pp. 2114–2115.

Douglas, S. C., A. Cichocki, and S. Amari, 1997a, "Quasi-Newton filtered-error and filtered-regressor algorithms for adaptive equalization and deconvolution," presented at *IEEE Workshop Signal Processing Adv. Wireless Comm.*, Pairs, April 1997, pp. 109–112.

Douglas, S. C., A. Cichocki, and S. Amari, 1997b, "Multichannel blind separation and deconvolution of sources with arbitrary distributions," in *Neural Networks for Signal Processing, Proc. of the 1997 IEEE Workshop (NNSP-97)*, Sept. 1997, (New York: IEEE Press) pp. 436–445.

Douglas, S. C., A. Cichocki, and S. Amari, July 9, 1998, "Bias removal for blind source separation with noisy measurements," *Electronics Lett.*, vol. 34, no. 14, pp. 1379–1380.

Edeleman, A., T. Arias, and S. T. Smith, in press, "The geometry of algorithms with orthogonality constraints" SIAM 1999.

Girolami, M., and C. Fyfe, Jan. 1997, "Temporal model of linear anti-Hebbian learning," *Neural Processing Lett.*, vol. 4, no. 3, pp. 1–10.

Girolami, M., and C. Fyfe, 1997, Extraction of independent signal sources using a deflationary exploratory projection pursuit network with lateral inhibition. In *IEE Proceedings on Vision on, Image and Signal Processing Journal*, vol. 14, pp. 299–306.

Girolami, M., A. Cichocki, and S. Amari, 1998, A common neural network model for unsupervised exploratory data analysis and independent component analysis, *IEEE Trans. Neural Networks*, vol. 9, no. 6, pp. 1495–1501.

Gorokhov, A., P. Loubaton, and E. Moulines, 1996, "Second order blind equalization in multiple input multiple output FIR systems: A weighted least squares approach," *Proc. IEEE Int. Conf. Acoust., Speech, Signal Processing*, Atlanta, GA, vol. 5, May 1996, pp. 2415–2418.

Gurelli, M. I., and C. L. Nikias, Jan. 1995, "EVAM: An eigenvector-based algorithm for multichannel bind deconvolution of input colored signals," *IEEE Trans. Signal Processing*, vol. SP-43, no. 1, pp. 134–149.

Haykin, S., 1992, *Blind equalization formulated as a self organized learning process, ASILOMAR*, 1992, Pacific Groves, pp. 346–350.

Haykin, S., ed., 1994, *Blind Deconvolution* (Englewood Cliffs, NJ: Prentice-Hall).

Haykin, S., 1996, *Adaptive Filters Theory* (Englewood Cliffs, NJ: Prentice-Hall).

Haykin, S., 1999, *Neural Networks*, 2nd ed. (Englewood Cliffs, NJ: Prentice-Hall).

Hua, Y., March 1996, "Fast maximum likelihood for blind identification of multiple FIR channels," *IEEE Trans. Signal Processing*, vol. SP-44, no. 3, pp. 661–672.

Hyvärinen, A., 1998, "Independent component analysis in the presence of noise: A maximum likelihood approach," *Proc. of Int. Workshop on Independence and Artificial Neural Networks*, ed. C. Fyfe, February 9–10, 1998, Tenerife, Spain, pp. 32–38.

Hyvarinen, A., and E. Oja, 1997, "One-unit learning rules for independent component analysis," in *Advances in Neural Information Processing System 9* (NIPS*96), (Cambridge, MA: MIT Press) pp. 480–486.

Jahn, O., A. Cichocki, A. Ioannides, and S. Amari, 1998, Identification and elimination of artifacts from MEG Signals using efficient Independent Components Analysis, in *Proc. of 11th Int. Conference on Biomagentism* BIOMAG-98, Sendai, Japan (Aug. 1998) in *Recent Advances in Biomagnetism*, Tohoku University Press, Seudai, Japan 1999, T. Yashimoto et al. eds., pp. 224–226.

Inouye, Y., 1997, "Criteria for blind deconvolution of multichannel linear time-invariant systems of non-minimum phase," in eds. T. Katayama and S. Sugimoto, *Statistical Methods in Control and Signal Processing*, (New York: Dekker) pp. 375–397.

Inouye, Y., and K. Hirano, June 1997, "Cumulant-based blind identification of linear multi-input multi-output systems driven by colored inputs," *IEEE Trans. on Signal Processing*, vol. 45, no. 6, pp. 1543–1552.

Inouye, Y., and T. Sato, 1997, "On-line algorithms for blind deconvolution of multichannel linear time-invariant systems," *Proc. IEEE Signal Processing Workshop on Higher-Order Statistics*, pp. 204–208.

Johnson, Jr., C. R., Jan. 1991, "Admissibility in blind adaptive channel equalization," *IEEE Control Systems Mag.*, vol. 11, pp. 3–15.

Jutten, C., and J. Herault, 1991, "Blind separation of sources, Part I: An adaptive algorithm based on neuromimetic architecture," *Signal Processing*, vol. 24, pp. 1–20.

Jutten, C., H. L. Nguyen Thi, E. Dijkstra, E. Vittoz, J. Caelen, 1991, "Blind separation of sources: an algorithm for separation of convolutive mixtures," *Proc. of Int. Workshop on Higher Order Statistics*, Chamrousse, France, July 1991, pp. 275–278.

Karhunen, J., 1996, "Neural approaches to independent component analysis and source separation," *Proc. European Symposium on Artificial Neural Networks*, ESANN-96 Bruges, Belgium, April 1996, pp. 249–266.

Karhunen, J., and J. Joutsensalo, 1993, "Learning of robust principal component subspace," *IJCNN'93. Proc. of 1993 Int. Joint Conf. on Neural Networks*, vol. 3, Nagoya, Japan, 1993, pp. 2409–2412.

Karhunen, J., A. Cichocki, W. Kasprzak, and P. Pajunen, April 1997, "On neural blind separation with noise suppression and redundancy reduction," *Int. J. Neural Systems*, vol. 8, no. 2, pp. 219–237.

Karhunen, J., L. Wang, and R. Vigario, 1995, "Nonlinear PCA type approaches for source separation and independent component analysis," *Proc. ICNN'95*, Perth, Australia, Nov.–Dec. 1995, pp. 995–1000.

Kasprzak, W., A. Cichocki, and S. Amari, Nov. 1997, "Blind source separation with convolutive noise cancellation," *J. Neural Computing and Applications*, vol. 3, no. 6, pp. 127–141.

Lambert, R. H., 1996, Multi-channel blind deconvolution: FIR matrix algebra and separation of multi-path mixtures, Ph.D. Thesis, Elec. Eng., Univ. of Southern California.

Lee, T. W., A. J. Bell, and R. Lambert, 1997, "Blind separation of delayed and convolved sources," *Advances in Neural Information Processing Systems 9*, (Cambridge MA: MIT Press) pp. 758–764.

Lewicki, M. S., and T. Sejnowski, 1998, "Learning nonlinear overcomplete representation for efficient coding," NIPS, vol. 10, pp. 815–821.

Liu, R.-W., and G. Dong, May 1997, "Fundamental theorem for multiple-channel blind equalization," *IEEE Trans. Circuits and Systems*, vol. 44, pp. 472–473.

Ljung, L., and T. Söderström, 1983, *Theory and Practice of Recursive Identification* (Cambridge, MA: MIT Press).

Macchi, O., and E. Moreau, 1997, Self-adaptive source separation part I: convergence analysis of a direct linear network controlled by Herault-Jutten algorithm." *IEEE Trans. Signal Processing*, vol. 45, pp. 918–926.

MacKay, D. J. C., 1996, "Maximum likelihood and covariant algorithms for independent component analysis," internal report, Cavendish Laboratory, Cambridge University.

Makeig, S., A. Bell, T.-P. Jung, and T. J. Sejnowski, 1996, "Independent component analysis in electro-encephalographic data," in M. Mozer et al., eds., *Advances in Neural Information Processing Systems 8* (Cambridge, MA: MIT Press) pp. 145–151.

Maluche, Z., and O. Macchi, 1997, "Adaptive separation of unknown number of sources," *Proc. IEEE Workshop on Higher Order Statistics*, Los Alamitos, CA, pp. 295–299.

Matsuoka, K., M. Ohya, and M. Kawamoto, 1995, "A neural net for blind separation of nonstationary signals," *Neural Networks*, vol. 8, no. 3, pp. 411–419.

Moreau, E., and O. Macchi, Jan. 1996, "High order contrasts for self-adaptive source separation," *Int. J. Adaptive Control and Signal Processing*, vol. 10, no. 1, pp. 19–46.

Moulines, E., J. F. Cardoso, and E. Gassiat, 1997, "Maximum likelihood for blind separation and deconvolution of noisy signals using mixture models," *Proc. Int. Conf. Acoust. Speech, Signal Processing, ICASSP-97*, Munich, Germany, 1997, pp. 3617–3620.

Murata, N., K. Mueller, A. Ziehle, and S.-I. Amari, 1997, "Adaptive on-line learning in changing environments," *Advances in Neural Information Processing Systems 9*, (Cambridge MA: MIT Press) pp. 312–318.

Nadal, J. P., and N. Parga, 1997, "ICA: Conditions on cumulants and information theoretic approach," *Proc. European Symposium on Artificial Neural Networks*, ESANN'97, Bruges, April 16–17 1997, pp. 285–290.

Nguyen Thi, H. L., and C. Jutten, 1995, "Blind source separation for convolved mixtures," *Signal Processing*, vol. 45, no. 2, pp. 209–229.

Oja, E., 1997, "The nonlinear PCA learning rule in independent component analysis" *Neurocomputing*, vol. 17, pp. 25–45.

Oja, E., and J. Karhunen, Nov. 1995, "Signal separation by nonlinear Hebbian learning," in M. Palaniswami et al., eds., *Computational Intelligence—A Dynamic System Perspective*, (New York: IEEE Press) pp. 83–97.

Oja, E., J. Karhunen, L. Wang, and R. Vigario, 1995, "Principal and independent components in neural networks—recent developments," *Proc. 7th Italian Workshop on Neural Networks (WIRN-95)*, Vietri sul Mare, Italy, May 1995, pp. 20–26.

Papadias, C. B., and A. Paulraj, June 1997, "A constant modulus algorithm for multi-user signal separation in presence of delay spread using antenna arrays," *IEEE Signal Processing Lett.*, vol. 4, no. 6, pp. 178–181.

Pearlmutter, B. A., and L. C. Parra, 1996, "A context-sensitive generalization of ICA," *Proc. of Int. Conf. on Neural Information Processing, ICONIP-96*, Hong Kong, Sept. 24–27, 1996, pp. 151–156.

Pham, D. T., P. Garrat, and C. Jutten, 1992, "Separation of a mixture of independent sources through a maximum likelihood approach," *Proc. EUSIPCO-92*, pp. 771–774.

Platt, J. C., and F. Faggin, 1992, "Networks for the separation of sources that are superimposed and delayed," *NIPS* 1992, vol. 4, pp. 730–737.

Sabala, I., Sept. 1998, Multichannel blind deconvolution/separation of statistically independent signals, Ph.D. thesis, Warsaw University of Technology (supervisor A. Cichocki).

Sabala, I., A. Cichocki, and S. Amari, 1998, "Relationships between instantaneous blind source separation and multichannel blind deconvolution," *Proc. Int. Joint Conference on Neural Networks*, Alaska USA, 1998, pp. 148–152.

Sato, Y., June 1975, "A method of self-recovering equalization for multilevel amplitude-modulation systems," *IEEE Trans. Communic.*, vol. 23, pp. 679–682.

Sato, Y., May 1994, "Blind equalization and blind sequence estimation," *IEICE Trans. Communications*, vol. E77-B, no. 5, pp. 545–556.

Shalvi, O., and E. Weinstein, March 1990, "New criteria for blind deconvolution of non-minimum phase systems," *IEEE Trans. Inform. Theory*, vol. 36, pp. 312–321.

Shalvi, O., and E. Weinstein, March 1993, "Super-exponential methods for blind deconvolution," *IEEE Trans. on Signal Processing*, vol. 39, pp. 504–519.

Shalvi, O., and E. Weinstein 1994, "Universal method for blind deconvolution," *Blind Deconvolution*, S. Haykin, ed., (Englewood Cliffs, NJ: Prentice-Hall) pp. 121–180.

Shynk J. J., and R. P. Gooch, March 1996, "The constant modulus array for co-channel signal copy and direction finding," *IEEE Trans. on Signal Processing*, vol. 44, no. 3, pp. 652–660.

Silva, F. M., and L. B. Almeida, 1991, "A distributed decorrelation algorithm," in E. Gelenba, ed., *Neural Networks, Advances and Applications*, (Amsterdam: North-Holland) pp. 145–163.

Slock, D. T. M., 1994, "Blind joint equalization of multiple synchronous mobile users oversampling and/or multiple antennas," *Proc. 28th Asilomar Conf.*, Pacific Grove, CA, Oct. 31–Nov. 2, 1994, pp. 1154–1158.

Thawonmas, R., A. Cichocki, and S. Amari, Sept. 1998, "A cascade neural network for blind signal extraction without spurious equilibria," *IEICE Trans. Fundamentals*, vol. E81-A, no. 9, pp. 1833–1846.

Torkkola, K., 1996a, "Blind separation of delayed sources based on information maximization," *Proc. IEEE ICASSP*, Atlanta, GA, vol. 4, 1996, pp. 3509–3513.

Torkkola, K., 1996b, "Blind separation of convolved sources based on information maximization," *Proc. IEEE Workshop on Neural Networks and Signal Processing*, Kyoto, Japan, Sept. 1996, pp. 423–432.

Tong, L., Y. Inouye, and R.-W. Liu, 1993, "Waveform preserving blind estimation of multiple independent sources," *IEEE Trans. on Signal Processing*, vol. 41, pp. 2461–2470.

Tong, L., R.-W. Liu, V. C. Soon, and J. F. Huang, 1991, "Indeterminacy and identifiability of blind identification," *IEEE Trans. Circuits and Systems*, pp. 409–509.

Tong, L., G. Xu, and T. Kailath, March 1994, "Blind identification and equalization based on second-order statistics: A time domain approach," *IEEE Trans. Inform. Theory*, vol. IT-40, no. 2, pp. 340–349.

Treichler, J. R., and M. G. Larimore, April 1985, "New processing techniques based on constant modulus adaptive algorithms," *IEEE Trans. on Acoustic, Speech and Signal Processing*, vol. ASSP-33, no. 2, pp. 420–431.

Tugnait, J. K., 1996, "Blind equalization and channel estimation for multiple-input multiple-output communications systems," *Proc. IEEE Int. Conf. Acoust., Speech, Signal Processing*, Atlanta, GA, May 1996, vol. 5, pp. 2443–2446.

Van der Veen, A.-J., S. Talvar, and A. Paulraj, Jan. 1997, "A subspace approach to blind space-time signal processing for wireless communication systems," *IEEE Trans. on Signal Processing*, vol. 45, no. 1, pp. 173–190.

Van Gerven, S., March 1996, Adaptive noise cancellation and signal separation with applications to speech enhancement, Ph.D. thesis, Catholic University Leuven, Leuven, Belgium.

Weinstein, E., M. Feder, and A. Oppenheim, 1993, "Multichannel signal separation by decorrelation," *IEEE Trans. on Speech and Audio Processing*, vol. 1, no. 4, pp. 405–413.

Widrow, B., and E. Walach, 1996, *Adaptive Inverse Control* (Upper Saddle River, NJ: Prentice-Hall PTR).

Xu, G., H. Liu, L. Tong, and T. Kailath, Dec. 1995, "A least-squares approach to blind channel identification," *IEEE Trans. Signal Processing*, vol. SP-43, no. 12, pp. 2982–2983.

Xu, L., C.-C. Cheung, J. Ruan, and S. Amari, 1997a, "Nonlinearity and separation capability: Further justification for the ICA algorithm with a learned mixture of parametric densities," *Proc. European Symposium on Artificial Neural Networks, ESANN'97*, Bruges, April 16–17, 1997, pp. 291–296.

Xu, L., C. C. Chung, H. H. Yang, and S. Amari, 1997b, "Independent component analysis by the information–theoretic approach with mixture of parametric densities," *Proc. of IEEE Int. Conf. on Neural Networks*, June 9–12, 1997, Houston, TX, USA, vol. III, pp. 1821–1826.

Yang, H. H., and S. Amari, 1997, "Adaptive on-line learning algorithms for blind separation—maximum entropy and minimum mutual information," *Neural Computation*, no. 9, pp. 1457–1482.

Yang, H. H., S. Amari, and A. Cichocki, 1998, "Information-theoretic approach to blind separation of sources in non-linear mixture," *Signal Processing*, vol. 64, no. 3, pp. 291–300.

Yellin, D., and E. Weinstein, Jan. 1996, "Multichannel signal separation: methods and analysis," *IEEE Trans. Signal Processing*, vol. 44, pp. 106–118.

Zhang, L., and A. Cichocki, 1998a, "Blind deconvolution and equalization using state-space models," in *Neural Networks for Signal Processing VIII*, (New York: IEEE Press) pp. 123–131.

Zhang, L., and A. Cichocki, 1998b, "Blind separation deconvolution of sources using canonical stable space model," *Proc. of the 1998 Int. Symposium on Nonlinear Theory and its Applications, NOLTA-98*, 1998, pp. 927–930.

Zhang, L., and A. Cichocki, 1998c, "Blind separation of filtered source using state-space approach," *Neural Information Processing Systems, NIPS'98*, Denver, USA, Dec. 1998 in vol. 11, Cambridge MA: The MIT Press 1999, pp. 648–654.

4

ENTROPIC CONTRASTS FOR SOURCE SEPARATION: GEOMETRY AND STABILITY

Jean-François Cardoso

ABSTRACT

Algorithms for the blind separation of sources can be derived from the optimization of *contrast functions*. This chapter provides a view of contrast functions in the framework of information geometry; identifiability, consistency, and stability of the algorithms based on the optimization of contrast functions are discussed, as is the notion of relative gradient.

4.1 INTRODUCTION

The problem of *source separation* is to recover signals from their linear mixtures when neither the mixture is known nor the signals observed. For an introduction to this problem, the reader is referred to previous chapters.

The purpose of this chapter is to provide a geometric view of the *contrast functions*, which are objective functions (or criteria) for source separation (contrast functions are briefly reviewed in Section 4.2). Simple ideas from *information geometry* (reviewed in Section 4.3) provide a unifying view of various contrast functions and of their relationships (see Section 4.4). This chapter discusses both "regular" and "orthogonal" contrast functions. The latter enforce the decorrelation of the recovered signals.

Unsupervised Adaptive Filtering, Volume I, Edited by Simon Haykin.
ISBN 0-471-29412-8 © 2000 John Wiley & Sons, Inc.

This chapter also addresses the issue of *stability* of adaptive algorithms for source separations. Most on-line techniques rely on nonlinear functions to adaptively learn a separating (demixing) matrix. These functions must be adapted to the probability distribution of the sources to guarantee the stability of the stationary points. Simple stability conditions are spelled out in Section 4.6.

4.1.1   The Basic Source Separation Model

The simplest source separation model considers noise-free observations of as many linear mixtures as independent source signals. Thus the basic source-separation model is that of an $n \times 1$ vector $\mathbf{x}$ with the following structure:

$$\mathbf{x} = A\mathbf{s} \tag{4.1}$$

where A is an invertible $n \times n$ square matrix and the random $n \times 1$ vector $\mathbf{s}$ is distributed with probability density function (pdf) $r(\mathbf{s})$ with respect to the Lebesgue measure. Nothing is assumed about the "mixing matrix" A apart from its invertibility. Therefore, the model density of $\mathbf{x}$ relates to the density of $\mathbf{s}$ as

$$p(\mathbf{x}; A, r) = \frac{1}{|\det A|} r(A^{-1}\mathbf{x}) \tag{4.2}$$

In this chapter, the "source signals" are assumed to be zero-mean. The key ingredient in the source-separation model is that vector $\mathbf{s}$ has independent components: its pdf $r(\mathbf{s})$ factors as

$$r(\mathbf{s}) = \prod_{i=1,n} r_i(s_i) \tag{4.3}$$

where $r_1, \ldots, r_n$ are the densities for components or "sources" $1, \ldots, n$.

It must be emphasized that Eqs. (4.1) and (4.2) only define a statistical *model*. The observations may or may not obey this model. Actually, in many applications, such a simple model cannot be expected to hold. It is therefore important to understand what happens when the source-separation model is fit to data vectors that are not a linear superposition of independent components. In the following, we will not necessarily assume that "the model holds." When we do so, it will be specified. Of course, a good part of the theoretical analysis can only be carried out "under the model": this is the case of most of the material presented in Section 4.6 on stability analysis.

Given realizations of vector $\mathbf{x}$, source separation can be formulated as an estimation problem:

Problem 1. Compute on-line or off-line an estimate $\hat{A}$ of the mixing matrix A.

In this case, "natural estimates" of the sources can be obtained as $\hat{s}(t) = \hat{A}^{-1}\mathbf{x}(t)$ for $t = 1, \ldots, T$. It is, however, more common and maybe more illuminating to state the problem as:

Problem 2. Find an $n \times n$ "separating matrix" B such that its output $\mathbf{y} = B\mathbf{x}$ is an estimate of the source signals.

This point of view is of course essentially equivalent to the "estimation" point of view, but is more natural when it comes to finding adaptive separation algorithms that update a separating matrix $B(t)$ into $B(t+1)$ upon reception of each new realization $\mathbf{x}(t+1)$.

Both points of view stem from the source-separation model in that this model assumes the existence of a vector $\mathbf{s}$ of sources and of a mixing matrix. In this case, it makes sense to define the *global system* C associated with any matrix B:

$$C = BA$$

so that the relation between a particular value $\mathbf{s}$ of the source and its estimated value $\mathbf{y} = B\mathbf{x}$ via matrix B is

$$\mathbf{y} = B\mathbf{x} = BA\mathbf{s} = C\mathbf{s}$$

and the aim of source separation can be restated as making the global system C equal to the identity matrix I. Then, the notions of consistency and accuracy make sense: we must study the departure of C from the identity.

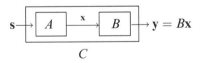

A third point of view is that of independent-component analysis (ICA) is more related in spirit to data analysis and in particular to principal-component analysis (PCA): matrix B is seen as a change of basis and one tries to solve the problem.

Problem 3. Find the basis B such that the entries of the transformed vector $\mathbf{y} = B\mathbf{x}$ are as independent as possible.

Here, the notion of maximal independence is not well defined yet, but it certainly relates to source separation, which is also supposed to recover a vector of independent components. Actually, the three problems just listed are intimately related: we shall see in particular that maximum-likelihood estimation of the parameters of the source-separation model is equivalent in a strong sense to the search for a vector **y** with the most independent entries.

4.1.2 Parameter of Interest and Nuisance Parameters

If the model holds and if we are interested in recovering the sources, that is, the vector **s**, it is sufficient to determine the value of the mixing matrix A. Clearly, this is the "parameter of interest" in source separation. However, in order to specify the probability distribution of **x**, it is necessary to also specify the distribution of **s** that amounts to specifying the source densities $r_1, \ldots, r_n$. These are "nuisance parameters" in statistical parlance and in the sense that we are not primarily interested in estimating these densities (especially since density estimation is a difficult task). Thus the unknown parameter is

$$\theta = (A, r) \tag{4.4}$$

whose first component (a matrix) is a parameter of interest and second component (a set of n univariate probability distributions) is a nuisance parameter. In generic inference problems, it is necessary to estimate nuisance parameters (whether or not one is interested in them) in order to consistently estimate the parameters of interest. Fortunately, this is not the case in source separation: we will see that (and explain why) it is possible not to estimate the source densities—using in their place (possibly roughly) "guessed" densities—so that source densities are not such a nuisance, after all! The price to pay for using more or less wild guesses about source densities is in terms of performance, not necessarily in terms of consistency. It is one of the objectives of this chapter to provide qualitative (geometric) and quantitative explanations (with some geometric flavor too) about these facts.

The geometric view proposed in this chapter is that of geometry in spaces of probability distributions (Amari 1985). It relies on the Kullback-Leibler (KL) divergence as a "pseudodistance" between probability distributions and on the related differential entropy.

For a probability distribution on $\mathbb{R}^n$ with probability density function f with respect to the Lebesgue measure, the *(differential) entropy* is denoted H_f. If g is the pdf of another distribution, the *KL divergence* between densities f and g is denoted $D_{f\|g}$. These are defined

$$H_f \overset{\text{def}}{=} -\int_{\mathbb{R}^n} \log(f(\mathbf{y})) f(\mathbf{y}) \, d\mathbf{y} \qquad D_{f\|g} \overset{\text{def}}{=} \int_{\mathbb{R}^n} \log\left(\frac{f(\mathbf{y})}{g(\mathbf{y})}\right) f(\mathbf{y}) \, d\mathbf{y} \tag{4.5}$$

whenever the integrals exist. A crucial property of the Kullback divergence is that $D_{f\|g} \geq 0$ with equality if and only if f and g agree f-almost everywhere.

4.1.3   Important Notation Convention

Since Kullback divergences are ubiquitous in this chapter, an "efficient" notation is needed. The following notation convention will be used: if $\mathbf{x}$ is a random variable, then $[\mathbf{x}]$ denotes the *probability distribution* of $\mathbf{x}$. For example, with this notation convention, if f (resp. g) is the pdf of a random vector $\mathbf{y}$ (resp. a random vector $\mathbf{z}$), we also denote

$$H_f = H[\mathbf{y}] \qquad D_{f\|g} = D[\mathbf{y}\|\mathbf{z}]$$

Some properties are more easily expressed with this notation. For instance, the fact that the Kullback divergence is invariant under invertible linear transforms reads:

$$D[\mathbf{y}\|\mathbf{z}] = D[T\mathbf{y}\|T\mathbf{z}] \tag{4.6}$$

for any square invertible matrix T and any random vectors $\mathbf{y}$ and $\mathbf{z}$. Other notations and symbols are summarized in Table 4.1.

Table 4.1   Notations and Symbols

Notations	
$\mathbf{u}, \mathbf{v}, \ldots$	Vectors in boldface, lowercase
u_i	The ith entry of vector $\mathbf{u}$
$.^\dagger$	Transpose of a matrix or vector
$\langle M_1 \| M_2 \rangle$	Euclidean scalar product between matrices M_1 and M_2
Symbols	
E	The expectation operator
$\hat{\cdot}$	A sample estimate
$\mathcal{N}(\mathbf{m}, R)$	Gaussian distribution with mean $\mathbf{m}$ and covariance matrix R
H_f	Shannon differential entropy of distribution f
$H[\mathbf{u}]$	Shannon differential entropy of the distribution of vector $\mathbf{u}$
$D_{f\|g}$	KL-divergence from distribution f to distribution g
$D[\mathbf{u}\|\mathbf{v}]$	KL-divergence from the distribution of vector $\mathbf{u}$ to the distribution of vector $\mathbf{v}$
$D\{R_1 \| R_2\}$	KL-divergence from $\mathcal{N}(0, R_1)$ to $\mathcal{N}(0, R_2)$
$\phi[\mathbf{u}]$	A function of the *distribution* of vector $\mathbf{u}$
$\mathbf{u}^P$	A vector with independent entries, each with the same distribution as the corresponding entry of $\mathbf{u}$
$\mathbf{u}^N$	A vector with a Gaussian distribution and the same covariance matrix as $\mathbf{u}$
$J[\mathbf{u}]$	Mutual information between the entries of vector $\mathbf{u}$: $J[\mathbf{u}] = D[\mathbf{u}\|\mathbf{u}^P]$
$N[\mathbf{u}]$	Negentropy of vector $\mathbf{u}$: $N[\mathbf{u}] = D[\mathbf{u}\|\mathbf{u}^N]$
$\mathcal{G}$	Gaussian manifold
$\mathcal{P}$	Product manifold

4.2 CONTRAST FUNCTIONS FOR SOURCE SEPARATION

Loosely put, a *contrast function* for source separation is a function of the *probability distribution* of **y**, with the following key property: a contrast function should reach its minimum value when B is a solution of the source-separation problem. In other words, using a contrast function turns the source-separation problem into an optimization problem.

4.2.1 Contrast Functions and Contrast Processes

The notion of contrast function is not specific to source separation; it is actually quite a general principle of statistical inference, which is briefly recalled below starting from an (over)simple example and leaving out technical details.

A Simple Example   Contrast functions and contrast processes can be introduced by a very simple example that is (apparently) not related to source separation. Let $X_T = (\mathbf{x}_1, \ldots, \mathbf{x}_T)$ denote a sample of T observations and consider the problem of estimating the mean of their distribution. One may consider

$$\frac{1}{T} \sum_{t=1}^{T} \|\mathbf{x}_t - \mathbf{m}\|^2 \tag{4.7}$$

as a measure of concentration of the data around a value **m** and obtain an estimate of the mean as the minimizer of this quantity:

$$\hat{\mathbf{m}} = \arg \min_{\mathbf{m}} \left(\frac{1}{T} \sum_{t=1}^{T} \|\mathbf{x}_t - \mathbf{m}\|^2 \right) \tag{4.8}$$

If the data points are independent realizations of a random variable, a sample average like Eq. (4.7) converges for large T to the ensemble average

$$\lim_{T \to \infty} \frac{1}{T} \sum_{t=1}^{T} \|\mathbf{x}_t - \mathbf{m}\|^2 = E\|\mathbf{x} - \mathbf{m}\|^2 \tag{4.9}$$

which depends on the distribution P_X of **x** and on the parameter **m**. Assume that **x** is distributed according to $\mathbf{x}_t = \mathbf{m}_* + \mathbf{v}_t$ where $\{\mathbf{v}_t\}$ is a zero-mean noise sequence. Then

$$E\|\mathbf{x} - \mathbf{m}\|^2 = E\|\mathbf{m}_* - \mathbf{m} + \mathbf{v}_t\|^2 = \|\mathbf{m}_* - \mathbf{m}\|^2 + \sigma^2 \tag{4.10}$$

where $\sigma^2 = \lim \frac{1}{T} \sum_t E\|\mathbf{v}_t^2\|$. This expression shows that the ensemble average

$E\|\mathbf{x} - \mathbf{m}\|^2$ is minimized in $\mathbf{m}$ by taking $\mathbf{m} = \mathbf{m}_*$. It is thus expected that the minimizer $\hat{\mathbf{m}}$ of the sample average, Eq. (4.7), is "close to the true value $\mathbf{m}_*$" when the sample average is close enough to the ensemble average, Eq. (4.9).

General Contrasts The notions of contrast function and contrast process offer a broad generalization of these ideas. Again, let $X_T = (\mathbf{x}_1, \ldots, \mathbf{x}_T)$ denote a sample of T observations and consider modeling it as a random quantity whose probability distribution depends on a parameter $\theta \in \Theta$. The minimum-contrast approach to estimating the parameter θ follows this course:

- A contrast process $U(X_T; \theta)$ is a real-valued function of the data sample and of the parameter θ of the parametric model. It is the analog of the sample mean-square deviation, Eq. (4.7), and is the basis for inferring θ: an estimate $\hat{\theta}$ is computed as the minimizer of the contrast process:

$$\hat{\theta} = \arg \min_{\theta} U(X_T; \theta) \qquad (4.11)$$

In the preceding example, $\hat{\theta}$ would just be the mean of the sample.

- If the data points in X_T are realizations (not necessarily independent) of a random variable with probability distribution P_X, we consider the case where the contrast process $U(X_T; \theta)$ converges to a deterministic limit denoted $\phi(P_X; \theta)$:

$$\lim_{T \to \infty} U(X_T; \theta) \stackrel{\text{def}}{=} \phi(P_X; \theta) \qquad (4.12)$$

which depends for each θ on the distribution P_X of $\mathbf{x}$. This limit $\phi(P_X; \theta)$ is called the *contrast function* associated with the contrast process. It is the analog of the ensemble mean-square error, Eq. (4.9).

- Let $p(\mathbf{x}; \theta)$ be a parametric model of distributions of $\mathbf{x}$. When the model holds, that is, when the parametric model $p(\cdot; \theta)$ does contain the actual distribution P_X of the data, then there is a "true value" θ_* of the parameter, that is, $P_X(\cdot) = p(\cdot; \theta_*)$. In such a case, the value of the contrast function $\phi(P_X; \theta)$ is a function of θ_* and of θ, which is denoted by

$$\mathscr{C}(\theta_*; \theta) \stackrel{\text{def}}{=} \phi(p(\cdot; \theta_*), \theta) \qquad (4.13)$$

It is the analog of $\|\mathbf{m}_* - \mathbf{m}\|^2 + \sigma^2$ in Eq. (4.10). A contrast function is said to be valid if it satisfies

$$\theta_1 \neq \theta_2 \Rightarrow \mathscr{C}(\theta_1; \theta_2) > \mathscr{C}(\theta_1; \theta_1) \qquad (4.14)$$

Thus, in the framework of minimum-contrast estimation, the *design* of the estimators is nicely split into two steps of a different nature:

1. Find a contrast function that is minimum at the true value of the parameter, that is, such that property (4.14) holds.
2. Find a contrast process converging to the contrast function as in Eq. (4.12).

In this two-step approach, the questions of identifiability are addressed by studying the contrast functions, while stochastic effects (finite sample size, convergence) are addressed by studying the associated contrast process.

4.2.2   The Likelihood Contrast: Kullback Matching

This section recalls how the classic method of maximum-likelihood (ML) parameter estimation is associated with a specific contrast function.

Consider a simple but quite general setting (not necessarily describing source separation) where T observations $\mathbf{x}(1), \ldots, \mathbf{x}(T)$ of a vector $\mathbf{x}$ are available and are to be modeled as independent realizations of a random vector distributed according to some parametric density $p(\mathbf{x}; \theta)$. In the ML approach, the adjustable vector of parameters θ is estimated by maximizing the probability of the data set. This probability, seen as a function of θ, is called the *likelihood* and is given in our setting by

$$l_T(X_T; \theta) = \prod_{t=1}^{T} p(\mathbf{x}(t); \theta) \tag{4.15}$$

Maximizing $l_T(\theta)$ is equivalent to minimizing minus the normalized log-likelihood, which we take as our contrast process:

$$U_{\mathrm{ML}}(X_T; \theta) \stackrel{\mathrm{def}}{=} -\frac{1}{T} \log l_T(\theta) = -\frac{1}{T} \sum_{t=1}^{T} \log p(\mathbf{x}(t); \theta) \tag{4.16}$$

When T grows, the sample average converges (under mild assumptions) to the ensemble average, that is, the expectation of $-\log p(\mathbf{x}; \theta)$:

$$U_{\mathrm{ML}}(X_T; \theta) \rightarrow -\mathrm{E} \log p(\mathbf{x}; \theta) \tag{4.17}$$

Denoting P_X the distribution of the observed vector $\mathbf{x}$, we have

$$-\mathrm{E} \log p(\mathbf{x}; \theta) = -\int P_X(\mathbf{x}) \log p(\mathbf{x}; \theta) \, d\mathbf{x} \tag{4.18}$$

$$= \int P_X(\mathbf{x}) \log \frac{P_X(\mathbf{x})}{p(\mathbf{x}; \theta)} \, d\mathbf{x} - \int P_X(\mathbf{x}) \log P_X(\mathbf{x}) \, d\mathbf{x} \tag{4.19}$$

$$= D_{P_X(\cdot) \| p(\cdot; \theta)} + H_{P_X} \tag{4.20}$$

where the Kullback divergence D and the Shannon entropy H are defined at Eqs. (4.5). Note that the entropy H_{P_X} does not depend on the parameter θ, so that the likelihood contrast process, Eq. (4.16), provides, up to this constant entropic term, a sample-based version of the divergence $D_{P_X(\cdot)\|p(\cdot\,;\theta)}$ between the distribution $P_X(\cdot)$ of the observations and the family of distributions $p(\cdot\,;\theta)$ in the model.

Therefore ML estimation can be understood as matching in the Kullback divergence a parametric family of distributions to the distribution of the observations. Up to an additive constant, the contrast function associated with the ML principle is

$$\phi_{\mathrm{ML}}(P_X,\theta) = D_{P_X\|p(\cdot\,;\theta)} \tag{4.21}$$

In practice, of course, one does not need to estimate the data distribution and then evaluate the Kullback divergence to the model: rather, the log-likelihood, Eq. (4.16), is the natural sample estimate (up to a constant entropic term) of this divergence.

4.2.3 The Likelihood Contrast for Source Separation

When applied to source separation, the general notion of contrast function takes a specific form, which stems from the fact the model of interest is a transformation model: the observations are modeled as resulting from a linear but otherwise unknown invertible transform of a vector of unobserved sources.

It will be convenient to denote by $\underline{s}$ a vector of independent components distributed with the product density $r(\cdot)$ of Eq. (4.3). With this convention, the nuisance parameter r is equivalently represented by $\underline{s}$ and the source separation model for $\mathbf{x}$ is compactly written as $\mathbf{x} = A\underline{s}$. According to the previous section subsumed by Eq. (4.21), ML estimation corresponds to minimizing the Kullback divergence between the distribution of observed values of $\mathbf{x}$ and the (model) distribution of $A\underline{s}$ by adjusting matrix A and the density r of $\underline{s}$. Thus the ML contrast function is the Kullback mismatch:

$$\phi_{\mathrm{ML}} = D[\mathbf{x}\|A\underline{s}] \tag{4.22}$$

and we must stress again that this conclusion is reached whether or not there exist some matrix A and some vector $\underline{s}$ of independent components such that the distribution of $\mathbf{x}$ is the distribution of $A\underline{s}$ or, in other words, whether or not the model of independent components holds. Now, for any invertible matrix A, we have

$$D[\mathbf{x}\|A\underline{s}] = D[A^{-1}\mathbf{x}\|A^{-1}A\underline{s}] = D[A^{-1}\mathbf{x}\|\underline{s}] \tag{4.23}$$

where the first equality stems from the invariance of the Kullback divergence under invertible transforms [Eq. (4.6)].

A key feature of the last expression in Eq. (4.23) is that it depends on A and on the distribution of the observed vector $\mathbf{x}$ *only* via $\mathbf{y} = A^{-1}\mathbf{x}$ (we will see that this fact is related to the notion of equivariance). Thus the ML principle is seen to correspond to minimizing the Kullback mismatch:

$$\phi_{\mathrm{ML}}[\mathbf{y}] = D[\mathbf{y}\|\underline{\mathbf{s}}] \tag{4.24}$$

between the distribution of $\mathbf{y} = A^{-1}\mathbf{x}$ and the distribution of a hypothetical source vector $\underline{\mathbf{s}}$. In principle, the minimization of this contrast should be conducted by optimizing over $\theta = (A, r)$, that is, over the hypothetical mixing matrix A (thus changing the distribution of $\mathbf{y}$) and on the hypothetical distribution of the sources (thus changing the distribution of $\underline{\mathbf{s}}$). Section 4.4 explains why minimizing $\phi_{\mathrm{ML}}[\mathbf{y}] = D[\mathbf{y}\|\underline{\mathbf{s}}]$ over the distribution of $\underline{\mathbf{s}}$, that is, over the nuisance parameter, is not necessary to obtain consistent estimates of A.

Orthogonal Contrasts   A standard approach to source separation is to require that the recovered source signals should be at least uncorrelated, that is, that they should verify

$$\mathrm{E}y_j y_j = 0 \qquad \text{for} \quad 1 \le i \ne j \le n \tag{4.25}$$

One may also decide to fix the scale of each recovered source signal by requiring it to have unit variance:

$$\mathrm{E}y_i^2 = 1 \qquad \text{for} \quad 1 \le i \le n \tag{4.26}$$

The scale-fixing conditions, Eq. (4.26), are consistent with the decorrelation conditions, Eq. (4.25), in the sense that both are based on second-order moments. These two sets of constraints can be summarized by the so-called "whiteness constraint":

$$\mathrm{E}\mathbf{y}\mathbf{y}^\dagger = I \tag{4.27}$$

Any vector $\mathbf{y}$ such that $\mathrm{E}\mathbf{y}\mathbf{y}^\dagger = I$ is said to be *spatially white*. One can consider the design of contrast functions $\phi[\mathbf{y}]$, which enforce decorrelation and are thus to be optimized under the whiteness constraint, Eq. (4.27). Such contrasts are called *orthogonal contrasts* in the following.

The term "orthogonal contrast" stems from the following idea. Let W be a "whitening matrix," that is, an $n \times n$ matrix such that $W\mathbf{x}$ is a white vector. Such a matrix can be easily found as a matrix square root of the inverse of the covariance matrix of $\mathbf{x}$, that is, $W = (\mathrm{E}\mathbf{x}\mathbf{x}^\dagger)^{-1/2}$. Denote by $\mathbf{w} = W\mathbf{x}$ a whitened (or "sphered") version of $\mathbf{x}$. Such a linear transform preserves the model of a linear mixture of independent components, and thus any source separation can be applied to $\mathbf{w}$ as well as to $\mathbf{x}$. If the output $\mathbf{y}$ is computed from w by a linear transform U, then the condition that $\mathbf{y} = U\mathbf{w}$ should also be white requires that matrix U should be an orthonormal matrix, that is, $UU^{\dagger} = I$, because it relates two white vectors.

The orthogonal approach may yield simpler algorithms because whitening is a simple procedure and it reduces the search for a separating matrix to the space of orthonormal matrices. It is also expected to provide some robustness because it amounts to using as much second-order information as possible in the construction of a separating matrix. There is, however, a price to pay in terms of achievable performance in the orthogonal approach [see Cardoso (1994) for a lower bound to performance induced by the whiteness constraint].

The minimization of an orthogonal contrast function $\phi[\mathbf{y}]$ is not necessarily implemented by prewhitening the data $\mathbf{x}$ into $\mathbf{w}$ and then optimizing over orthonormal matrices U a contrast process for $\mathbf{y} = U\mathbf{w}$: an alternative equivalent approach is to solve the constrained optimization problem:

$$\text{Minimize} \quad \phi[\mathbf{y}] \quad \text{under the constraint} \quad \mathbf{E}\mathbf{y}\mathbf{y}^{\dagger} = I$$

It is important that the orthogonal approach admits a formulation uniquely in terms of the distribution of $\mathbf{y}$ because it makes it clear that, even if implemented in a two-step approach, the solution of the optimization problem still is uniquely determined in terms of the distribution of $\mathbf{y}$.

Equivariance It is Comon (1994) who introduced the notion of contrast function in the field of source separation[1] as a function of the distribution of the output vector $\mathbf{y}$. Comon proposed using the mutual information between the entries of $\mathbf{y}$ (see below) as an objective, and gave an approximation to it defined in terms of the cumulants of $\mathbf{y}$. We just saw that the ML principle also leads to an objective function that depends only on A and $\mathbf{x}$ through the distribution of $\mathbf{y} = A^{-1}\mathbf{x}$ and requires matching the distribution of $\mathbf{y}$ to a target distribution. Finally, the whiteness constraint also is a constraint on the distribution of $\mathbf{y}$. As we shall see in Section 4.5, the fact that the objective can be specified in terms of the distribution of $\mathbf{y}$ is linked to the fact that the source-separation model is a *transformation model*, and to the fact that it is possible to design source-separation algorithms with a separating performance that does not depend on the value of the mixing matrix.

Besides maximizing a measure of independence between the entries of $\mathbf{y}$ or matching the distribution of $\mathbf{y}$ to a target with independent entries, other ideas for designing contrast, are based on the intuition that mixing independent variables tends to Gaussianize their distributions so that unmixing could be obtained by making the distribution of $\mathbf{y}$ as non-Gaussian as possible or by minimizing its entropy under appropriate constraints. Section 4.4 explores

---

[1] Comon's definition is unnecessarily restrictive, however, since it explicitly requires that a contrast function be invariant by permutation, excluding in particular the likelihood contrast.

the relationships between these ideas within the framework of information geometry.

Before looking at information geometry, we outline the cumulant-based version of contrast processes and contrast functions for source separation. For a random vector $\mathbf{z}$, denote $\mathrm{Cum}[\mathbf{z}]$, a $p \times 1$ vector made by stacking p cross-cumulants of $\mathbf{z}$ (the particular choice of these p cross-cumulants is not discussed here), and let $\widehat{\mathrm{Cum}}[\mathbf{z}]$ denote a sample estimate of these cumulants. Let us then consider the following contrast process:

$$U(X_T; \theta) = U(X_T, A) = \|\widehat{\mathrm{Cum}}[A^{-1}\mathbf{x}]\|^2 \qquad (4.28)$$

If $\widehat{\mathrm{Cum}}[\cdot]$ is a consistent estimate, then for any A the contrast process, Eq. (4.28), converges as

$$U(X_T, A) = \|\widehat{\mathrm{Cum}}[A^{-1}\mathbf{x}]\|^2 \rightarrow \|\mathrm{Cum}[A^{-1}\mathbf{x}]\|^2 = \|\mathrm{Cum}[\mathbf{y}]\|^2 \qquad (4.29)$$

This is a valid contrast function for source separation if it has property (4.14). When the model holds, we have $\mathbf{x} = A_1\mathbf{s}$ for some "true" mixing matrix A_1 and a vector $\mathbf{s}$ of independent components. In this case, for any invertible matrices A_1 and A_2:

$$U(X_T, A_2) \rightarrow \|\mathrm{Cum}[A_2^{-1}\mathbf{x}]\|^2 = \|\mathrm{Cum}[A_2^{-1}A_1\mathbf{s}]\|^2 \overset{\text{def}}{=} \mathscr{C}(A_1, A_2) \qquad (4.30)$$

That $\mathscr{C}(A_1, A_2)$ actually has property (4.14) is therefore equivalent to the property that $\|\mathrm{Cum}[C\mathbf{s}]\|^2 > \|\mathrm{Cum}[\mathbf{s}]\|^2$ for any matrix C different from the identity matrix I. Since $\mathbf{s}$ has independent entries by assumption, all its cross-cumulants vanish: $\mathrm{Cum}[\mathbf{s}] = 0$, so that property (4.14) also reads

$$C \neq I \Rightarrow \|\mathrm{Cum}[C\mathbf{s}]\|^2 > 0 \qquad (4.31)$$

We note that property (4.31) cannot hold as is because if C is any nonmixing matrix (that is, if it has one and only one nonzero entry in each row and each column), then $C\mathbf{s}$ has independent entries, so that $\mathrm{Cum}[C\mathbf{s}] = 0$, not only for $B = A^{-1}$, but also for all separating matrices. Of course, as far as source separation is concerned, it is sufficient to find a nonmixing global system. This type of contrast functions has been investigated, for instance, in Cardoso (1998), Cardoso and Souloumiac (1993), Comon (1994), Moreau and Macchi (1996), and many others.

4.3 SOME INFORMATION GEOMETRY OF RANDOM VECTORS

In preparation for discussing the geometry of entropic contrast functions, we recall in this section some basic notions of information geometry. An authoritative reference is Amari's book (Amari 1985).

4.3.1 Exponential Families of Probability Distributions

In statistical inference, *exponential families* of probability distributions have a special status and many "nice" properties. Consider $p_0(\mathbf{x})$ and $p_1(\mathbf{x})$, two probability density functions. The *exponential segment* between p_0 and p_1 is the one-parameter family of distributions with density

$$p_\beta(\mathbf{x}) = g(\beta) p_0^{1-\beta}(\mathbf{x}) p_1^{\beta}(\mathbf{x}) \qquad (4.32)$$

for $\beta \in [0, 1]$ where $g(\beta)$ is a normalizing constant ensuring that $\int p_\beta(\mathbf{x})\, d\mathbf{x} = 1$. When β varies from 0 to 1, the density p_β varies from p_0 to p_1 along the exponential segment. In some sense (see, for instance, the Pythagorean theorem below), such a segment is a straight line, and exponential families can be thought as being flat. A family of distributions is said to be an *exponential family* if it contains the exponential segment between any two of its members.

Two families of zero-mean probability distributions on $\mathbb{R}^n$ play an important role in source separation: the set of all Gaussian distributions and the set of all distributions with independent entries. These two sets form exponential families as seen next.

Gaussian Manifold The *Gaussian manifold* is defined as the set of all n-variate zero-mean normal distributions with positive covariance matrix. This family of distributions is denoted by $\mathscr{G}$. Any distribution in $\mathscr{G}$ is uniquely characterized by its covariance matrix R and has density $|2\pi R|^{-1/2} \exp - \frac{1}{2} \mathbf{x}^\dagger R^{-1} \mathbf{x}$. If p_0 and p_1 are two distributions in $\mathscr{G}$ with covariance matrices R_0 and R_1, respectively, it can be readily verified that their exponential mixture p_β, as defined by Eq. (4.32), is also a zero-mean normal distribution with covariance matrix R_β, given by

$$R_\beta^{-1} = (1 - \beta) R_0^{-1} + \beta R_1^{-1} \qquad (4.33)$$

Therefore $\mathscr{G}$ is a (finite-dimensional) exponential family.

Product Manifold Consider the set $\mathscr{P}$ of all probability distributions of vectors of $\mathbb{R}^n$ having independent entries, that is, those factoring as Eq. (4.3). If $r_0(\mathbf{s})$ and $r_1(\mathbf{s})$ are two distributions of independent components, they factor as in Eq. (4.3), so that $r_\beta(\mathbf{s})$ defined in Eq. (4.32) also exhibits the same factorization and thus belongs to $\mathscr{P}$. Therefore, $\mathscr{P}$ is an (infinite-dimensional) exponential family. Since every element in $\mathscr{P}$ is the product of its marginals, the family $\mathscr{P}$ is called the *product manifold*.

4.3.2 A Pythagorean Property in Distribution Space

In order to quantify how far a given probability distribution is from a given family of distributions (like the Gaussian manifold or the product manifold, for instance), we shall use the Kullback divergence. When the given family of

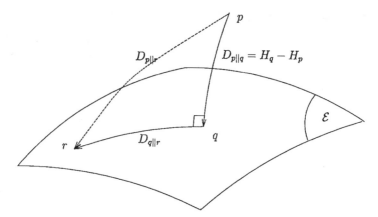

Figure 4.1 A Pythagorean theorem in distribution space: $D_{p\|r} = D_{p\|q} + D_{q\|r}$.

distributions is exponential, additional insight is gained from the following proposition, which is illustrated in Fig. 4.1.

Proposition 1. Let p be a probability distribution and let $\mathcal{E}$ be an exponential family of distributions. Let q be the distribution in $\mathcal{E}$ that is the closest to p in the sense that $\forall r \in \mathcal{E}$, $D_{p\|r} \geq D_{p\|q}$. Then,

$$\forall r \in \mathcal{E}, \qquad D_{p\|r} = D_{p\|q} + D_{q\|r}$$

and, in addition, the divergence from p to $\mathcal{E}$ is also given by

$$D_{p\|q} = H_q - H_p$$

This important property of exponential families, informally adapted from Amari (1985), is the analog in the space of distributions of the familiar Pythagorean theorem in Euclidean spaces. In the latter case, the projection Q of P on a linear subspace is such that $\|\mathbf{PR}\|^2 = \|\mathbf{PQ}\|^2 + \|\mathbf{QR}\|^2$ for any other point R of the subspace. In the space of distributions, the Kullback divergence plays the role of a squared Euclidean distance, the exponential family plays the role of a linear subspace, and q is the projection of p onto this subspace. The divergence between the distribution and its projection onto an exponential family defines the divergence between the probability distribution and the given exponential family.

4.3.3 Mutual Information and Negentropy

The Pythagorean theorem applied to the Gaussian manifold and to the product manifold gives rise to quantities that play a key role in the theory of source separation.

Mutual Information and the Product Manifold The degree of dependence between the components of a vector **y** may be quantified by the divergence from its distribution to the product manifold. Let us denote by $J[\mathbf{y}]$ the Kullback divergence from $[\mathbf{y}]$ to the closest point in $\mathscr{P}$:

$$J[\mathbf{y}] \stackrel{\text{def}}{=} \min_{\mathbf{z} \in \mathscr{P}} D[\mathbf{y}\|\mathbf{z}]. \tag{4.34}$$

It is well known [e.g., Cover and Thomas (1991)] and easily checked (for instance, using the Lagrange multiplier method) that this minimum is reached for the distribution of a vector that is denoted by $\mathbf{y}^P$ in the following and is characterized by the fact that its entries are independent and have the same distributions as the corresponding entries of **y**. This is

$$\mathbf{y}^P \sim \prod_{i=1}^{n} r_i(y_i)$$

where r_i is the distribution of the ith entry of **y**. With this notation, we have

$$J[\mathbf{y}] = D[\mathbf{y}\|\mathbf{y}^P]$$

This quantity is often called *mutual information*, since it appears as a natural extension of the mutual information between two random variables as classically defined in information theory.

Let **z** be a vector of independent components. Then, by Proposition 1, $D[\mathbf{y}\|\mathbf{z}] = D[\mathbf{y}\|\mathbf{y}^P] + D[\mathbf{y}^P\|\mathbf{z}]$. The first term in this equation is the mutual information of **y** and the second term can further be decomposed as

$$D[\mathbf{y}^P\|\mathbf{z}] = \sum_{i=1}^{n} D[y_i\|z_i]$$

because only product densities appear in $D[\mathbf{y}^P\|\mathbf{z}]$. Therefore, we have:

Proposition 2. The Kullback divergence from a vector **y** to a vector **z** of independent components verifies

$$D[\mathbf{y}\|\mathbf{z}] = J[\mathbf{y}] + \sum_{i=1}^{n} D[y_i\|z_i]$$

This proposition is illustrated in Fig. 4.2, and states that the divergence from any distribution $[\mathbf{y}]$ on R^n to a given distribution of independent components is made up of two terms: one measures how far **y** is from having independent components, and the second is the sum of the divergences between the marginal

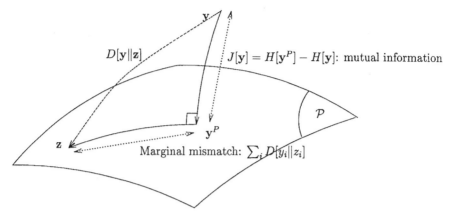

Figure 4.2   Decomposition of the divergence to a vector of independent components.

distributions of **y** and **z**. The first term says something about the joint distribution of **y**, while the second term ignores its joint distribution and only measures one-dimensional divergences between marginals.

Negentropy and the Gaussian Manifold   Like the definition of mutual information as a divergence from the product manifold, the non-Gaussianity of a zero-mean random vector **y** with covariance matrix R_y can be defined as the divergence from its distribution to the *Gaussian manifold*. Thus, the non-Gaussianity of a zero-mean vector **y** is defined as

$$N[\mathbf{y}] \overset{\text{def}}{=} \min_{\mathbf{z} \in \mathscr{G}} D[\mathbf{y}\|\mathbf{z}] \tag{4.35}$$

Not surprisingly, this minimum is found to be reached at the zero-mean Gaussian distribution with the same covariance matrix as **y**. Denoting by $\mathbf{y}^N$ such a Gaussian vector, we thus have, by Proposition 1,

$$N[\mathbf{y}] = D[\mathbf{y}\|\mathbf{y}^N] = H[\mathbf{y}^N] - H[\mathbf{y}] \tag{4.36}$$

Because the Gaussian manifold $\mathscr{G}$ is an exponential family, Proposition 1 again yields a decomposition of the divergence from the distribution of **y** to the distribution of any zero-mean Gaussian vector **n**. This is

$$D[\mathbf{y}\|\mathbf{n}] = D[\mathbf{y}\|\mathbf{n}^N] + D[\mathbf{y}^N\|\mathbf{n}] \tag{4.37}$$

Since the divergence $D[\mathbf{y}^N\|\mathbf{n}]$ is from one Gaussian vector to another, it can be expressed only in terms of the covariance matrices of **n** and of $\mathbf{y}^N$, the latter being equal to R_y by definition of $\mathbf{y}^N$. Incidentally, there is a close-form expression for $D[\mathbf{n}_1\|\mathbf{n}_2]$ when $\mathbf{n}_1$ and $\mathbf{n}_2$ are two zero-mean Gaussian vectors with

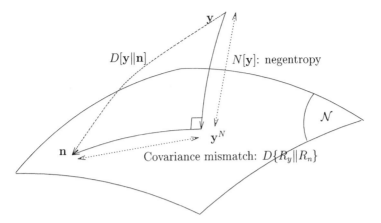

Figure 4.3 Decomposition of the divergence to a Gaussian vector.

covariance matrices R_1 and R_2:

$$D[\mathbf{n}_1\|\mathbf{n}_2] = D\{R_1\|R_2\} \stackrel{\text{def}}{=} \tfrac{1}{2}\,\text{trace}(R_2^{-1}R_1 - I) - \tfrac{1}{2}\log|\det R_2^{-1}R_1| \qquad (4.38)$$

In summary, we can state the following proposition as a corollary of Proposition 1:

Proposition 3. The divergence from the distribution of vector $\mathbf{y}$ with covariance matrix R_y to the distribution of a Gaussian vector $\mathbf{n}$ with covariance matrix R_n can be decomposed as

$$D[\mathbf{y}\|\mathbf{n}] = N[\mathbf{y}] + D\{R_y\|R_n\}$$

The first term measures the non-Gaussianity of $\mathbf{y}$ and does not depend on $\mathbf{n}$; the second term, which involves only second-order moments of $\mathbf{y}$ and $\mathbf{n}$, measures the deviation between their respective covariance matrices, as illustrated by Fig. 4.3.

It is interesting to note that the non-Gaussianity of a random vector as measured by its negentropy is invariant under invertible transforms. Indeed, if T is an invertible matrix, we have

$$N[T\mathbf{y}] = \min_{\mathbf{z} \in \mathscr{G}} D[T\mathbf{y}\|\mathbf{z}] \qquad (4.39)$$

$$= \min_{\mathbf{z} \in \mathscr{G}} D[\mathbf{y}\|T^{-1}\mathbf{z}] \qquad (4.40)$$

$$= \min_{\mathbf{z} \in \mathscr{G}} D[\mathbf{y}\|\mathbf{z}] \qquad (4.41)$$

$$= N[\mathbf{y}] \qquad (4.42)$$

where the first equality is by definition, the second one stems from the invariance, Eq. (4.6), of D under invertible transform, and the last one stems from the global invariance of $\mathscr{G}$ under any invertible linear transform.

4.4 INFORMATION GEOMETRY OF CONTRAST FUNCTIONS

This section offers a global geometric picture of the classic entropic contrast functions.

4.4.1 System Manifold

For a random $n \times 1$ vector $\mathbf{x}$, not necessarily made of independent components, the *system manifold* is defined as the set $\mathscr{S}$ of probability distributions that includes the distribution of $\mathbf{x}$ as well as the distributions of all vectors obtained by linear invertible transformations of $\mathbf{x}$. By this definition, the system manifold is completely determined by the distribution of $\mathbf{x}$, and it is the set of the distributions of $\mathbf{y} = B\mathbf{x}$ for all invertible matrices B, hence the name "system manifold," since this manifold is parameterized by the separating system B.

The definition of a contrast function ϕ for source separation amounts to assigning a numerical value $\phi[\mathbf{y}]$ to each point of the system manifold and the learning of a separating matrix B may be seen as a trajectory on the system manifold trying to descend the contrast function.

The learning of a separating matrix is illustrated at Fig. 4.4. The system manifold is represented as a 2-dimensional manifold (its actual dimensionality is larger, indeed: see below). The point labeled $\mathbf{x}$ is the distribution of the observations. As a separating system is updated, staring from $B_0 = I$, the distri-

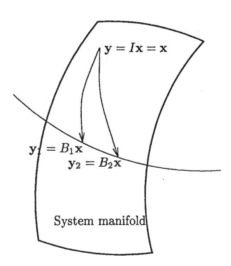

Figure 4.4 Two learning trajectories for two hypothetical source-separation algorithms.

bution of $y = Bx$ evolves along some trajectory on the system manifold. In this example, we show two trajectories starting from the same initial point (the distribution of x). Each trajectory corresponds to a particular learning rule. They end up at two different points that may be considered equivalent to y_1 and y_2 if they differ only by scale factors.[2]

The dimension of the system manifold *a priori* is equal to n^2, since this is the number of independent entries in an $n \times n$ matrix. However, the dimension is smaller if some linear transforms of x does not change its distribution. In particular, if x is a Gaussian vector with covariance matrix R, then vector Bx has the same distribution as vector x for $B = R^{1/2}UR^{-1/2}$, where $R^{1/2}$ denotes the symmetric square root of R and U is *any* orthonormal matrix (that is, $UU^\dagger = I$). In this case, the system manifold has only dimension $d = n^2 - (n(n-1)/2) = n(n+1)/2$ because there are $n(n-1)/2$ degrees of freedom in specifying an $n \times n$ orthonormal matrix. If the distribution of x "tends to a Gaussian distribution," the system manifold wraps around itself, loosing $n(n-1)/2$ of its dimensions. More generally, if x contains q Gaussian components, the dimension of the system manifold is easily seen to reduce to $d = n^2 - q(q-1)/2$. In the following, we assume that when the model holds, vector x does not contain more than one Gaussian component.

When the model holds, that is, when $x = As$ for some mixing matrix A and a vector s of independent components, the system manifold, which contains the distributions of all linear invertible transforms of x, contains in particular the distribution of $s = A^{-1}x$. It also contains the distribution of Λs for any diagonal matrix Λ with nonzero diagonal entries. These distributions form an n-dimensional submanifold of the system manifold that we call the *diagonal manifold*. Any algorithm driving the distribution of $y = Bx$ to the diagonal manifold provides a satisfactory solution to the source-separation problem, since all the distributions in the diagonal manifold are nothing but rescaled versions of the sources.

4.4.2 Blind Identifiability

From the statistical point of view, the first question to be asked about the source-separation model is that of identifiability: For a given invertible matrix A and a vector s of independent components, does there exist another invertible matrix $\tilde{A}$ such that $x = As$ has the same distribution as $\tilde{A}\tilde{s}$ for a source vector $\tilde{s}$ of independent components? If this is the case, matrix A cannot be uniquely identified and the source vector cannot be uniquely restored. Of course, this situation occurs with $\tilde{A} = A\Lambda$ and $\tilde{s} = \Lambda^{-1}s$ for any diagonal matrix Λ, so the actual identifiability question is rather the existence of a matrix $\tilde{A}$ such that the related $\tilde{s}$ is not a rescaled version of s (possibly with permuted entries).

The identifiability issue can be rephrased by noting that if $As = \tilde{A}\tilde{s}$, then

---

[2] This figure is purely illustrative: the two trajectories have been drawn arbitrarily and do not correspond to any actual source separation algorithm.

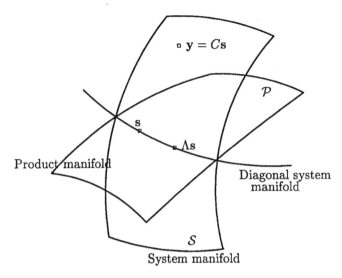

Figure 4.5 Identifiability.

$\tilde{\mathbf{s}} = \tilde{A}^{-1} A\mathbf{s} = C\mathbf{s}$ with $C = \tilde{A}^{-1} A$. Thus, one can equivalently ask: What are the invertible matrices C such that $C\mathbf{s}$ has independent components? Therefore, the question of identifiability can finally be rephrased in geometric terms: What is the intersection between the system manifold (the set of distributions of $C\mathbf{s}$ for all invertible C) and the product manifold (the set of all distributions with independent components)? Blind identifiability is granted if the intersection occurs only for matrices C which are nonmixing, that is, if it has one and only one nonzero entry in each row and each column.

Assume that the model holds: $\mathbf{x} = A\mathbf{s}$ for some invertible matrix A and a vector $\mathbf{s}$ of independent components. Then $\mathbf{y} = B\mathbf{x} = BA\mathbf{s} = C\mathbf{s}$ where matrix $C = BA$ denotes the "global system" obtained by concatenating the mixing and the unmixing matrix. The distribution of $\mathbf{s}$ belongs to the system manifold and to the product manifold. A pictorial representation is given by Fig. 4.5, which shows both these manifolds and their intersection, which includes the distribution of $\mathbf{s}$ and a rescaled version $\Lambda\mathbf{s}$ Which other distributions belong to the intersection? Any such distribution would be the distribution of an invertible linear transform of $\mathbf{s}$ with independent components. We know [from Comon (1994), for instance] that if $\mathbf{s}$ has at most on Gaussian entry, then the entries of $\mathbf{y} = C\mathbf{s}$ remain independent if and only if C is nonmixing. This is the very property that makes blind source separation possible. Its geometric expression is that the product manifold and the system manifold intersect only at those points parameterized by nonmixing matrices.

Characterizing the indeterminations of the source separation problem with unknown source distributions is equivalent to characterizing the intersection between the product manifold and the system manifold. The fact that source separation is possible even when the source distributions are unknown is equivalent to the statement that the intersection reduces to nonmixing matrices.

4.4.3 Fitting the Mixture

This subsection illustrates the behavior of the likelihood contrast $\phi_{ML}[y]$ defined by Eq. (4.24) when it is optimized with respect to the mixing matrix only, that is, when we content ourselves with a fixed assumption about the distribution of the sources. We consider the case where x actually is a mixture of independent components, that is, $x = As$ and the entries $s_1, \ldots, s_n$ are independent with unknown densities $q_1, \ldots, q_n$. The target source densities (the "model densities" or "hypothesized densities") are denoted by $r_1, \ldots, r_n$ and, as in Eq. (4.24), we denote by $\underline{s}$ a random vector with the product density, Eq. (4.3). As seen earlier, the likelihood contrast associated to this model is

$$\phi_{ML}[y] = D[y\|\underline{s}]$$

Even though it is not necessarily assumed in the following that the source model is correct (that is, s is distributed as $\underline{s}$), we start with this most favorable case where the model densities $r_1, \ldots, r_n$ are equal to the true densities $q_1, \ldots, q_n$ of the sources so that

$$\phi_{ML}[y] = D[y\|\underline{s}] = D[y\|s] = D[Cs\|s]$$

The Kullback divergence then actually reaches its lowest possible value (that is, zero) at $C = I$ for which $D[Cs\|s] = D[s\|s] = 0$. If the sources have identical distributions, this minimum is also reached when C is any permutation matrix; if the sources are symmetrically distributed, the sign of each row of C can also be changed. All these cases correspond to nonmixing matrices C, which are "far away" from $C = I$: they can be ignored in the vicinity of $C = I$. Figure 4.6 complements Fig. 4.5 by showing the trajectories of a hypothetical source-separation algorithm for three different starting points, all ending at the true distribution of the source vector, at which point we have $D[y\|s] = D[s\|s] = 0$.

More interesting is the case when the model distribution $\underline{s}$ (the target distribution) used in defining ϕ_{ML} is incorrect. Figure 4.7 now shows the assumed distribution at a point labeled $\underline{s}$ lying on the product manifold, since $\underline{s}$ is a vector of independent components, but not lying on the system manifold, since a wrong hypothesis for the source distributions is assumed here.

Consider the distributions of the diagonal submanifold: these are scaled versions of the true distribution of s, that is, they are the distributions of all vectors Λs where $\Lambda = \text{Diag}(\lambda_1, \ldots, \lambda_n)$. Among the distributions of the diagonal submanifold, the closest to the target distributions in the sense of minimizing the Kullback divergence $D[\Lambda s\|\underline{s}]$ is obtained with the scale factors $\tilde{\lambda}_1, \ldots, \tilde{\lambda}_n$

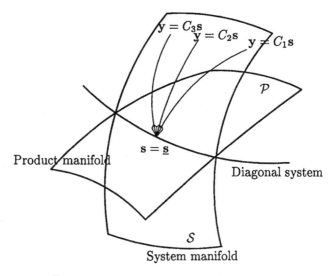

Figure 4.6   Attracting the distribution of **y** to **s**.

solutions of

$$\tilde{\lambda}_i = \arg \min_{\lambda} D[\lambda s_i \| \underline{s}_i] \tag{4.43}$$

because the divergence $D[\Lambda \mathbf{s} \| \underline{\mathbf{s}}]$, being between vectors of independent entries, decomposes as

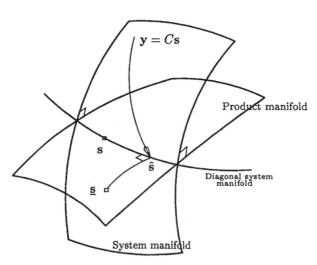

Figure 4.7   Minimizing the divergence from the output to a wrong hypothesized distribution.

$$D[\Lambda \mathbf{s}\|\underline{\mathbf{s}}] = \sum_{i=1}^{n} D[\lambda s_i\|\underline{s}_i] \tag{4.44}$$

Thus, for each source, the scaling factor $\tilde{\lambda}_i$ brings the distribution of $\lambda_i s_i$ as close as possible to the assumed model density r_i. Let us then denote $\tilde{s}_i$ the rescaled sources:

$$\tilde{s}_i = \tilde{\lambda}_i s_i \qquad \tilde{\mathbf{s}} = [\tilde{s}_1, \dots, \tilde{s}_n] = \tilde{\Lambda} \mathbf{s} \tag{4.45}$$

with $\tilde{\Lambda} = \mathrm{Diag}(\tilde{\lambda}_1, \dots, \tilde{\lambda}_n)$. Among all rescaled versions of $\mathbf{s}$, the source vector $\tilde{\mathbf{s}}$ is the closest to the target distribution:

$$D[\tilde{\mathbf{s}}\|\underline{\mathbf{s}}] = \min_{C \text{ diagonal}} D[C\mathbf{s}\|\underline{\mathbf{s}}] \tag{4.46}$$

and, lying on the system manifold, it is also the best distribution that can be obtained from source separation in terms of matching the target distribution. Since source separation is not concerned with the scales of the recovered signals, vector $\tilde{\mathbf{s}}$ is a solution to the source-separation problem.

In practice, there is no means of constraining the separating matrix B to be such that $C = BA$ is diagonal; indeed, it is the objective of source separation to achieve such a result. Therefore, it is necessary to determine if the minimization of $D[C\mathbf{s}\|\underline{\mathbf{s}}]$ *without* constraining C to be a diagonal matrix also occurs for $C = \tilde{\Lambda}$. If this is the case, then $\phi_{\mathrm{ML}}[\mathbf{y}]$ indeed is a valid contrast for source separation for the particular distribution of $\mathbf{s}$ and for the particular target $\underline{\mathbf{s}}$.

Thus the question to be asked is: Is the closest point $\mathbf{y} = C\mathbf{s}$ to $\underline{\mathbf{s}}$ on the system manifold identical to $\tilde{\mathbf{s}} = \tilde{\Lambda}\mathbf{s}$ even when C is not constrained to be diagonal? If this is the case, then

$$\tilde{\Lambda} = \arg \min_{C \text{ diagonal}} D[C\mathbf{s}\|\underline{\mathbf{s}}] = \arg \min_{C \text{ invertible}} D[C\mathbf{s}\|\underline{\mathbf{s}}] = \arg \min_{\mathbf{y}=C\mathbf{s}} \phi_{\mathrm{ML}}[\mathbf{y}] \tag{4.47}$$

so that the minimization of $\phi_{\mathrm{ML}}[\mathbf{y}]$ leads to source separation even though a wrong target distribution is used in the ML contrast.

The geometrical answer to this question is more easily pictured by drawing only a section of Fig. 4.7. This section is sketched at Fig. 4.8 and is done in such a way that it contains the target $\underline{\mathbf{s}}$, the rescaled source vector $\tilde{\mathbf{s}} = \tilde{\Lambda}\mathbf{s}$, and a generic point $\mathbf{y} = C\mathbf{s}$ on the system manifold. The figure suggests that, as desired, the point $\mathbf{y} = C\mathbf{s}$ of the system manifold that is the closest to the target $\underline{\mathbf{s}}$ is indeed $\tilde{\mathbf{s}}$. Figure 4.8 suggests this property, because the sections of two manifolds have been drawn as intersecting at a right angle.

Analogous to Euclidean geometry, consider the right panel of Fig. 4.8 for which

$$\|\mathbf{RP}\|^2 = \|\mathbf{RQ} + \mathbf{QP}\|^2 = \|\mathbf{RQ}\|^2 + 2\mathbf{RQ} \cdot \mathbf{QP} + \|\mathbf{QP}\|^2 \tag{4.48}$$

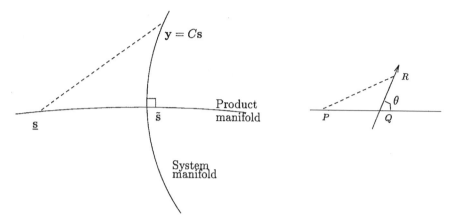

Figure 4.8   Minimizing the distance to a wrong target distribution.

and consider $\|\mathbf{RP}\|^2$ as a function of point $\mathbf{R}$. If vector $\mathbf{QR}$ is constrained to be orthogonal to $\mathbf{QP}$, then $\mathbf{RQ} \cdot \mathbf{QP} = 0$ so that $\|\mathbf{RP}\|$ is minimized for $\mathbf{R} = \mathbf{Q}$. Under such an orthogonal constraint, we have

$$\|\mathbf{RP}\|^2 = \|\mathbf{RQ}\|^2 + \|\mathbf{QP}\|^2 \tag{4.49}$$

and in particular

$$\|\mathbf{RP}\|^2 = \|\mathbf{QP}\|^2 + O(\|\mathbf{RQ}\|^2) \tag{4.50}$$

which is sufficient to show that the function $\mathbf{R} \rightarrow \|\mathbf{PR}\|^2$ is stationary at point $\mathbf{R} = \mathbf{Q}$ when the orthogonal constraint $\mathbf{PQ} \cdot \mathbf{QR} = 0$ is enforced.

Equation (4.49) is the Pythagorean theorem in Euclidean space, and Eq. (4.50) is a weaker version expressing orthogonality. This weaker version has a direct analog in distribution space for the source-separation problem when $\mathbf{y}$ is on the system manifold in the vicinity of $\tilde{\mathbf{s}}$: then $\mathbf{y}$ is related to $\tilde{\mathbf{s}}$ by an infinitesimal linear transform, that is, $\mathbf{y} = (I + \mathcal{E})\tilde{\mathbf{s}}$ where $\mathcal{E}$ is an infinitesimal $n \times n$ matrix. Then

$$\phi_{\mathrm{ML}}[\mathbf{y}] = D[\mathbf{y}\|\underline{\mathbf{s}}] = D[(I + \mathcal{E})\tilde{\mathbf{s}}\|\underline{\mathbf{s}}] = D[\tilde{\mathbf{s}}\|\underline{\mathbf{s}}] + O(\|\mathcal{E}\|^2)$$
$$= \phi_{\mathrm{ML}}[\tilde{\mathbf{s}}] + O(\|\mathcal{E}\|^2) \tag{4.51}$$

The last equalities are proved in Section 4.6 [where an explicit expression is given for the $O(\|\mathcal{E}\|^2)$ term]. It is the analog of Eq. (4.50), and shows that $\phi_{\mathrm{ML}}[\mathbf{y}]$ is stationary over the system manifold at point $\mathbf{y} = \tilde{\mathbf{s}} = \hat{\Lambda}\mathbf{s}$. It does express indirectly (just as Eq. (4.50) does in Euclidean space) the fact that the system manifold and the product manifold intersect at a right angle. The

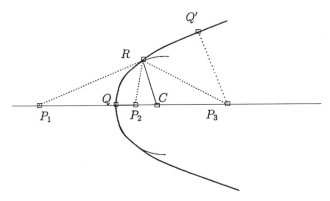

Figure 4.9 Local stability and curvature: point C is located at the center of curvature of the smooth curve at Q. The distance from a generic point R of the smooth curve to P_1 or P_2 is minimum at Q. The distance from R to P_3 shows a local *maximum* at Q and is minimum at some point Q' because P_3 is located beyond the center of curvature.

orthogonality of the system manifold and of the product manifold at $\tilde{s}$ (and actually at any rescaled version of s) can be established more directly [see Amari and Cardoso (1997) for this point and for other consequences of manifold orthogonality].

In summary, even if a wrong target $\underline{s}$ is selected in the source model, the likelihood contrast $\phi_{\mathrm{ML}}[\mathbf{y}] = D[\mathbf{y}\|\underline{s}]$ remains stationary at a point $\tilde{s}$, which is a rescaled version $\tilde{s} = \Lambda s$ of the source vector distribution with the scaling factors in Λ given at Eq. (4.43) as a best Kullback fit. This is a very fortunate circumstance: it means that selecting a wrong target distribution (or, in the jargon of Section 4.1, fixing the nuisance parameter to a wrong value) affects the stationary points of the likelihood contrast only by changing the scale of each source. The geometric origin of this property is the orthogonality between nuisance parameter and the parameter of interest, that is, the fact that the product manifold and the system manifold intersect at a right angle along the diagonal manifold.

4.4.4 Stability

The fact that the likelihood contrast $\phi_{\mathrm{ML}}[\mathbf{y}] = D[\mathbf{y}\|\underline{s}]$ is stationary at a point $\mathbf{y} = \tilde{\Lambda}s$ is not enough to claim that any target distribution $\underline{s}$ could be used in defining a likelihood contrast: it is also necessary to ensure that this stationary point further is a *maximum* of the likelihood and not a minimum or a saddle point; in these latter cases, starting from a neighborhood of $\tilde{s}$ and maximizing the likelihood from there would drive $\mathbf{y}$ away from $\tilde{s}$: this is the so-called *stability problem*.

The stability problem has a simple interpretation in the information geometric framework. In Fig. 4.9 a smooth curve intersects a straight line orthog-

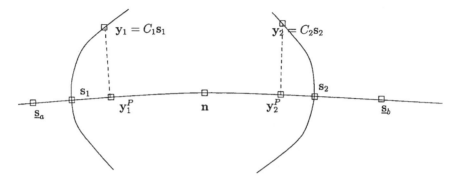

Figure 4.10   Stability: being on the right side of the Gaussian distribution.

onally at point Q. Point C is the center of curvature of the curve at point Q. Points P_1, P_2, and P_3 lie on a straight line and are located with respect to Q and C as indicated in the figure. Let R be a point on the smooth curve and consider the functions $\phi_i(R) \stackrel{\text{def}}{=} \|\mathbf{RP}_i\|$ for $i = 1, 2, 3$. Functions ϕ_1, ϕ_2, and ϕ_3 are stationary at point $R = Q$, because the intersection at Q is orthogonal. However, this stationary point Q is a (local) *minimum* of ϕ_1 and ϕ_2 and a (local) *maximum* of ϕ_3, because P_3 lies beyond the center of curvature: the closest point to P_3 on the smooth curve is not at Q but somewhere else, at a point labeled Q' in the figure.

The information-geometric analog of Fig. 4.9 is given in Fig. 4.10. Before proceeding further, it must be stressed again that the analogy cannot be fully developed because the space of probability distributions is not Euclidean and because the system manifold is multidimensional, so a proper center of curvature cannot be defined as in Fig. 4.9. This said, some idea can still be gained from Fig. 4.10 on which *two* system manifolds have been represented: one generated from a source distribution labeled s_1 with a generic point labeled $\mathbf{y}_1 = C s_1$ and similarly for a second system manifold generated from another source vector labeled s_2. The idea here is to represent with a single picture the sub-Gaussian case and the super-Gaussian case. Super-Gaussian (resp. sub-Gaussian) distributions have heavier (resp. lighter) tails than the Gaussian distribution. Thus vector s_1 could be a vector of independent entries with, say, Laplace distributions (tails decreasing as $e^{-|s|}$, that is, more slowly than the Gaussian), while s_2 could be a vector of uniformly distributed entries. In the figure, a smooth line is drawn that runs through s_1 and s_2 along the product manifold, that is, this line is a submanifold of the product manifold. We assume that this line goes through the Gaussian distribution at a point labeled $\mathbf{n}$. Of course, if any type of segment is drawn from s_1 to s_2, it will not (in general) include the Gaussian distribution: this assumption is made only for the sake of developing the geometric analogy; still the Gaussian distribution is somewhere between the super-Gaussian s_1 and the super-Gaussian s_2, even if it is not necessarily on any kind of straight line joining them.

Figure 4.10 shows the two system manifolds as curving in the direction of the Gaussian $\mathbf{n}$ rather than curving away from it. This is not arbitrary and it can be understood as follows. Consider the projections $\mathbf{y}_1^P$ and $\mathbf{y}_2^P$ of $\mathbf{y}_1 = C_1 \mathbf{s}_1$ and $\mathbf{y}_2 = C_2 \mathbf{s}_2$ onto the product manifold. Vectors $\mathbf{y}_1^P$ and $\mathbf{y}_2^P$ have independent entries by construction. The distributions of these entries, each resulting from a mixture (by matrices C_1 or C_2) of independent components, are expected to be closer to the normal distribution than the entries of the unmixed source vectors $\mathbf{s}_1$ and $\mathbf{s}_2$; hence, the bending of the systems manifolds in the direction of the Gaussian rather than in the opposite direction.

Figure 4.10 also suggests that the normal distribution lies at the center of curvature of the two system manifolds. Some consistency can be given to such a suggestion in the case of orthogonal contrasts (see below); for the time being, we will not take it too seriously. If the source vector to be unmixed is, say, $\mathbf{s} = \mathbf{s}_1$ in Fig. 4.10, then by analogy with Fig. 4.9, an appropriate target distribution is $\mathbf{s}_a$ (or any distribution on the same side of the curvature as $\mathbf{s}_a$) or $\mathbf{s}_1$ (indeed) or any distribution between $\mathbf{s}_1$ and $\mathbf{n}$. The same could be said *mutatis mutandis* on the other side, that is, when $\mathbf{s} = \mathbf{s}_2$. This suggests the following rule of thumb: in order to make $\phi_{\mathrm{ML}}[\mathbf{y}] = D[\mathbf{y}|\underline{\mathbf{s}}] = D[C \mathbf{s}|\underline{\mathbf{s}}]$ behave properly, that is, to show a minimum at a nonmixing point, the target $\underline{\mathbf{s}}$ should be taken "on the same side of the Gaussian as the distribution of the source vector." This is a rephrasing of a similar rule of thumb: for processing mixtures of sub-Gaussian (resp. super-Gaussian) sources, use a sub-Gaussian (resp. super-Gaussian) source model. This statement is just as vague as any other rule of thumb and cannot be much more explicit for at least two reasons. First, there is no clear-cut definition of sub-Gaussianity and super-Gaussianity that would allow *any* probability distribution to be classified in either category. Second, the rule leaves out the case when both heavy-tailed and light-tailed entries are simultaneously present in the source vector.

In any case, one should certainly try to use target distributions that are as close as possible to the true source distributions. However, the actual stability criteria, which are spelled out in Section 4.6, are not directly expressed in terms of the closeness of probability distributions, but rather in terms of the closeness of the associated score functions (see below).

4.4.5 Fitting Everything: The Mutual Information Contrast

In the introductory section, the parameters of the source-separation model were decomposed as $\theta = (A, r)$, with the mixing matrix A being the parameter of interest and the distribution r of the sources being a nuisance parameter. These parameters also appear separately in the contrast function $\phi_{\mathrm{ML}}[\mathbf{y}] = D[\mathbf{y}\|\underline{\mathbf{s}}]$ associated with the ML principle: fitting the parameter of interest affects the distribution of $\mathbf{y} = B\mathbf{x} = A^{-1}\mathbf{x}$, while fitting the nuisance parameter affects, by definition, the distribution of $\underline{\mathbf{s}}$. So far, we have considered only fitting the parameter of interest, using a fixed assumption $\underline{\mathbf{s}}$ for the nuisance parameter. However, there is no reason not to also apply the ML principle to fitting the source distributions. Actually, fitting the full source-separation model in the ML sense corresponds to minimizing $\phi_{ML}[\mathbf{y}] = D[\mathbf{y}\|\underline{\mathbf{s}}]$ by adjusting both $\mathbf{y}$ and $\underline{\mathbf{s}}$. According to Proposition 2, the likelihood contrast can be decomposed as

$$D[\mathbf{y}\|\underline{\mathbf{s}}] = J[\mathbf{y}] + \sum_{i=1}^{n} D[y_i\|\underline{s}_i] \qquad (4.52)$$

in which each individual source target enters in only one term of the preceding sum. Thus, for a fixed $\mathbf{y}$, the ML principle leads, quite naturally, to the minimization of $D[y_i\|\underline{s}_i]$ for $i = 1, \ldots, n$. The minimization of $D[y_i\|\underline{s}_i]$ with respect to the distribution of $\underline{s}_i$ is achieved when this distribution is equal to the distribution of y_i, at which $D[y_i\|\underline{s}_i]$ reaches its smallest value, namely zero. In this case, the ML contrast reduces to the first term on the right-hand side of Eq. (4.52), that is, the mutual information $J[\mathbf{y}]$ between the entries of $\mathbf{y}$. Therefore, when applied to the full model (that is, an optimization over both the nuisance parameter and the parameter of interest), the ML principle is associated with a contrast function that is precisely the mutual information $J[\mathbf{y}]$ as originally proposed by Comon (1994) without reference to the ML principle.[3] By taking the marginal distributions of $\mathbf{y}$ as its target distributions, this contrast function creates, in some sense, its own adaptive targets. This is illustrated in Figure 4.11, which is a 2-dimensional section through the system manifold and the product manifold and shows the decomposition of the likelihood contrast according to Eq. (4.52). In this figure, the target distribution $\underline{\mathbf{s}}$ is (maybe exaggeratedly) wrong.

We just saw that, when $\mathbf{y}$ is kept fixed, minimizing $D[\mathbf{y}\|\underline{\mathbf{s}}]$ with respect to $\underline{\mathbf{s}}$ amounts to estimating the distribution of each source based on the distribution of the corresponding entry of $\mathbf{y}$. This suggests source-separation schemes for minimizing the mutual information $J[\mathbf{y}]$ in which the likelihood contrast would be minimized by alternatively minimizing with respect to the nuisance parameters and with respect to the parameter of interest, that is, alternating between (1) source separation based on a fixed assumption about the source distributions, and (2) a reestimation of the source distributions based on the current marginal distribution of the vector $\mathbf{y}$ of recovered sources. Such a process is illustrated in Fig. 4.12. Several options can be considered for implementing such a scheme: one is to estimate the marginal densities of $\mathbf{y}$ using kernel estimates [for instance, see Pham (1996)], simpler generic parametric models [for instance, see Perlmutter (1996)]. However, as will be seen below, when the optimization with respect to $\mathbf{y}$ is conducted by a gradient technique, one does not need to estimate the source densities themselves, but rather the associated score functions [defined by Eq. (4.80)]. This is addressed, for instance, in Pham and Garat (1997) and in Bell and Sejnowski (1995).

How much is the achievable separation performance affected when the

---

[3] Actually, Comon considered the optimization of $J[\mathbf{y}]$ under the whiteness constraint.

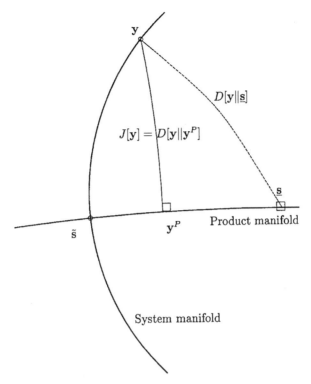

Figure 4.11 Kullback divergence to a fixed target source distribution **s** or to the adaptive target $\mathbf{y}^P$, as suggested by the ML principle.

source distributions also have to be estimated? Maybe surprisingly, there is—at least asymptotically (that is, for large enough data sets)—no impact [Amari and Cardoso (1997)]. In essence, this is due to the orthogonality between the system manifold and the product manifold. Therefore, one can look forward to designing source-separation/estimation algorithms that perform equally well whether the source distributions are known in advance or are estimated from the data.

4.4.6 On a Kullback Sphere: Geometry of Orthogonal Contrasts

This section considers the effect of whitening on the information geometry landscape: it deals with *orthogonal* contrast functions, that is, those contrast functions $\phi[\mathbf{y}]$ to be minimized under the whiteness constraint $\mathbf{E}\mathbf{y}\mathbf{y}^\dagger = I$. Accordingly, the geometric description is restricted to the space of probability distributions of zero-mean random vectors whose covariance is the identity matrix. Note that in this case, there is only *one* Gaussian distribution, namely $\mathcal{N}(0, I)$: the process of whitening the observations has the effect of shrinking the whole Gaussian manifold to the single point, $\mathcal{N}(0, I)$. In the following, **n**

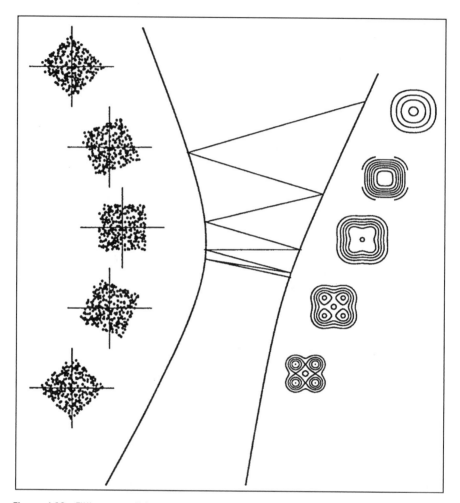

Figure 4.12 Fitting everything: alternate minimizations with respect to the mixing matrix and to the source distribution by successive projections on the system manifold and the product manifold (or a submanifold of it).

denotes a zero-mean white Gaussian vector so distributed (**n** was previously used to denote a Gaussian vector with an arbitrary covariance matrix).

Under the whiteness constraint, the system manifold has a specific geometric property. Consider the Kullback divergence from a point **y** of the system manifold to the (unique) Gaussian distribution **n**. As an instance of the Pythagorean theorem applied to the Gaussian manifold we have

$$D[\mathbf{y}\|\mathbf{n}] = D[\mathbf{y}\|\mathbf{n}^N] + D[\mathbf{y}^N\|\mathbf{n}] \qquad (4.53)$$

which is a repeat of Eq. (4.37), where $\mathbf{y}^N$ is a Gaussian vector with the same covariance matrix as **y**. Since **y** is constrained to be white, the distribution of $\mathbf{y}^N$ is by definition equal to $\mathcal{N}(0, I)$, so that $D[\mathbf{y}^N\|\mathbf{n}] = 0$. Hence, the second term on the right-hand side of Eq. (4.53) is equal to zero. The first term $D[\mathbf{y}\|\mathbf{n}^N]$ is the negentropy $N[\mathbf{y}]$ of **y** as seen in Eq. (4.36). Recalling that negentropy is invariant under invertible transform [Eq. (4.39)], we also have $N[\mathbf{y}] = N[B\mathbf{x}] = N[\mathbf{x}]$. It follows that under the whiteness constraint

$$D[\mathbf{y}\|\mathbf{n}] = N[\mathbf{x}] = \text{constant} \tag{4.54}$$

for any white **y** on the system manifold. This expression shows that the system manifold lies entirely under the whiteness constraint, on a Kullback sphere centered at the standardized Gaussian distribution $\mathbf{n} \sim \mathcal{N}(0, I)$, and with a radius equal to the negentropy $N[\mathbf{x}]$ of the observed vector. This remark is valid whether or not the model holds. If the model does hold, however, that is, if $\mathbf{x} = A\mathbf{s}$, then the radius of the Kullback sphere is also given by the negentropy $N[\mathbf{s}]$ of the source vector. We also note that for Gaussian observations, the negentropy is equal to 0: the system manifold under whiteness constraint reduces to a single point!

Regarding the geometry of the likelihood contrast $\phi_{\text{ML}}[\mathbf{y}] = D[\mathbf{y}|\mathbf{s}]$ with a fixed target distribution, there is little to add with respect to previous sections when the whiteness constraint is enforced except that, as we just saw, the standardized Gaussian distribution is, in some sense, the center of a Kullback sphere including the system manifold (this does *not* mean, however, that **n** can be considered as a center of curvature for the system manifold in the true sense of the term).

The geometry of the likelihood contrast when the target source distributions are optimized, that is, as discussed earlier, the geometry of the mutual information contrast $J[\mathbf{y}]$, is more interesting under the whiteness constraint. Again, we can apply the Pythagorean theorem [Proposition 1] by projecting **y** onto the product manifold and using the standardized Gaussian as the third vertex of the triangle. This is

$$D[\mathbf{y}\|\mathbf{n}] = D[\mathbf{y}\|\mathbf{y}^P] + D[\mathbf{y}^P\|\mathbf{n}] \tag{4.55}$$

On the left-hand side of Eq. (4.55), thanks to the whiteness constraint, we recognize as discussed earlier, the (constant) negentropy $N[\mathbf{y}] = N[\mathbf{x}]$, which is also the radius of the Kullback sphere. The first term on the right-hand side of Eq. (4.55) is the mutual information $J[\mathbf{y}] = D[\mathbf{y}\|\mathbf{y}^P]$ of **y**. The second term on the right-hand side of Eq. (4.55) is a Kullback divergence between two vectors, each having independent entries so that it decomposes as the sum of the divergences between entries

$$D[\mathbf{y}^P\|\mathbf{n}] = \sum_{i=1}^{n} D[y_i\|n_i] \tag{4.56}$$

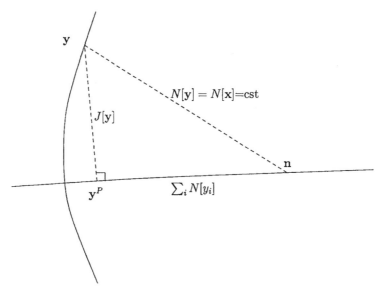

Figure 4.13 Mutual information and maximum marginal negentropy under the whiteness constraint.

Now, each entry n_i of **n** has unit variance by definition and each entry y_i of **y** also has unit variance by the whiteness constraint. Thus, $D[y_i \| n_i]$ is the negentropy of y_i, that is, $D[y_i \| n_i] = N[y_i]$. Hence, we have identified the length of the last edge of the triangle as the sum of the negentropies of the marginal distributions of **y**. Summing up, the whiteness constraint entails this decomposition:

$$J[\mathbf{y}] + \sum_{i=1}^{n} N[y_i] = N[\mathbf{y}] = N[\mathbf{x}] = \text{constant radius} \qquad (4.57)$$

This is illustrated in Fig. 4.13, which shows that bringing **y** as close as possible to the product manifold is equivalent to bringing its entries as far away as possible from the Gaussian when measuring the former by the mutual information $J[\mathbf{y}]$ and the latter by the sum of the marginal negentropies. Indeed the sum of these criteria is just the constant radius of the Kullback sphere equal to the negentropy $N[\mathbf{x}]$ of the observed vector **x**.

There is another formulation for these two equivalent contrasts. Specializing Eq. (4.36) to the case of scalar variables, we have for each output,

$$N[y_i] = H[y_i^N] - H[y_i] \qquad (4.58)$$

where y_i^N is a Gaussian variable with the same variance as y_i. Since the variance of y_i is constant under the whiteness constraint, the entropy $H[y_i^N]$ is also constant. Therefore, combining Eq. (4.57) and Eq. (4.58), we have

$$J[\mathbf{y}] = -\sum_{i=1}^{n} N[y_i] + c_1 = \sum_{i=1}^{n} H[y_i] + c_2 \tag{4.59}$$

where c_1 and c_2 are constant under the whiteness constraint. Equation (4.59) shows that the following ideas are equivalent under the whiteness constraint when they are expressed in terms of the information-theoretic criteria considered in this section:

- Maximize the independence between the entries of $\mathbf{y}$,
- Maximize the non-Gaussianity of the entries of $\mathbf{y}$,
- Minimize the entropies of the entries of $\mathbf{y}$.

Minimizing mutual information was first considered in the ICA approach of Comon (1994). The minimum entropy idea was introduced by Donoho (1981) in the context of blind deconvolution. It must be stressed that for source separation, the plain entropy $H[\mathbf{y}]$ of the output is not a valid orthogonal contrast function, since it is *constant* under the whiteness constraint; rather, the ML principle suggests minimizing the sum of the *marginal* entropies.

4.4.7 Spurious Minima

When the model holds and the distribution of the source vector $\mathbf{s}$ is known, it seems very reasonable to use the plain-likelihood contrast $\phi_{\mathrm{ML}}[\mathbf{y}] = D[\mathbf{y}\|\mathbf{s}]$ to drive a source-separation algorithm, especially since it is known from inference theory that the ML estimate will then have the lowest possible estimation variance (at least for large enough data sets). However, most of the nice properties of the ML estimation are only local: the ML contrast function has no particular reason to be well behaved away from the separating points. In particular, one may wonder if $\phi_{\mathrm{ML}}[\mathbf{y}]$ is guaranteed to be free of local minima at mixing points, even in the case when the true source distributions are used. This is like asking about the global shape of $\phi_{\mathrm{ML}}[\mathbf{y}] = D[\mathbf{y}\|\mathbf{s}] = D[C\mathbf{s}\|\mathbf{s}]$ as a function of the global system matrix C.

It is actually possible to find simple instances of source distributions such that the map $C \rightarrow D[C\mathbf{s}\|\mathbf{s}]$ shows local minima at mixing points. The idea for constructing such a counterexample is to think in terms of distribution matching. Since $D[\mathbf{y}\|\mathbf{s}]$ measures the closeness of the distribution of $\mathbf{y}$ to the distribution of $\mathbf{s}$, a "bumpy" likelihood landscape arises when the source distributions are multimodal.

The simplest example of ill-convergence that we have been able to build along these lines is illustrated at Fig. 4.14. This is a case with $n = 2$ identically distributed sources, with trimodal densities $q_1(s_1)$ and $q_2(s_2)$ having modes at

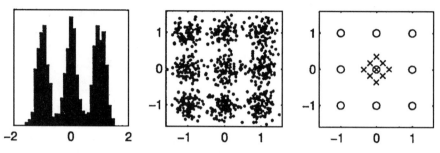

Figure 4.14 Trapping the ML estimator when the true source distributions are used: an example with multimodal distributions (see text).

points $-1, 0$, and 1:

$$q_{1,2}(s) = \frac{g(s-1) + g(s) + g(s+1)}{3} \tag{4.60}$$

where $g(s)$ is a Gaussian kernel: $g(s) = (2\pi\sigma^2)^{-1/2} \exp - s^2/2\sigma^2$. In our example, we set $\sigma = 0.2$, which corresponds to the histogram in the left panel of Fig. 4.14. The center panel shows the (sample) joint distribution of $\mathbf{s} = [s_1, s_2]^\dagger$, with $9 = 3 \times 3$ modes. Using the score functions that correspond to the density, Eq. (4.60), the ML algorithm is found to converge to the identity when initialized close enough to it. For many other initial points, however, the separating system converges to a mixing solution. These bad attractors are such that the output distribution (the distribution of $\mathbf{y}$) is as depicted in the right panel of Fig. 4.14. In this panel, the circles are located at the modes of the distribution of $\mathbf{s}$ and the crosses are located at the modes of the distribution of $\mathbf{y}$ after convergence. Therefore, the figure shows that the global system after convergence C_∞ is a mixing matrix. We find numerically that $C_\infty \approx 0.25 \times R_{\pi/4}$, where $R_{\pi/4}$ denotes the $\pi/4$ rotation matrix. Clearly, there are other bad attractors similar to C_∞ that are deduced from it by permutations and changes of signs. The reason why such mixing matrices are local attractors can be easily understood as follows.

Consider the square pattern made up by the nine modes of the joint distribution of s_1 and s_2. Let us call the four modes at the vertices of the square corresponding to the patterns $\{(1,1), (-1,1), (1,-1), (-1,-1)\}$ the *outer modes*, the four modes on the middle of the edges of the square corresponding to the pattern $\{(0,1), (0,-1), (1,0), (-1,0)\}$ the *inner modes*, and the mode at the center of the square corresponding to $\{(0,0)\}$ the *central mode*. A correct convergence of the algorithm brings the inner modes of $\mathbf{y}$ onto the inner modes of $\mathbf{s}$, and similarly for the outer modes, thus yielding a perfect match between the distribution of $\mathbf{y}$ and the distribution of $\mathbf{s}$. The example of ill-convergence shown on the right panel of Fig. 4.14 corresponds to an imperfect matching of these distributions. The ML estimator is fooled because, thanks to a small scaling factor of about 0.25, vector **y** is shrunk enough for all its nine modes to fit under the central mode of the distribution of **s**. The value ~ 0.25 of the scaling appears to be selected by the ML estimator for the nine modes of **y** to (roughly) represent the central mode of **s**. Now, for the same scaling factor, all nine modes of **y** would sit just as comfortably under the central mode of **s** for any rotation angle. The ML estimator selects a $\pi/4$ rotation or any rotation of $\pi/4$ modulo $\pi/2$ because such a rotation of **y** gives a maximum overlap with the distribution of **s**. The outer modes of **y** play a more important role in the overlap than the inner modes of **y** because they are closer to the inner modes of **s**; similarly, the inner modes of **s** play a more important role in the overlap than the outer modes of **s** because they sit closer to the origin where the distribution of **y** is concentrated. They therefore interact more strongly with the modes of the scaled-down vector **y**. This said, it should be clear that the rotations that cause a maximum overlap are those bringing the outer modes of **y** closer to the inner modes of **s**; and these are the rotations of $\pi/4$ modulo $\pi/2$.

4.5 USING THE GROUP STRUCTURE: EQUIVARIANCE AND RELATIVE GRADIENT

The source-separation model is a *transformation model* in the sense that the distributions for **x** in the model are obtained as a result of transforming the distribution of a hypothetical source vector **s** of independent components. More specifically, these transformations are all the linear invertible transformations of an $n \times 1$ vector. This set of transformations can be represented by all the invertible $n \times n$ matrices that form a group, with the group operation being the usual matrix multiplication.

When an estimator in a transformation model is compatible with the structure of the transformation group, it is called *equivariant*. This section investigates what "compatible" means in the case of source separation, and why it is a desirable property; it leads to the definition of a specific notion of gradient and to adaptive algorithms that combine simplicity and efficiency.

4.5.1 Equivariance

The notion of equivariance is classic in statistical inference [see, for instance, Lehmann and Casella (1998)], but it is usually only considered in the context of batch estimation. Its definition in such a context is recalled below in the case of source separation, and we explain why equivariant estimators display a "uniform performance property." The following section extends the idea of equivariance to adaptive algorithms.

Equivariance of Batch Algorithms   Denote by $\hat{A}$ an estimate of A obtained from a batch of T samples $\mathbf{x}_1, \ldots, \mathbf{x}_T$ using a particular estimator of the mixing matrix. This estimator is called equivariant if, for any invertible $n \times n$ matrix M, it returns the estimated value $M\hat{A}$ when applied to the transformed data $M\mathbf{x}_1, \ldots, M\mathbf{x}_T$. In other words, a transformation of the data results in a similar transformation of the estimated parameter. It is in this sense that the estimator is "compatible with the transformation group."

If an estimator of the mixing matrix A is equivariant and if the source signals are estimated by applying the inverse of the estimated mixture to the observations, that is, if $\hat{\mathbf{s}} = \mathbf{y} = \hat{A}^{-1}\mathbf{x}$, it is straightforward to check that the *exact* same source signals are obtained whether the procedure is applied to $\mathbf{x}_1, \ldots, \mathbf{x}_T$ or to any (invertible) transform $M\mathbf{x}_1, \ldots, M\mathbf{x}_T$ of it. In this sense, equivariant estimators for source separation have *uniform performance* in the mixing matrix, that is, the quality of the recovery of source signals does not depend at all on the particular matrix A applied to the sources.

Equivariance via Contrast Functions   Estimators based on the optimization of contrast functions are equivariant because they are defined uniquely in terms of the output vector $\mathbf{y}$. This is easily seen as follows. Let B_* denote a separating matrix obtained as the result of minimizing a contrast function $\phi[\mathbf{y}]$:

$$B_* = \arg \min_B \phi[B\mathbf{x}] \tag{4.61}$$

and denote by $\mathbf{y}_*$ the corresponding estimate of the source vector: $\mathbf{y}_* = B_*\mathbf{x}$. Next consider an arbitrary invertible matrix M and solve the same minimization problem on a new data vector $\tilde{\mathbf{x}}$ obtained from $\mathbf{x}$ by multiplication by M, that is, $\tilde{\mathbf{x}} = M\mathbf{x}$. The minimization of the contrast function for the new vector is obtained at a value $\tilde{B}$ of the separating matrix, which is determined by

$$\tilde{B} = \arg \min_B \phi[B\tilde{\mathbf{x}}] = \arg \min_B \phi[BM\mathbf{x}] \tag{4.62}$$

The preceding equation shows that, by Eq. (4.61), the minimum is reached for $\tilde{B}M = B_*$. Therefore the source signals estimated from the transformed vector $M\mathbf{x}$ are $\tilde{\mathbf{y}} = \tilde{B}\tilde{\mathbf{x}} = \tilde{B}M\mathbf{x} = B_*\mathbf{x} = \mathbf{y}_*$, that is, they are identical to those estimated from the original vector $\mathbf{x}$. This is, of course, a direct consequence of defining the estimation procedure only in terms of the distribution of the output vector $\mathbf{y}$. Thus equivariance is granted by the use of contrast functions; in practice, the contrast function is not available, but the approximating contrast process is, that is, the estimate of the contrast function based on a finite set of samples. However, the same reasoning as just given applies: as long as the contrast process itself is a function of $\mathbf{y}$ only, its minimization yields an equivariant estimate. This remains true for minimization under a constraint provided the constraint is also defined only in terms of the output samples (as is the case of the whiteness constraint). A simple example of equivariance loss is when matrix B is constrained to have a fixed diagonal. Such a constraint is sometimes used to fix the scales of the output vector $\mathbf{y}$, but it is bad practice, because the resulting procedure is not equivariant, so its behavior becomes more difficult to understand and to control, since it would depend on the particular mixture to be inverted.

Equivariance of Adaptive Algorithms In the following, we consider on-line algorithms that learn a separating matrix B_t according to the following rule:

$$B_{t+1} = (I - \mu_t G(\mathbf{y}_t))B_t \qquad (4.63)$$

The main reason for preferring such a learning rule over other types of updates is that is extends to adaptive algorithms the uniform performance property of batch algorithms discussed earlier.

Assume that the model holds when $\mathbf{x} = A\mathbf{s}$. Being interested in separating sources, we consider the trajectory of the global system $C_t = B_t A$. The trajectory is simply given by right-multiplying Eq. (4.63) by A, yielding

$$C_{t+1} = (I - \mu_t G(C\mathbf{s}_t))C_t \qquad (4.64)$$

where we have used the fact that $\mathbf{y} = B\mathbf{x} = BA\mathbf{s} = C\mathbf{s}$. The key observation here is that the trajectory of the global system, being governed by Eq. (4.64), depends only on the source sequence $\{\mathbf{s}_t\}$ and on the initial point C_0 for the global system. Therefore, the particular value of the mixing matrix only influences the starting point $C_0 = B_0 A$, but is otherwise completely irrelevant to the trajectory of C_t. Thus, the behavior of the algorithm can be studied independently of the value of A. Since the performance of the algorithm in terms of source separation also depends only on the global system (because this is the quantity that determines the accuracy of separation), the learning rule Eq. (4.63) can be said to display uniform (with respect to the mixing matrix) performance (Cardoso and Laheld 1996). The good behavior of this type of updates was also noted in Cichocki et al. (1994). These types of algorithms are shown below to derive from gradient-descent techniques, provided the gradient is properly defined by taking the group structure into account.

4.5.2 Relative Gradient

A well-established technique of adaptive signal processing is the (stochastic) gradient technique. In the case of source separation, it could be applied to the minimization of the contrast functions just introduced. However, as will be seen below, a naive implementation of gradient descent does not lead to equivariant learning rules in the form of Eq. (4.63). The gradient approach has to be rethought: it does yield equivariant algorithms, provided an appropriate definition of the gradient is chosen. Here, not surprisingly, an "appropriate definition" means a definition based on the fact that the unknown parameter is an invertible square matrix and thus belongs to a multiplicative group.

Let us then introduce a few definitions and notations related to matrices. The Euclidean scalar product between a matrix M and a matrix N of the same dimensions is denoted by

$$\langle M|N \rangle \overset{\text{def}}{=} \sum_{ij} M_{ij} N_{ij} = \text{trace}\{M^\dagger N\} \tag{4.65}$$

It corresponds to the Frobenius norm $\| \cdot \|_F$ defined by

$$\|M\|_F^2 = \langle M|M \rangle \overset{\text{def}}{=} \sum_{ij} M_{ij}^2 \tag{4.66}$$

The derivative of a function $f(M)$ of an $m \times n$ matrix M is the $m \times n$ matrix $[\partial f(B)/\partial B]$ with entries given by

$$\left[\frac{\partial f(B)}{\partial B} \right]_{ij} \overset{\text{def}}{=} \frac{\partial f(B)}{\partial b_{ij}} \tag{4.67}$$

When it exists, it can be used in a first-order approximation of f around point M:

$$f(B + \delta B) = f(B) + \left\langle \frac{\partial f(B)}{\partial B} \middle| \delta B \right\rangle + o(\|\delta B\|) \tag{4.68}$$

The matrix of derivatives $[\partial f(B)/\partial B]$ is called the *Euclidean gradient* of f at M because it results from seeing the set of $m \times n$ matrices as a linear space with the Euclidean scalar product, Eq. (4.65). Thus, this definition does not try to express the fact that matrices can form a multiplicative group; actually, it cannot do so, since all these definitions apply to matrices that are not necessarily square or invertible.

Relative Gradient   We are only concerned with variations of an $n \times 1$ vector $\mathbf{y}$ under linear transforms. Any linear transform of $\mathbf{y}$ can be written as

$$\mathbf{y} \leftarrow (I + \mathscr{E})\mathbf{y} = \mathbf{y} + \mathscr{E}\mathbf{y} \tag{4.69}$$

where $\mathscr{E}$ is an $n \times n$ matrix. This transform is an infinitesimal transform (that is, close to the identity transform) when $\mathscr{E}$ is an infinitesimal matrix. Consider a function $\phi[\mathbf{y}]$ of the distribution of $\mathbf{y}$. If it is regular enough, its first-order variation can be characterized by

$$\phi[(I + \mathscr{E})\mathbf{y}] = \phi[\mathbf{y}] + \langle \nabla\phi[\mathbf{y}]|\mathscr{E} \rangle + o(\|\mathscr{E}\|) \tag{4.70}$$

where $\nabla\phi[\mathbf{y}]$ is a matrix depending on the distribution of $\mathbf{y}$, whose entries are by definition

$$\{\nabla\phi[\mathbf{y}]\}_{ij} = \frac{\partial\phi[\mathbf{y} + \mathcal{E}\mathbf{y}]}{\partial\mathcal{E}_{ij}}\bigg|_{\mathcal{E}=0} \tag{4.71}$$

This matrix is called the *relative gradient* of ϕ. In the source-separation problem, it is closely related (Cardoso 1998) to Amari's natural gradient (Amari 1998) discussed in Chapter 2 of this volume. It is interesting to note that the relative gradient of a contrast function can be defined by referring only to the distribution of $\mathbf{y}$, that is, without reference to a separating matrix B. Equation (4.75) gives a simple example.

Like the Euclidean gradient, Eq. (4.67), the relative gradient can also be defined for a function $f(B)$ of a (square) matrix by

$$\{\nabla f(B)\}_{ij} = \frac{\partial f((I + \mathcal{E})B)}{\partial\mathcal{E}_{ij}}\bigg|_{\mathcal{E}=0} \tag{4.72}$$

In such a case, if $f(B)$ is defined as $f(B) = \phi[B\mathbf{x}]$, then $\nabla f(B) = \nabla\phi[B\mathbf{x}]$. It is also straightforward to relate the relative gradient to the Euclidean gradient: comparing the first-order expansions, Eqs. (4.68) and (4.70), shows that

$$\nabla f(B) = \frac{\partial f(B)}{\partial B} B^{\dagger} \tag{4.73}$$

A simple example is the relative gradient of a measure of whiteness. We defined a "contrast function for whitening" as

$$\phi_w[\mathbf{y}] \stackrel{\text{def}}{=} D[\mathbf{y}^N \| \mathbf{n}] = D\{R_y \| I\} \tag{4.74}$$

where, as in Section 4.3, $\mathbf{y}^N$ is a Gaussian vector with the same covariance matrix R_y as $\mathbf{y}$ and $\mathbf{n}$ is a white Gaussian vector. As indicated, it is also equal to $D\{R_y \| I\}$, where the divergence between covariance matrices $D\{\cdot\|\cdot\}$ is defined by Eq. (4.38). Then $\phi_w[\mathbf{y}] \geq 0$ with equality if and only if $R_y = I$, that is, $\mathbf{y}$ is spatially white. The relative gradient of $\phi_w[\mathbf{y}]$ if found (Cardoso and Laheld 1996) to be

$$\nabla\phi_w[\mathbf{y}] = E\mathbf{y}\mathbf{y}^{\dagger} - I \tag{4.75}$$

and is canceled only if $E\mathbf{y}\mathbf{y}^{\dagger} = I$, indeed. For later use, we note that $\nabla\phi_w[\mathbf{y}]$ is a symmetric matrix. This is actually an instance of a more general property of the relative gradient, which is stated without proof.

Proposition 4. If for some random vector $\mathbf{y}$, a function $f[\mathbf{y}]$ is rotationally invariant, that is, if $f[\mathbf{y}] = f[U\mathbf{y}]$ for any orthonormal matrix U, then $\nabla f[\mathbf{y}] = \nabla f[\mathbf{y}]^{\dagger}$.

In other words, the relative gradient of a rotationally invariant function is a symmetric matrix.

Stationary Points   Since we restrict ourselves to invertible matrices, it is clear from Eq. (4.73) that the Euclidean gradient and the relative gradient of functions of matrices cancel at the same points. The stationarity of a contrast function as a function of the separating matrix $f(B) = \phi[Bx]$ may be expressed as $(\partial f(B)/\partial B) = 0$ or, equivalently, as $\nabla \phi[\mathbf{y}] = 0$.

The stationary points of an orthogonal contrast function can also be expressed straightforwardly in terms of $\nabla \phi[\mathbf{y}]$. In a relative update, vector $\mathbf{y}$ undergoes a small change by left multiplication by $I + \mathcal{E}$ where $\mathcal{E}$ is a small matrix. By definition, such a transformation is orthonormal if $(I + \mathcal{E})(I + \mathcal{E})^\dagger = I$, which implies that $\mathcal{E} + \mathcal{E}^\dagger = -\mathcal{E}\mathcal{E}^\dagger = O(\|\mathcal{E}\|_F^2) = o(\|\mathcal{E}\|_F)$. Thus, infinitesimal rotations are of the form $I + \mathcal{E}$, with $\mathcal{E}$ being an infinitesimal skew-symmetric matrix: $\mathcal{E}^\dagger = -\mathcal{E}$. Therefore, the stationarity of $\phi[\mathbf{y}]$ under the whiteness constraint is equivalent to the condition $\langle \nabla \phi[\mathbf{y}] | \mathcal{E} \rangle = 0$ for any skew-symmetric matrix $\mathcal{E}$. This implies that the skew-symmetric part of $\nabla \phi[\mathbf{y}]$ should be equal to zero. Therefore, a stationary point of $\phi[\mathbf{y}]$ under the whiteness constraint verifies

$$\nabla \phi_w[\mathbf{y}] = 0 \quad \text{and} \quad \nabla \phi[\mathbf{y}] - \nabla \phi[\mathbf{y}]^\dagger = 0 \qquad (4.76)$$

The first condition is the whiteness condition, and is expressed as the cancellation of a symmetric matrix; the second condition is the cancellation of a skew-symmetric matrix.

Relative Gradient of the Likelihood Contrast   The relative gradient of the likelihood contrast $\phi_{\text{ML}}[\mathbf{y}] = D[\mathbf{y}|\mathbf{s}]$ is obtained as follows. The target density $r(\cdot)$ of $\underline{\mathbf{s}}$ is associated with a score function $\varphi(\mathbf{s})$:

$$\varphi(\mathbf{s}) \overset{\text{def}}{=} -\frac{\partial \log r(\mathbf{s})}{\partial \mathbf{s}} \qquad (4.77)$$

This is the $n \times 1$ column vector made of the partial derivative of $-\log r(\mathbf{s})$, that is, the ith entry of $\varphi(\mathbf{s})$ is equal to $-(d \log r(\mathbf{s})/ds_i)$. This allows the first-order differential of the model log-density to be written as

$$\log r(\mathbf{s} + d\mathbf{s}) = \log r(\mathbf{s}) - \varphi(\mathbf{s})^\dagger d\mathbf{s} \qquad (4.78)$$

Thanks to the product form, Eq. (4.3), of the density for independent components, the score function for the n-variate model density $r(\mathbf{s})$ has the special form

$$\varphi(\mathbf{s}) = [\varphi_1(s_1), \dots, \varphi_n(s_n)]^\dagger \qquad (4.79)$$

where the scalar functions φ_i are the score functions for the distribution of each individual source, that is,

$$\varphi_i(\mathbf{s}) \overset{\text{def}}{=} -\frac{d \log r_i(s_i)}{ds_i} = -\frac{r_i'(s_i)}{r_i(s_i)} \qquad i = 1, \ldots, n \qquad (4.80)$$

Thus, the assumption of source independence corresponds to a simple property of the score φ: *the ith entry of $\varphi(\mathbf{s})$ depends only on s_i.* Simple calculus [see, e.g., Pham and Garat (1997)] yields the relative gradient of $\phi_{\text{ML}}[\mathbf{y}]$:

$$\nabla \phi_{\text{ML}}[\mathbf{y}] = \mathrm{E} G_\varphi(\mathbf{y}) \qquad (4.81)$$

where the vector-to-matrix mapping G_φ is defined as

$$G_\varphi(\mathbf{y}) \overset{\text{def}}{=} \varphi(\mathbf{y})\mathbf{y}^\dagger - I \qquad (4.82)$$

The meaning of this particular form is discussed at Section 4.6.

Relative Gradient of the Mutual Information The relative gradient of the mutual information contrast has a similar form: recall that $\phi_{\text{ML}}[\mathbf{y}] = D[\mathbf{y}|\mathbf{s}]$, and the value of this divergence is the mutual information $J[\mathbf{y}]$ when the target $\mathbf{s}$ is taken to be $\mathbf{y}^P$, which is a vector of independent entries with the same marginal distributions as $\mathbf{y}$. Therefore

$$\nabla J[\mathbf{y}] = \mathrm{E} G_{\varphi_y}(\mathbf{y}) \qquad (4.83)$$

where G_{φ_y} is defined as G_φ, except that the score functions $\varphi_1, \ldots, \varphi_n$ are those associated with the marginal densities of $\mathbf{y}$ rather than with the fixed target densities $r_1, \ldots, r_n$, which appear in Eq. (4.80).

4.5.3 Relative-Gradient Algorithms

We are now ready to describe relative-gradient algorithms to optimize contrast functions.

Relative-Gradient Descent Formally, a relative-gradient descent of $\phi[\mathbf{y}]$ consists in changing $\mathbf{y}$ into $(I + \mathscr{E})\mathbf{y}$, with $\mathscr{E}$ being a "small" matrix in the direction opposite to the gradient. Indeed, according to Eq. (4.70), taking $\mathscr{E} = -\mu \nabla \phi[\mathbf{y}]$ changes the value of $\phi[\mathbf{y}]$ into

$$\phi[(I - \mu \nabla \phi[\mathbf{y}])\mathbf{y}] = \phi[\mathbf{y}] + \langle \nabla \phi[\mathbf{y}] \| - \mu \nabla \phi[\mathbf{y}] \rangle + o(\mu)$$

$$= \phi[\mathbf{y}] - \mu \| \nabla \phi[\mathbf{y}] \|_F^2 + o(\mu) \qquad (4.84)$$

Therefore such a change guarantees a decrease in ϕ for a small enough positive step μ until the gradient is brought to zero. An abstract version of a relative-gradient descent with step size μ is thus

$$\mathbf{y} \leftarrow (I - \mu \nabla \phi[\mathbf{y}]) \mathbf{y} \tag{4.85}$$

For on-line learning of a separating matrix B, the variation $\mathbf{y} \leftarrow (I + \mathcal{E})\mathbf{y}$ corresponds to changing a separating matrix B into $(I + \mathcal{E})B$. Hence, in terms of the separating matrix, the learning rule, Eq. (4.85), is

$$B \leftarrow (I - \mu \nabla \phi[\mathbf{y}]) B \tag{4.86}$$

Note that the relative-gradient update could also be obtained, via the regular Euclidean gradient approach by using a local coordinate system. Here "local" means that the coordinate system is changed at each step. Indeed, this is exactly what is achieved if B_{t+1} is parameterized by $\mathcal{E}$ according to $B_{t+1} = (I + \mathcal{E})B_t$.

Relative-Gradient Optimization of Orthogonal Contrasts   The minimization of an orthogonal contrast function could be implemented in two stages: a whitening stage to enforce whiteness, and a rotation stage to minimize the contrast function while preserving whiteness. Each of these stages may resort to a relative-gradient technique, but these two stages are better combined in a simple one-stage relative update, as we examine next.

Let $\phi_w[\mathbf{y}]$ be a function measuring how far $\mathbf{y}$ is from being white. It could be the function defined by Eq. (4.74) or any other measure of whiteness. For the time being, it is only assumed to be differentiable and rotationally invariant. This is a natural property to assume for a measure of whiteness, since the whiteness property itself is invariant under rotations. A whitening matrix can be learned by descending along the relative gradient of ϕ_w, that is, according to

$$\mathbf{y} \leftarrow (I - \mu \nabla \phi_w[\mathbf{y}]) \mathbf{y} \tag{4.87}$$

Not much is said as long as ϕ_w is not specified, but we recall that by Proposition 4, the relative gradient $\nabla \phi_w[\mathbf{y}]$ is a symmetric matrix.

Now consider the rotation stage. Under the orthogonality constraint, the formal learning rule, Eq. (4.85), becomes

$$\mathbf{y} \leftarrow \left(I - \mu \frac{\nabla \phi[\mathbf{y}] - \nabla \phi[\mathbf{y}]^\dagger}{2} \right) \mathbf{y} \tag{4.88}$$

involving only the skew-symmetric part, $(\nabla \phi[\mathbf{y}] - \nabla \phi[\mathbf{y}]^\dagger)/2$, of the relative gradient in order to satisfy, at least at first order, the orthogonality constraint.

One could think of implementing separate realizations of a whitening stage and of a rotation stage. It is a better idea to combine them in a single relative update rule:

$$y \leftarrow \left(I - \mu\left\{\nabla\phi_w[y] + \frac{\nabla\phi[y] - \nabla\phi[y]^\dagger}{2}\right\}\right)y \tag{4.89}$$

The stationary points for such a rule are characterized by the cancellation of the matrix between curly braces in Eq. (4.89), which implies the cancellation of both its symmetric and skew-symmetric parts. Therefore a stationary point of Eq. (4.89) verifies the stability conditions, Eqs. (4.76), because $\phi_w[y]$ is symmetric and thus has no skew-symmetric part. This double condition thus expresses that y is white and that the contrast is stationary under the whiteness constraint.

Algorithms   Earlier we described an "abstract" gradient algorithm to minimize a generic contrast function with or without the whiteness constraint. Next we describe specific stochastic relative gradient descent algorithms for source separation.

The relative gradient of $\phi_w[y]$ defined by Eq. (4.74) is seen in Eq. (4.75) to be the expectation of the vector-to-matrix function: $y \rightarrow yy^\dagger - I$, and the relative gradient of ϕ_{ML} is the expectation of the vector-to-matrix function $y \rightarrow G_\varphi(y) = \phi(y)y^\dagger - I$. Therefore stochastic-gradient algorithm are straightforwardly obtained by deleting the expectation operator in front of the gradient. Recalling the equivalence between a linear transform of y and the update of a separating matrix as summarized by Eqs. (4.85) and (4.86), we obtain for the minimization of the likelihood contrast the following algorithm

$$B_{t+1} = (I - \mu_t G_\varphi(y_t))B_t \tag{4.90}$$

which is just the rule given by Eq. (4.63), with $G(y) = G_\varphi(y)$.

For the optimization under the whiteness constraint, one uses the rule derived from Eq. (4.89), which again yields an equivariant algorithm in the form of Eq. (4.63) with $G(y) = G_\varphi^\circ(y)$ defined as

$$G_\varphi^\circ \overset{\text{def}}{=} yy^\dagger - I + \frac{\varphi(y)y^\dagger - y\varphi(y)^\dagger}{2} \tag{4.91}$$

4.6   ASYMPTOTIC STABILITY

This section complements the geometric description of likelihood contrasts and the result of their minimization with or without the whiteness constraint. In particular, the issue of stability that was only quantitatively discussed in the geometrical framework can now be fully resolved by computing the second-order variations of contrast functions at stationary points, which amounts to characterizing the curvature of the system manifold.

4.6.1 Local Stability Conditions

For a given target distribution $\underline{s}$, the contrast function $\phi_{ML}[y] = D[y\|\underline{s}]$ is stationary with respect to all (relative) linear variations of y if its (relative) gradient cancels at point y. According to Eqs. (4.81) and (4.82), this condition is

$$EG_\varphi(y) = 0 \qquad \text{or equivalently} \qquad E\varphi(y)y^\dagger = I \qquad (4.92)$$

These matrix-valued conditions can be split into n^2 scalar conditions; the diagonal and off-diagonal terms have different interpretations. The diagonal terms, for $i = 1, \ldots, n$, are $[EG_\varphi(y)]_{ii} = 0$, or

$$E\varphi_i(y_i)y_i = 1 \qquad (4.93)$$

Only the ith entry of y enters in the (i, i) condition, which appears as *fixing the scale* of the ith output. The off-diagonal terms are $[EG_\varphi(y)]_{ij} = 0$ for $i \neq j$, that is,

$$E\varphi_i(y_i)y_j = 0 \qquad (4.94)$$

These conditions have a simple interpretation: at any stationary point of $\phi_{ML}[y]$, the jth output y_j is uncorrelated with a nonlinear function of the i-output y_i for all pairs $i \neq j$. This nonlinear function φ_i is the score function (4.80) associated with the target probability distribution for the ith source. It is in general nonlinear because linear scores are associated with Gaussian distributions. Actually, if all the score functions were linear, then the (i, j)th stationarity condition would boil down to the simple decorrelation between y_i and y_j, and thus would be identical with the (j, i)th stationarity condition, leaving too small a number of conditions to uniquely determine a separating point.

Under the whiteness constraint, a different set of conditions appears. These conditions are summarized by the matrix equation $EG_\varphi^\circ[y] = 0$. Of course, the diagonal terms boil down to fixing the output scale by requesting unit variance:

$$Ey_i^2 = 1 \qquad (4.95)$$

while the conditions enforced by the off-diagonal terms are

$$Ey_iy_j = 0 \qquad \text{and} \qquad E\varphi_i(y_i)y_j - E\varphi_i(y_j)y_i = 0, \qquad 1 \leq i < j \leq n \quad (4.96)$$

These characterizations of the stationary points of $\phi_{ML}[y]$ with or without the whiteness constraint are obtained *without* any assumption about the distribution of the observed vector x. In particular, the stationarity condition holds whether or not x actually is a linear combination of independent components.

If this is the case, however, it is readily verified that separating matrices exist (with a proper scaling) and are stationary points because the entries of $\mathbf{y}$ then are independent so that $E\varphi_i(y_i)y_j = 0$ for $i \neq j$ for zero-mean signals. Therefore all the off-diagonal stationary conditions are verified at any separating point regardless of the scale of the entries of $\mathbf{y}$. The output scales are determined by the diagonal terms, either by the familiar unit-variance condition (4.95) when the whiteness constraint is in force, or by condition (4.93) when it is not. One can verify that this last condition corresponds precisely to the rescaling factors defined at Eq. (4.45).

Thus, in spite of a possibly incorrect specification of target source distributions, the contrast $D[\mathbf{y}\|\underline{s}]$ still has stationary points that are also separating points.

The picture is, however, far from being complete. First, the *variability* of the estimates obtained from a finite batch of data or by on-line minimization of $\phi_{\mathrm{ML}}[\mathbf{y}]$ also depends on choosing a good target distribution, or, equivalently, good score functions $\varphi_1, \ldots, \varphi_n$. Second, nothing is said of other possible, not necessarily nonmixing, stationary points of the contrast function. Third, the nonmixing stationary points characterized earlier are not necessarily *minima*. Of course, they are expected to be minima if the target distribution is not too different from the distribution of the sources, but it is not clear how wrong we can afford to be in our guess of the source distribution. The geometric presentation of Section 4.4 suggested that the target distribution should be on the same side of the Gaussian as the true source distribution, this rule of thumb being based on an argument of curvature. This curvature, or rather the second-order variation of the contrast function, can be computed at a stationary point when the model holds (it does not make much sense to compute it when the model does not hold) thanks to the independence between the source signals. The following proposition is adapted from Pham and Garat (1997).

Proposition 5. The second-order variation of the contrast $\phi_{\mathrm{ML}}[\mathbf{y}]$ at a separating stationary point (where the first-order variation is identically zero) is given by

$$\phi_{\mathrm{ML}}[(I + \mathscr{E})\mathbf{y}] = \phi_{\mathrm{ML}}[\mathbf{y}] + \frac{1}{2}\sum_{i=1,n}\delta_i D + \frac{1}{2}\sum_{1 \leq i < j \leq n}\delta_{ij}D + o(\|\mathscr{E}\|^2) \qquad (4.97)$$

where $\delta_i D$ and $\delta_{ij}D$ have the following quadratic forms in $\mathscr{E}$:

$$\delta_i D = \mathscr{E}_{ii}^2(E\varphi_i(y_i)y_i + E\varphi_i'(y_i)y_i^2) \qquad (4.98)$$

$$\delta_{ij}D = \begin{bmatrix} \mathscr{E}_{ij} \\ \mathscr{E}_{ji} \end{bmatrix}^\dagger \begin{bmatrix} E\varphi_i'(y_i)Ey_j^2 & E\varphi_i(y_i)y_i \\ E\varphi_j(y_j)y_j & E\varphi_j'(y_j)Ey_i^2 \end{bmatrix} \begin{bmatrix} \mathscr{E}_{ij} \\ \mathscr{E}_{ji} \end{bmatrix} \qquad (4.99)$$

The various moments of y_i and y_j that appear in the proposition are those obtained when $\mathbf{y}$ is the output of a separating matrix, so that each y_i is a rescaled version of some entry of the source vector. Besides the fact that there is a possible permutation of the sources at the output, there is also a possible rescaling of each source, the scale of the ith output being fixed by the condition (4.93). Thus, the nonlinear moment $E\varphi_i(y_i)y_i$ could be replaced by 1 in Eqs. (4.98) and (4.99), but this general form will be useful later.

Stability of Regular Contrasts To have a local minimum, and thus local stability, the quantities in Eqs. (4.98) and (4.99) must be positive for any $\mathcal{E}$. The ith quadratic form $\delta_i D$ of Eq. (4.98) depends only on $\mathcal{E}_{ii}$: it thus describes the stability with respect to the scaling of the ith output. The form is positive if

$$E\varphi_i(y_i)y_i + E\varphi_i'(y_i)y_i^2 > 0 \tag{4.100}$$

We note that this condition is met regardless of the distribution of y_i if φ_i is an increasing odd function.

The other stability conditions of Eq. (4.99) appear to be pairwise conditions. They govern the stability with respect to mixing the ith source with the jth source. Working out the eigenvalues of the 2×2 matrix in Eq. (4.99) reveals that they depend on the following nonlinear moments:

$$\xi_i \stackrel{\text{def}}{=} E\varphi_i'(y_i)Ey_i^2 - E\varphi_i(y_i)y_i \tag{4.101}$$

and that the positivity of $\delta_{ij}D$ yields the following stability condition.

Proposition 6. A separating stationary point of $\phi_{\mathrm{ML}}[\mathbf{y}]$ is stable with respect to a pair (i, j) of sources if and only if

$$1 + \xi_i > 0 \qquad 1 + \xi_j > 0 \qquad (1 + \xi_i)(1 + \xi_j) > 1 \tag{4.102}$$

Thus in the (ξ_i, ξ_j) plane, the stability domain for a given pair (i, j) of sources is bounded by the arc of hyperbola defined by Eq. (4.102) and displayed in Fig. 4.15.

Stability Under the Whiteness Constraint Local stability conditions under the whiteness constraint are more easily found because the infinitesimal transform $I + \mathcal{E}$ must be orthonormal. As seen earlier, this implies that $\mathcal{E}$ is restricted to be skew-symmetric: $\mathcal{E}^\dagger = -\mathcal{E}$ so that $\mathcal{E}_{ii} = 0$ and $\mathcal{E}_{ij} = -\mathcal{E}_{ji}$ for $i \neq j$. Therefore the condition regarding $\delta_i D$ is void and the positivity of the variation of $\delta_{ij}D$ is guaranteed if and only if

$$\begin{bmatrix} 1 \\ -1 \end{bmatrix}^\dagger \begin{bmatrix} E\varphi_i'(y_i)Ey_j^2 & E\varphi_i(y_i)y_i \\ E\varphi_j(y_j)y_j & E\varphi_j'(y_j)Ey_i^2 \end{bmatrix} \begin{bmatrix} 1 \\ -1 \end{bmatrix} > 0 \tag{4.103}$$

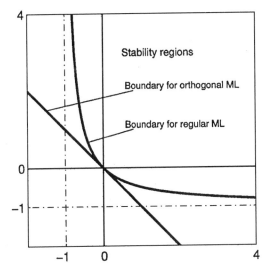

Figure 4.15 Stability domains in the (ξ_i, ξ_j) plane for a given pair (i, j) of sources.

Using definition (4.101) and elementary algebra, condition (4.103) yields the following proposition.

Proposition 7. Under the whiteness constraint, a stationary point of $\phi_{\mathrm{ML}}[\mathbf{y}]$ is stable with respect to a pair (i, j) of sources if and only if

$$\xi_i + \xi_j > 0 \tag{4.104}$$

This stability condition is identical to that given in Cardoso and Laheld (1996) for the corresponding adaptive algorithm. The local stability is more easily met under the whiteness constraint, since Eq. (4.104) defines, in the (ξ_i, ξ_j) plane, a half-plane of stability that contains the stability domain of the unconstrained contrast (see Fig. 4.15). It is not very surprising that the whiteness makes local stability easier to achieve because about half the stability is provided by the second-order constraint of having a white output, without any particular requirement regarding the source distributions.

Note that the stability conditions with and without the whiteness constraint may not be as similar as it seems, because the value of each moment ξ_i is affected by the scale of the corresponding output. This effect may be difficult to predict unless φ_i is an homogeneous function of y_i.

4.6.2 Geometric Interpretation of the Stability Conditions

Sections 4.3 and 4.4 gave a quick introduction to some global aspects of information geometry and a geometric interpretation of contrast functions.

The exposition was restricted to a "global picture" without trying to describe local features. Actually, such a basic notion as "tangent planes" in information geometry was not even discussed. Thus the idea of orthogonality between manifolds has remained vague. This final section gives just a hint of these ideas by showing how the stability conditions are related to the notion of tangent vector. As a benefit, the rule of thumb that stability requires that the target distribution be on the right side of the Gaussian will be given a quantitative meaning.

In information geometry, given a parametric family of probability densities (a manifold of distributions), the tangent vector at a given point may be represented as a random variable obtained by differentiating the log-density with respect to one of the parameters. The tangent plane at a given point (that is, for a given density in the parametric family) is the set of all linear combinations of these tangent vectors obtained by differentiating the log-density with respect to all parameters. This linear set of random variables can be equipped with a scalar product that is just the expectation of the product of any two random variables of the tangent plane.

Turning back to the source-separation problem, denote by q_i the density of y_i at the ith output of a separating matrix: this is a possibly rescaled version of one of the sources. We can define two simple (one-parameter) families of distributions based on q_i. The translation family is the set of densities in the form $q_i(y_i - m)$ for all values of m; these are the densities for the translated variables $y_i + m$. Similarly, the scale family contains the densities in the form $|a|^{-1} r_i(y_i/a)$ for all positive scales a; these are the densities of the scaled variables ay_i. The "translation score" and the "scale score" are the log-derivative of the densities of the translated and scaled variable at points $m = 0$ and $a = 1$, respectively. One finds

$$
\left. \frac{d \log q_i(y_i - m)}{dm} \right|_{m=0} = \psi_i(y_i),
$$

$$
\left. \frac{d \log |a|^{-1} q_i(y_i/a)}{da} \right|_{a=1} = \psi_i(y_i) y_i - 1
$$

(4.105)

where the function ψ_i is defined similarly to φ_i,

$$
\psi_i(y_i) \stackrel{\text{def}}{=} -\frac{q_i'(y_i)}{q_i(y_i)}
$$

(4.106)

Regarding stability, the similarity between true source distribution and target distribution depends on the similarity between the true scores ψ_i and the target scores φ_i, as examined next. If y_i is a zero-mean Gaussian variable, then $\psi_i(y_i) = y_i/\mathrm{E}y_i^2$. Since (non)-Gaussianity is so important in source separation, we decompose the translation score for a non-Gaussian variable in two terms: the first term is the score for the best Gaussian approximation to y_i, that is, the zero-mean Gaussian variable with the same variance, and the second term is the remainder. In symbols, this decomposition is

$$\psi_i(y_i) = \frac{y_i}{Ey_i^2} + \psi_i^{\perp}(y_i) \qquad \text{where} \qquad \psi_i^{\perp}(y_i) \overset{\text{def}}{=} \psi_i(y_i) - \frac{y_i}{Ey_i^2} \qquad (4.107)$$

We use the notation $\psi_i^{\perp}(y_i)$ to stress that this is an orthogonal decomposition: the variable $\psi_i^{\perp}(y_i)$ and the linear part y_i/Ey_i^2 are uncorrelated, as can be readily checked using the fact that $E\psi_i(y_i)y_i = 1$. Another view of $\psi_i^{\perp}(y_i)$ is that it is minus the log-derivative of the density of y_i *with respect to the Gaussian distribution with variance* Ey_i^2. Indeed, if this density is denoted $q_i^{\perp}$, we have by definition

$$q_i(y_i) = q_i^{\perp}(y_i) \frac{1}{\sqrt{2\pi}\sigma_i} \exp - \frac{y_i^2}{2\sigma_i^2}$$

with $\sigma_i^2 = Ey_i^2$ and it is readily checked that $\psi_i^{\perp}(y_i) = -(d/dy_i)\log q_i^{\perp}(y_i)$; hence, the following property.

Proposition 8. $\psi_i^{\perp}(y_i) = 0$ if and only if y_i is normally distributed.

The stability conditions of separating matrices have been seen to depend on the signs of the pseudokurtosis $\xi_1, \ldots, \xi_n$. In particular, the quarter plane $\{\xi_i > 0, \xi_j > 0\}$ is a stability domain with or without the whiteness constraint, and this domain decouples the stability conditions in the sense that a sourcewise (as opposed to pairwise) sufficient stability condition is that ξ_i for all i. Thus the sign of ξ_i is the crucial quantity to control in order to ensure local stability. It appears that each ξ_i (and in particular its sign), is directly related to the nonlinear parts of the score functions $\psi_i^{\perp}(y_i)$. Indeed, integration by parts yields:

$$\xi_i = E\{\psi_i^{\perp}(y_i)\varphi_i^{\perp}(y_i)\}E\{y_i^2\} \qquad (4.108)$$

where similarly to definition (4.107), we have defined the nonlinear part of the target score function:

$$\varphi_i^{\perp}(y_i) \overset{\text{def}}{=} \varphi_i(y_i) - \frac{y_i}{Ey_i^2} \qquad (4.109)$$

Therefore, local stability is guaranteed if the nonlinear (non-Gaussian) parts of the true score and the nonlinear part of the target score are positively correlated. This is the precise quantitative meaning to be given to the rule of thumb that the true distribution and the target distribution should be, for the sake of stability, on the same side of the Gaussian.

Similarly, the stability condition with respect to scale (4.100) can also be shown to be the condition that scale scores make an acute angle: one can establish

$$\mathrm{E}\varphi_i'(y_i)y_i^2 + \mathrm{E}\varphi_i(y_i)y_i = \mathrm{E}\{(\psi_i(y_i)y_i - 1)(\varphi_i(y_i)y_i - 1)\} \qquad (4.110)$$

so that the stability condition for scale amounts to a positive correlation between true-scale score and target-scale score.

ACKNOWLEDGMENTS

Part of this work was initiated during a visit at the Riken Institute. I gratefully acknowledge Professor S. I. Amari for support and inspiration and suggesting that I look into the geometry of blind source separation. I also thank M. Basseville and A. Hyvärinen for comments on a preliminary version of this material.

This chapter is partly based on the paper "Entropic Contrasts for Source Separation" presented at a NIPS*96 workshop organized by A. Cichocki.

REFERENCES

Amari, S.-I., 1985, *Differential-Geometrical Methods in Statistics*, Number 28 in Lecture Notes in Statistics, Springer-Verlag.

Amari, S.-I., 1998, "Natural gradient works efficiently in learning," *Neural Computation*, vol. 10, pp. 251–276.

Amari, S.-I., and J.-F. Cardoso, November 1997, "Blind source separation—semiparametric statistical approach," *IEEE Trans. on Signal Processing*, vol. 45, no. 11, pp. 2692–2700, special issue on neural networks.

Bell, A. J., and T. J. Sejnowski, 1995, "An information-maximisation approach to blind separation and blind deconvolution," *Neural computation*, vol. 7, no. 6, pp. 1004–1034.

Cardoso, J.-F., September 1994, "On the performance of orthogonal source separation algorithms," *Proc. EUSIPCO*, Edinburgh, pp. 776–779.

Cardoso, J.-F., 1999, High-order contrasts for independent component analysis, *Neural Computation*, vol. 11, no. 1, pp. 157–192.

Cardoso, J.-F., 1998, "Learning in manifolds: the case of source separation," *Proc. SSAP '98*.

Cardoso, J.-F., and B. Laheld, December 1996, "Equivariant adaptive source separation," *IEEE Trans. on Signal Processing*, vol. 44, no. 12, pp. 3017–3030.

Cardoso, J.-F., and A. Souloumiac, December 1993, "Blind beamforming for non Gaussian signals," *IEE Proceedings-F*, vol. 140, no. 6, pp. 362–370.

Cichocki, A., R. Unbehauen, L. Moszczynski, and E. Rummert, 1994, "A new on-line adaptive learning algorithm for blind separation of source signals," *Proc. ISANN*, pp. 406–411.

Comon, P., April 1994, "Independent component analysis, a new concept?" *Signal Processing*, vol. 36, no. 3, pp. 287–314, special issue on higher-order statistics.

Cover, T., and J. Thomas, 1991, *Elements of Information Theory*, New-York: Wiley.

Donoho, D., 1981, On minimum entropy deconvolution, in *Applied Time-Series Analysis II* (New York: Academic Press), pp. 565–609.

Lehmann, E. L., and G. Casella, 1998, *Theory of Point Estimation*, Springer texts in statistics, Springer.

Moreau, E., and O. Macchi, Jan. 1996, "High order contrasts for self-adaptive source separation," *Int. J. of Adaptive Control and Signal Processing*, vol. 10, no. 1, pp. 19–46.

Pearlmutter, B. A., and L. C. Parra, 1996, "A context-sensitive generalization of ICA," in *Int. Conf. on Neural Information Processing*, Hong Kong.

Pham, D.-T., Nov. 1996, "Blind separation of instantaneous mixture of sources via an independent component analysis," *IEEE Trans. on Signal Processing*, vol. 44, no. 11, pp. 2768–2779.

Pham, D.-T., and P. Garat, July 1997, "Blind separation of mixture of independent sources through a quasi-maximum likelihood approach," *IEEE Trans. Signal Processing*, vol. 45, no. 7, pp. 1712–1725.

5

BLIND SOURCE SEPARATION: MODELS, CONCEPTS, ALGORITHMS, AND PERFORMANCE

Pierre Comon and Pascal Chevalier

ABSTRACT

Blind source separation has recently become of increasing interest. For instance, it is now often considered as a means to exploit the spatial diversity in antenna array processing when source signals and array response are unknown, yielding more powerful processing schemes in digital communications, radar, and sonar.

Another instance is the so-called independent-component analysis (ICA), which can be viewed as a general-purpose tool taking the place of the principal-component analysis (PCA), which means it is applicable in a wide range of problems, including data analysis. Instances of this versatile framework are pointed out later.

This chapter is a survey of the problem, encompassing algebraic and statistical tools, concrete numerical algorithms, and performance analysis. It includes a thorough, up-to-date bibliography. The authors think that the blind source-separation algorithms may be implemented in the near future in operational systems on a much larger scale. Improvements to be carefully studied include, in particular, the ability to detect and extract more sources than sensors.

Unsupervised Adaptive Filtering, Volume I, Edited by Simon Haykin.
ISBN 0-471-29412-8 © 2000 John Wiley & Sons, Inc.

5.1 SOURCE-SEPARATION PROBLEMS

After a brief description of possible channel and signal models, the goals of source separation are introduced. The optimality of some filtering structures for several operational contexts is then discussed. Finally, the state of the art of the main blind algorithms is surveyed.

5.1.1 Modeling

Convolutive Model It is assumed throughout this chapter that P source signals, generally random, propagate through a linear deterministic channel, are corrupted by additive noise, and are received by an array of K sensors. Thus we assume that the mathematical observation model for the signals $x_i(n)$ received on the ith sensor of the array, $1 \le i \le K$, is described by

$$x_i(n) = \sum_{j=1}^{P} a_{ij}(n, \tau) * m_j(\tau) + v_i(n) \tag{5.1}$$

where $a_{ij}(n, \tau)$ is the impulse response linking sensor i with source j at time n, $*$ is the convolution operator, $m_i(\tau)$ is the ith source signal, and $v_i(n)$ is the noise contribution on the ith sensor. It is useful to rewrite this model equation in a compact form as

$$\mathbf{x}(n) = \mathbf{A}(n, \tau) * \mathbf{m}(\tau) + \mathbf{v}(n) \tag{5.2}$$

where vectors $\mathbf{x}(n)$ and $\mathbf{m}(\tau)$ are of dimension K and P, respectively, and matrix $\mathbf{A}(n, \tau)$ is of dimension $K \times P$. In addition, the zero-mean noise $\mathbf{v}(n)$ is assumed to be statistically independent of the source vector, $\mathbf{m}(\tau)$. This is the general model assumed in the antenna array processing literature, and this characterizes a noisy, convolutive, and possibly time-varying multiple-output mixture of sources (it is multiple-output from the point of view of the channel, due to the multisensor reception).

In Eq. (5.2), all variables can take their values either from the complex field or from the real field. In the following, we assume that the variables are complex, unless otherwise specified, which means that model (5.2) describes the analytical properties of the signals at the output of the sensors. For signals modulating a carrier frequency, the model (5.2) generally describes the complex envelopes of the signals at the output of the sensors. In the latter case, however, a residual carrier may still affect the sources after the baseband conversion, as is generally the case for passive listening applications. In these conditions, it is generally preferable to use the following model:

$$\mathbf{x}(n) = \mathbf{A}(n, \tau) * \mathbf{m}^c(\tau) + \mathbf{v}(n) \tag{5.3}$$

where $\mathbf{m}^c(\tau)$ is the vector whose components are the signals $m_k^c(n) = m_k(n)e^{j(\Delta\omega_k t + \phi_k)}$, with $j^2 = -1$,[1] where $\Delta\omega_k$ denotes the frequency-carrier mismatch and ϕ_k the carrier phase of source k. In the absence of the carrier residual, Eq. (5.3) matches Eq. (5.2). This model covers a wide range of applications, including digital communications, passive listening, sonar, radar, and acoustics (Chaumette et al. 1993; Deville and Andry 1996; D'Urso et al. 1997; Haykin 1984; Proakis 1995).

In the case of stationary mixtures, the impulse-response matrix $\mathbf{A}(n, \tau)$ does not vary with time, and $\mathbf{A}(n, \tau) = \mathbf{A}(\tau)$. The potential stationarity of the sources is characterized by the constancy, with respect to the time parameter n, of the statistical properties of $\mathbf{m}^c(n)$.

Static Model   A simplified model is that of instantaneous channels, for which a single propagation path links each source to every sensor: $a_{ij}(n, \tau) = a_{ij}(n)\delta(\tau - \tau_{ij}(n))$. The quantity $\tau_{ij}(n)$ denotes the relative time delay of the path linking source j to sensor i, taking sensor 1 as a reference ($\tau_{1j} = 0$). Each component of model (5.1) can be written as

$$x_i(n) = \sum_{j=1}^{P} a_{ij}(n)m_j(n - \tau_{ij}(n)) + v_i(n)$$

An especially interesting case in digital communications is the narrow-band (NB) sources. This case is encountered when the signal bandwidth B and the maximal sensor spacing d satisfy the inequality $\pi Bd/c \ll 1$, where c denotes the wave celerity. NB sources propagating through a stationary instantaneous channel lead to the observation model (Monzingo and Miller 1980):

$$\mathbf{x}(n) = \mathbf{Am}^c(n) + \mathbf{v}(n) \tag{5.4}$$

where matrix $\mathbf{A}$ is the matrix of the so-called source-steering vectors, containing all the information about the array of sensors and the direction of arrival of the sources. This model, which characterizes a stationary, noisy, and instantaneous multiple-output mixtures of NB sources is called *static*. It already has been used in many applications, such as radar, sonar, radio communications, passive listening, and nuclear power plants. Additionally, model (5.4) can be encountered in other contexts where variable n does not necessarily denote a discrete time index, in particular in data analysis (Caroll et al. 1980), sparse coding (Hyvärinen et al. 1999), or numerical complexity (Kruskal 1977).

Lastly, for stationary channels and sources, notice that the Fourier transform of model (5.2) obviously leads to a static model for any fixed-frequency bin.

---

[1] The square root of unity is denoted by the dotless j.

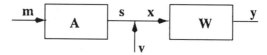

Figure 5.1   Filter **W** aims at compensating for the effect of channel **A**.

NB Sources and Channels with Propagation Multipaths   A particular case of practical interest is the case of NB sources propagating through channels with a finite number of delayed multipaths. The vector impulse response of the vector channel associated with source i can then be written as

$$\mathbf{a}_i(n, \tau) = \sum_{k=1}^{M_i} \alpha_{ik}(n) \exp(-2j\pi f_0 \tau_{ik}(n)) \delta(\tau - \tau_{ik}(n)) \mathbf{a}_{ik}(n)$$

and the contribution of source i to vector $\mathbf{x}(n)$ is given by $\mathbf{a}_i(n, \tau) * m_i^c(\tau)$, where $\mathbf{a}_{ik}(n)$ is the steering vector of path k from source i at time n; M_i is the number of propagation paths associated with source i; $\alpha_{ik}(n)$ and $\tau_{ik}(n)$ are the attenuation and the propagation delay associated with path k from source i at time n, respectively, and f_0 is the carrier frequency. Note that if all the propagation paths are considered as different sources, then, for stationary channels, one ends up with a model similar to Eq. (5.4) instead of considering the convolutive character of the mixture.

5.1.2   Goals, Filtering Structures, and Optimality

Goals   The general goal of source separation is to process or to filter the observation vector over a given observation duration so as to (1) separate the contributions of the different sources impinging on the array, and (2) extract the information contained in each of these sources.

The kind of information about the sources we want to extract depends on the application and on the nature of the sources. Indeed, we may want, for instance,

- To recover, from the vectors $\mathbf{x}(n)$, the complex signals associated with each source of interest. In this case, the goal of source separation is to generate, at each time n, an output vector $\mathbf{y}(n)$ corresponding to a good estimate of the vector $\mathbf{m}^c(n)$, up to scale and permutation factors. This estimation problem is typical of the analog transmission or the passive listening contexts. Since ordering or weighting of sources is of little importance, the source estimates may be defined to within a permutation matrix, $\mathbf{P}$, and a diagonal matrix, Λ. In other words, the goal of source separation is to find a filter $\mathcal{W}$ to be applied to the observations such that the following approximation holds in some sense:

$$\mathbf{y}(n) = \mathcal{W}(\mathbf{x}(n), n_0 \leq n \leq n_0 + n_1) \approx \mathbf{P}\Lambda\mathbf{m}^c(n) \qquad (5.5)$$

- To generate, from the vectors $\mathbf{x}(n)$, an estimate of the symbols transmitted by the sources of interest when the latter are digitally modulated. In this case, source separation is no longer an estimation problem, but actually corresponds to one of decision. This process requires the *a priori* knowledge or estimation of the symbol duration of each source. This is typical of the digital radio-communications field and in particular of mobile digital cellular networks.
- To detect the presence or the absence of the sources, corresponding to different echoes, in radar or sonar applications. In this case, the problem of source separation is a multiple target detection problem.

In all cases, if we denote by P' the number of sources of interest for the receiver, the process of source separation jointly performs the separation of the P' sources of interest, the rejection of the $P - P'$ interfering sources and, for convolutive mixtures, the deconvolution, called *equalization* for digital sources, of the different propagation channels. For the sake of simplicity and without restricting the generality, it will be subsequently assumed that all sources are interesting, that is, $P = P'$.

Filtering Structures and Optimality   Whatever the goal, the structure of the optimal separators directly depends

- On the chosen ideal optimization criterion [likelihood, contrast, symbol, or bit error rate after decision, or receiver operator characteristic (ROC) curves]
- On the assumptions about the nature (random, deterministic, narrow-band, or wide-band) and the statistical properties (Gaussian or not, circular or not, stationary or not, continuous or discrete) of sources and noise. See Section 5.1.3 about taxonomy.
- On the kind of mixture (convolutive or static).

For example, in the presence of non-Gaussian noise or interferences, it is well-known that the optimal structures for filtering are no longer linear but become nonlinear. The interest in using particular nonlinear structures [e.g., Volterra (Schetzen 1980)], instead of linear ones for the array processing of non-Gaussian signals, has already been investigated in several papers (Chevalier and Picinbono 1996; Souloumiac et al. 1993), both in detection and estimation contexts.

On the other hand, for noncircular signals (Picinbono 1994), the optimal structures consist of jointly filtering the observation vector $\mathbf{x}(n)$ and its complex conjugate, $\mathbf{x}(n)^*$, and generating output vectors of the form $\mathscr{W}\{\mathbf{x}(n), \mathbf{x}(n)^*\}$ (Chevalier 1996b; Picinbono and Chevalier 1995b). The gain in performance obtained by using these kinds of structures instead of linear ones in noncircular contexts has been investigated in (Chevalier 1996b; Comon, Grellier, and Mourrain 1998; Grellier and Comon 1998a; Picinbono and Chevalier

1995a,b) for estimation, identification, equalization, detection and array processing respectively.

Moreover, for nonstationary sources, the optimal filtering structures become time varying. They become quasi- or polyperiodic (Gardner 1993) for (quasi)-cyclostationary sources (digital modulations) (Gardner et al. 1987). The interest of polyperiodic structures for filtering in array processing contexts has been investigated recently in (Chevalier and Maurice 1997, 1998).

Besides, for wide-band signals or convolutive mixtures, the optimal structures are spatiotemporal (Paulraj and Papadias 1997). The spatial structures of filtering become optimal only for instantaneous mixtures of temporally white NB sources (Pipon et al. 1997).

Chosen Filtering Structures   In the following, we consider only linear and time-invariant (TI) filtering structures, the most widely investigated in the open literature, and shall restrict ourselves to finite-duration impulse-response (FIR) filters for practical reasons [it is understood that an infinite-duration impulse response (IIR) can be approximated by a long FIR]. Of course, for non-Gaussian, noncircular or nonstationary contexts, these structures obviously become suboptimal, but suboptimality does not necessary mean lack of practical usefulness.

On the other hand, for instantaneous mixtures of NB sources, it is sufficient to consider only spatial structures for separation. In this case, a linear source separator is characterized by a $K \times P$ matrix $\mathbf{W}$, outputting the $P \times 1$ vector $\mathbf{y}(n)$ defined by

$$\mathbf{y}(n) = \mathbf{W}^H \mathbf{x}(n) \qquad (5.6)$$

where $(^H)$ denotes Hermitian transposition. On the other hand, any mixture of wide-band sources needs space–time structures for separation. This also holds true for convolutive mixtures of any type of sources. In these cases, a source separator is characterized by a $K \times P$ matrix of FIR filters, $\mathbf{W}(n)$, outputting the $P \times 1$ vector $\mathbf{y}(n)$ defined by

$$\mathbf{y}(n) = \mathbf{W}(n)^H * \mathbf{x}(n) \qquad (5.7)$$

5.1.3 Taxonomy

It is difficult to classify all the possible models and algorithms related to source separation. However, the latter are closely linked to the assumptions made about sources, channel, and noise. In this subsection, we attempt to list most of these assumptions.

In model (5.3), it is clear that mixture $\mathbf{A}(n, \tau)$ and sources $\mathbf{m}(n)$ cannot be determined uniquely, even if sources are assumed to be statistically independent (Comon 1994a; Tong et al. 1991). In fact, let $\check{\mathbf{A}}$ denote the Fourier transform of $\mathbf{A}$; then the pair $(\check{\mathbf{A}}, \mathbf{m})$ can be replaced by $(\check{\mathbf{A}}\check{\mathbf{\Lambda}}^{-1}, \check{\mathbf{\Lambda}}\mathbf{m})$, for any invertible diagonal filter $\check{\mathbf{\Lambda}}(z)$. This inherent indeterminacy needs to be fixed by additional hypotheses, and this can be done in various manners, as now summarized.

Hypotheses on the Sources m(n)

HS1. *Sources $m_i(n)$ are statistically independent.* For approaches based on statistics of order r, the independence should be understood at order r only.

HS2. *Sources are known to have different second-order spectra.* This hypothesis can be used when the mixture is instantaneous and constant.

HS3. *The number of sources, P, is strictly smaller than the number of sensors, K.* This hypothesis is necessary in most approaches resorting to second-order statistics.

HS4. *The sources are known during part of the time.* This hypothesis is the case in many radio communication links, for example, in the global system for mobile communications (GSM), because they utilize training sequences.

HS5. *Some sources are almost completely known.* This hypothesis is the case for pilot sources (e.g., in UMTS).

HS6. *Sources have constant modulus,* for example, M-ary phase shift keying (M-PSK) with rectangular pulse function.

HS7. *Sources have discrete distribution,* for example, M-PSK, GMSK, or QAM16.

HS8. *Sources are non-Gaussian.*

HS9. *Sources are stationary or (quasi)-cyclostationary.*

HS10. *Each source $m(n)$ is independently and identically distributed (i.i.d.), that is, strongly white.* For instance, a linearly modulated digital source sampled at the symbol rate.

HS11. *Sources are white at order 2,* that is, only second-order spectra are considered. This hypothesis is weaker than HS10.

HS12. *Sources have unit variance.* With HS11 (or HS10), this condition shows that the source spectral matrix is constant and equal to unity.

Hypotheses on the Filters A, W, and $C = W^H * A$

HF0. *Filters $\mathbf{A}$, $\mathbf{W}$, and $\mathbf{C}$ are time-invariant*

HF1. *Diagonal entries of $\check{\mathbf{W}}(z)$ are constant and equal to 1.* This means: Diag $\check{\mathbf{W}}(z) = I, \forall z$.

HF2. *Columns of $\check{\mathbf{C}}(z)$ are normalized.* This means: Diag $\check{\mathbf{C}}(z)^H \check{\mathbf{C}}(z) = I, \forall z$.

HF3. *$\check{\mathbf{C}}(z)$ is normalized over the whole spectrum:* Diag $\oint \check{\mathbf{C}}(z)^H \check{\mathbf{C}}(z) \, dz = I$. This is weaker than HF2.

HF4. *Matrix $\check{\mathbf{A}}(z)$ is full column rank for all z.* In particular, if $\check{\mathbf{A}}(z)$ is FIR, then it admits a FIR left inverse (i.e., it is minimum phase).

HF5. *$\check{\mathbf{A}}(z)$ is a FIR with known degrees.*

HF6. *The rows $\check{\mathbf{A}}(z)$ are FIR with strictly decreasing orders.*

Properties Due to Preprocessing of the Data $\mathbf{x}(n)$

PX1. *Every observation $x_i(n)$ is of unit variance.*

PX2. *Every component $x_i(n)$ is whitened at order 2.* If (HS11) and (HS12) are jointly assumed, then $\text{Diag}\,\breve{\mathbf{W}}(z)\breve{\mathbf{W}}(z)^H = I, \forall z$, and rows of $\breve{\mathbf{W}}(z)$ are normalized. For instance, lossless filters satisfy this property.

PX3. *The vector process $\mathbf{x}(n)$ is whitened at order 2, both spatially and spectrally.* That is, $E\{\mathbf{x}(z)\mathbf{x}(z)^H\} = I$. With (HS11) and (HS12), this yields that $\breve{\mathbf{W}}(z)\breve{\mathbf{W}}(z)^H = I$, for all z. This is stronger than (PX2).

PX4. *Rows of $\breve{\mathbf{W}}(z)$ are globally normalized over the whole spectrum.* This means that $\text{Diag} \oint \breve{\mathbf{W}}(z)\breve{\mathbf{W}}(z)^H dz = I$. This hypothesis is weaker than (PX2).

Hypotheses on the Noise $\mathbf{v}(n)$

HN1. *The noise $\mathbf{v}(n)$ is Gaussian.*

HN2. *The noise $\mathbf{v}(n)$ is non-Gaussian.*

HN3. *The noise $\mathbf{v}(n)$ is temporally white.*

HN4. *The noise $\mathbf{v}(n)$ is spatially uncorrelated.*

HN5. *The noise is spatially correlated with known spatial coherency* $\mathbf{B}$: $\mathbf{R}_v = E\{\mathbf{v}(n)\mathbf{v}(n)^H\} = \eta_v \mathbf{B}$, *where η_v is unknown and trace $\{\mathbf{B}\} = K$.*

Some important concepts and definitions have to be introduced at this stage. As already stated earlier, source separation is a filtering problem that includes separation, rejection, deconvolution, or equalization. However, this process may, in some cases, require the *a priori* estimation of the matrix $A(n, \tau)$. This operation is referred to as channel identification.

From the point of view of the channels, the sources are the inputs and the sensor observations are the outputs. With these definitions, we can consider the following cases:

- SISO (single input – single output) systems: $P = 1$ and $K = 1$
- SIMO (single input – multiple output) systems: $P = 1$ and $K > 1$
- MIMO (multiple input – multiple output) systems: $P > 1$ and $K > 1$

When *a priori* information such as training sequences (HS4) or direction of arrival is available, the process of source separation or channel identification is said to be *informed*. On the other hand, when no *a priori* information is available about the sources, the channels, or the wavefront, the latter processing is referred to as *blind*. When the input is known part of the time, the processing is termed *semiblind*. Informed processing can be efficiently performed from the second-order statistics of the data (Hannan and Deistler 1988), with possibly noise in both inputs and outputs (Soderstrom and Stoica 1989).

The source separator is said to be second order (SO) if only the SO statistics of the data are exploited. It is called higher order (HO) otherwise. This chapter focuses essentially on blind techniques, and mostly HO. A general review is given before concentrating on static models.

5.1.4 Blind Separators of Static Mixtures

If the sources can be distinguished by their time coherency (HS2), that is, their second-order spectrum, then they can be separated with the help of second-order statistics. But of course this is a very rare situation, which might be encountered, for instance, when sources are not sampled at the symbol rate. The situation has been addressed in a few papers (Belouchrani et al. 1993; Fety 1988; Van Gerven and Van Compernolle 1995; Lacoume et al. 1997; Tong et al. 1991).

If one cannot take advantage of any source time coherency, then it is sufficient to consider the model $x = Am + v$, where x and m are random variables. In order to recover vector m from realizations of the sole observation of vector x, the components m_i are assumed to be statistically independent (HS1). This problem is now referred to as independent-component analysis (ICA) (Comon 1994a).

With no additional assumption (e.g., on the structure of the mixing matrix A), one must resort to high-order statistics (HOS) of the observed vector x, and one can only have access to an equivalence class of solutions, in which every solution is deduced from the others by a permutation and a scale factor (Cao and Liu 1996; Comon 1994a; Picinbono and Chevalier 1995a). The first algorithms that appeared in the literature were iterative (Bar-Ness et al. 1982; Comon et al. 1991; Jutten and Herault 1991), and their numerical behavior has been studied in detail (Deville 1996; Macchi and Moreau 1997; Sorouchyari 1991).

On the other hand, their behavior remained incompletely understood for a few years. Then methods *explicitly* using second- and higher-order moments of the data allowed one to show that there was indeed some statistical foundation in the problem (Cardoso 1989; Comon 1989b; Lacoume and Ruiz 1992). Further improvements and understanding led to more refined algorithms (Cardoso and Souloumiac 1993; Comon 1994a; Delfosse and Loubaton 1995), whose performance is now beginning to be known rather accurately both for stationary (Cardoso 1994, 1998b; Chevalier 1995b) or (quasi) cyclostationary (Ferreol and Chevalier 1998a) independent sources, and in the presence of potentially correlated multipaths (Chevalier et al. 1997). Performance analysis in real operational contexts include Chaumette et al. (1993) in radar, or Chevalier et al. (1999) in radio communications.

Besides, approximate maximum-likelihood (ML) approaches have been developed (Gaeta and Lacoume 1990; Pham and Garat 1997), but require the noise to be Gaussian. A deeper analysis of approaches founded on the ML principle can be found in Cardoso (1998a). The maximum entropy principle proposed in Bell and Sejnowski (1995) can be seen to belong to this family (Cardoso 1997).

If sources have discrete distributions (HS7), specific techniques can be used in order to improve on the performances (Talwar et al. 1996; Comon, Grellier, and Mourrain 1998). For instance, only the cardinality of the support is assumed to be known in Gamboa (1995), whereas sources are assumed to be binary phase shift keying (BPSK) modulated in (Van der Veen 1997). If only the constancy of the modulus of the inputs is assumed (HS6), as in Van der Veen and Paulraj (1996), then the discrete character of the sources is not fully exploited.

Very few papers have investigated the use of nonlinear, widely linear, or time-varying source separators, although optimal in non-Gaussian, noncircular, or nonstationary contexts, as described in Section 5.1. However, some initial studies have been carried out on nonlinear separators (Krob and Benidir 1993).

If there are more sources than sensors, then either specific additional assumptions need to me made (Cao and Liu 1996; Tong 1996), or HOS have to be used without prewhitening by second-order statistics (Cardoso 1992; Comon and Mourrain 1996; DeLathauwer et al. 1994, 1996; Emile et al. 1998; Comon 1998; Laheld and Cardoso 1994), in order to identify the mixture. The source extraction itself then requires assumptions on the source statistics such as (HS7), as in Comon (1998). The number of sources that can be detected is limited both by the order of the moments that are utilized and by the noise level. An increase in the number of sensors, K, is just shifting this upper bound upward. For instance, in the absence of noise, Viterbi-like algorithms can theoretically extract an arbitrarily large number of sources. In practice, this is limited by the large numerical complexity, so that suboptimal algorithms need to be used (Seshadri; 1994).

5.1.5 Blind Separators of Convolutive Mixtures

Let's now review some blind identification and deconvolution algorithms. Some of them explicitly exploit the discrete character of source distributions, and as such, should be considered as blind equalizers (cf. Section 5.1).

Single-Output Systems The SISO blind equalization is an old problem since it can be traced back to the 1970s with the works of Sato and Godard (1975 and 1980, respectively). Benveniste probably first applied the qualifier "blind" to such approaches that cannot access the input (Benveniste and Goursat 1984). The framework has been analyzed in detail by Donoho (1981), who established a real hierarchy within optimization criteria (later called *contrasts*). See also Benveniste and Goursat (1984) and Ding and Kennedy (1993) for convergence issues.

Kurtosis maximization algorithms, such as Shalvi and Weinstein (1990) and Souloumiac et al. (1993) can be seen to enter Donoho's framework. Bellini proposed using the cross-correlation between observations and a nonlinear function of them (Bellini and Rocca 1990); he thus implicitly uses HOS. See, for example, Proakis (Nikias 1991; Proakis and Nikias 1991) for partial surveys.

In many applications of practical value, the sources take their values on the complex unit circle (HS6). This observation leads to the so-called constant-modulus algorithms (CMAs) (Agee 1986; Fijalkow et al. 1997; Godard 1980; Hilal and Duhamel 1992; Treichler and Agee 1983).

When sources are discrete (HS7), then CMA is suboptimal, even if the values that sources are allowed to take lie on the unit circle (except in the presence of carrier mismatch). This is the idea of decision directed and Bussgang techniques (Haykin 1994; Macchi 1995; Macchi and Eweda 1984), or more sophisticated ones (Gassiat 1995; da Rocha and Macchi 1994). For instance in (Mayrargue 1994), the CM property is combined with oversampling to yield an algebraic solution. Additional knowledge can also be exploited, such as the modulation memory (Grellier and Comon 1998a), in order to access closed-form solutions, which are computationally less costly compared to the well-known Viterbi algorithm (Forney 1972; Seshadri 1994), and possibly in decision-directed mode when the channel is unknown (Proakis 1995).

Another direction consists of assuming that the input is distributed so as to minimize the Fisher information (Vuattoux and Le Carpentier 1996; Zivojnovic 1995). Parameter estimation is then granted some robustness.

Multiple-Output Systems Let's now turn to the multiple-output case. A FIR channel cannot be deconvolved with a FIR equalizer in the SISO case. But in multiple-output (SIMO or MIMO) cases, it becomes possible under mild hypotheses, for example, (HF4). Multiple outputs can be obtained either by using several sensors, possibly exploiting the polarization, by oversampling, or by collecting blocks of data at different periods of time; one then refers to spatial, polarization, time, or delay diversities.

Subspace methods resorting to second-order statistics include (Gesbert et al. 1997; Giannakis and Halford 1997; Moulines et al. 1995; Slock 1994a; Tong et al. 1994; Tsatsanis and Giannakis 1997; Deneir et al. 1997) in the SIMO case, and (Abed-Meraim et al. 1997a; Ding 1996; Gorokhov and Loubaton 1997; Ng et al. 1998), in the MIMO case. They are based on fractional sampling, induced cyclostationarity, and/or knowledge of the transmit pulse shape. Channel lengths must be known in order to equalize the channel. If they are equal, a postprocessing by an ICA is necessary in order to identify $\mathbf{W}(0)$. The problem of spatially colored noise is addressed in Abed-Meraim et al. (1994) and Fijalkow and Loubaton (1995). Note the possibility of applying subspace methods in the semi-blind context as well (Kristensson et al. 1996).

An alternative approach is the one of linear prediction. If sources are i.i.d. (HS10), it suffices to identify the nonmonic MIMO model and to compute the residuals. Again, this requires the additional use of ICA to extract the inputs (Abed-Meraim et al. 1997b; Comon 1989a; Gorokhov and Loubaton 1998; Slock 1994b; Swami et al. 1994). Note that the whiteness of the sources was not required in the subspace-based methods just mentioned. In Lindgren and Van der Veen (1996) a second-order algebraic method is proposed when sources have different spectra. Identifiability results are also available when channels have more poles than zeros (Broman et al. 1999).

The previous methods (especially subspace-based) are in general usable only if the number of outputs K strictly exceeds the number of inputs (HS3). If this is not the case, then the additional use of HOS is always necessary. The oldest idea consisting of identifying a linear model, and of computing the residues, is in principle still applicable. This will yield the sources, up to some indeterminacies, if they are indeed i.i.d. and independent; a rather complete survey can be found in Swami et al. (1994).

The extension of SISO algorithms to MIMO cases also led to iterative algorithms such as (Gorokhov and Cardoso 1997; Inouye and Habe 1995; Lindgren et al. 1995; Nguyen-Thi and Jutten 1995; Swindlehurst et al. 1995; Touzni et al. 1998; Van der Veen et al. 1995; Yellin and Porat 1993), some methods being based on multispectra (Yellin and Weinstein 1994). But their iterative character often raises the issue of uncertain convergence, and the existence of spurious stationary points. In Lindgren et al. (1995), the convergence of a 2-input 2-output algorithm inspired from Jutten and Herault (1991) is studied. A criterion of the Godard type is proposed in Tugnait (1997) and allows the extraction of sources one by one. The discrete character of the sources can also be exploited to build MIMO closed-form equalizers (Grellier and Comon 1998a). A quite accessible partial survey can be found in Duhamel (1997).

A recent direction of research concerns the way to handle non-Gaussian noise with unknown statistics, in particular via contrasts (Comon 1994a, 1996; Moreau and Pesquet 1997). The idea is to devise optimization criteria that would have as maxima only acceptable solutions, and that would be asymptotically insensitive to Gaussian noise. For instance, contrasts based only on fourth-order cumulants are being sought. A definition based on (HF3) has been proposed in Comon (1996), but no numerical algorithm is yet available.

5.2 STATISTICAL TOOLS

In this section, the main statistical tools that are used for devising powerful blind separators are briefly presented (see Chapter 1 for complements).

5.2.1 Standardization, Spatial Prewhitening

For various reasons, it is often convenient to standardize the data. The standardized random variable $\tilde{X} = \mathbf{T}X$ associated with X is defined (nonuniquely) as a linear function of the data having a unit covariance (Kendall and Stuart 1977; McCullagh 1987). For instance, if $\mathbf{R}_x = E\{XX^H\}$ denotes the *circular* covariance matrix of X, $\mathbf{T}^{-1}$ can be any square root of $\mathbf{R}_x$, if the latter is full rank. In practice, one often takes for $\tilde{X}$ the matrix of right-singular vectors of the data matrix $\mathbf{X}$ associated with nonzero singular values (Comon 1994a). To be more precise, denote by $\mathbf{X} = \mathbf{V}\Sigma_o\tilde{\mathbf{X}}$ the "economical" Singular Value Decomposition (SVD) of the data matrix $\mathbf{X}$, where $\tilde{\mathbf{X}}$ is of full rank r and Σ_o is $K \times r$ diagonal real positive. Then, one can choose $\mathbf{T} = \Sigma_o^+\mathbf{V}^H$, where Σ_o^+ is the pseudoinverse of the diagonal rectangular matrix Σ_o.

This operation is also sometimes called *spatial whitening*. If the spatial noise coherency $\mathbf{B}$ is known, one can use the noiseless covariance $\mathbf{R}_s = \mathbf{R}_x - \mathbf{R}_v$ instead of $\mathbf{R}_x$. This process is classic in antenna array processing, and yields orthogonal steering vectors (cf. Section 5.2.4). In the scalar case, the standardization merely reduces to division by the standard deviation.

5.2.2 Basic Statistical Tools

Characterization of a Complex Random Vector A random vector X is entirely characterized by its probability density function $f_X(\mathbf{x})$, possibly in the sense of Dirac distributions. Alternatively, the first and second characteristic functions of X, $\Phi_X(\mathbf{u}) = \mathrm{E}\{e^{j\Re(\mathbf{u}^H\mathbf{x})}\}$, and $\Psi(\mathbf{u}) = \log(\Phi_X(\mathbf{u}))$, completely characterize the random variable X; here, $\mathrm{E}\{\cdot\}$ denotes expectation. Then the Taylor expansion of $\Psi(\mathbf{u})$ [resp. of $\Phi(\mathbf{u})$] about the origin yields the cumulants (resp. moments) of increasing orders (Kendall and Stuart 1977), assuming that they exist. For these reasons, the random vector X can also be characterized by its moments or its cumulants, when they are finite.

Moment and Cumulant Examples Let A, B, C, D be *zero-mean* scalar random variables with finite moments up to order 4. Then one defines the joint cumulants, which roughly correspond to a kind of centered version of joint moments. For zero-mean variables, moments and cumulants coincide up to order 3, but at order 4, the following relation is satisfied, and it will be sufficient for our purpose:

$$\mathrm{Cum}\{A, B, C, D\} = \mathrm{E}\{ABCD\} - \mathrm{E}\{AB\}\mathrm{E}\{CD\} - \mathrm{E}\{AC\}\mathrm{E}\{BD\}$$
$$- \mathrm{E}\{AD\}\mathrm{E}\{BC\} \quad (5.8)$$

Let X be a random variable of dimension K. If X takes its values in the complex field, it is useful to distinguish between the three kinds of cumulants that can be formed (the others being deducible by permutation or complex conjugation):

$$\mathscr{C}_{ijk\ell, X} = \mathrm{Cum}\{X_i, X_j, X_k, X_\ell\}$$
$$\mathscr{C}_{ijk, X}^\ell = \mathrm{Cum}\{X_i, X_j, X_k, X_\ell^*\} \quad (5.9)$$
$$\mathscr{C}_{ij, X}^{k\ell} = \mathrm{Cum}\{X_i, X_j, X_k^*, X_\ell^*\}$$

Note that when variables are complex conjugated, their index appears as a superscript. For reasons that will soon become apparent, $\mathscr{C}_{ij,X}^{k\ell}$ is sometimes referred to as the fourth-order circular cumulant (Lacoume et al. 1997). Moments and cumulants of a standardized variable $\tilde{X}$ are often called standardized moments and cumulants of X.

The cumulants are often more attractive than the moments since the logarithm function associated with the second characteristic function transforms the multiplicative properties of the moments into additive ones for the cumulants. This is particularly useful for statistically independent variables, as is discussed later.

Now, let us consider a component X_1 of a zero-mean scalar complex random variable X, and denote by $\mathscr{C}_{1,x}^1$ its variance. Then we may define the standardized fourth-order (FO) moment as

$$\mu_{11\tilde{x}}^{11} = \mathrm{E}\{|X_1|^4\}/(\mathscr{C}_{1,x}^1)^2$$

From the previous definitions, the corresponding standardized cumulant is

$$\mathscr{K}_{11,x}^{11} \overset{\text{def}}{=} \mathscr{C}_{11\tilde{x}}^{11} = \mathrm{Cum}\{X_1, X_1, X_1^*, X_1^*\}/(\mathscr{C}_{1,x}^1)^2 \qquad (5.10)$$

The latter standardized cumulant is generally referred to as the kurtosis.[2] It can be shown (Lacoume et al. 1997; Cardoso 1997) by applying a Cauchy-Schwarz inequality that $\mu_{11\tilde{x}}^{11} \geq 1$. As a consequence, we also have: $\mathscr{K}_{11,x}^{11} \geq -1 - |\mathrm{E}\tilde{X}_1^2|^2 \geq -2$. This minimal value is reached for BPSK distributions.

5.2.3 Properties

Circularity   The assumption of circularity of a random vector X, defined and discussed in Picinbono (1994), is very often used in the literature to simplify the developments and especially to introduce the probability density function of a complex Gaussian vector (Goodman 1963). Besides, for random processes or signals, this concept has some links with the property of stationarity (Picinbono 1994). However, in practice, numerous random processes are noncircular, in particular in the context of digital radio communications. This property has to be taken into account in the estimation of the HOS of the data (Ferreol and Chevalier 1998a,b).

A random vector X is said to be circular if and only if its probability density function (pdf) is invariant to rotation, which means that $f_X(\mathbf{x}) = f_X(e^{j\theta}\mathbf{x})$ whatever the value of θ (Lacoume et al. 1997; Picinbono 1994). As a consequence, for a circular vector X of dimension K and for $1 \leq i_k, j_k \leq K$, we obtain if $p \neq r$:

---

[2] In some early works, the standardized moment also sometimes received the appellation of *kurtosis*. In the present paper, the kurtosis will designate exclusively the standardized cumulant.

$$\mathrm{Cum}\{X_{i_1}, \ldots, X_{i_p}, X_{j_1}^*, \ldots, X_{j_r}^*\} = 0 \quad \text{and} \quad \mathrm{E}\{X_{i_1}, \ldots, X_{i_p} X_{j_1}^*, \ldots, X_{j_r}^*\} = 0$$

$$(5.11)$$

The random vector X is said to be the qth-order circular if the previous property is satisfied for all the indices $1 \leq i_k, j_k \leq K$ such that $p + r = q$.

It is shown in Picinbono (1994) that the analytical signal and the complex envelope of a NB stationary real signal are necessarily circular up to an order that increases as the signal bandwidth decreases, which relates the concept of circularity to a physical property. Thus, as digitally modulated signals (omnipresent in radio-communications contexts) are not stationary but (quasi)-cyclostationary, they may be noncircular. For example, BPSK signals are $2p$th-order noncircular, $\forall p$. Nevertheless, a noncircular cyclostationary signal (e.g., BPSK) sampled at the symbol rate, becomes stationary but still keeps a non-circular behavior. On the other hand, a quarternary phase-shift keying (QPSK) signal is second-order circular but $4p$th-order noncircular. More generally (M-PSK) signals are Mpth-order noncircular (Lacoume et al. 1997).

Gaussian Case   A complex random vector X is said to be Gaussian if and only if the real random vector obtained by concatenation of its real and imaginary parts is a Gaussian vector (Goodman 1963). If X is zero-mean complex Gaussian, its odd moments are zero, and its even moments can be deduced from the SO moments; as a consequence, all the cumulants of order 3 or more are zero.

Because complex Gaussian variables are completely characterized by their first- and SO moments, the concept of circularity of a complex Gaussian vector is reduced to the concept of SO circularity. Thus, a complex Gaussian vector X is circular if and only if $\mathrm{E}\{XX^T\} = 0$. The FO standardized moment of a circular complex Gaussian variable is thus equal to 2, whereas it is equal to 3 for real variables.

Statistical Independence P random vectors X_i, $1 \leq i \leq P$ are said to be statistically independent if and only if the joint pdf $f(\mathbf{x}_1, \mathbf{x}_2, \ldots, \mathbf{x}_P)$ of the P-uplet $(X_1, X_2, \ldots, X_P)$ is equal to the product of the marginal ones $f_{X_i}(\mathbf{x}_i)$, that is,

$$f(\mathbf{x}_1, \mathbf{x}_2, \ldots, \mathbf{x}_P) = \prod_{i=1}^{P} f_{X_i}(\mathbf{x}_i) \tag{5.12}$$

If the P random vectors X_i, $1 \leq i \leq P$ are statistically independent, then for any family of functions $\{g_i\}$: $\mathrm{E}\{\prod_i g_i(\mathbf{x}_i)\} = \prod_i \mathrm{E}\{g_i(\mathbf{x}_i)\}$. In particular, if random variables $\mathbf{x}_i$ are scalar, if the pdf of $\mathbf{x}_1$ is even, and if g_1 is an odd function, then

$$\mathrm{E}\left\{\prod_{i=1}^{P} g_i(\mathbf{x}_i)\right\} = 0$$

This last property is the core of some iterative source separators, such as that of Jutten and Herault (1991). On the other hand, if two random vectors X and Y are statistically independent, then all the cross-cumulants between X and Y are zero:

$$\text{Cum}\{X_{i_1}, \ldots, X_{i_p}, Y_{j_1}, \ldots, Y_{j_q}\} = 0 \qquad (5.13)$$

for any set of indices compatible with the dimensions of X and Y. Note that this property is at the heart of numerous FO source separators (Cardoso and Souloumiac 1993) or (Comon 1994a).

5.2.4 Statistics of Sensor Measurements

From now on, we shall restrict our attention mainly to model (5.4), valid for static mixtures of NB sources.

SO Statistics   The SO statistics of the data are completely described by the so-called circular correlation $\mathbf{R}_x(n, \tau) = \mathrm{E}\{\mathbf{x}(n + \tau/2)\mathbf{x}(n - \tau/2)^{\mathrm{H}}\}$, and non-circular correlation $\mathbf{C}_x(n, \tau) = \mathrm{E}\{\mathbf{x}(n + \tau/2)\mathbf{x}(n - \tau/2)^{T}\}$ of $\mathbf{x}(n)$, which, from (5.4), can be written as

$$\mathbf{R}_x(n, \tau) = \mathbf{A}\mathbf{R}_{mc}(n, \tau)\mathbf{A}^{\mathrm{H}} + \mathbf{R}_v(n, \tau) = \mathbf{R}_s(n, \tau) + \mathbf{R}_v(n, \tau) \qquad (5.14)$$

$$\mathbf{C}_x(n, \tau) = \mathbf{A}\mathbf{C}_{mc}(n, \tau)\mathbf{A}^{T} + \mathbf{C}_v(n, \tau) = \mathbf{C}_s(n, \tau) + \mathbf{C}_v(n, \tau) \qquad (5.15)$$

where $\mathbf{R}_{mc}(n, \tau) = \mathrm{E}\{\mathbf{m}^c(n + \tau/2)\mathbf{m}^c(n - \tau/2)^{\mathrm{H}}\}$, $\mathbf{C}_{mc}(n, \tau) = \mathrm{E}\{\mathbf{m}^c(n + \tau/2) \cdot \mathbf{m}^c(n - \tau/2)^{T}\}$, and with a similar notation for $\mathbf{v}$.

For nonstationary sources or noise, these matrices depend on the time index n. In particular, for (quasi)-cyclostationary sources, the first and second correlation matrices of both $\mathbf{m}^c(n)$ and $\mathbf{x}(n)$ have a Fourier series expansion showing off the cyclic frequencies contained in $\mathbf{m}^c(n)$ and $\mathbf{x}(n)$ (Gardner 1987; Gardner et al. 1987). Finally, for stationary sources and noise, the previous correlation matrices no longer depend on parameter n. Note that most of the HO source separators developed for instantaneous mixtures of NB sources exploit only the information contained in the correlation matrices of the data for $\tau = 0$.

On the other hand, for circular observation, the second (noncircular) correlation matrix is zero. This occurs, for example, if noise and sources impinging on the array are stationary, but it does not necessarily occur when digital sources are sampled at the symbol rate. Indeed, sampling at the symbol rate makes the source stationary, but can preserve the noncircularity of the source (e.g., BPSK).

Thus, for stationary complex sources and observations $\mathbf{x}(n)$, the SO statistics that are exploited by most of the HO separators of instantaneous mixtures of NB sources are given by, with $\mathbf{R}_x$ standing for $\mathbf{R}_x(\tau = 0)$,

$$\mathbf{R}_x = \mathbf{A}\mathbf{R}_{mc}\mathbf{A}^H + \mathbf{R}_v \overset{\text{def}}{=} \mathbf{R}_s + \mathbf{R}_v \tag{5.16}$$

$$\mathbf{C}_x = \mathbf{A}\mathbf{C}_{mc}\mathbf{A}^T + \mathbf{C}_v \overset{\text{def}}{=} \mathbf{C}_s + \mathbf{C}_v \tag{5.17}$$

FO Statistics We have seen that the FO cumulants $\mathscr{C}_{ijk\ell,X}$ and $\mathscr{C}^\ell_{ijk,X}$ are zero for circular variables; hence the name of *noncircular cumulants*. Thus, the cumulant-based FO separators of static mixtures must exploit the FO information contained in the circular FO cumulants, $\mathscr{C}^{j\ell}_{ik,X}$. For practical purposes, this four-way array is sometimes stored in matrix form. One possibility consists of assigning the rows to indices i and j of $\mathscr{C}^{j\ell}_{ik,X}$, and the columns to indices k and ℓ. As a result, the so-called quadricovariance matrix, $\mathbf{Q}_x(n)$, is Hermitian, and

$$\mathbf{Q}_x(n) = [\mathbf{A} \otimes \mathbf{A}^*]\mathbf{Q}_{mc}(n)[\mathbf{A} \otimes \mathbf{A}^*]^H + \mathbf{Q}_v(n) \tag{5.18}$$

Remark that this storing preserve one symmetry ($Q_{k\ell ij} = Q_{ijk\ell}$), but may loose the two others present in the original array (Comon and Cardoso 1990) (namely $Q_{\ell kji} = Q^*_{ijk\ell}$ and $Q_{jik\ell} = Q_{ijk\ell}$) after matrix manipulations. For stationary observations, the quadricovariance matrix does not depend on time index n, and for a Gaussian noise $\mathbf{Q}_v(n)$ is zero.

Statistical Estimators For given statistical properties of the observation vector $\mathbf{x}(n)$, the SO and FO cumulant estimators have to be chosen so as to generate asymptotically unbiased estimates with a covariance tending to zero as the observation time increases. In fact, it is not always possible to implement such estimators, since some properties of ergodicity or cycloergodicity (Boyles and Gardner 1983) are required. For stationary and ergodic observations, the classic sample estimates enjoy the latter properties (Hannan 1970; Kendall and Stuart 1977).

However, even in the presence of (quasi)-cyclostationary and cycloergodic sources, the sample estimators generally yield biased estimates of the FO statistics of the data (Ferreol and Chevalier 1998a,b), and more powerful estimators are often preferred. A discussion on this matter can be found in Ferreol and Chevalier (1998a,b).

5.2.5 Approximation of the Mutual Information

Mutual Information By definition, the components of the random vector X are independent if and only if their joint distribution deflates into the product of the marginal ones, Eq. (5.12). A quite natural measure of statistical independence is thus a divergence between $f_X(\mathbf{u})$ and $\prod_i f_{X_i}(u_i)$. Assuming the Kullback divergence leads to the mutual information as an independence criterion (Blahut 1987; Cardoso 1998a; Comon 1994a; Kendall and Stuart 1997). See also Chapter 1 for more details.

Contrary to what we could expect, $I(f_X)$ is not invariant to linear transforms. In order to see this, just consider the mutual information of a Gaussian variable with pdf Φ_X and covariance $\mathbf{R}$: $I(\Phi_X) = \frac{1}{2}\log(\det^{-1}\mathbf{R}\prod R_{ii})$. The information $I(\Phi_X)$ is invariant only under transforms of the form $\Lambda\mathbf{P}$, where Λ is diagonal and $\mathbf{P}$ is a permutation, unless the covariance matrix $\mathbf{R}$ is diagonal. Such transforms are sometimes referred to as *generalized permutations*, or more generally *trivial transforms* (Cao and Liu 1996; Comon 1996).

Negentropy   Now, given any random variable X with pdf f_X, denote by Φ_X the Gaussian pdf with the same mean and covariance matrix. One defines the negentropy of X as

$$J(f_X) = \int f_X(\mathbf{u})\log\frac{f_X(\mathbf{u})}{\Phi_X(\mathbf{u})}\, d\mathbf{u}$$

The negentropy is a measure of departure from normality, since it can be seen to be equal to the divergence $D_{f_X\|\Phi_X}$. Negentropy is invariant by unitary change of coordinates, as is the classic differential entropy (Blahut 1987), but it is also invariant to any invertible transform.

It is interesting to write the mutual information as a combination of negentropies, as pointed out in Comon (1992, 1994a):

$$I(f_X) = I(\Phi_X) + J(f_X) - \sum_i J(f_{X_i}) \tag{5.19}$$

From the preceding properties, it can be shown that mutual information is invariant to scale change (i.e., any diagonal invertible transform). But this formula also emphasizes that three terms can contribute to creating some statistical dependence between components X_i. First $I(\Phi_X)$ is a SO contribution, and can be eliminated by standardization (cf. Section 5.2). The two remaining terms are contributions of higher orders. Yet, after standardization, only unitary transforms are allowed in order to maintain the first term at zero. And we have just seen that $J(f_X)$ is invariant under such transforms. So the only term that can be utilized in order to measure the dependence between the components of a unit variance random vector is the third one.

Edgeworth Expansion   In practice, a negentropy is not easy to estimate; in order to decrease its computational cost, it has been suggested that it be approximated by a finite expansion. The so-called *type A* Edgeworth expansions aim at approximating densities about the Gaussian density with the same mean and variance (Kendall and Stuart 1977, 6.49; McCullagh 1987). And this is precisely what we are doing when utilizing negentropy. The difference with Gram-Charlier expansions lies in the ordering of terms, which is arbitrary in Gram-Charlier, whereas it is consistent with the Central Limit theorem in Edgeworth. For all these reasons, an Edgeworth expansion of negentropies has been re-tained in Comon (1994a) as an optimization criterion, whereas a Gram-Charlier expansion has been chosen in Gaeta and Lacoume (1990) and Lacoume and Ruiz (1992).

A simple but painful calculation has shown (Comon 1992, 1994a) that if X is scalar standardized, then

$$J(f_X) \approx \frac{1}{12}\mathscr{C}_{(3)}^2 + \frac{1}{48}\mathscr{C}_{(4)}^2 + \frac{7}{48}\mathscr{C}_{(3)}^4 - \frac{1}{8}\mathscr{C}_{(3)}^2\mathscr{C}_{(4)} \qquad (5.20)$$

where $\mathscr{C}_{(3)} = \mathrm{Cum}\{X, X, X\}$ and $\mathscr{C}_{(4)} = \mathrm{Cum}\{X, X, X, X\}$. If the distribution of X is skew, then the approximation $J(f_X) \approx \frac{1}{12}\mathscr{C}_{(3)}^2$ can be sufficient. On the other hand, if it is symmetric, then $J(f_X) \approx \frac{1}{48}\mathscr{C}_{(4)}^2$ holds true. As discussed in the next section, these two criteria have been successfully proposed for practical use.

5.2.6 Contrasts

Contrasts have served as optimization criteria in SISO blind deconvolution problems (Donoho 1981), in blind source separation (Comon 1994a), and more recently in MIMO blind deconvolution (Comon 1996; Moreau and Pesquet 1997). In practice, the mutual information, Eq. (5.19) is difficult to utilize. Therefore, more practical criteria are sought, and it is desired that they possess suitable properties.

Let $\mathscr{P}$ be a set of multidimensional random processes and $\mathscr{H}$ a set of authorized transformations operating on it. A contrast Υ on $(\mathscr{P}, \mathscr{H})$ is a mapping associating any element of $\mathscr{H} \cdot \mathscr{P}$ with a real number, and satisfying three properties:

C1. Υ is invariant to scale change: $\Upsilon(\mathbf{y}) = \Upsilon(\Lambda\mathbf{y})$, for all diagonal constant matrices Λ of $\mathscr{H}$.

C2. If $\mathbf{y}$ has independent components, then $\Upsilon(H * \mathbf{y}) \leq \Upsilon(\mathbf{y})$, for all $H \in \mathscr{H}$.

C3. Equality occurs in the foregoing if and only if H is trivial. The set of trivial transforms (e.g., the set of generalized permutations) defines how much the contrast is a discriminant.

In order to separate sources, it suffices to search $\mathscr{H}$ for the transform $\mathbf{W}$ maximizing $\Upsilon(\mathbf{y})$, where $\mathbf{y} = \mathbf{W}^H * \mathbf{x}$ is the output of the separator and $\mathbf{x}$ the observation.

Examples Take for $\mathscr{P}$ the set of non-Gaussian standardized random variables with finite marginal circular cumulants of order r, $\mathscr{C}_{(r),y}$, $r > 2$, and for $\mathscr{H}$ the set of unitary matrices. Then $\Upsilon_{\alpha,r} = \sum_i |\mathscr{C}_{(r),y_i}|^\alpha$ are contrasts for any $r > 2$ and $\alpha > 0$ (Comon 1994a). In the same framework, another criterion has been used, for $r = 4$; $\Upsilon_{\mathrm{JADE}} = \sum_{ik\ell} |\mathscr{C}_{ij,y}^{i\ell}|^2$ (Cardoso and Souloumiac 1993), and recently has been shown to be asymptotically equivalent to $\Upsilon_{2,4}$ in the presence of Gaussian noise and provided model (5.4) is exactly satisfied.

In the convolutive scalar case, take for $\mathscr{P}$ the set of scalar random processes with finite cumulants of order r, and $\mathscr{H}$ the set of nonzero filters. Then $\Upsilon_{\alpha,r} = |\mathscr{K}_{(r),y}|^\alpha$ are contrasts, for any $r > 2$ and $\alpha > 0$, where $\mathscr{K}_{(r),y}$ is the standardized cumulant of the output $y(n)$ (e.g., the kurtosis if $r = 4$). Trivial filters are then those whose impulse response is zero everywhere except at one lag.

In the multidimensional case, one can take for $\mathscr{P}$ the set of non-Gaussian processes of dimension K with finite cumulants of order r, and $\mathscr{H}$ a certain class of linear filters with finite norm (Comon 1996; Moreau and Pesquet 1997). Then $\Upsilon_{\alpha,r} = \sum_i |\mathscr{K}_{(r),y_i}|^\alpha$ are again contrasts, if $\mathscr{H}$ are standardized cumulants. Trivial filters are formed as the product of a permutation by a diagonal matrix of scalar trivial filters.

5.3 ULTIMATE SEPARATORS

In Section 5.1, we reviewed a number of problems and models that can be encountered. In Section 5.2, we introduced statistical tools potentially required to implement blind source separators. In this section, separators that are optimal with respect to some performance criteria are analyzed. The qualifier *ultimate* refers to the fact that (1) separators are optimal in a certain sense, and (2) channel and source statistics are supposed to be perfectly known (e.g., ultimate performances sometimes coincide with asymptotic ones). One could also attempt to bound ultimate performances by looking for the easiest channel possible (Grellier and Comon 1998b), but this is beyond the scope of the present section.

As was pointed out in Section 5.1, the nature of ultimate separators is closely linked to the goal we want to reach, to the assumptions we make about sources and noise, and to performance criteria. See the references cited therein for further details.

In this section, we concentrate on static mixtures of NB sources, and only investigate the problem of recovering the complex source inputs by means of a spatial linear and TI filter, $\mathbf{W}$. Under these assumptions, the sensor observations satisfy model (5.4), and the goal is summarized by expression (5.5), where the filtering operation is defined by Eq. (5.6).

5.3.1 Performance Criteria

In the context presented previously, source separators require the introduction of a performance criterion. Several performance criteria have already been used by numerous authors. Among these criteria, we can cite the $\Lambda\mathbf{P}$–invariant gap between true mixture matrix $\mathbf{A}$ and its blind estimation $\hat{\mathbf{A}}$, based on L^1 and L^2 norms (Comon 1994a). The main advantage of the latter criterion is that it does not require the estimation of the best permutation matrix $\mathbf{P}$, thus allowing reduced complexity. The drawback is that it does not quantitatively characterize the restoration quality of a given source, but only allows comparisons.

Another criterion is the residual source interference rate at each output (Cardoso 1994). Its inconvenience is that it does not take into account the presence of background noise, and requires an explicit estimation of **P**.

A good performance criterion should, if computational complexity is ignored:

- Be able to characterize quantitatively the restitution quality of each of the P sources by an arbitrary source separator;
- Allow the performance comparison of two arbitrary separators for the restitution of each source;
- Take into account all the physical parameters that may alter the process of source separation (P, K, background noise, relative power of the sources, sources-to-noise ratio, source distributions, angular separation between sources, etc.)
- Be related to an ideal performance criterion, such as ROC curves or (BER) after demodulation, depending on the application.

Such a criterion at the output of a source separator was introduced in Chevalier (1995b) and is based on the signal-to-interference-plus-noise ratio (SINR) of each source at the separator outputs. To introduce it, let us call *interference for source* k all the sources j with $j \neq k$. Then we define the SINR of source k at output i of separator **W** by the following expression

$$\text{SINR}_k(\mathbf{w}_i) = \sigma_k^2 \frac{|\mathbf{w}_i^H \mathbf{a}_k|^2}{\mathbf{w}_i^H \mathbf{R}_{v_k} \mathbf{w}_i} \tag{5.21}$$

where $\mathbf{w}_i$ is the ith column of **W**; $\mathbf{a}_k$ is the kth column of **A**; $\mathbf{R}_{v_k}$ is the temporal mean (for nonstationary sources) of the correlation matrix of the noise plus interference for source k, defined by $\mathbf{R}_{v_k} = \mathbf{R}_x - \sigma_k^2 \mathbf{a}_k \mathbf{a}_k^H$; and σ_k^2 is the input power of source k, defined as the temporal mean (for nonstationary sources) of the variance $\mathrm{E}\{|m_k^c(n)|^2\}$.

With these definitions, for each source k, we define the maximum SINR of source k at the outputs of **W**, denoted by $\text{SINRM}_k(\mathbf{W})$, by the maximum value of the quantities $\text{SINR}_k(\mathbf{w}_i)$, when i varies from 1 to P. Finally, the performance index of the separator **W** is defined by the P-tuplet

$$\mathscr{M}(\mathbf{W}) = (\text{SINRM}_1(\mathbf{W}), \text{SINRM}_2(\mathbf{W}), \dots, \text{SINRM}_P(\mathbf{W})) \tag{5.22}$$

We verify that **W** and **WΛP** have the same performance $\mathscr{M}(\mathbf{W})$, which is meaningful, since they belong to the same equivalence class (Cao and Liu 1996; Comon 1994a). Moreover, we say that a separator **W**1 is better than a separator $\mathbf{W}2$ for the restitution of the source k in a given context if $\text{SINRM}_k(\mathbf{W}_1) > \text{SINRM}_k(\mathbf{W}2)$ for this context.

5.3.2 Ultimate Separators

Spatial Matched-Filter Source Separator   The ultimate NB source separator is the separator that gives the best performance for the restitution of each source. It maximizes, over all the possible separators $\mathbf{W}$, the quantities $\text{SINRM}_k(\mathbf{W})$ for each source k. It implements a spatial matched filter (SMF) to each source, and is defined by the following expression

$$\mathbf{W}_{\text{SMF}} = \mathbf{R}_x^{-1}\mathbf{A}\mathbf{\Lambda}\mathbf{P} \tag{5.23}$$

Its output vector is given by

$$\mathbf{y}_{\text{SMF}}(n) = \mathbf{W}_{\text{SMF}}^H\mathbf{x}(n) = \mathbf{P}^H\mathbf{\Lambda}^H\mathbf{A}^H\mathbf{R}_x^{-1}\mathbf{x}(n) \tag{5.24}$$

which gives, by using the model (5.4)

$$\mathbf{y}_{\text{SMF}}(n) = \mathbf{P}^H\mathbf{\Lambda}^H\{\mathbf{A}^H\mathbf{R}_x^{-1}\mathbf{A}\mathbf{m}^c(n) + \mathbf{A}^H\mathbf{R}_x^{-1}\mathbf{v}(n)\} \tag{5.25}$$

We deduce from the latter expression that in the presence of noise, the exact restitution of the sources is not possible. Moreover, as the matrix $\mathbf{A}^H\mathbf{R}_x^{-1}\mathbf{A}$ has no reason to be diagonal, in most situations, an interference residue exists at each output of the optimal separator. As a consequence, interferences are not completely canceled, except when the matrix $\mathbf{A}^H\mathbf{R}_v^{-1}\mathbf{A}$ is diagonal or when noise is absent (Chevalier 1999). A detailed description of the properties and performance of the optimal NB source separator $\mathbf{W}_{\text{SMF}}$ is presented in Chevalier (1999). The optimal performance vector is eventually given by

$$\mathscr{M}(\mathbf{W}_{\text{SMF}}) = (\sigma_1^2\mathbf{a}_1^H\mathbf{R}_{v1}^{-1}\mathbf{a}_1, \sigma_2^2\mathbf{a}_2^H\mathbf{R}_{v2}^{-1}\mathbf{a}_2, \ldots, \sigma_P^2\mathbf{a}_P^H\mathbf{R}_{vP}^{-1}\mathbf{a}_P) \tag{5.26}$$

Weighted Least-Square Source Separator   The NB source separator, optimal with respect to criterion (5.26), that completely nulls the interferences at each output, corresponds to the weighted least-square (WLS) source separator (Chevalier 1999), denoted by $\mathbf{W}_{\text{WLS}}$. This separator generates the WLS estimate of $\mathbf{m}^c(n)$, that is, the estimate minimizing

$$\Upsilon_{\text{WLS}}(\mathbf{m}^c(n)) = (\mathbf{x}(n) - \mathbf{A}\mathbf{m}^c(n))^H\mathbf{B}^{-1}(\mathbf{x}(n) - \mathbf{A}\mathbf{m}^c(n)) \tag{5.27}$$

The WLS equivalence class contains the separators defined by

$$\mathbf{W}_{\text{WLS}} = \mathbf{B}^{-1}\mathbf{A}(\mathbf{A}^H\mathbf{B}^{-1}\mathbf{A})^{-1}\mathbf{\Lambda}\mathbf{P} \tag{5.28}$$

whose output vectors are given by

$$\mathbf{y}_{\mathrm{WLS}}(n) = \mathbf{W}_{\mathrm{WLS}}^{H}\mathbf{x}(n) = \mathbf{P}^{H}\boldsymbol{\Lambda}^{H}(\mathbf{A}^{H}\mathbf{B}^{-1}\mathbf{A})^{-1}\mathbf{A}^{H}\mathbf{B}^{-1}\mathbf{x}(n) \qquad (5.29)$$

which, by using model (5.4) gives

$$\mathbf{y}_{\mathrm{WLS}}(n) = \mathbf{P}^{H}\boldsymbol{\Lambda}^{H}\{\mathbf{m}^{c}(n) + (\mathbf{A}^{H}\mathbf{B}^{-1}\mathbf{A})^{-1}\mathbf{A}^{H}\mathbf{B}^{-1}\mathbf{v}(n)\} \qquad (5.30)$$

where $\mathbf{R}_v = \eta_v \mathbf{B}$, with η_v corresponding to the mean power of the noise per sensor, and $\mathbf{B}$ being the matrix of the noise spatial coherence, such that trace $\{\mathbf{B}\} = K$. The WLS source separator corresponds to the optimal one whenever the latter completely nulls the interference, that is, when the matrix $\mathbf{A}^{H}\mathbf{B}^{-1}\mathbf{A}$ is diagonal or in the absence of noise. The properties and performance of the WLS source separator are presented in Chevalier (1999).

5.3.3 Ultimate Performance

The computation of the optimal separator performance index allows us to understand the way in which physical parameters such as the signal-to-noise ratio (SNR), the angular separation of the sources, the noise spatial coherence, the number of sources and sensors, or the array geometry, influence the output performance. Besides, these optimal performance indices represent an upper bound to the performance achievable by any linear spatial TI blind NB source separator.

Two-Source Case In the two-source case ($P = 2$), the SINRM$_1$ at the output of the optimal separator is given by

$$\mathrm{SINRM}_1(\mathbf{W}_{\mathrm{SMF}}) = \varepsilon_1 \left\{ 1 - \frac{\varepsilon_2}{1 + \varepsilon_2} |\alpha_{12}|^2 \right\} \qquad (5.31)$$

where

$$\varepsilon_k = \frac{\sigma_k^2}{\eta_v} \mathbf{a}_k^{H}\mathbf{B}^{-1}\mathbf{a}_k \qquad (5.32)$$

and where α_{12} is the spatial correlation coefficient of sources 1 and 2, defined by

$$\alpha_{12} = \frac{\mathbf{a}_1^{H}\mathbf{B}^{-1}\mathbf{a}_2}{(\mathbf{a}_1^{H}\mathbf{B}^{-1}\mathbf{a}_1)^{1/2}(\mathbf{a}_2^{H}\mathbf{B}^{-1}\mathbf{a}_2)^{1/2}} \qquad (5.33)$$

This coefficient is a function of the noise spatial coherence matrix, the angular separation between the sources, the array geometry, the number and the kind of sensors, the polarization of the sources and sensors. Note that the quantity $\text{SINRM}_2(\mathbf{W}_{\text{SMF}})$ is obtained from Eq. (5.31) by exchanging indices 1 and 2.

Equation (5.31) shows off all the physical parameters that control the optimal source-separator performance. It shows in particular that the SINRM_1 at the output of $\mathbf{W}_{\text{SMF}}$ is always proportional to the input SNR of source 1. Besides, this SINRM_1 is maximum and equal to ε_1 in the absence of interference ($\varepsilon_2 = 0$) or when the two sources are orthogonal in the $\mathbf{B}^{-1}$ metric ($\alpha_{12} = 0$). In these latter situations the output SINRM_1 is, for spatially white noise and identical sensors, proportional to the number of sensors K.

Nevertheless, in the general case, the presence of a second source (interference) degrades the performance, and the output SINRM_1 becomes a decreasing function of ε_2 and $|\alpha_{12}|^2$, which means in particular that the output SINRM_1 decreases when the input SNR of source 2 increases or when the angular separation of the two sources decreases beyond the 3-dB beam width of the array (for sensors not diversely polarized). In particular when $|\alpha_{12}|^2$ is maximum and equal to 1, which is the case when the two sources come from the same direction and have the same polarization (if the sensors are diversely polarized), the output SINRM_1 is equal to the input 1, and the source separator is inoperative.

On the other hand, for a strong interference source ($\varepsilon_2 \gg 1$), as long as $|\alpha_{12}|^2$ is not too close to unity, Eq. (5.31) can be written approximately as

$$\text{SINRM}_1(\mathbf{W}_{\text{SMF}}) \approx \varepsilon_1\{1 - |\alpha_{12}|^2\} \tag{5.34}$$

This expression no longer depends upon the input SNR of the interference source, since the latter is rejected under the output background noise level, which solely controls the total output noise power. Thus, in the presence of a strong interference not too close to source 1, the SINRM_1 at the output of $\mathbf{W}_{\text{SMF}}$ is mainly controlled by the maximal SINRM_1, ε_1, that can be reached at the output of $\mathbf{W}_{\text{SMF}}$. It is also controlled by the square modulus of the spatial correlation coefficient α_{12}, which gathers all the physical parameters controlling the optimal source-separator performance for source 1, except the input SNR1 controlling ε_1.

Note that the SINRM_1 at the output of the WLS source separator is exactly given by the right-hand side of Eq. (5.34), whatever the power and angular separation of the sources. This shows that in all cases, $\text{SINRM}_1(\mathbf{W}_{\text{WLS}}) \leq \text{SINRM}_1(\mathbf{W}_{\text{SMF}})$, except for $\mathbf{B}^{-1}$-orthogonal sources, or in the absence of noise, where the two quantities coincide. In fact, the performances at the output of the optimal and the WLS separators are not very different in most cases, except for weak sources that are angularly close (Chevalier 1999).

General Case of P Sources   The previous results can be generalized to an arbitrary number of sources, provided that $P \leq K$. The details of this generalization are presented in Chevalier (1999), where it is shown in particular that an increase in the number of sources necessarily decreases the output performance, as illustrated in Fig. 5.2.

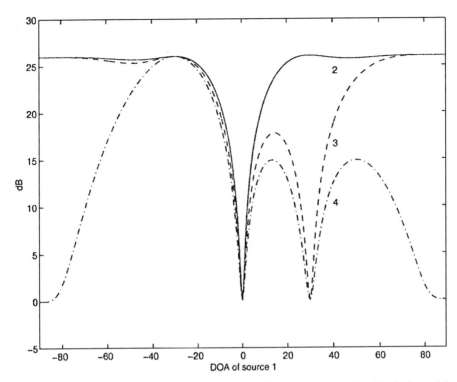

Figure 5.2   SINRM$_1$(**W**$_{SMF}$) as a function of θ_1, for $K = 4$ sensors and for $P = 2$, 3, and 4 sources.

This figure shows the variations of SINRM$_1$ at the output of **W**$_{SMF}$ as a function of θ_1, the direction of arrival (DOA) of source 1, for several values of the number of sources and for a uniform linear array of $K = 4$ sensors, assuming that the input SNR of all sources is equal to 20 dB. The DOAs of sources 2, 3, and 4 are such that $\theta_2 = 0, \theta_3 = 30$, and $\theta_4 = 90$ degrees. Note the degradation in performance as the number of sources increases and as the angular separation between source 1 and the others decreases.

5.4   A CLASS OF OFF-LINE NUMERICAL ALGORITHMS

In this section we concentrate on two families of block algorithms aimed at separating P sources from K sensor measurements, $P \le K$. For reasons of space, their (obvious) adaptive implementation is not discussed.

5.4.1   Unitary Transforms

As pointed out in Section 5.2, spatial whitening is often applied to the data in a first stage. The effect is twofold. First, it allows for a dimensionality reduction if the rank is deficient. And second, the data are standardized so that their variance is unity and their cross-correlation is zero; this removes the first term present in the mutual information, $I(\Phi_X)$, in Eq. (5.19).

As a consequence, the only transformations that we are allowed to apply to standardized data are those that preserve the SO covariance, that is, unitary transformations. Yet any unitary matrix U can be split, in a nonunique manner, into a product of so-called Givens plane rotations, and a diagonal matrix Δ of unit modulus entries:

$$U = \prod_{i<j} G[i, j]\Delta, \qquad G[i, j] = \begin{pmatrix} I & \vdots & O & \vdots & O \\ \cdots & c & \cdots & s & \cdots \\ O & \vdots & I & \vdots & O \\ \cdots & -s^* & \cdots & c & \cdots \\ O & \vdots & O & \vdots & I \end{pmatrix} \qquad (5.35)$$

These matrices differ from identity only in their (i, i), (i, j), (j, i), and (j, j) coordinates, and can always be required to have a real cosine, c. Conversely, any mixing matrix A can be factored into a product $A = TU\Delta$, where U is unitary and can be decomposed into a product of Givens rotations.

We are looking for a unitary matrix U, such that $Y = U\tilde{X}$ has a maximal contrast. Now consider the following class of contrast functions, already introduced in Section 5.2.6:

$$\Upsilon_{\alpha, r} = \sum_i |\mathscr{C}_{(r), Y_i}|^\alpha$$

Yet, because of the multilinear property of cumulants (McCullagh 1987), $\mathscr{C}_{(r), Y_i}$ $= \sum_{k\ell mn} U_{ik} U_{i\ell} U^*_{im} U^*_{in} \mathscr{K}^{mn}_{k\ell, X}$. As a consequence, the matrix U should maximize the following criterion (taking $r = 4$ for clarity):

$$\Upsilon_{\alpha, 4} = \sum_i \sum_{k\ell mn} U_{ik} U_{i\ell} U^*_{im} U^*_{in} |\mathscr{K}^{mn}_{k\ell, X}|^\alpha \qquad (5.36)$$

It turns out that this complicated expression takes quite a simple form if Givens rotations are considered, as explained in the next subsection. This is one of the major advantages of applying a sequence of plane rotations, instead of estimating a sequence of full unitary matrices.

5.4.2 The Case of Two Sources After Prewhitening

Define the Givens rotation:

$$G = \frac{1}{\sqrt{1 + \theta\theta^*}} \begin{pmatrix} 1 & \theta \\ -\theta^* & 1 \end{pmatrix}$$

where $\theta = s/c$ is the complex tangent of the rotation angle, whose modulus is required to lie within the interval $(-\pi/2, \pi/2]$. Estimating the Givens rotation consists solely of computing parameter θ. Thus the optimization problem reduces to a 1-dimensional search. But looking more carefully at the optimization criterion reveals that stationary points can be obtained by mere polynomial rooting techniques.

Real Case To be more concrete, consider the case of real mixtures, and let $\xi = \theta - 1/\theta$. Then the stationary values of contrasts $\Upsilon_{2,i}$, $i = 3, 4$, can be shown to satisfy $\omega_{2,i}(\xi) = 0$, where (Comon 1992, 1994a)

$$\Upsilon_{2,3}(\theta) = \left(\theta + \frac{1}{\theta}\right)^{-3} \sum_{i=1}^{3} a_i(\theta^i - (-\theta)^{-i}); \qquad \omega_{2,3}(\xi) = d_2\xi^2 + d_1\xi - 4d_2; \quad (5.37)$$

$$\Upsilon_{2,4}(\xi) = (\xi^2 + 4)^{-2} \sum_{i=0}^{4} b_i\xi^i; \qquad \omega_{2,4}(\xi; g) = \sum_{i=0}^{4} c_i\xi^i \qquad (5.38)$$

The values of parameters a_i, b_i, c_i, d_i are given functions of the standardized data cumulants, $\mathcal{K}_x$, and can be found in Comon (1994b).[3] As a consequence, the absolute maxima of $\Upsilon_{2,3}$ and $\Upsilon_{2,4}$ can be found solely by analytical means. The relation mapping θ to ξ being bijective, one obtains θ from ξ by rooting $\theta^2 - \xi\theta - 1 = 0$. The two roots belong to the same equivalence class (i.e., the corresponding Givens rotations derive from each other by a factor of the form ΛP).

Similarly, for $\Upsilon_{1,i}$, $i = 3, 4$, the way to obtain the absolute maxima analytically is even simpler. In fact, one can show that, up to a sign (the common sign of source standardized cumulants):

$$\Upsilon_{1,3}(\theta) = (1 + \theta^2)^{-3/2}(\alpha\theta^3 + \beta\theta^2 + \gamma\theta + \delta);$$

$$\omega_{1,3}(\xi) = \beta\theta^3 + (2\gamma - \alpha)\theta^2 + (\delta - 2\beta)\theta - \gamma$$

$$\Upsilon_{1,4}(\xi) = (\xi^2 + 4)^{-1}(a\xi^2 + 4b\xi + 2c); \qquad (5.39)$$

$$\omega_{1,4}(\xi) = b\xi^2 + (c - 2a)\xi - 4b$$

where $\alpha = \mathcal{K}_{222,x} - \mathcal{K}_{111,x}$; $\beta = \mathcal{K}_{112,x} + \mathcal{K}_{122,x}$; $\gamma = \mathcal{K}_{112,x} - \mathcal{K}_{122,x}$; $\delta = \mathcal{K}_{222,x} + \mathcal{K}_{111,x}$; and $a = \mathcal{K}_{1111,x} + \mathcal{K}_{2222,x}$; $b = \mathcal{K}_{1222,x} - \mathcal{K}_{1112,x}$; $c = 6\mathcal{K}_{1122,\tilde{x}} + a$. The rooting of $\omega_{1,3}(\theta)$ is easy, and two roots correspond to equivalent solutions maximizing $\Upsilon_{1,3}(\theta)$, opposite and inverse of each other. Likewise, the rooting of $\omega_{1,4}(\xi)$ does not raise any difficulty; one root corresponds to an absolute maximimum of $\Upsilon_{1,4}(\xi)$; to select it, it suffices to compute

---

[3] Downloadable from www.i3s.unice.fr/~comon.

the contrast at those two values. Then, as earlier, rooting $\theta^2 - \xi\theta - 1 = 0$ leads to two equivalent solutions for θ.

For real data and mixtures, the number of operations is thus extremely small, and is dominated by the computation of the cumulants of the standardized observations, $\tilde{x}_1$ and $\tilde{x}_2$. If N denotes the data length, this represents only $O(8\,N)$ floating-point operations for the 5 FO source cumulants if $\Upsilon_{1,4}$ is used (Comon 1994a), or $O(6\,N)$ for the 4 third-order ones if $\Upsilon_{1,3}$ is used. The numerical complexity for separating more than two sources is studied in a later section.

Complex Case   In the complex case, the same reasoning holds true if one defines the auxiliary variable $\xi = \theta - 1/\theta^*$ (Chaumette et al. 1993), or equivalently $\theta\theta^* - \xi\theta^* - 1 = 0$. But the rooting involves a larger computational burden, since variable ξ is complex.

For this reason, the analytical maximization of contrast $\Upsilon_{1,4}$ has been studied in by Comon and Moreau (1997). In fact, because the cumulants are not squared, the degree of the polynomials is decreased by half (i.e., down to degree 2). It has been shown that stationary values $Re^{j\phi}$ of $\Upsilon_{1,4}(\xi)$ satisfy the system of two equations below:

$$b\cos(\phi + \beta)(R^2 - 4) + (2d\cos(2\phi + 2\beta + \mu) + f - 2a)R = 0 \quad (5.40)$$

$$b\sin(\phi + \beta)R + 2d\sin(2\phi + 2\beta + \mu) = 0 \quad (5.41)$$

where parameters a, b, d, f, μ, β are given functions of the data standardized cumulants $\mathcal{K}_x$, and can be found in Comon and Moreau (1997);[4] the maximization of $\Upsilon_{1,4}$ can then be completed entirely analytically with reduced computational cost (rooting of 1 cubic and 2 quadratics). In addition, it has been observed that the solution obtained with $\Upsilon_{1,4}(\xi)$ has a slightly smaller variance than $\Upsilon_{2,4}(\xi)$, which confirms the theory (Comon et al. 1997).

Noiseless Case   If the noise is Gaussian, and for long integration lengths, N, everything happens as if noise were absent. In such a case, there is a simpler and more comprehensive expression for the optimal separator. In fact, the cumulants of the observations $\tilde{x}_1$ and $\tilde{x}_2$ can be directly expressed as linear functions of the source cumulants. More precisely, let the output of the separator be expressed as

$$\mathbf{y} = \begin{pmatrix} \cos\alpha & \sin\alpha\, e^{j\varphi} \\ -\sin\alpha\, e^{-j\varphi} & \cos\alpha \end{pmatrix} \begin{pmatrix} \tilde{x}_1 \\ \tilde{x}_2 \end{pmatrix}$$

Then it can be shown (Comon 1989b; Lacoume et al. 1997), by eliminating the output marginal cumulants in the linear system, and by forcing the cross-

---

[4] Paper and MATLAB codes can be downloaded from www.i3s.unice.fr/~comon.

cumulants of **y** to zero, that

$$\frac{\mathcal{K}_{12,x}^{22} - \mathcal{K}_{11,x}^{12}}{\mathcal{K}_{12,x}^{12}} = -2\cot(2\alpha)e^{j\varphi}$$

In other words, parameters α and φ can be identified directly from three standardized cross-cumulants of the observation. If $\mathcal{K}_{12,x}^{12}$ is close to zero, then the ratio is ill-conditioned. Fortunately, in that case, it can be computed by either $\mathcal{K}_{11,x}^{11}/\mathcal{K}_{11,x}^{12}$ or $\mathcal{K}_{22,x}^{22}/\mathcal{K}_{12,x}^{22}$.

Contrary to what one would expect, this noiseless solution still behaves surprisingly well when estimation errors affect the observation cumulants, $\mathscr{C}_{ij,\tilde{x}}^{k\ell}$ (in particular for short data records, e.g., 200 samples). On the other hand, in the presence of significant non-Gaussian noise (or interferences) with unknown distribution, the solutions given in the two previous subsections generally exhibit much better performance.

5.4.3 The Contrast Maximization Algorithms

The contrast maximization algorithm (CoM) introduced in Comon (1992, 1994a) consists of proceeding pairwise, and processes each pair with the help of one of the closed-form solutions presented in the previous subsection.

Given a $K \times N$ data matrix, **X**, we first initialize the separating matrix **W** to the $P \times K$ whitening matrix **T** and the estimated source matrix **Y** to the standardized data matrix **X̃**. Then we sweep the $P(P-1)$ possible pairs of rows (i, j) in **Y**, $1 \le i < j \le K$. For each of them, we process the $2 \times N$ submatrix **Y**$[i, j]$ formed of rows i and j of **Y** by applying one of the separation algorithms described in Section 5.4.2. The result obtained is an estimate of the separating matrix **G**$[i, j]$ and an estimate of the two sources, **G**$[i, j]$**Y**$[i, j]$. Then we update the outputs as **Y** $\leftarrow$ **G**$[i, j]$**Y** and the separating matrix as **W** $\leftarrow$ **WG**$[i, j]$. The application of the $P(P-1)$ possible Givens rotations constitutes a sweep. In practice, it is necessary to run several sweeps (an order of $2 + \sqrt{K}$ is reasonable) (Comon 1994a), just as in the classic matrix diagonalization techniques (Garbow et al. 1997; Golub and van Loan 1989). At the end of the last sweep, the $P \times N$ matrix **Y** contains the estimate of the P sources, and the relation **Y** = **WX** is satisfied.

In the sequel, we use CoM1 (resp. CoM2) to refer to the CoM algorithm maximizing $\Upsilon_{1,4}$ (resp. $\Upsilon_{2,4}$). Contrasts $\Upsilon_{1,3}$ and $\Upsilon_{2,3}$ will not be subsequently utilized, since they involve third-order cumulants, which are often ill-conditioned.

5.4.4 The JADE Algorithms

Principle Because the transform **U** to be found is constrained to be unitary, the maximization of $\Upsilon_{2,4}$ in Eq. (5.36) is equivalent to minimizing the sum of the squared moduli of all cross-cumulants (Comon 1992, 1994a). On the other hand, Cardoso and Souloumiac (1993) proposed maximizing $\Upsilon_{\text{JADE}} = \sum_{ik\ell} |\mathscr{C}_{ij,z}^{i\ell}|^2$, which is equivalent to minimizing the sum of the squared moduli of the cumulants with distinct first and second indices (cf. Section 5.2.6). In such a criterion, not all cross-cumulants are meant to be cancelled, even if in practice the effect is often achieved (Chevalier 1995b).

The ICA can be seen as a tensor diagonalization problem (Comon 1994b; Comon and Cardoso 1990), and tools similar to those utilized for matrix diagonalization can be thought of (Cardoso and Comon 1990). But in order to do that, the symmetry structure must be broken (only one of the three symmetries is imposed; see our remark in Section 5.2.4). This is what we observe in criterion Υ_{JADE}, whereas the symmetry structure is preserved in $\Upsilon_{2,4}$. In return, efficient matrix eigenvalue decomposition (EVD) algorithms are at our disposal (Cardoso 1989).

The joint approximate diagonalization of eigenmatrices (JADE) algorithm introduced by Cardoso and Souloumiac (1993) is a trade-off in the sense that it still operates on matrices, but performs better than a mere EVD of a contracted version of the original tensor, as in Cardoso (1989). The idea is to try to approximately diagonalize several matrices, instead of attempting to diagonalize the tensor itself (Cardoso and Comon 1996; Comon 1994a). The core of the algorithm includes the proper choice of P' matrices, and the joint diagonalization technique.

Choice of a Set of Matrices   Any tensor of order 4 can be considered (in a nonunique fashion) as an operator acting on matrices. For instance, given $C_{ik}^{j\ell}$, one can map any matrix $\mathbf{M}$ to the matrix $\Omega(\mathbf{M})$ defined as $\Omega(\mathbf{M})_{ij} = \sum_{k\ell} C_{ik}^{j\ell} M_{k\ell}$. In the complex case, one can then relate a Hermitian operator to the tensor of circular cumulants. As a linear Hermitian operator, it then admits an EVD. For the sake of convenience, one can call its P^2 eigenelements *eigenmatrices*. The idea is to select the P' dominant eigenmatrices. Originally, these eigenmatrices were weighted by the eigenvalues λ_p with which they were associated (Cardoso and Souloumiac 1993). This will be referred to as the JADE2 algorithm. Alternatively, we can weight them by $\sqrt{\lambda_p}$ instead. The corresponding algorithm has a behavior different from that of JADE2, and later will be called JADE1.

Joint Diagonalization   Given a set of P' square matrices $\mathbf{M}(m)$, $1 \leq m \leq P'$, the algorithm seeks a unitary matrix $\mathbf{V}$ such that $\text{off}(\mathbf{V}) = \sum_m |\text{Diag}(\mathbf{V}^H \mathbf{M}(m)\mathbf{V}|^2$ is minimized. In order to do this, the algorithm described in Cardoso and Souloumiac (1993) proceeds pairwise, similarly to the Jacobi techniques proposed earlier (Comon 1989b). Due to the special form of the optimization criterion, however, the calculation of each Givens rotation is carried out in a specific way (Cardoso and Souloumiac 1993, 370).

Comments   The JADE algorithms have some limitations, besides the fact that they break symmetry. In fact, contrary to CoM, they cannot be applied to sta-

Table 5.1   Polynomial Rooting Involved in the Three Basic Source-Separation Block Algorithms

	Real	Complex
CoM2	1 quartic + 1 quadratic	1 higher degree + 1 quadratic
CoM1	2 quadratics	1 cubic + 2 quadratics
JADE	Eigenmatrices + 1 3 × 3 EVD	Eigenmatrices + 1 3 × 3 EVD

tistics of odd orders (e.g., skewness), nor to tensors lacking Hermitian symmetry (e.g., noncircular FO cumulant). On the other hand, the matrix formalism makes them easier to understand.

5.4.5   Numerical Complexity

Let us call CoM1 (resp. CoM2) the algorithm maximizing $\Upsilon_{1,4}$ (resp. $\Upsilon_{2,4}$). Then, the separation of two sources from a 2-dimensional mixture by any of the four algorithms CoM1, CoM2, JADE1, or JADE2 involves polynomial rooting, as given in Table 5.1.

Now it is possible to evaluate rather accurately the computational complexity in terms of the number of floating-point operations (flops) as a function of the number of sources P, the number of sensors K, and the number of available samples N. Here, one multiplication followed by one addition are combined into one flop; complexities are given below in complex flops.

With these conventions, we can evaluate the numerical complexity of the standardization of a $K \times N$ data matrix:

$$\min\left[\frac{1}{2}NK^2 + \frac{4}{3}K^3, NK^2\right] + PKN$$

The execution of the JADE1 or JADE2 algorithms demands the additional computational complexity defined below:

$$3Nt + \min\left[8P^5 + 24P^3, \frac{4}{3}P^6\right] + 2IP^4 + \frac{21}{2}IP^3 + 37IP^2 + NP^2$$

whereas for CoM1 the extraneous computations take

$$\min\left[3Nt + 2ItP^2 + NP^2, \frac{13}{2}INP^2\right] + 2IP^3 + 16IP^2$$

complex flops, with $t = \frac{1}{8}P^4 + \frac{1}{4}P^3$, $I = \text{round}(\sqrt{P}) + 2$, and for $P > 2$. These calculations assume that integers $\{P, K, N\}$ satisfy $i^2 \gg i$, that is, the simplify-

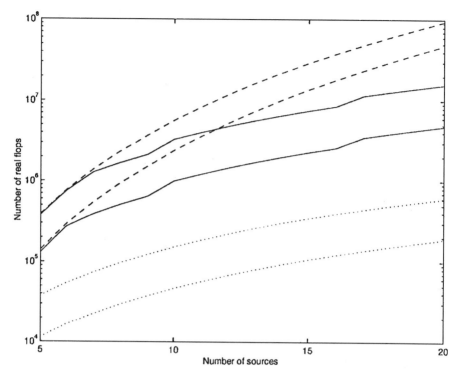

Figure 5.3 Complexity in number of flops as a function of P. Dotted line: data standardization alone; dashed line: JADE and solid line: CoM1, both without including standardization. The complexity is given for $K = P$ ranging from 5 to 20 sources, and for data lengths of 300 and 1000 samples.

ing approximations may be questionable for small values. The complexities vary with P as shown in Fig. 5.3. In particular, we observe that the complexity of CoM1 improves compared to JADE2 when P increases.

5.5 BLIND SOURCE-SEPARATOR PERFORMANCE

In this section we present some results of the performance of the methods CoM1, CoM2, JADE1, and JADE2 described in Section 5.4. The separator outputs described in Section 5.4 do not correspond to the blind estimates of optimal separator outputs of Section 5.3. In this section, an SMF is constructed, $\mathbf{W} = \mathbf{R}_X^{-1}\hat{\mathbf{A}}$, matched to the blind estimate of the mixing matrix.

A performance analysis of methods CoM2 and JADE2 is presented in Chevalier (1995a,b), while the same kind of analysis is presented in Comon et al. (1997) for methods CoM1 and JADE1.

Summary Numerous computer simulations associated with multiple scenarios of sources have enabled us to draw some conclusions on the behavior of CoM1, CoM2, JADE1, and JADE2.

More precisely, in the presence of Gaussian noise with a known spatial coherence matrix, the blind separators CoM2, JADE1, and JADE2 are equivalent in the steady state and implement the optimal separator, regardless of the power and the angular separation of the sources, provided that $P \leq K$ and that at most one of the sources is Gaussian. Next, if all source kurtoses have the same sign, the separator CoM1 also becomes equivalent to CoM2, JADE2, and JADE1 in steady state and thus becomes optimal. This is no longer true when source kurtoses have different signs, in which case the separator CoM1 may become suboptimal, and may even fail to separate the sources (Comon et al. 1997).

Moreover, in the preceding favorable conditions, as long as the number of sources is not overestimated, the four separators JADE1, JADE2, CoM1, and CoM2 have practically the same performance regardless of the observation duration, provided that no source is Gaussian. In these cases, the convergence speed of the output SINRM of the sources toward optimality depends on the SNR (see Fig. 5.5) and on source kurtosis (Chevalier 1995a,b). For instance, this convergence slows down when the source kurtoses decrease to zero (see Fig. 5.6). A consequence of this result is that a passband filtering of the data, or a fast Fourier transform (FFT), implemented before the source separator, might reduce the convergence speed of the algorithms or may even prevent it from separating the sources (Chevalier 1995a,b). In fact, this can be explained by a tendency to Gaussianity if the bandwidth of the filter is much smaller than the source bandwidth.

Nevertheless, in the presence of a Gaussian source and for a finite observation duration, the performance of JADE1 and CoM1 are equivalent to each other, and degrade with respect to that of JADE2 and CoM2, which are also equivalent to each other. This loss in performance of JADE1 and CoM1 becomes greater as SNR increases and as observation duration decreases (Comon 1997).

On the other hand, an overestimation of the number of sources slows down the convergence of the four blind methods. Nevertheless, the decrease in performance is much worse for JADE1 and JADE2 than for CoM1 and CoM2 (Chevalier 1995a).

In the presence of several Gaussian sources, the four methods fail in separating the Gaussian sources but still succeed, under the favorable conditions described previously, in separating the non-Gaussian sources (Chevalier 1995a).

For noise with an unknown coherence matrix, the performance of the four methods become generally suboptimal, which does not necessarily mean inefficient (Chevalier 1995a).

Last, in the presence of non-Gaussian noise, the performances of the four methods rapidly degrade as soon as the absolute value of the noise kurtosis becomes greater than a threshold that increases with SNR and with source kurtosis (Fig. 5.4). Nevertheless, the robustness of the four methods with re-

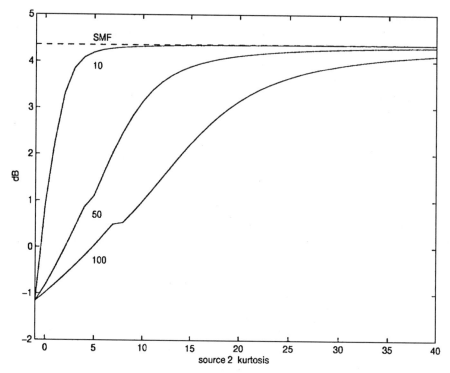

Figure 5.4   SINRM₁ as a function of source two kurtosis, and for various noise kurtosis: 10, 50, 100. The horizontal line is the SINRM₁ of the SMF.

spect to a non-Gaussian noise is not the same and method CoM1 seems the most robust, followed by the three others, which seem less and equally robust.

Computer Results   To illustrate the preceding results, in this section we present some computer simulations.

Figure 5.4 illustrates the behavior of CoM2 in the presence of non-Gaussian noise. It shows the variations of $SINRM_1$ at the output of CoM2 as a function of the kurtosis of source 2, $\mathcal{K}_2$, for different values of the noise kurtosis, $\mathcal{K}_v$, when the array of four sensors received two QPSK sources; the SNR is 0 dB and the directions of arrival are 0 and 15 degrees, respectively. Note the improvement in performance of CoM2 as $\mathcal{K}_2$ increases with respect to $\mathcal{K}_v$. The same behavior is observed for JADE2. These results have been computed with exact statistics (no actual data have been generated). Therefore, they can be considered as asymptotic performances.

Figure 5.5 illustrates the behavior of CoM2 and JADE2 in the presence of a spatially white Gaussian noise, for two QPSK sources sampled at the symbol rate, angularly separated by 20 degrees; the SNR ranges from −10 dB to 30 dB. The SINRM₁ at the output of CoM2 and JADE2 are plotted as a function of the

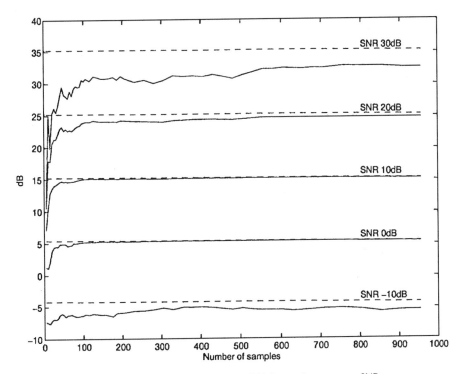

Figure 5.5 SINRM$_1$ as a function of N, for various source SNRs.

number of independent snapshots of noise and sources. Note the quick convergence of the output SINRM$_1$ as long as the steady-state output SINRM$_1$ remains below 15 dB, and note the decreasing convergence speed as the output SINRM$_1$ increases above 15 dB.

Figure 5.6 shows the influence of the source kurtoses on the convergence speed of JADE2 and CoM2. It shows the variations of the SINRM$_1$ at the output of JADE2 or CoM2 as a function of the number of independent snapshots for several values of the source kurtoses, for two sources having an SNR of 10 dB, and an angular separation of 20 degrees. Note the decreasing convergence speed of SINRM$_1$ as the kurtosis decreases.

5.6   CONCLUSIONS AND PROSPECTIVE APPLICATIONS

The advantages of blind techniques compared to informed ones are that they require neither array calibration nor precise information on the sources. We have shown that approaches based on FO statistics are powerful, and that they allow the processing of a wide class of signals, in both radar and communications, among other areas.

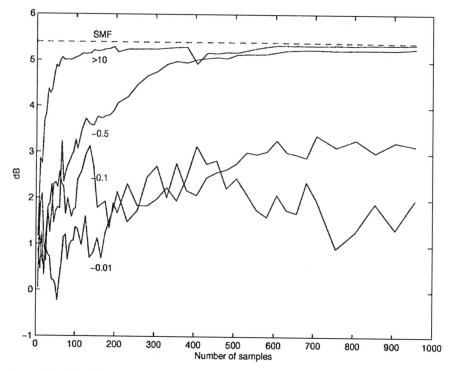

Figure 5.6   SINRM₁ as a function of N for various values of source kurtoses, taken equal.

The authors are convinced that source-separation techniques, and especially the so-called independent-component analysis (ICA), will play an increasingly important role in the scientific community, not only for blind identification or deconvolution of multivariate channels, but also in data analysis, which is intrinsically static. In addition, they feel that ICA will be utilized in a wider variety of operational applications.

The present chapter is an attempt to gather the main results concerning source-separation concepts, tools, and performance measures, with an emphasis on off-line algorithms solving the ICA problem. In fact, because of the extensive computing power now available, the performances do not seem to be limited by computational costs, but by the convergence speed of adaptive algorithms. Therefore, block methods should be preferred to fully recursive ones, in particular in time varying environments.

More precisely, on average, block approaches require a number of arithmetic operations of the same order as stochastic-gradient adaptive approaches, but their convergence is always better, by construction, in a stationary environment. In addition, they are better matched to data arriving in bursts, like in GSM, even in the presence of time-varying channels. Therefore, we think that block algorithms will develop further in the coming years.

We did not look much at convolutive mixtures, mainly because of space, but also because the subject is not yet completely mature, though we did point out a number of results available in the literature. We emphasized that ICA plays a central role in blind deconvolution problems. These problems are important for various reasons. From the point of view of source separation in communications, multichannel equalization avoids network saturation by decreasing the number of apparent sources (in static separation, multipaths are viewed as distinct sources).

Finally, the problem of extracting more sources than sensors might also become crucial in some antenna array processing problems, for example, cellular networks, because of the increasing number of users and because of the decreasing resources per user. This problem can be addressed by using high-order statistics exclusively, that is, without resorting to spatial prewhitening.

ACKNOWLEDGMENT

This work, initiated while the author was with the Eurecom Institute, has been supported in part by the CNRS Telecommunications program TL97104. We thank Dr. A. Swami for his careful reading of the manuscript.

REFERENCES

Abed-Meraim, K., P. Loubaton, and E. Moulines, 1994, "Subspace method for blind identification of multichannel FIR filters in noise field with unknown spatial covariance," in *Asilomar conference*, California.

Abed-Meraim, K., P. Loubaton, and E. Moulines, Mar. 1997a, "A subspace algorithm for certain blind identification problems," *IEEE Trans. Inform. Theory*, pp. 499–511.

Abed-Meraim, K., E. Moulines, and P. Loubaton, Mar. 1997b, "Prediction error method for second-order blind identification," *IEEE Trans. Signal Processing*, vol. 45, no. 3, pp. 694–705.

Agee, B. G., Apr. 7–11 1986, "The Least Squares CMA: a new technique for for rapid correction of constant modulus signals," in *Proc. ICASSP*, Tokyo, pp. 953–956.

Amari, S., 1998, "Natural gradient works efficiently in learning," *Neural Computation*, vol. 10, pp. 251–276.

Bar-Ness, Y., J. W. Carlin, and M. L. Steinberger, Nov. 1982, "Bootstrapping adaptive interference cancelers: Some practical limitations," *Proc. The Globecom. Conference*, Miami, pp. 1251–1255, paper No F3.7.

Bell, A. J., and T. J. Sejnowski, Nov. 1995, "An information-maximization approach to blind separation and blind deconvolution," *Neural Computation*, vol. 7, no. 6, pp. 1129–1159.

Bellini, S., and F. Rocca, 1990, "Asymptotically efficient blind deconvolution," *Signal Processing*, vol. 20, pp. 193–209.

Belouchrani, A., K. Abedmeraim, J. F. Cardoso, and E. Moulines, 1993, "Second-order blind separation of correlated sources," *Proc. Int. Conf. Digital Signal Processing,* Cyprus, pp. 346–351.

Benveniste, A., and M. Goursat, Aug. 1984, "Blind equalizers," *IEEE Trans. Communic.,* vol. 32, no. 8, pp. 871–883.

Blahut, R. E., 1987, *Principles and Practice of Information Theory,* (Reading, MA: Addison-Wesley).

Boyles, R. A., and W. A. Gardner, Jan. 1983, "Cycloergodic properties of discrete parameters non stationary stochastic processes," *IEEE Trans. Inform. Theory,* vol. 39, no. 1.

Broman, H., U. Lindgren, H. Sahlin, and P. Stoica, Feb. 1999, "Source separation: a TITO system identification approach," special issue on blind separation and deconvolution, *Signal Processing,* Elsevier, vol. 73, no. 2.

Cao, X. R., and R. W. Liu, Mar. 1996, "General approach to blind source separation," *IEEE Trans. Signal Processing,* vol. 44, no. 3, pp. 562–570.

Cardoso, J. F., 1989, "Source separation using higher order moments," *Proc. ICASSP* Glasgow, pp. 2109–2112.

Cardoso, J. F., 1992, "Iterative techniques for blind source separation using only fourth order cumulants," *Proc. EUSIPCO,* Brussels, Belgium.

Cardoso, J. F., Sept. 1994, "On the performance of source separation algorithms," *Proc. EUSIPCO,* Edinburgh, pp. 776–779.

Cardoso, J. F., Apr. 1997, "Infomax and Maximum Likelihood for source separation," *IEEE Signal Processing Letters,* vol. 4, no. 4, pp. 112–114.

Cardoso, J. F., 1998a, "Blind signal separation: statistical principles," *Proc. IEEE,* Special issue on blind identification and estimation, To appear.

Cardoso, J. F., 1998b, "High-order contrasts for independent component analysis," *Neural Computation,* to appear.

Cardoso, J. F., and P. Comon, 1990, "Tensor-based independent component analysis," *Proc. European Signal Processing Conf. EUSIPCO,* Barcelona, Spain, September 18–21 1990, pp. 673–676.

Cardoso, J. F., and P. Comon, May 1996, "Independent component analysis, a survey of some algebraic methods," *Proc. ISCAS Conference,* Atlanta, vol. 2, pp. 93–96.

Cardoso, J. F., and A. Souloumiac, Dec. 1993, "Blind beamforming for non-Gaussian signals," *IEE Proceedings—Part F,* vol. 140, no. 6, pp. 362–370, special issue on applications of high-order statistics.

Caroll, J. D., S. Pruzansky, and J. B. Kruskal, Mar. 1980, "Candelinc: A general approach to multidimensional analysis of many-way arrays with linear constraints on parameters," *Psychometrika,* vol. 45, no. 1, pp. 3–24.

Chaumette, E., P. Comon, and D. Muller, Dec. 1993, "An ICA-based technique for radiating sources estimation; application to airport surveillance," *IEE Proceedings—Part F,* vol. 140, no. 6, pp. 395–401, special issue on applications of high-order statistics.

Chevalier, P., Sept. 1995a, "Méthodes aveugles de filtrage d'antennes," *Revue d'Electronique et d'Electricité, SEE,* no. 3, pp. 48–58.

Chevalier, P., 1995b, "On the performance of higher order blind sources separation methods," in *IEEE-ATHOS Workshop on Higher-Order Statistics,* Begur, Spain, 12–14 June 1995, pp. 30–34.

Chevalier, P., May 1996a, "Optimal array processing for non stationary signals," *ICASSP*, Atlanta, vol. 5, pp. 2868–2871.

Chevalier, P., 1996b, "Optimal time invariant and widely linear spatial filtering for radiocommunications," *Proc. EUSIPCO 96*, Trieste, Italy, Sept. 10–13 1996, pp. 559–562.

Chevalier, P., Feb. 1999, "Optimal separation of independent narrow-band sources—concept and performance," *Signal Processing*, Elsevier, vol. 73, no. 1, special issue on blind separation and deconvolution.

Chevalier, P., and A. Maurice, Apr. 1997, "Constrained beamforming for cyclostationary signals," *ICASSP*, Munich, pp. 3789–3792.

Chevalier, P., and A. Maurice, 1998, "Blind and informed cyclic array processing for cyclostationary signals," *Eusipco*, Rhodes, Greece, Sept. 8–11 1998, pp. 1645–1648.

Chevalier, P., and B. Picinbono, Oct. 1996, "Complex linear-quadratic systems for detection and array processing," *IEEE Trans. Signal Processing*, vol. 44, no. 10, pp. 2631–2634.

Chevalier, P., V. Capdevielle, and P. Comon, 1997, "Behavior of higher order blind source separation methods in presence of correlated multipaths," *IEEE SP Workshop on HOS*, Banff, Canada, July 21–23 1997, pp. 363–367.

Chevalier, P., V. Capdevielle, and P. Comon, 1999, "Performance of HO blind source separation methods: Experimental results on ionospheric HF links," *ICA99, IEEE Workshop on Indep. Comp. Anal. and Blind Source Separation*, Aussois, France, Jan. 11–15 1999, pp. 443–448.

Comon, P., 1989a, "Separation of sources using high-order cumulants," *SPIE Conf. Adv. Alg. Archi. Signal Processing*, San Diego, Ca, August 8–10 1989, pp. 170–181, vol. Real-time signal processing XII.

Comon, P., 1989b, "Separation of stochastic processes," *Proc. Workshop on Higher-Order Spectral Analysis*, Vail, Colorado, June 28–30 1989, IEEE-ONR-NSF, pp. 174–179.

Comon, P., 1991, "Independent component analysis," *Proc. Int. Signal Processing Workshop on Higher-Order Statistics*, Chamrousse, France, July 10–12 1991, pp. 111–120. Republished in *Higher-Order Statistics*, J. L. Lacoume ed., Elsevier, 1992, pp. 29–38.

Comon, P., Apr. 1994a, "Independent component analysis, a new concept?", *Signal Processing*, Elsevier, vol. 36, no. 3, pp. 287–314, special issue on higher-order statistics.

Comon, P., 1994b, "Tensor diagonalization, a useful tool in signal processing," in *IFAC-SYSID, 10th IFAC Symposium on System Identification*, M. Blanke, T. Soderstrom, eds., Copenhagen, Denmark, July 4–6 1994, vol. 1, pp. 77–82, invited session.

Comon, P., July 1996, "Contrasts for multichannel blind deconvolution," *Signal Processing Lett.*, vol. 3, no. 7, pp. 209–211.

Comon, P., 1998, "Blind channel identification and extraction of more sources than sensors," in *SPIE Conference*, San Diego, July 19–24 1998, pp. 2–13, keynote address.

Comon, P., and J. F. Cardoso, 1990, "Eigenvalue decomposition of a cumulant tensor with applications," in *SPIE Conf. Adv. Signal Processing Alg.*, San Diego, CA, July 10–12 1990, vol. 1348, Arch. and Implem., pp. 361–372.

Comon, P., O. Grellier, and B. Mourrain, 1998, "Blind channel Identification with MSK inputs," in Asilomar Conference, Pacific Grove, CA, Nov. 1–4, pp. 1569–1573.

Comon, P., and E. Moreau, 1997, "Improved contrast dedicated to blind separation in communications," *ICASSP*, Munich, April 20–24 1997, pp. 3453–3456.

Comon, P., and B. Mourrain, Sept. 1996, "Decomposition of quantics in sums of powers of linear forms," *Signal Processing*, Elsevier, vol. 53, no. 2, pp. 93–107, special issue on high-order statistics.

Comon, P., P. Chevalier, and V. Capdevielle, 1997, "Performance of contrast-based blind source separation," *SPAWC-IEEE Signal Processing Advances in Wireless Communic.*, Paris, April 16–18 1997, pp. 345–348.

Comon, P., C. Jutten, and J. Herault, July 1991, "Separation of sources, part II: Problems statement," *Signal Processing*, Elsevier, vol. 24, no. 1, pp. 11–20.

DeLathauwer, L., B. DeMoor, and J. Vandewalle, Sept. 1994, "Blind sources separation by higher-order singular value decomposition," in *Proc. EUSIPCO*, Edinburgh, pp. 175–178.

DeLathauwer, L., B. DeMoor, and J. Vandewalle, June 1996, "Independent component analysis based on higher-order statistics only," in *8th IEEE SP Workshop on Statistical Signal and Array Processing*, Corfu, Greece, pp. 356–359.

Delfosse, N., and P. Loubaton, 1995, "Adaptive blind separation of independent sources: a deflation approach," *Signal Processing*, vol. 45, pp. 59–83.

Deneire, L., J. Ayadi, and D. Slock, 1997, "Subspace Fitting without Eigendecomposition," in *Proc. DSP'97*, Santorini, Greece, July 1997.

Deville, Y., June 1996, "A unified stability analysis of the Herault-Jutten source separation neural network," *Signal Processing*, Elsevier, vol. 51, no. 3, pp. 229–233.

Deville, Y., and L. Andry, Oct. 1996, "Application of blind source separation techniques to multi-tag contactless identification systems," *IEICE Trans. Fund. Electr. Com. Comput. Sci.*, vol. E79-A, no. 10, pp. 1694–1699.

Ding, Z., June 1996, "An outer-product decomposition algorithm for multichannel blind identification," in *8th IEEE SP Workshop SSAP*, Corfu, Greece, pp. 132–135.

Ding, Z., and R. A. Kennedy et al., Jan. 1993, "Local convergence of the Sato blind equalizer and generalizations under practical constraints," *IEEE Trans. Inform. Theory*, vol. 39, no. 1.

Donoho, D., 1981, "On minimum entropy deconvolution," in *Applied time-series analysis II*, (New York: Academic Press) pp. 565–609.

Duhamel, P., 1997, "Blind multivariate equalization," in *DSP'97, 13th Int. Conf. Digital Signal Processing*, Santorini, Greece, July 2–4, 1997.

D'Urso, G., P. Prieur, and C. Vincent, 1997, "Blind identification methods applied to Electricite de France civil works and power plants monitoring," *IEEE SP Workshop on HOS*, Banff, Canada, July 21–23 1997, pp. 82–86.

Emile, B., P. Comon, and J. Leroux, July 1998, "Estimation of time delays with fewer sensors than sources," *IEEE Trans. Signal Processing*, vol. 46, no. 7, pp. 2012–2015.

Feder, M., A. V., Oppenheim, and E. Weinstein, Feb. 1989, "ML noise cancellation using the EM algorithm," *IEEE Trans. Acoust. Speech Signal Processing*, vol. 37, no. 2, pp. 204–216.

Ferreol, A., and P. Chevalier, 1998a, "On the fourth-order cumulants estimation for the HO blind separation of cyclostationary sources," in *ICASSP*, Seattle, May 12–15 1998, pp. 2313–2316.

Ferreol, A., and P. Chevalier, 1998b, "On the limits of current higher order blind source separation methods for cyclostationary sources," *IEEE Trans. Signal Processing*, submitted.

Fety, L., 1988, Methodes de Traitement d'Antenne Adaptees aux Radio-communications, Doctorat, ENST.

Fijalkow, I., and P. Loubaton, 1995, "Identification of rank one rational spectral densities from noisy observations: a stochastic realization approach," *Systems and Control Lett.*, vol. 24, pp. 201–205.

Fijalkow, I., A. Touzni, and J. R. Treichler, Jan. 1997, "Fractionally spaced equalization using CMA: Robustness to channel noise and lack of disparity," *IEEE Trans. Signal Processing*, vol. 45, no. 1, pp. 56–66.

Forney, G. D., May 1972, "Maximum-likelihood sequence estimation of digital sequences in the presence of intersymbol interference," *IEEE Trans. Inform. Theory*, vol. 18, no. 3.

Gaeta, M., and J. L. Lacoume, 1990, "Source separation without a priori knowledge: the maximum likelihood solution," *Proc. EUSIPCO*, Barcelona, Spain, pp. 621–624.

Gamboa, F., 1995, "Separation of sources having unknown discrete supports," in *IEEE-ATHOS Workshop on Higher-Order Statistics*, Begur, Spain, 12–14 June 1995, pp. 56–60.

Garbow, B. S., et al., 1977, *EISPACK*, vol. 51 of *Lecture Notes in computer sciences*, (New York: Springer Verlag).

Gardner, W. A., Jan. 1993, "Cyclic wiener filtering: Theory and method," *IEEE Trans. Communic.*, vol. 41, no. 1, pp. 151–163.

Gardner, W. A., W. A. Brown III, and C. K. Chen, June 1987, "Spectral correlation of modulated signals: Part II—digital modulation," *IEEE Trans. Communic.*, vol. 35, no. 6, pp. 595–601.

Gassiat, E., 1995, "Blind deconvolution of discrete linear systems perturbed with additive noise," in *IEEE-ATHOS Workshop on Higher-Order Statistics*, Begur, Spain, 12–14 June 1995, pp. 305–309.

Van Gerven, S., and D. Van Compernolle, July 1995, "Signal separation by symmetric adaptive decorrelation: Stability, convergence, and uniqueness," *IEEE Trans. Signal Processing*, vol. 43, no. 7, pp. 1602–1612.

Gesbert, D., P. Duhamel, and S. Mayrargue, Sept. 1997, "On-line blind multichannel equalization based on mutually referenced filters," *IEEE Trans. Signal Processing*, vol. 45, no. 9, pp. 2307–2317.

Giannakis, G. B., and S. D. Halford, Sept. 1997, "Blind fractionally spaced equalization of noisy FIR channels: Direct and adaptive solutions," *IEEE Trans. Signal Processing*, vol. 45, no. 9, pp. 2277–2292.

Godard, D., Nov. 1980, "Self recovering equalization and carrier tracking in two dimensional data communication systems," *IEEE Trans. Communic.*, vol. 28, no. 11, pp. 1867–1875.

Golub, G. H., and C. F. Van Loan, 1989, *Matrix computations* (Baltimore, MD: The John Hopkins University Press).

Goodman, N. R., 1963, "Statistical analysis based on certain multivariate complex normal distributions," *Annals Math. Stat.*, vol. 34, pp. 152–177.

Gorokhov, A., and J. F. Cardoso, Apr. 1997, "Equivariant blind deconvolution of MIMO-FIR channels," *Proc. SPAWC International Workshop*, Paris.

Gorokhov, A., and P. Loubaton, Sept. 1997, "Subspace based techniques for blind separation of convolutive mixtures with temporally correlated sources," *IEEE Trans. Cir. Syst.*, vol. 44, pp. 813–820.

Gorokhov, A., and P. Loubaton, 1998, "Blind identification of MIMO-FIR systems: a generalized linear prediction approach," to appear in *Signal Processing*.

Grellier, O., and P. Comon, 1998a, "Blind separation and equalization of a channel with MSK inputs," in *SPIE Conference*, San Diego, July 19–24 1998, pp. 26–34, invited session.

Grellier, O., and P. Comon, 1998b, "Performance of blind discrete source separation," in *Eusipco*, Rhodes, Greece, Sept. 8–11, 1998, vol. IV, pp. 2061–2064, Invited session.

Hannan, E. J., 1970, *Multiple Time Series* (New York: Wiley).

Hannan, E. J., and M. Deistler, 1988, *The Statistical Theory of Linear Systems* (New York: Wiley).

Haykin, S., ed., 1984, *Array Signal Processing* (Prentice Hall).

Haykin, S., 1994, *Blind Equalization*, Information and System Sciences. (Prentice-Hall).

Hilal, K., and P. Duhamel, Aug. 1992, "A convergence study of the constant modulus algorithm leading to a normalized-CMA and a block normalized-CMA," *Proc. European Signal Processing Conf. EUSIPCO*, Brussels, pp. 135–138.

Hyvärinen, A., P. Hoyer, and E. Oja, 1999, "Denoising of non gaussian data by independent component analysis and sparse coding," in *ICA99, IEEE Workshop on Indep. Comp. Anal. and Blind Source Separation*, Aussois, France, Jan 11–15, pp. 485–489.

Inouye, Y., and T. Habe, May 1995, "Blind equalization of multichannel linear time-invariant systems," *The Institute of Electronics Information and Communication Engineers*, vol. 24, no. 5, pp. 9–16.

Jutten, C., and J. Herault, 1991, "Blind separation of sources, part I: An adaptive algorithm based on neuromimetic architecture," *Signal Processing, Elsevier*, vol. 24, no. 1, pp. 1–10.

Kendall, M., and A. Stuart, 1977, *The Advanced Theory of Statistics, Distribution Theory*, vol. 1, C. Griffin.

Kristensson, M., B. Ottersten, and D. T. M. Slock, 1996, "Blind subspace identification of a BPSK communication channel," *Proc. 30th Asilomar Conf. Sig. Syst. Comp.*, Pacific Grove, CA.

Krob, M., and M. Benidir, 1993, "Blind identification of a linear-quadratic mixture," *Proc. IEEE SP Workshop on Higher-Order Stat.*, Lake Tahoe, USA, pp. 351–355.

Kruskal, J. B., 1977, "Three-way arrays: Rank and uniqueness of trilinear decompositions," *Linear Algebra and Applications*, vol. 18, pp. 95–138.

Lacoume, J. L., and P. Ruiz, Dec. 1992, "Separation of independent sources from correlated inputs," *IEEE Trans. Signal Processing*, vol. 40, no. 12, pp. 3074–3078.

Lacoume, J. L., P. O. Amblard, and P. Comon, 1997, *Statistiques d'ordre supérieur pour le traitement du signal*, Collection Sciences de l'Ingénieur. Masson.

Laheld, B., and J. F. Cardoso, Sept. 1994, "Adaptive source separation without pre-whitening," in *Proc. EUSIPCO*, Edinburgh, pp. 183–186.

Lindgren, U., and A. J. Van der Veen, June 1996, "Source separation based on second order statistics—an algebraic approach," *Proc. 8th IEEE SSAP Workshop*, Corfu, pp. 324–327.

Lindgren, U., T. Wigren, and H. Broman, Dec. 1995, "On local convergence of a class of blind separation algorithms," *IEEE Trans. Signal Processing*, vol. 43, no. 12, pp. 3054–3058.

Macchi, O., 1995, *Adaptive Processing* (New York: Wiley).

Macchi, O., and E. Eweda, Mar. 1984, "Convergence analysis of self-adaptive equalizers," *IEEE Trans. Inf. Theory*, vol. 30, no. 2, pp. 161–176.

Macchi, O., and E. Moreau, Apr. 1997, "Self-adaptive source separation. part I: Convergence analysis of a direct linear network controlled by the Herault-Jutten algorithm," *IEEE Trans. Signal Processing*, vol. 45, no. 4, pp. 918–926.

Mayrargue, S., Apr. 1994, "A blind spatio-temporal equalizer for a radio-mobile channel using the CMA," in *ICASSP*, Adelaide, vol. III, pp. 317–320.

McCullagh, P., 1987, *Tensor Methods in Statistics*, Monographs on Statistics and Applied Probability, (New York: Chapman and Hall).

Monzingo, R. A., and T. N. Miller, 1980, *Introduction to Adaptive Arrays*, (New York: Wiley).

Moreau, E., J. C. Pesquet, June 1997, "Generalized contrasts for multichannel blind deconvolution of linear systems," *IEEE Signal Processing Lett.*, vol. 4, no. 6, pp. 182–183.

Moulines, E., P. Duhamel, J. F. Cardoso, and S. Mayrague, Feb. 1995, "Subspace methods for the blind identification of multichannel FIR filters," *IEEE Trans. Signal Processing*, vol. 43, no. 2, pp. 516–525.

Ng, B. C., D. Gesbert, and A. Paulraj, 1998, "A semi-blind approach to structured channel equalization," *Proc. ICASSP*, Seattle, May 12–15 1998.

Nguyen-Thi, H. L., and C. Jutten, 1995, "Blind source separation for convolutive mixtures," *Signal Processing*, vol. 45, no. 2, pp. 209–229.

Nikias, C. L., 1991, "Blind deconvolution using high-order statistics," *Proc. Int. Signal Processing Workshop on Higher-Order Stat.*, Chamrousse, France, pp. 155–162.

Paulraj, A. J., and C. Papadias, Nov. 1997, "Space-time processing for wireless communications," *IEEE Signal Processing Mag.*, vol. 14, no. 6, pp. 49–83.

Pham, D. T., and P. Garat, July 1997, "Blind separation of mixture of independent sources through a quasi-maximum likelihood approach," *IEEE Trans. Signal Processing*, vol. 45, no. 7, pp. 1712–1725.

Picinbono, B., Dec. 1994, "On circularity," *IEEE Trans. Signal Processing*, vol. 42, no. 12, pp. 3473–3482.

Picinbono, B., and P. Chevalier, 1995a, "Extensions of the minimum variance method," *Signal Processing*, Elsevier, vol. 49, pp. 1–9.

Picinbono, B., and P. Chevalier, Aug. 1995b, "Widely linear estimation with complex data," *IEEE Trans. on Signal Processing*, vol. 43, no. 8, pp. 2030–2033.

Pipon, F., P. Chevalier, P. Vila, and D. Pirez, Apr. 1997, "Practical implementation of a multichannel equalizer for a propagation with ISI and CCI—application to a GSM link," *Proc. IEEE Vehic. Tech. Conf.*, Phoenix, pp. 889–893.

Proakis, J. G., 1995, *Digital Communications*, 3rd ed. (New York: McGraw-Hill).

Proakis, J. G., and C. L. Nikias, 1991, "Blind equalization," in *SPIE Adaptive Signal Processing*, vol. 1565, pp. 76–88.

Da Rocha, C., and O. Macchi, Apr. 1994, "A novel self adaptive recursive equalizer with unique optimum for QAM," in *ICASSP*, Adelaide, vol. III, pp. 481–484.

Sato, Y., June 1975, "A method of self recovering equalization for multilevel amplitude-modulation systems," *IEEE Trans. Communic.*, vol. 23, pp. 679–682.

Schetzen, M., 1980, *Volterra and Wiener Theory and Non Linear Systems* (New York: Wiley).

Seshadri, N., Mar. 1994, "Joint data and channel estimation using fast blind trellis search techniques," *IEEE Trans. Communic.*, vol. 42, pp. 1000–1011.

Shalvi, O., and E. Weinstein, Mar. 1990, "New criteria for blind deconvolution of non-minimum phase systems," *IEEE Trans. Inform. Theory*, vol. 36, no. 2, pp. 312–321.

Slock, D. T. M., Apr. 1994a, "Blind fractionally-spaced equalization, perfect-reconstruction filter banks and multichannel linear prediction," *Proc. ICASSP 94 Conf.*, Adelaide, Australia.

Slock, D. T. M., 1994b, "Blind joint equalization of multiple synchronous mobile users using oversampling and/or multiple antennas," *Proc. 28th Asilomar Conf. Sig. Syst. Comp.*, Pacific Grove, CA, Oct. 31–Nov. 2 1994, pp. 1154–1158.

Soderstrom, T., and P. Stoica, 1989, *System Identification* (New York: Prentice-Hall).

Sorouchyari, E., July 1991, "Separation of sources, part III: Stability analysis," *Signal Processing*, Elsevier, vol. 24, no. 1, pp. 21–29.

Souloumiac, A., P. Chevalier, and C. Demeure, May 1993, "Improvement in non Gaussian jammers rejection with a non linear spatial filter," in *Proc. ICASSP*, Minneapolis, vol. V, pp. 670–673.

Swami, A., and G. Giannakis, and S. Shamsunder, Apr. 1994, "Multichannel ARMA processes," *IEEE Trans. Signal Processing*, vol. 42, no. 4, pp. 898–913.

Swindlehurst, A. L., S. Daas, and J. Yang, Dec. 1995, "Analysis of a decision directed beamformer," *IEEE Trans. Signal Processing*, vol. 43, no. 12, pp. 2920–2927.

Talwar, S., M. Viberg, and A. Paulraj, May 1996, "Blind estimation of multiple co-channel digital signals arriving at an antenna array: Part I, algorithms," *IEEE Trans. Signal Processing*, pp. 1184–1197.

Tong, L., Sept. 1996, "Identification of multichannel MA parameters using higher-order statistics," *Signal Processing*, Elsevier, vol. 53, no. 2, pp. 195–209, special issue on high-order statistics.

Tong, L., R. Liu, and V. C. Soon, May 1991, "Indeterminacy and identifiability of blind identification," *IEEE Trans. Cir. Syst.*, vol. 38, no. 5, pp. 499–509.

Tong, L., G. Xu, and T. Kailath, Mar. 1994, "Blind identification and equalization based on second-order statistics: a time domain approach," *IEEE Trans. Inform. Theory*, vol. 40, no. 2, pp. 340–349.

Touzni, A., I. Fijalkow, M. Larimore, and J. R. Treichler, 1998, "A globally convergent approach for blind MIMO adaptive deconvolution," *ICASSP*, Seattle, May 12–15 1998.

Treichier, J. R., and B. G. Agee, Apr. 1983, "A new approach to multipath correction of constant modulus signals," *IEEE Trans. Acoust. Speech Signal Processing*, vol. 31, no. 4, pp. 459–471.

Tsatsanis, M. K., and G. B. Giannakis, Apr. 1997, "Cyclostationarity in partial response signaling: A novel framework for blind equalization," in *ICASSP*, Munich, pp. 3597–3600.

Tugnait, J., Jan. 1992, "Comments on 'New criteria for blind deconvolution of nonminimum phase systems,'" *IEEE Trans. Inform. Theory*, vol. 38, no. 1, pp. 210–213.

Tugnait, J. K., Jan. 1997, "Blind spatio-temporal equalization and impulse response estimation for MIMO channels using a Godard cost function," *IEEE Trans. Signal Processing*, vol. 45, no. 1, pp. 268–271.

van der Veen, A. J., Apr. 1997, "Analytical method for blind binary signal separation," *IEEE Trans. Signal Processing*, vol. 45, no. 4, pp. 1078–1082.

van der Veen, A. J., and A. Paulraj, May 1996, "An analytical constant modulus algorithm," *IEEE Trans. Signal Processing*, vol. 44, no. 5, pp. 1136–1155.

van der Veen, A. J., S. Talwar, and A. Paulraj, May 1995, "Blind estimation of multiple digital signals transmitted over FIR channels," *IEEE Signal Processing Lett.*, vol. 2, no. 5, pp. 99–102.

Vuattoux, J. L., and E. Le Carpentier, 1996, "Efficient ARMA parameter estimation on non Gaussian processes by minimization of the Fisher information under cumulant constraints," in *Proc. ICASSP*, Atlanta, Georgia, May 7–10 1996, pp. 218–221.

Yellin, D., and B. Porat, 1993, "Blind identification of FIR systems excited by discrete-alphabet inputs," *IEEE Trans. Signal Processing*, vol. 41, no. 3, pp. 1331–1339.

Yellin, D., and E. Weinstein, Aug. 1994, "Criteria for multichannel signal separation," *IEEE Trans. Signal Processing*, vol. 42, no. 8, pp. 2158–2168.

Zivojnovic, V., 1995, "Higher-order statistics and Huber's robustness," in *IEEE-ATHOS Workshop on Higher-Order Statistics*, Begur, Spain, 12–14 June 1995, pp. 236–240.

6

INFORMATION THEORY, INDEPENDENT-COMPONENT ANALYSIS, AND APPLICATIONS

Anthony J. Bell

ABSTRACT

Independent-component analysis (ICA) is a linear transform of multivariate data designed to make the resulting random vector as statistically independent (factorial) as possible. There are several different perspectives on the problem. Here we discuss it from an information-theoretic point of view, through an understanding of the entropy maximization (or Infomax) algorithm, and its relationship to maximum-likelihood approaches.

Insights about both the algorithm and the structure of particular data sets have been produced by applying the algorithm in real-world situations. For example, ICA can be used to produce a set of basis vectors for natural signals that are sensitive to their phase structure. In the case of natural images, the ICA basis vectors are edges. In the second part of the chapter, we show the results on decomposing images, sounds, and biomedical recordings (electroencephalographic and magnetic resonance data). We believe these are just the first of many such applications.

Unsupervised Adaptive Filtering, Volume I, Edited by Simon Haykin.
ISBN 0-471-29412-8 © 2000 John Wiley & Sons, Inc.

6.1 INTRODUCTION

Independent-component analysis (ICA) is an attractive subject for a number of reasons. First, although it is a linear model, and linear models have been analyzed extensively for many years, the solution to the ICA problem does not appear in any standard textbook. Second, it is a conceptually simple problem that introduces a wealth of interesting technical material—higher-order statistics, information theory, sparse coding, projection pursuit, and information geometry, to name a few.

But the third and main reason has to be the applications. In this chapter, as well as outlining the conceptual issues, we will also take a close look at applications on real-world data where the results have been not only impressive but also instructive. The results have helped us to understand the algorithm better. We believe that ICA can be used profitably to transform any multivariate data, as long as there is some question as to the meaning of those data. As we show, insights can be derived both from the transformed data, and from the way the data are transformed. This distinguishes ICA from fixed-basis signal-processing techniques such as spectral analysis and wavelet transforms. In ICA, the data themselves tell us how they should be transformed.

After the following short historical review, we follow a discursive path in introducing the basic ideas in Sections 6.2.1 and 6.2.2. The approach taken throughout is the information-theoretic one sometimes called Infomax/maximum-likelihood. Some justification is required in identifying Infomax with ICA, which we tackle in Section 6.2.5. The relationships with maximum likelihood are covered in Section 6.2.6. The applications we discuss are finding independent basis sets for images, and the brain-probing biomedical data produced by electroencephalographic (EEG) recordings and magnetic resonance imaging. We close with a discussion of problems and approaches in extending ICA to transforms that are nonlinear and/or stochastic (Section 6.4).

The problem was introduced by Herault and Jutten in 1986 (Herault and Jutten 1986). The results of their algorithm were poorly understood and led to Comon's 1994 paper defining the problem (Comon 1994), and to his solution using fourth-order statistics. As early as 1992 Pham et al. (Pham et al. 1992) proposed a maximum-likelihood approach that formed the basis of Cardoso and Laheld 1996 EASI method (Cardoso and Laheld 1996). These methods are very close to the information-theoretic approach proposed in Bell and Sejnowski (1995). An increase in interest took place in the neural-networks community, with major contributions from Amari and colleagues (Amari 1997; Amari et al. 1996). We refer the reader to the rest of this book for better histories than this.

6.2 THEORY OF ICA

6.2.1 Filters and Basic Functions

We start by reviewing some basic concepts. We are given an ensemble of P data points, $x_1, \ldots, x_P$, in an N-dimensional space. Our job is to transform this data set into a new affine coordinate system that satisfies some criterion. Our new data will be written as

$$\mathbf{u} = \mathbf{Wx} + \mathbf{w} \tag{6.1}$$

where $\mathbf{W}$ is an $N \times N$ matrix and $\mathbf{w}$ is an $N \times 1$ vector. This is an *affine* transform (as opposed to just a *linear* one) because we shift the origin of the data with $\mathbf{w}$ as well as transforming the space with $\mathbf{W}$.

The rows of the matrix $\mathbf{W}$ correspond to N *filters* because each creates a separate output of the vector $\mathbf{u}$, thus

$$u_i = \sum_{j=1}^{N} W_{ij} x_j + w_j \tag{6.2}$$

But we can turn this picture on its head by imagining that the data set, $\mathbf{x}$, is formed by another affine transform that we get by solving Eq. (6.1) for $\mathbf{x}$:

$$\mathbf{x} = \mathbf{W}^{-1}\mathbf{u} - \mathbf{W}^{-1}\mathbf{w} \tag{6.3}$$

The columns of the matrix $\mathbf{W}^{-1}$ are called *basis vectors* because they provide a way of talking about how the data are formed, sometimes called a *generative model*. Let's say $\mathbf{A} = \mathbf{W}^{-1}$ and $\mathbf{a} = -\mathbf{W}^{-1}\mathbf{w}$, and using $\mathbf{A}^j$ to denote one of the columns of $\mathbf{A}$, then we can see that each data point is formed out of a set of N coefficients, u_j, each representing the degree to which basis vector $\mathbf{A}^j$ is present in the data:

$$\mathbf{x} = \sum_{j=1}^{N} \mathbf{A}^j u_j + \mathbf{a} \tag{6.4}$$

We will use both the *basis vector* and the *filter* way of talking in future sections.

6.2.2 Decorrelation and Independence

What affine transform $\{\mathbf{W}, \mathbf{w}\}$ or, equivalently, what basis, $\{\mathbf{A}, \mathbf{a}\}$, should we choose? Signal processing provides us with some ready-made basis vectors, such as the Fourier basis, which is an orthogonal-basis set, or various forms of orthogonal or nonorthogonal wavelet transforms. But rather than using a fixed basis for all signals, it is more interesting to allow the statistics of the data themselves determine what basis we use to describe them. In this subsection, primarily to clarify the relations with ICA, we look at several ways of doing this using the covariance matrix of the data.

The lowest-order statistics are the first-order statistics (the mean vector), $\langle \mathbf{x} \rangle$, and the second-order statistics (the covariance matrix), $\langle \mathbf{xx}^T \rangle$. All other statistics are known as higher-order statistics.

A common procedure, sometimes known as *sphering* or *whitening*, is to transform the data so that they are zero-mean, unit-variance, and decorrelated:

$$\langle \mathbf{u} \rangle = \mathbf{0} \quad \text{(the zero vector)} \tag{6.5}$$

$$\langle \mathbf{uu}^T \rangle = \mathbf{I} \quad \text{(the identity matrix)} \tag{6.6}$$

Plugging Eq. (6.1) into Eqs. (6.5) and (6.6) gives (after some matrix manipulations)

$$\mathbf{w} = -\langle \mathbf{Wx} \rangle \tag{6.7}$$

$$\mathbf{W}^T \mathbf{W} = (\langle \mathbf{xx}^T \rangle - \langle \mathbf{x} \rangle \langle \mathbf{x} \rangle^T)^{-1} \tag{6.8}$$

Thus, $\mathbf{W}$ is to be calculated from the inverse of the covariance matrix of $\mathbf{x}$. However, Eq. (6.8) is underconstrained, consisting of $(N^2 + N)/2$ equations in N^2 variables, so there are multiple decorrelating solutions. We now consider three ways of constraining the solution.

Principal-Component Analysis (The Orthogonal solution) Principal-component analysis (PCA) is calculated from the eigenvectors of the covariance matrix, which are columns of an orthogonal matrix $\mathbf{E}$, satisfying

$$\mathbf{EDE}^{-1} = \langle \mathbf{xx}^T \rangle - \langle \mathbf{x} \rangle \langle \mathbf{x} \rangle^T \tag{6.9}$$

where $\mathbf{D}$ is a diagonal matrix with the eigenvalues on the diagonal. We can define $\mathbf{W}_P$, a decorrelating transform coming from this is, as follows:

$$\mathbf{W}_P = \mathbf{D}^{1/2} \mathbf{E}^T \tag{6.10}$$

What is special about this transform is that $\mathbf{W}_P$ is an orthogonal matrix: $\mathbf{W}_P^T = \mathbf{DW}_P^{-1}$. (If $\mathbf{D}$ were the identity matrix, it would be an ortho*normal* transform.) Constraining the matrix to be orthogonal is artificial, which is why PCA is not a good way to find structure in data, though it is useful for dimensionality reduction when we wish to minimimize the mean-squared error of reconstruction.

Zero-Phase Decorrelation (The Symmetrical solution) To show that PCA is not the only way to constrain $\mathbf{W}$, we introduce a transform we call zero component analysis (ZCA). This is obtained by constraining $\mathbf{W}$ to be symmetrical ($\mathbf{W}^T = \mathbf{W}$), in which case Eq. (6.8) is solved by a transform, $\mathbf{W}_Z$, satisfying

$$\mathbf{W}_Z = (\langle \mathbf{xx}^T \rangle - \langle \mathbf{x} \rangle \langle \mathbf{x} \rangle^T)^{-1/2} \tag{6.11}$$

In general, this is a nonorthogonal decorrelator that is called *zero-phase* because it decorrelates spatial (or temporal) data while leaving its phase structure intact, an example being the spatial whitening of an image.

Independent-Component Analysis (The "Independent-as-Possible" Solution) In contrast to the two preceding methods, which have artificial constraints on the matrix $\mathbf{W}$, ICA finds its constraints in the data. As well as decorrelating data, it attempts to remove higher-order correlations, ultimately aiming for statistical independence, in which the joint probability distribution of the data factorizes into the product of its marginal (univariate) distributions, as follows:

$$f(\mathbf{u}) = \prod_{i=1}^{N} f_i(u_i). \tag{6.12}$$

There are three facts that are useful to remember about this.

1. Statistical independence is nothing to do with linear independence.
2. Decorrelation is nothing to do with orthogonality.

In both cases, the former quantity is statistical, to do with the random variables, and the latter is to do with the vectors making up the matrix $\mathbf{W}$. Decorrelation and orthogonality are sometimes conflated because in PCA and other eigenvector-based methods, the two properties co-occur.

3. Decorrelation $\overset{\times}{\underset{\longleftarrow}{\longrightarrow}}$ Independence.

The only case where decorrelation implies independence is when the underlying distribution, $f(\mathbf{u})$, is multivariate Gaussian: a distribution that is completely specified by first- and second-order statistics.

Another difference between independence and decorrelation is that decorrelation is always achievable (when the covariance matrix is full rank), while independence can only be achieved if the data have really been formed by linear mixing together of independent variables. Since this is, in general, not the case, ICA must be satisfied with the goal of finding a $\mathbf{W}$ that produces as-independent-as-possible u_i. An information-theoretic measure of how independent our outputs are is

$$I(\mathbf{u}) = E\left[\log\frac{f(\mathbf{u})}{\prod_{i=1}^{N} f_i(u_i)}\right] \left(= \int f(\mathbf{u}) \log\frac{f(\mathbf{u})}{\prod_{i=1}^{N} f_i(u_i)} d\mathbf{u}\right) \tag{6.13}$$

where $E = [\cdot]$ denotes expected value. This *redundancy* measure has the value of 0 when the distribution $f(\mathbf{u})$ factorizes as in Eq. (6.12). For the case of $N = 2$, the redundancy measure gives the *mutual information* of two variables, $I(u_1, u_2)$, which can be seen as a higher-order analog of decorrelation.

6.2.3 Information Maximization

We have discussed decorrelation and independence, but we have not yet shown how to perform ICA, because it is a more difficult problem—more difficult not only because true independence is impossible in general to achieve, but also because it is difficult to minimize the independence cost function in Eq. (6.13).

To approach the ICA problem, we change direction altogether and, following the presentation in Bell and Sejnowski (1995), show how to solve an easier problem: the problem of information maximization, or *Infomax*. The reader who would prefer a more leisurely and intuitive exposition of the material below is referred to our earlier article. The close relationships between the Infomax approach and the maximum-likelihood approach of Pham et al. (1992) will be discussed in Section 6.2.6.

Infomax, for our purposes, addresses the problem of maximizing the mutual information, $I(\mathbf{y}, \mathbf{x})$, between the input random vector, $\mathbf{x}$, and an invertible nonlinear transform of it, $\mathbf{y}$, given by

$$\mathbf{y} = \mathbf{g}(\mathbf{u}) = \mathbf{g}(\mathbf{Wx} + \mathbf{w}) \tag{6.14}$$

Here, the degrees of freedom in the maximization are the $N \times N$ matrix $\mathbf{W}$, the $N \times 1$ "bias"-vector, $\mathbf{w}$, and $\mathbf{g}$, an invertible and bounded nonlinear vector function, the components of which are $g_i(u_i) = g_i(\sum_j w_{ij} x_j + w_i)$. As we will see, it was the achievement of Nadal and Parga (1994) to notice that maximizing $I(\mathbf{y}, \mathbf{x})$ with respect to all three of $\mathbf{W}, \mathbf{w}$, and $\mathbf{g}$ could be seen as a way of approaching ICA. (They did not, however, propose a maximization algorithm.) But for now, we will consider the maximization only with respect to $\mathbf{W}$ and $\mathbf{w}$.

Because this is a deterministic (noiseless) mapping, maximizing $I(\mathbf{y}, \mathbf{x})$ is the same thing as maximizing the output entropy, $H(\mathbf{y})$ (Nadal and Parga 1994). This is because the final term in

$$I(\mathbf{y}, \mathbf{x}) = H(\mathbf{y}) - H(\mathbf{y}|\mathbf{x}) \tag{6.15}$$

drops out when we differentiate with respect to $\mathbf{W}$. Thus we could equally well call the technique "entropy maximization," as some do. Our motivation in choosing the Infomax name were partly to establish a connection with earlier information-theoretic, unsupervised learning studies by Linsker (1992) and Atick (1992), but also to avoid assocation with "maximum entropy" techniques (Nailong 1997) as used, for example, in spectral estimation, which were not really related to our methods. In addition, when the mapping is noisy, Infomax is actually what we want to do: since we will wish to minimize the $H(\mathbf{y}|\mathbf{x})$ term caused by the noise.

A simple relation holds (Papoulis 1984) between the probability distributions of the input and output:

$$f(\mathbf{y}) = \frac{f(\mathbf{x})}{|J|} \tag{6.16}$$

where $|\cdot|$ denotes absolute value, and J is the determinant of the Jacobian matrix: $J = \det([\partial y_i/\partial x_j]_{ij})$.

Since the output entropy is $H(\mathbf{y}) = -E[\log f(\mathbf{y})] = -\int f(\mathbf{y}) \log f(\mathbf{y}) \, d\mathbf{y}$, we can substitute Eq. (6.16) into it to obtain

$$H(\mathbf{y}) = E[\log|J|] + H(\mathbf{x}) \tag{6.17}$$

The input entropy, $H(\mathbf{x})$, is not affected by the parameters that we are learning, so we can write a learning rule for $\mathbf{W}$ as follows:

$$\Delta \mathbf{W} \propto \frac{\partial H(\mathbf{y})}{\partial \mathbf{W}} = E\left[\frac{\partial}{\partial \mathbf{W}} \log|J|\right] \tag{6.18}$$

The first thing to note here is that in the *stochastic-gradient* method, we can update $\mathbf{W}$ every time step (ignoring the expected value operator, $E[\cdot]$). The second thing to note is that the $\log|J|$ term, in the case of Eq. (6.14) decomposes as follows:

$$\log|J| = \log \det \mathbf{W} + \sum_{i=1}^{N} \log|y_i'| \tag{6.19}$$

where $|y'|$ is the absolute value of the slope $(\partial y/\partial u)$ of the output y. Differentiating this with respect to $\mathbf{W}$ gives our stochastic gradient learning rule [full derivations in Bell and Sejnowski (1995)]:

$$\Delta \mathbf{W} \propto \frac{\partial}{\partial \mathbf{W}} \log|J| = \mathbf{W}^{-T} + \Phi(\mathbf{u})\mathbf{x}^T \tag{6.20}$$

Here "$-T$" is shorthand for inverse transpose and the vector function $\Phi(\mathbf{u})$ has elements as follows:

$$\phi_i(u_i) = \frac{\partial y_i'}{\partial y_i} = \frac{\partial}{\partial u_i}\log|y_i'| \tag{6.21}$$

The elements of this vector depend on the choice of nonlinear functions, $\mathbf{g}$, in Eq. (6.14). Any invertible functions can be chosen for $\mathbf{g}$, but common choices are the logistic sigmoid, $y = (1 + e^{-u})^{-1}$, or the hyperbolic tangent, $y = \tanh(u)$, for which $f(u)$ evaluates to $1 - 2y$ and $-2y$, respectively. Many other examples appear in Bell and Sejnowski (1995).

Similar reasoning leads to the rule for adjusting the bias vector:

$$\Delta\mathbf{w} \propto \frac{\partial}{\partial\mathbf{w}}\log|J| = \Phi(\mathbf{u}) \tag{6.22}$$

Equations (6.20)–(6.22) thus represent stochastic-gradient learning rules for Infomax.

6.2.4 Natural (or Relative) Gradient

In practice we never use Eqs. (6.20)–(6.22) anymore. This is because Amari et al. (1996), based on earlier experiments by Cichocki et al. (1994), proposed multiplying the entropy gradient in Eq. (6.20) by $\mathbf{W}^T\mathbf{W}$ to get a simpler rule, one that avoids inverting $\mathbf{W}$ at every learning step:

$$\Delta\mathbf{W} \propto \frac{\partial\log|J|}{\partial\mathbf{W}}\mathbf{W}^T\mathbf{W} = (\mathbf{I} + \Phi(\mathbf{u})\mathbf{u}^T)\mathbf{W} \tag{6.23}$$

Amari's information-geometrical explanation (Amari 1997) makes the case that

1. Postmultiplying the (matrix) gradient of Eq. (6.20) by $\mathbf{W}^T\mathbf{W}$ has the same effect as
2. Premultiplying the (vector[1]) gradient by the inverse of the Fisher information matrix; equivalently, the inverse of the metric tensor of the vector space formed by the components of the matrix $\mathbf{W}$.

The goal in both cases is to take into account the non-Euclidean nature of distances in the $\mathbf{W}$-space that results from $\mathbf{W}$ being a matrix. However, (1) and (2) cannot be exactly the same because the inverse of the Fisher information matrix for Infomax (using Vec to convert matrices to column vectors):

$$E\left[\text{Vec}\left(\frac{\partial\log|J|}{\partial\mathbf{W}}\right)\text{Vec}\left(\frac{\partial\log|J|}{\partial\mathbf{W}}\right)^T\right]^{-1} \tag{6.24}$$

---

[1] The vector gradient is just the $N \times N$ matrix gradient turned into a N^2-length column vector.

is data-dependent, while the postmultiplication $\mathbf{W}^T\mathbf{W}$ in (1) only becomes so as $\mathbf{W}$ learns. MacKay (1996) outlines some approximations whereby $\mathbf{W}$ could be seen to acquire metric information during learning. Cardoso approaches equivariance from the point of view of invariant group action (Cardoso and Laheld 1996; Cardoso 1995). And in Cardoso and Amari (1998), a noninformation-geometrical account is given that has the useful side effect of defining a relative-gradient rule for the bias update [compare with Eq. (6.22)]:

$$\Delta\mathbf{w} \propto \Phi(\mathbf{u}) + (\mathbf{I} + \Phi(\mathbf{u}))\mathbf{w} \qquad (6.25)$$

Chapter 2 of this volume presents a detailed discussion of natural gradient, so we return to the relationship of Infomax to ICA.

6.2.5 Infomax and ICA

The Infomax objective of maximizing the entropy of the the nonlinearly transformed outputs, $\mathbf{y}$, is not the same as the ICA objective of minimizing the statistical dependence between the linear outputs, $\mathbf{u}$. But intuitively, we can see that they are going to be related: the less redundancy between the elements of a random vector, the more information (entropy) they can carry together.

The redundancy objective:

$$I(\mathbf{u}) = E\left[\log\frac{f(\mathbf{u})}{\prod_{i=1}^{N} f_i(u_i)}\right] \qquad (6.26)$$

is the negative of the Kullback-Leibler distance of the output distribution from an independent distribution with the same marginal distributions. It is related to the entropy as follows (see also Yang and Amari (1997)):

$$H(\mathbf{y}) = -I(\mathbf{u}) + E\left[\sum_i \log\frac{|y_i'|}{f_i(u_i)}\right] \qquad (6.27)$$

Proof of Eq. (6.27). Because of

$$f(\mathbf{y}) = \frac{f(\mathbf{u})}{\det([\partial y_i/\partial u_i]_{ij})} = \frac{f(\mathbf{u})}{\prod_i |y_i'|} \qquad (6.28)$$

we can write

$$H(\mathbf{y}) = -E[\log f(\mathbf{y})] \qquad (6.29)$$

$$= -E[\log f(\mathbf{u})] + E\left[\log\prod_i |y_i'|\right] \qquad (6.30)$$

$$= -E\left[\log \frac{f(\mathbf{u})}{\prod_i f_i(u_i)}\right] - E\left[\log \prod_i f_i(u_i)\right] + E\left[\log \prod_i |y_i'|\right] \quad (6.31)$$

$$= -I(\mathbf{u}) + E\left[\sum_i \log \frac{|y_i'|}{f_i(u_i)}\right] \quad (6.32)$$

Thus if $|y_i'| = f_i(u_i)$ for all i, then maximizing the entropy of $\mathbf{y}$ is exactly equivalent to ICA (minimizing the redundancy of $\mathbf{u}$). This means that the gs can be seen as *estimators* for the form of the source cumulative distribution functions (cdf's). (We say "form" because scaling and shifting the gs only has the effect of scaling and shifting our source estimates. The scale and shift of the true sources are unrecoverable.) When the estimators are accurate, the learning rules in Eqs. (6.20)–(6.22) will perform ICA. We can regard the $\log(|y_i'|/f(u_i))$ terms as "interference" terms because they have the potential to interfere with Infomax performing ICA.

What if these interference terms are not zero? One approach is to notice, as did Nadal and Parga (1994), that allowing flexibility in the g_i can allow the $|y_i'|$ to approach the $f(u_i)$, thus reducing the interference terms in Eq. (6.27) and making Infomax converge to ICA. On-line estimation of the gs has been pursued by several authors [Pearlmutter and Parra (1996), Obradovic and Deco (1998)].

Another approach (which we took) was to run Eqs. (6.20)–(6.22) on various types of data and see to what extent the interference terms actually interfered. What emerged was a remarkable pattern of robustness to misestimation of the gs. The robustness took roughly the form that if the sources were all super-Gaussian,[2] then any nonlinearity whose derivative $g'(u)$ was a super-Gaussian probability density function (pdf) would suffice to separate the sources. The converse was also true—sub-Gaussian $g'(u)$'s separated sub-Gaussian sources. For example, using the logistic sigmoid for $g(u)$, we were able to separate several hundred super-Gaussian sources (Bell and Sejnowski 1997).

The closest that anyone has come to explaining this is the stability analysis of Amari et al. (1997). Using the Hessian of the gradient, they looked at conditions for the separating solution to be a stable equilibrium, and even proposed an on-line Hessian-estimation scheme that makes the correct solution stable. Most helpfully, they proved that using logistic sigmoids for $g(u)$ will separate all super-Gaussians.

A slightly simpler, slightly less elegant, but eminently practical scheme was devised by Girolami (1996) and followed up by Lee et al. (1997b, 1998). Given the robustness to misestimation, why devote huge machinery to on-line estimation of complicated nonlinearities or Hessians? Why not instead just look to see whether a given output, u_i, seems to be super- or sub-Gaussian and switch

---

[2] A Gaussian has kurtosis 0, a super-Gaussian has a positive kurtosis, is usually peaky and long-tailed, and a sub-Gaussian has negative kurtosis and is often flat or bimodal.

the nonlinearity accordingly? They termed their approach "extended ICA." It is simply Eq. (6.23) with the $\phi_i(u_i)$'s defined as follows (with $k(u_i)$ meaning the online-estimated kurtosis of the distribution $f_i(u_i)$):

$$\phi_i(u_i) = \begin{cases} -\tanh(u_i) - u_i & \text{if } k(u_i) > 0 \text{ (super-Gaussian)} \\ \tanh(u_i) - u_i & \text{if } k(u_i) < 0 \text{ (sub-Gaussian)} \end{cases} \tag{6.33}$$

Using Eq. (6.34), we can check to see exactly what prior on the source distribution is implied by the $\phi_i(u_i)$'s in Eq. (6.33). There are many other super-/sub-Gaussian pairs of functions that we could choose from. One possible failing of this method is instability when many sources are close to Gaussian. Training noise could cause the random switching of the earlier back and forth from the super- and sub-Gaussian regimes.

In a final note, one might well ask: Why not tackle ICA [$I(\mathbf{u})$ in Eq. (6.26)] directly, instead of doing Infomax [$H(\mathbf{y})$ in Eq. (6.26)]? The answer is that while the gradient of $H(\mathbf{y})$ is easily estimated, in practice it is impossible to estimate the gradient of $I(\mathbf{u})$. An attempt can be made by, for example, approximating $I(\mathbf{u})$ with a Gram-Charlier series expansion [as was done by Yang and Amari (1997)]. This will give rise to a learning rule identical to ours [Eq. (6.20)], but with a set of estimating nonlinearities $\Phi(\mathbf{u})$, derived from the Gram-Charlier approximation. Such approximations always fail at some point, and their failure might not always be understood, except by converting the Gram-Charlier-derived $\Phi_i(u_i)$'s back into earlier source estimates via the inverse procedure to Eq. (6.21):

$$\hat{f}(u_i)[= g_i'(u_i)] = e^{\int \phi_i(u_i)\, du_i} \tag{6.34}$$

(and since this is indefinite integration, we can use the integration constant to ensure that $\hat{f}(u_i)$ integrates to 1). In fact, performing this operation on the fourth-order Gram-Charlier $\phi_i(u_i)$ in (Yang and Amari 1997) leads to a bimodal sub-Gaussian $\hat{f}(u_i)$, which explains their success in separating sub-Gaussian sources and their failure in separating super-Gaussian sources.

So even in the case of an ICA method focused on $I(\mathbf{u})$, it is helpful to use Infomax/maximum-likelihood concepts to help explain performance. We have added "maximum likelihood" here, because, as the next section explains, Infomax considered as a density estimation problem is exactly a maximum-likelihood procedure.

6.2.6   Infomax and Maximum Likelihood

When the functions, $\mathbf{g}$, are sigmoids going from 0 to 1 (as with the logistic), it has the form of a cdf and $y'(x)$ can be interpreted as a pdf on x. Similarly, in the multivariate case, $|J(\mathbf{x})|$ has the form of a pdf on the vector $\mathbf{x}$. Now consider the quantity $H(\mathbf{y})$ we are maximizing. When the $\mathbf{g}$ go from 0 to 1, $\mathbf{y}$ is constrained in the unit hypercube, and the highest entropy occurs when $\mathbf{y}$ is uniformly distributed in this volume—in other words, when $f(\mathbf{y}) = 1$ in the support of $\mathbf{y}$. Looking at Eq. (6.16), we see that in this case, $|J(\mathbf{x})| = f(\mathbf{x})$. Therefore in maximizing $H(\mathbf{y})$, we are attempting to make $|J(\mathbf{x})|$ into an estimate that is as close as possible to $f(\mathbf{x})$. The estimate is parameterized by $\mathbf{W}, \mathbf{w}$, and $\mathbf{g}$, and may be written formally as

$$\hat{f}(\mathbf{x}|\mathbf{W}, \mathbf{w}, \mathbf{g}) = (\det \mathbf{W}) \prod_{i=1}^{N} |y_i'| \tag{6.35}$$

This reasoning was used by Roth and Baram (1996) in their independent derivation of these results. Thus our learning rule in Eq. (6.20) can be rewritten as

$$\Delta \mathbf{W} \propto \frac{\partial}{\partial \mathbf{W}} \log \hat{f}(\mathbf{x}|\mathbf{W}, \mathbf{w}, \mathbf{g}) \tag{6.36}$$

and similarly for $\mathbf{w}$ (and even $\mathbf{g}$). Equation (6.36) is exactly the form of a density estimation algorithm maximizing a parameterized log-likelihood function. This explains the connection between the Infomax and maximum-likelihood approaches to source separation, which were proposed first by Pham in 1992 (Pham et al. 1992) and elaborated more recently in Belouchrani and Cardoso (1995), Cardoso and Laheld (1996), Pearlmutter and Parra (1996), and Pham (1996). The Infomax/maximum-likelihood connection is explored more fully by Cardoso (1997). In essence, the two approaches are the same. Thinking in terms of maximum likelihood allows natural generalizations to more complex likelihood functions (Pearlmutter and Parra 1996) and likelihood functions with hidden variables, in other words, expectation-maximization (EM) algorithms, as proposed by Belouchrani and Cardoso (1994, 1995) and Attias (1998). On the other hand, thinking in information-theoretic terms helps to keep us closer to the essential goal of ICA, which is to reduce the redundancy measure $I(\mathbf{u})$ in Eq. (6.26). Either way, it is an interesting and useful insight that by maximizing the entropy gained in a bounded invertible nonlinear transformation of a random vector $\mathbf{x}$, we implicitly calculate a density estimate on $\mathbf{x}$.

6.2.7   Other Approaches

Before moving on to applications, it is important to point out that the theory described in the previous section has been restricted to a class of algorithms that falls within the Infomax/maximum-likelihood family. An extensive integrative analysis of this family of methods can be found in Lee et al. (1998). As well as the issues discussed here, relations are drawn to the negentropy maximization approach of Girolami (1996), the nonlinear PCA methods of Karhunen and Joutsensalo (1994), and Bussgang approaches. The earlier fourth-order method of Comon et al. (1991) is also described. Comon's paper and those of Herault and Jutten (1986) and Jutten and Herault (1991) essentially defined the ICA problem. Another useful fourth-order method is Cardoso's JADE algorithm, which has a proven performance (Cardoso 1998).

Another extremely useful method is Hyvärinen's "FastICA" (1997). Like Girolami's negentropy method, this is one of a family of the projection pursuit (Huber 1985) style of methods that utilize the fact that the independent components are bound (by the Central Limit theorem) to be those projections whose $f_i(u_i)$'s have the greatest possible distance from the Gaussian distribution. What makes FastICA special is that the independent components are found one at a time, a process called *deflation*, which is commonly used in projection pursuit. This makes FastICA particularly useful for very high-dimensional data, when only a fraction of the independent components are desired. Use has been made of this by Van Hateren and Van der Schaaf (1998) (see Section 6.3.1) in finding 1728-dimensional independent spatiotemporal basis functions of natural movies. On that note, we move on to applications.

6.3 APPLICATIONS OF ICA

The concept of ICA has been more or less synonymous with that of blind source separation (BSS) since the early papers of Herault and Jutten (1986). The algorithms are often the same (though see Section 6.3.5), but there is a very subtle difference when it comes to applications. The difference is perhaps best captured in the phrase "Let's run the data through ICA and see what happens." That is, it became rewarding (and fun) to apply ICA when we had no clear concept of what a "source" was, quite unlike the situation usually encountered in signal processing. This happened repeatedly when I was working at the Salk Institute in 1994–1997, and caused all kinds of epistemological debates about whether, for example, what an electroencephalographic (EEG) expert called a "brain source" had anything to do with how ICA decomposed EEG data. We decided finally (I think) that the minimal state-of-affairs was that ICA could be used without any reference at all to any source model. It would then be up to the expert (typically the person who gathered the data) to judge in what ways, if at all, the result was interesting.

Thus a whole domain of subjectivity was opened up to us. Fortunately, experimental scientists are familiar with the process of constructing interpretations from data, and they will use whatever tools help form (or justify) those interpretations. The applications we will concentrate on here all involve natural data where there is no clear source model. Thus, much of this section is taken up with the subjective knowledge necessary to understand the significance of what ICA tells us about the particular data involved.

6.3.1 Natural Images

From the point of view of computer vision and computational neuroscience, perhaps the most interesting result [see Bell and Sejnowski (1997)] was the ICA

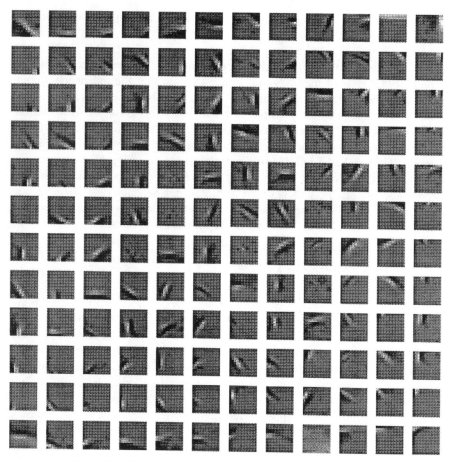

Figure 6.1 The independent basis functions of natural images.

basis vectors obtained for a data set of small image patches drawn from natural images. These basis vectors consisted of oriented, localized Gabor-type functions (see Figure 6.1), sometimes refered to as *edges* (though an edge is really something to do with object boundaries).

The reason this was interesting was because both the classic experiments of Hubel and Wiesel (1968, 1974) on orientation-selective neurons in visual cortex, and several decades of theorizing about feature detection in vision (Marr and Hildreth 1980) had left open the question most succinctly phrased by Barlow and Tolhurst (1992): "Why do we have edge detectors?" In other words, were there any coding principles that could predict the formation of localized, oriented receptive fields?

Barlow was the first to propose that our visual cortical feature detectors might be the end result of a *redundancy reduction* process (Barlow 1989; Atick 1992), in which the activation of each feature detector is supposed to be as statistically independent from the others as possible. Several authors proposed Hebbian unsupervised feature-learning algorithms based on second-order statistics (Linsker 1992; Miller 1988; Oja 1989; Sanger 1989; Földiák 1990; Atick and Redlich 1993), but as we discussed in Section 6.2.2, second-order statistics do not constrain the solution enough. In particular, the PCAs of natural images are Fourier filters ranked in frequency, quite unlike oriented localized filters (Hancock et al. 1992).

It wasn't until the work of Field (1987, 1994) and Intrator (1992) that progress on this question was made. Arguing for the importance of sparse, or "minimum entropy," coding (Barlow 1994), in which each feature detector is activated as rarely as possible, these authors set the scene for feature-learning algorithms with a "projection pursuit" (Huber 1985) flavor, the most successful of which has been Olshausen and Field's (1996a,b) demonstration of the self-organization of local, oriented receptive fields using a sparseness criterion.

By identifying sparseness with super-Gaussianity (which Olshausen and Field implicitly did), we can readily see, based on the discussion in Section 6.2.5, why an Infomax/ICA net with the logistic nonlinearity for its $g_i(u_i)$'s would produce the filters that produced the sparsest activation distributions when passed over the images. These distributions, furthest from Gaussian on the super-Gaussian side, were the most likely to be as statistically independent as possible, through the Central Limit theorem argument that any mixture of two independent distributions would produce a distribution that was closer to Gaussian. The fact that the basis functions produced by ICA, shown in Figure 6.1, were so similar to those published by Olshausen and Field (1996a) helped to connect the sparseness/projection pursuit arguments back up with the original information-theoretic arguments of Barlow. In addition, the ICA network and algorithm was more computationally tractable and conceptually simpler than the networks based on sparseness (see Olshausen (1996) for a detailed analysis of the connections between the two).

The assumption implicit in both approaches has been that the first layer of visual processing should attempt to invert the simplest possible image-formation process, in which the image is formed, just as in Eq. (6.4), by linear superposition of basis vectors (columns of $\mathbf{A}$), each activated by independent (or sparse) causes, s_i:

$$\mathbf{x} = \mathbf{As} \tag{6.37}$$

Learning the inverse of $\mathbf{A}$, called $\mathbf{W}$ in Eqs. (6.21)–(6.23), and then inverting $\mathbf{W}$, gives the basis vectors shown in Figure 6.1. Full implementational details are in Bell and Sejnowski (1997). In this case, $\mathbf{A}$ is a 144×144 matrix, each column of which contains the pixels we have plotted as 12×12 images in Figure 6.1. Linear superposition of these basis vectors, with independent weighting coefficients, s_i, corresponds to the image-formation process.

To evaluate the sparseness of the independent-component distributions, we calculated their kurtoses and compared them with those given by PCA and

ZCA (see Section 6.2.2) transforms. The results were 10.0 for ICA, 3.7 for PCA, and 4.5 for ZCA, showing that of the decorrelating transforms studied, ICA produced the sparsest distributions, as we might expect. This could well turn out to be an important fact for adaptive image-/video-compression systems in the future.

On might object, correctly, that because we have used the logistic sigmoid for $g_i(u_i)$, we are biasing the net away from finding *sub*-Gaussian independent components of the images (see Section 6.2.5). But experiments with the extended ICA algorithm of Eq. (6.33) have shown that even when the network can find sub-Gaussian components, none were found (Te-Won Lee, personal communication). This reinforces a pattern we have observed with natural data—that the vast majority of naturally occurring independent components are super-Gaussian. Even in the case of EEG signals (see Section 6.3.3), the only sub-Gaussian components came from sinusoidal 60-Hz electrical line noise, a man-made signal.

Since our study, impressive results have been obtained by Van Hateren and Van der Schaal (1998). Using a much larger and better calibrated image database, and employing Hyvärinen and Oja's FastICA method (1997) they found 18×18-oriented basis functions, and compared their properties in detail with receptive fields of primate visual cortex. Perplexed (as we were) that the spatial frequency distribution of the ICA basis functions was clustered around the high available frequencies (unlike the broader band of frequencies seen in monkey cortex), they argued that to truly compare ICA bases with the properties of monkey neurons, the temporal response properties of the cells had to be included. This led to a heroic effort by Van Hateren and Ruderman (1998), who used a Cray computer, to calculate some of the 1728 spatiotemporal basis functions of $12 \times 12 \times 12$ "natural movie patches," each consisting of 12 frames. The resulting spatiotemporal bases were localized, oriented, and moving perpendicular to their orientation direction, just as in monkey visual cortex. And significantly, there were many more with the much lower spatial frequency required to match the primate data.[3]

6.3.2 Natural Sounds

The next natural candidate to submit to the ICA-basis treatment was audio data (Bell and Sejnowski 1996). Randomly selecting windows of length N in a raw audio signal, we can ask what N independent bases compose the signal. The results are less inspiring than for image data. The analysis of audio data has been dominated for so long by the Fourier basis set that one might hope for something different. But ICA on a sufficiently long sample of speech data gives bases that are sinusoidal and of different frequencies and phases. They are roughly orthogonal and not qualitatively different from those found by PCA, and thus second-order statistics seem to be dominant in the case of speech data.

---

[3] Movies of the basis functions may be viewed on the Web at http://hlab.phys.rug.nl/index.html.

To demonstrate the ability of ICA to find nonorthogonal bases related to the higher-order statistics of the sounds, we have to analyze audio signals where the structure does not lie mostly in the frequencies present, but also in the phase structure.

It is worth explaining this in more detail, for the same arguments apply to the ICA image-bases results reported earlier. If we Fourier transform a signal of length N, we get N complex numbers. The squares of their lengths form the power spectrum, and their angles form the phase spectrum. If we take the power spectrum and Fourier transform *it*, we get the signal's autocorrelation function, which is exactly the second-order statistics of the signal. Thus, by a process of elimination, since statistics of all orders are sufficient to reconstruct the signal, all the statistics of order higher than two (the higher-order statistics) can be seen to carry the same information as the phase spectrum.

In Bell and Sejnowski (1996), we deliberately chose to analyze a signal with both time–local and frequency–local structure. A series of short musical notes were generated by tapping on a tooth with a fingernail and altering the structure of the oral cavity. Each note consisted of a broad-band click followed by a decaying sinusoidal envelope at a frequency dictated by the shape of the mouth. The resulting ICA basis vectors split into two types: brief time–local functions corresponding to the clicks, and more temporally extended frequency–local functions corresponding to the envelopes of the notes. Given that the timing of a note and its frequency are generally independent variables, such a decomposition is quite sensible. On the same data, PCA produced bases with no such interesting phase structure.

6.3.3 Electroencephalographic Data: Decomposition and Artifact Removal

EEG data are produced by spatial arrays of tiny electrodes implanted on the surface of the scalp. What makes it interesting is how the signal changes with the cognitive state of the subject. For example, if you have a lot of activity at 10 Hz (called alpha waves), you are probably asleep. Although the electrodes pick up quite correlated signals, different signals in different electrodes are enough to suggest to experimenters spatial differences in brain activity. With ICA, however, we can go further, and learn spatial filters across the N electrodes, which pull out independent activity. The form of the spatial filter tells us what parts of the scalp are most responsible for the activity.

For EEG analysis, the strength of the N EEG signals at each time t, provides an input, $\mathbf{x}$, to the ICA network. The ICA transform for that timepoint is [as in Eq. (6.1)] given by $\mathbf{u} = \mathbf{Wx}$. After learning on all timepoints, the columns of $\mathbf{W}^{-1}$ are used to give the "scalp map" (see Figure 6.2). For a given independent component, this is an interpolated function expressing the spatial projection of the independent component onto the electrodes.

What is measured in EEG are the brain's electric fields, which propagate linearly (by volume conduction) with negligible time delays (Nunez 1981). This

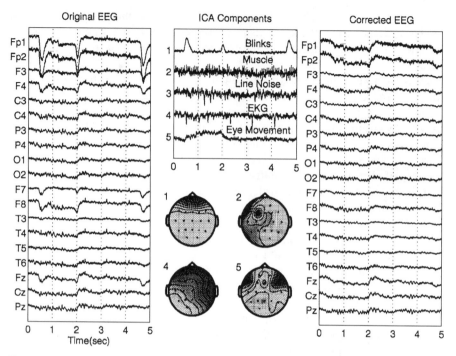

Figure 6.2 A 5-s portion of the EEG time series (left), ICA components accounting for eye movements, cardiac signals, and line noise sources (center), and the EEG signals "corrected" for artifacts by removing the five components (right). (Figure reprinted from Jung et al. (1997).)

makes it perfect for ICA analysis. If there were 14 electrical dipoles in the brain, each with independently fluctuating charges, and 14 noiseless electrodes on the scalp, then the dipole signals could be perfectly recovered by ICA. This is, of course, not the case. But the results are interesting nonetheless.

The first results of ICA on EEG were obtained by Makeig et al. (1995), with followup studies in Jung et al. (1998) and Makeig et al. (1997). The basic results are as follows:

1. As shown in Figure 6.2, what can appear as correlated alpha-wave activity across the electrodes, can be ICA-decomposed into prominent rhythms in different components with different time courses.

2. Many of the ICA outputs were easy to identify with artifacts known to contaminate the brain-wave data. In Figure 6.2 [from Jung et al. (1998)], five of these are displayed, corresponding to eyeblinks, localized scalp muscle movements, 60-Hz electrical line noise, heartbeat, and a horizontal eye movement. Both their time courses and their scalp maps help support these interpretations. The EEG data can then be cleansed of these artifacts, as in Figure 6.2, by zeroing the columns in $\mathbf{W}^{-1}$ corresponding to the artifacts, and reconstructing the data thus: $\mathbf{x} = \mathbf{W}^{-1}\mathbf{u}$. Using the

extended ICA algorithm of Eq. (6.33) was more successful on these data than using ICA with a fixed nonlinearity like the logistic, since it enabled isolation of the sub-Gaussian sinusoidal line noise.

The artifact removal result is important for EEG researchers. Previous methods aimed at artifact reduction used regression to remove the artifact from the data. But such methods relied on having clean reference signals, such as the electro-oculogram (EOG) for eye movements, when there was known to be cross-contamimation of the EEG and the reference signal. An additional advantage is that with ICA many different artifacts can be removed in one analysis.

What about the rest of the independent components? The answer to this question must be given by the expert on the data. The decomposition of cross-electrode alpha-wave activity into alpha activity occurring in different compo-nents at different times, is very suggestive, expecially when it follows a sequence across the scalp. But in the absence of further information about what the subject is doing, it is difficult to interpret. However, in another study (Makeig et al. 1997) applying ICA to event-related potential (ERP) data,[4] the spatio-temporal decomposition is easier to interpret. The subject (whose alertness was being monitored) sometimes failed and sometimes succeeded in responding to the event (a tone). Of the 14 ICA-transformed ERP waveforms, 3 showed "bumps" that only occurred when the subject succeeded in responding to the event (a tone). Another 4 showed bumps when the subject failed to respond. Another 3 accounted for the brain's phasic response to a 39-Hz background click, thus removing this response from the event-related activity. In all cases, scalp maps could be constructed to investigate the scalp localization of the dif-ferent event-related responses.

This kind of study is quite common in studies on cognition and human per-formance. EEG and ERP experimenters design experiments in which the sub-ject is exposed to stimuli with a variety of modalities and varying parameters, for example, sounds and lights in different places. What ICA opens up for them is the tantalizing possibility that these modalities and parameters will be re-flected in different activity in different independent components, thus enabling greater localization in both space and time, of differentiated brain functions. The applications for both diagnosis and the probing of brain function are clear.

In the next section, we turn to another example, this time a brain probe with a much finer spatial grain, magnetic resonance imaging (MRI).

6.3.4 Decomposing fMRI Signals

Functional magnetic resonance imaging (fMRI) monitors humans during per-formance of psychomotor tasks and produces a 3-dimensional picture of their brain activity with a spatial resolution of about 5 mm^2 and a temporal resolu-

---

[4] If, during recording of EEG, we expose the subject to some event repeatedly (in this case, we play a tone), and afterward we take all the EEG windows centered around the times when the tone played, and average them, then we have the ERP [see Hillyard and Pictor (1980)].

tion of about 2 s. It utilizes the different magnetic properties of oxygenated and deoxygenated blood, and the fact that increased neural activity locally deoxygenates the blood with a stereotypical latency of 5–8 s. However, task-related changes in the fMRI signal are heavily corrupted by machine noise, subtle subject movements, and heart and breathing rhythms. To counter this, experimenters typically ask the subject to alternate a control task with the main task, in a sequence of task–block cycles. Then the average fMRI picture when the control task is being done, is subtracted from that when the main task is being done, to yield a brain image supposedly representing the crucial difference between the tasks. Unfortunately, this throws away any data on transiently task-related activations such as the subject learning the task, or any other variations in the subject's performance, arousal, attention, or effort.

In the color-naming experiment reported in McKeown et al. (1998), the control task and main task each last 40 s and alternate for two 6-min trials.[5] fMRI data collection led to a sequence of 146 brain images, each with 10,000 voxels (volume pixels). This was suitable for ICA analysis with each input data vector, $\mathbf{x}$, consisting of the time course of a single voxel across the experiment. An ICA weight matrix, $\mathbf{W}$, trained on these data with the logistic nonlinearity, $g(u)$, transforms the data into a series of 146 independent, and sparsely activated "brain-maps," exactly the kind of images that are useful for localizing brain function. The columns of the inverse matrix, $\mathbf{W}^{-1}$, in this case represent the time courses of the activations of these sparse brain maps.

The most striking result of this study is shown in Figure 6.3c. In all six trials (three subjects performing the experiment twice each), a single one of these time courses turned on and off in an exact 40-s cycle, corresponding to the alternation of the tasks. There was even a phase delay corresponding to the known 5–8-s latency of the blood-flow signal. Neither PCA nor a fourth-order ICA method did as good a job of finding one time course corresponding to the task cycle, as shown in Figure 6.3a,b. When the independent components are ranked in terms of their strength in the original signal, this time course was ranked between fourteenth and forty-first out of the 146, so it was by no means a strong signal in the original data. The brain map associated with it contained activations in relevant areas of the brain corresponding mainly to vision and visual association, but also some in a motor area, and some in prefrontal cortex.

Frontal cortical activation was also seen for another one of the 146 time courses (not shown), one that was on for the first cycle of the main task, and not thereafter. Frontal cortex is associated with, among other things, stimulus novelty, working memory, and visual–spatial attention. It seems reasonable to explain the transiently task-related nature of this component by saying that the subject is learning the task on the first cycle.

---

[5] The control task was to covertly say ("think") the name of a colored rectangle being presented, while the main task was to covertly say the color of some presented text that spelled the name of another color. For example, the word "green" might be presented in blue text, and the subject should covertly say "blue."

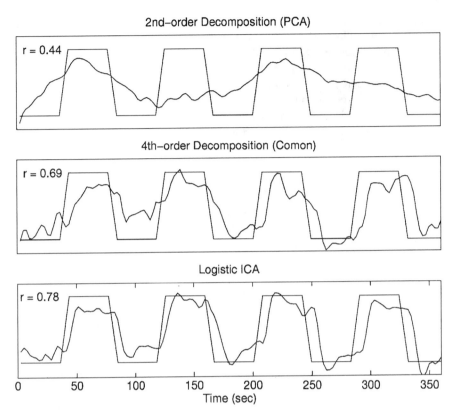

Figure 6.3 Comparison of 3 algorithms on the fMRI data. For each we display the time course (independent basis vector) most correlated with the reference function (a cyclic function representing the alternation of the control and main tasks). *r* denotes the correlation coefficient between the time course and the reference function.

In each case, in focusing on individual time courses, we are looking at on the order of 1/146 of the brain activation data. Thus the vast majority of the possibly confusing and irrelevant brain activation is stripped away for us, because its brain maps are statistically independent from the ones that concern us. Looking further at the time-courses of these other components, McKeown et al. (1998a,b) identified some that coincided with slow and fast head movements. These could be used for artifact removal, as with the EEG data.

6.3.5 Blind Separation of Audio Data

We will not cover in any detail the use of ICA to separate audio signals. The separation of artifically mixed audio signals was the main application reported in Bell and Sejnowski (1995). But separation of signals mixed in a real room (with echos and time delays) is a much more difficult problem, requiring multipath deconvolution to be added to the basic ICA methodology. Several studies have addressed this problem (Cichocki et al. 1997; Lambert and Bell 1997; Lambert 1996; Lee et al. 1997a; Torkkola 1996; Yellin and Weinstein 1996) with varying degrees of success, and a number of them are described in this book.

However, it is not even clear that by adding deconvolution to ICA we obtain the best approach to this problem. For example, Ehlers and Schuster (1997) have produced very good results using second-order statistics only. Their method extends Molgedey and Schuster's work (1994) using time-delayed decorrelation to separate artificially mixed audio signals. These second-order methods [see also references in Ehlers and Schuster (1997)] exploit the temporal correlations in audio signals to constrain the solution, and they are not strictly ICA methods, although they could perhaps be improved by including higher-order statistics. Another study (Ngo and Bhadkamkar 1998) that exploits the time delays in a real microphone situation (rather than attempting to cancel the delays), also produces a very good performance with a purpose-built two-microphone recording device.

6.4 DISCUSSION AND CONCLUSION

For a technique that maps data linearly and deterministically into a new coordinate system, the obvious extensions are to make it nonlinear[6] and stochastic. Theoretical problems abound here.

First, if the nonlinearity used in the mapping is sufficiently powerful, then our probability density estimate, $\hat{f}(\mathbf{x}|\mathbf{W})$ as in Eq. (6.36), can be warped in any way by the mapping. This can create independent variables in any direction desired. There will literally be an infinite number of ICA solutions.

Thus the mapping nonlinearity must be constrained in some way. Lee et al. (1997c) consider the case where each element of the data vector $\mathbf{x}$ is passed through a (possibly flexible) nonlinearity before standard ICA unmixing occurs. This is designed, for example, to invert microphone nonlinearities. Other methods (Lin and Cowan 1997; Pajunen 1996) use self-organizing maps and vector quantization to produce nonlinear coordinate axes in data.

Another possibility is to consider the constrained nonlinearity of an $N \rightarrow N \rightarrow N$ multilayer ICA network where we map $\mathbf{x}$ into $\mathbf{y}$ as before $(\mathbf{y} = \mathbf{g}(\mathbf{u}), \mathbf{u} = \mathbf{W}\mathbf{x})$, but we add an extra layer, which we distinguish by the $\sim$ overscript, giving $\tilde{\mathbf{y}} = \tilde{\mathbf{g}}(\tilde{\mathbf{u}}), \tilde{\mathbf{u}} = \tilde{\mathbf{W}}\mathbf{y}$. The Infomax learning rules for such a system display nice properties [compare with Eqs. (6.20)–(6.22)]:

$$\Delta\tilde{\mathbf{W}} \propto \tilde{\mathbf{W}}^{-T} + \tilde{\Phi}(\tilde{\mathbf{u}})\mathbf{y}^T \tag{6.38}$$

$$\tilde{f}_i(\tilde{u}_i) = \frac{\partial \tilde{y}_i'}{\partial \tilde{y}_i} \tag{6.39}$$

---

[6] This a nonlinearity in our mapping of $\mathbf{x}$ to $\mathbf{u}$, not to be confused with the nonlinearity $g(u)$ used as a source cdf estimator.

$$\Delta \mathbf{W} \propto \mathbf{W}^{-T} + \Phi(\mathbf{u}, \tilde{\mathbf{u}}, \tilde{\mathbf{W}})\mathbf{x}^T \tag{6.40}$$

$$\phi_i(u_i, \tilde{\mathbf{u}}, \tilde{\mathbf{W}}) = \frac{\partial y_i'}{\partial y_i} + y_i' \sum_{j=1}^{N} \tilde{w}_{ji} \tilde{f}_i(\tilde{u}_i) \tag{6.41}$$

The top layer of weights, $\tilde{\mathbf{W}}$, learns as before, and the bottom layer, $\mathbf{W}$, learns with a modified set of nonlinearities, Φ, which contain information back-propagated from the outputs.

Unfortunately, for the interesting case of an $N \to M \to N$ system where $M \neq N$ (presumably $M > N$), very complex rules result. And no one has thought of a use yet for the tractable $M = N$ case.

The problem with these methods is that the number of independent components in the data could be much less than the dimensionality of our data vectors. Consider the case of a data set of 12×12 images where each one contains only a rectangle of "on" pixels. The four independent variables are the position and the dimensions of the rectangle, and they form a curved 4-dimensional subspace in the 144-dimensional data space. Amazingly, no one has solved the problem of blindly extracting these four variables from the data, though the unsupervised learning literature in the neural-network field has devoted much attention lately to the inversion of such "nonlinear generative models," sometimes called nonlinear factor analysis after the classic work on factor analysis (Everitt 1984). An example of such a network is the Helmholtz machine (Dayan et al. 1995). One possible avenue is to consider noncommutative transformation groups (such as Lie groups) that can span the data manifold in a global way. Embryonic efforts in this direction can be seen in Nordberg (1994) and Rao and Ruderman (1999). Another approach is to abandon all hope of finding a global coordinate system, and to approximate the data manifold with a lot of local surfaces (Bregler and Omohundro 1995; Kambhatla and Leen).

Some progress has been made, however, in extending ICA to stochastic models, where as well as estimating the sources, some other "hidden" variables have to be taken into account. Perhaps the most up-to-date proposals on this topic are given by Attias (1998). In his independent factor analysis (IFA) networks, the EM algorithm (Dempster 1977; Neal and Hinton 1998) is used in an ICA setting to estimate source densities, mixing matrix, $\mathbf{W}$, and the parameters of corrupting noise. From this, an optimal nonlinear estimator can be built to reconstruct the sources. EM is basically maximum-likelihood density estimation with some hidden parameters that have to be estimated as well, so it is a natural extension of the maximum-likelihood view of ICA outlined in Section 6.2.6.

ACKNOWLEDGMENTS

The work reported here was done while I was in Terry Sejnowski's laboratory at the Salk Institute. Terry vigorously promoted and encouraged the work on ICA. The result is that

I am able to report here the work of a whole interdisciplinary team effort, with a constant flow of surprises and ideas coming from my colleagues, in particular, Martin McKeown, Scott Makeig, Tzzy-Ping Jung and Te-Won Lee, whose results appear throughout this chapter.

REFERENCES

Amari, S.-I., 1997, "Natural gradient works efficiently in learning," *Neural Computation*, vol. 10, pp. 251–276.

Amari, S.-I., T.-P. Chen, and A. Cichocki, 1997, "Stability analysis of adaptive blind source separation," *Neural Networks*, vol. 10, no. 8, pp. 1345–1351.

Amari, S., A. Cichocki, and H. H. Yang, 1996, "A new learning algorithm for blind signal separation," *Advances in Neural Information Processing Systems 8*, (Cambridge, MA: MIT Press).

Atick, J. J., 1992, "Could information theory provide an ecological theory of sensory processing?" *Network* vol. 3, pp. 213–251.

Atick, J. J., and A. N. Redlich, 1993, "Convergent algorithm for sensory receptive field development," *Neural Computation*, vol. 5, pp. 45–60.

Attias, H., 1998, "Independent factor analysis," *Neural Computation*, vol. 10, no. 6, pp. 1373–1425.

Barlow, H. B., 1989, "Unsupervised learning," *Neural Computation*, vol. 1, pp. 295–311.

Barlow, H. B., and D. J. Tolhurst, 1992, "Why do you have edge detectors?" *Optical Society of America: Technical Digest*, vol. 23, p. 172.

Bell, A. J., and T. J. Sejnowski, 1995, "An information maximization approach to blind separation and blind deconvolution," *Neural Computation*, vol. 7, pp. 1129–1159.

Bell, A. J., and T. J. Sejnowski, 1996, "Learning the higher-order structure of a natural sound," *Network: Computation in Neural Systems*, vol. 7, p. 2.

Bell, A. J., and T. J. Sejnowski, 1997, "The 'Independent Conponents' of natural scenes are edge filters," *Vision Research*, vol. 37, pp. 3327–3338.

Belouchrani, A., and J.-F. Cardoso, 1994, "Maximum likelihood source separation for discrete sources," *Proc. EUSIPCO*, Edinburgh, Sept. 1994, pp. 768–771.

Belouchrani, A., and J.-F. Cardoso, 1995, "Maximum likelihood source separation by the expectation-maximization technique: Deterministic and stochastic implementation," *Proc. Intern. Symp. on Nonlinear Theory and Applications (NOLTA)*, Las Vegas, Dec. 1995.

Bregler, C., and S. M. Omohundro, 1995, "Nonlinear image interpolation using manifold learning," *Advances in Neural Information Processing Systems 7*, (Cambridge, MA: MIT Press).

Cardoso, J.-F., 1995, "The invariant approach to source separation," *Proc. NOLTA*, pp. 55–60.

Cardoso, J.-F., 1997, "Infomax and maximum likelihood for blind source separation," *IEEE Signal Processing Lett.*, vol. 4, no. 4, pp. 112–114.

Cardoso, J.-F., 1998, "High-order contrasts for independent component analysis, *Neural Computation*, submitted.

Cardoso, J.-F., and S.-I. Amari, 1998, "Maximum likelihood source separation: equi-variance and adaptivity," *Proc. SYSID'97, 11th IFAC symposium on system identification*, Fukuoka, Japan.

Cardoso, J.-F., and B. Laheld, 1996, "Equivariant adaptive source separation," *IEEE Trans. on Signal Processing*.

Cichocki, A., S.-I. Amari, and J. Cao, 1997, "Neural network models for blind separation of time-delayed and convolved signals," *Japanese IEICE Trans. on Fundamentals*, vol. E-82-A, no. 9.

Cichocki, A., R. Unbehauen, and E. Rummert, 1994, "Robust learning algorithm for blind separation of signals," *Electronics Lett.*, vol. 30, no. 17, pp. 1386–1387.

Comon, P., 1994, "Independent component analysis, a new concept?" *Signal Processing*, vol. 36, pp. 287–314.

Comon, P., C. Jutten, and J. Herault, 1991, "Blind separation of sources, part II: problems statement, *Signal Processing*, vol. 24, pp. 11–21.

Obradovic, D., and G. Deco, 1998, "Information maximization and independent component analysis: Is there a difference?" *Neural Computation*, vol. 10, no. 8, pp. 2085–2103.

Dayan, P., G. Hinton, R. Neal, and R. Zemel, 1995, "The Helmholtz machine," *Neural Computation*, vol. 7, pp. 889–904.

Dempster, A. P., N. M. Laird, and D. B. Rubin, 1997, "Maximum likelihood from incomplete data via the EM algorithm," *J. Royal Statistical Society B*, vol. 39, pp. 1–38.

Ehlers, F., and H. G. Schuster, 1997, "Blind separation of convolutive mixtures and an application in automatic apeech recognition in noisy environment," *IEEE Trans. on Signal Processing*, vol. 45, no. 10, pp. 2608–2611.

Everitt, B., 1984, *An Introduction to Latent Variables* (London: Chapman & Hall).

Field, D. J., 1987, "Relations between the statistics of natural images and the response properties of cortical cells," *J. Opt. Soc. Am. A*, vol. 4, no. 12, pp. 2370–2393.

Field, D. J., 1994, "What is the goal of sensory coding?" *Neural Computation*, vol. 6, pp. 559–601.

Girolami, M., 1996, "Negentropy and kurtosis as projection pursuit indices provide generalised ICA algorithms," in *Advances in Neural Information Processing Systems 9*.

Hancock, P. J. B., R. J. Baddeley, and L. S. Smith, 1992, "The principal components of natural images," *Network*, vol. 3, pp. 61–72.

Herault, J., and C. Jutten, 1986, "Space or time adaptive signal processing by neural network models," in J. S. Denker, ed., *Neural Networks for Computing: AIP Conference Proceedings 151*, New York: (American Institute for Physics).

Hillyard, S., and T. Picton, 1980, "Electrophysiology of cognition," in *Handbook of Psychology—The Nervous System V*.

Hubel, D. H., and T. N. Wiesel, 1968, "Receptive fields and functional architecture of monkey striate cortex," *J. Physiol.*, vol. 195, pp. 215–244.

Hubel, D. H., and T. N. Wiesel, 1974, "Uniformity of monkey striate cortex: a parallel relationship between field size, scatter, and magnification factor," *J. Comp. Neurol.*, vol. 158, pp. 295–306.

Huber, P. J., 1985, "Projection pursuit," *Ann. Stat.*, vol. 13, pp. 435–475.

Hyvärinen, A., and E. Oja, 1997, "A fast fixed-point algorithm for independent component analysis," *Neural Computation*, vol. 9, pp. 1483–1492.

Intrator, N., 1992, "Feature extraction using an unsupervised neural network," *Neural Computation*, vol. 4, pp. 98–107.

Jung T.-P., C. Humphries, T.-W. Lee, S. Makeig, M. J. McKeown, V. Iragui, and T. J. Sejnowski, 1998, "Extended ICA removes artifacts from electroencephalographic recordings," *Advances in Neural Information Processing Systems (NIPS)*, vol. 10 (Cambridge, MA: MIT Press).

Jutten, C., and J. Herault, 1991, "Blind separation of sources, part I: an adaptive algorithm based on neuromimetic architecture," *Signal Processing* vol. 24, pp. 1–10.

Kambhatla, N., and T. K. Leen, "Dimension reduction by local principal component analysis," *Neural Computation*, vol. 9, p. 1493.

Karhunen, J., and J. Joutsensalo, 1994, "Representation and separation of signals using non-linear PCA type learning," *Neural Networks*, vol. 7, no. 1, pp. 113–127.

Lambert, R. H., 1996, Multi-channel blind deconvolution: FIR matrix algebra and separation of multipath mixtures, Ph.D. Thesis, Elec. Eng., Univ. of Southern California.

Lambert, R. H., and A. J. Bell, 1997, "Blind separation of multiple speakers in a multipath environment, submitted to *ICASSP '97*, Munich.

Lee, T.-W., A. J. Bell, and R. Lambert, 1997a, "Blind separation of delayed and convolved sources," in *Advances in Neural Information Processing Systems 9* (Cambridge, MA: MIT Press).

Lee, T.-W., M. Girolami, and T. J. Sejnowski, 1997b, Independent component analysis using an extended infomax algorithm for mixed sub-Gaussian and super-Gaussian sources, *Neural Computation*, submitted.

Lee, T.-W., B. Koehler, and R. Orglmeister, 1997c, "Blind separation of nonlinear mixing models," *IEEE Workshop on Neural Nets for Signal Processing*, Florida, USA.

Lee, T.-W., M. Girolami, A. J. Bell, and T. J. Sejnowski, 1998, "A unifying information-theoretic framework for independent component analysis," *Int. J. on Mathematical and Computer Modeling*, in press.

Lewicki, M., and T. J. Sejnowski, 1997, "Learning nonlinear overcomplete representations for efficient coding," in *Advances in Neural Information Processing Systems 10*.

Linsker, R., 1988, "Self-organization in a perceptual network," *Computer*, vol. 21, pp. 105–117.

Linsker, R., 1989, "An application of the principle of maximum information preservation to linear systems," in *Advances in Neural Information Processing Systems 1*, D. S. Touretzky, ed., Morgan-Kauffman.

Lin, J., and J. Cowan, 1997, "Faithful representation of separable input distributions, *Neural Computation*, vol. 9, no. 6, pp. 1305–1320.

MacKay, D., 1996, "Maximum likelihood and covariant algorithms for independent component analysis," *Tech. Report*, University of Cambridge, Cavendish Lab.

Makeig, S., A. J. Bell, T.-P. Jung, and T. J. Sejnowski, 1995, "Independent component analysis of electroencephalographic data," in M. Mozer et al. eds. *Advances in Neural Information Processing Systems 8*, (Cambridge, MA: MIT Press).

Makeig, S., A. J. Bell, T.-P. Jung, D. Ghahremani, and T. J. Sejnowski, 1997, "Blind separation of auditory evoked potentials into independent components," *Proc. Natl. Acad. Sci. USA*, vol. 94, pp. 10797–10984.

Marr, D., and E. Hildreth, 1980, "Theory of edge detection," *Proc. R. Soc. Lond. Ser. B*, vol. 207, pp. 187–217.

McKeown, M. J., S. Makeig, G. G. Brown, T.-P. Jung, S. S. Kinderman, A. J. Bell, and T. J. Sejnowski, 1998a, "Analysis of fMRI data by blind separation into independent spatial components," *Human Brain Mapping*, vol. 6, pp. 160–188.

McKeown, M. J., T.-P. Jung, S. Makeig, G. G. Brown, S. S. Kinderman, T.-W. Lee, and T. J. Sejnowski, 1998b, "Spatially independent activity patterns in functional magnetic resonance imaging data during the Stroop color-naming task," *Proc. Natl. Acad. Sci. USA*, vol. 95, pp. 803–810.

Miller, K. D., 1988, "Correlation-based models of neural development," in M. Gluck and D. Rumelhart, eds., *Neuroscience and Connectionist Theory*, (Hillsdale, NJ: Lawrence Erlbaum) pp. 267–353.

Molgedey, L., and H. G. Schuster, 1994, Separation of independent signals using time-delayed correlations, *Phys. Rev. Letts.* vol. 72, no. 23, pp. 3634–3637.

Nadal, J.-P., and N. Parga, 1994, "Non-linear neurons in the low noise limit: a factorial code maximises information transfer," *Network*, vol. 5, pp. 565–581.

Nailong, W., 1997, *The Maximum Entropy Method* (Berlin: Springer).

Neal, R., and G. E. Hinton, 1998, A view of the EM algorithm that justifies incremental, sparse and other variants, in M. I. Jordan, ed., *Learning in Graphical Models*.

Ngo, T., and N. A. Bhadkamkar, 1998, "Adaptive blind separation of audio sources by a physically compact device using second-order statistics," *Proc. of ICA '99*, Aussois, France.

Nordberg, K., 1994, "Signal representation and processing using operator groups, *Technical Report, Dissertation no. 366*, Dept. of Electrical Engineering, Linköping University.

Nunez, P., 1981, *Electric Fields of the Brain* (New York: Oxford).

Olshausen, B. A., 1996, "Learning linear, sparse, factorial codes," *MIT AI-memo No. 1580*, AI-lab, MIT.

Olshausen, B. A., and D. J. Field, 1996a, "Emergence of simple-cell receptive fieldproperties by learning a sparse code for natural images," *Nature*, vol. 381, pp. 607–609.

Olshausen, B. A., and D. J. Field, 1996b, "Natural image statistics and efficient coding," *Network: Computation in Neural Systems*, vol. 7, no. 2.

Olshausen, B. A., and D. J. Field, 1997, "Sparse coding with an overcomplete basis set: a strategy employed by V1?" *Vision Research*, vol. 37, pp. 3311–3325.

Pajunen, P., 1996, Nonlinear independent component analysis by self-organising maps, *Proc. Int. Conf. on Artif. Neural Networks (ICANN) 1996*.

Papoulis, A., 1984, *Probability, Random Variables and Stochastic Processes*, 2nd ed., (New York: McGraw-Hill).

Pearlmutter, B. A., and L. C. Parra, "A context-sensitive generalization of ICA, *Proc. ICONIP '96*, Japan.

Pham, D. T., P. Garrat, and C. Jutten, 1992, "Separation of a mixture of independent sources through a maximum likelihood approach," in *Proc. EUSIPCO*, pp. 771–774.

Pham, D.-T., 1996, "Blind separation of instantaneous mixtures of sources via an independent component analysis, *IEEE Trans. on Signal Processing*, vol. 44, no. 11, pp. 2768–2779.

Rao, R. P. N., and D. L. Ruderman, 1999, "Learning lie groups for invariant visual perception," *Advances in Neural Information Processing Systems 11*.

Roth, Z., and Y. Baram, 1996, "Multidimensional density shaping by sigmoids," *IEEE Trans. on Neural Networks*, vol. 7, no. 5, pp. 1291–1298.

Sanger, T. D., 1989, Optimal unsupervised learning in a single-layer network, *Neural Networks*, vol. 2, pp. 459–473.

Schmidhuber, J., 1992, "Learning factorial codes by predictability minimisation, *Neural Computation* vol. 4, no. 6, pp. 863–887.

Torkkola, K., 1996, Blind separation of convolved sources based on information maximisation, *Proc. IEEE Workshop on Neural Networks and Signal Processing*, Kyota, Japan, Sept. 1996.

van Hateren, J. H., 1992, A theory of maximising sensory information, *Biol. Cybern.*, vol. 68, pp. 23–29.

van Hateren, J. H., and D. L. Ruderman, 1998, Independent component analysis of natural image sequences yields spatiotemporal filters similar to simple cells in primary visual cortex. *Proc. R. Soc. Lond. B*, in press.

van Hateren, J. H., and A. van der Schaaf, 1998, "Independent component filters of natural images compared with simple cells in primary visual cortex," *Proc. R. Soc. Lond. B*, vol. 265, pp. 359–366.

Yang, H., and S.-I. Amari, 1997, Adaptive on-line learning algorithms for blind separation—maximum entropy and minimum mutual information, *Neural Computation*, vol. 9, pp. 1457–1483.

Yellin, D., and E. Weinstein, 1996, Multichannel signal separation: methods and analysis, *IEEE Trans. on Signal Processing*, vol. 44, no. 1, pp. 106–118.

7

INFORMATION-THEORETIC LEARNING

Jose C. Principe, Dongxin Xu, and John W. Fisher III

7.1 INTRODUCTION

This chapter addresses the important issue of *extracting information directly from data*, which is at the core of the issue of learning from examples in both biological and artificial systems. The learning-from-examples scenario starts with a data set that globally conveys information about a real-world event, and the goal is to capture the information in the parameters of a learning machine. The information exists in a "distributed" mode in the data set, and appears "condensed" in the parameters of the learning machine after successful training. Learning in artificial neural networks and adaptive filters has used almost exclusively correlation (the L2 norm or mean-square error) as a criterion to compare the information carried by the signals and the response of the learning machine, but there is mounting evidence that correlation (a second-order moment) is a poor measure to ascertain the equivalence of information between the desired response and the output of the mapper. The fundamental issue is to find the appropriate methodology to study this "change in state" and elucidate the issues of designing systems that are capable of producing the transfer of information as efficiently as possible.

We propose to utilize information theory (IT) as the mathematical infrastructure, because it is the best possible approach to deal with manipulation of information (Shannon and Weaver 1949). In a 1948 classic paper Shannon laid

Unsupervised Adaptive Filtering, Volume I, Edited by Simon Haykin.
ISBN 0-471-29412-8 © 2000 John Wiley & Sons, Inc.

down the foundations of IT (Shannon 1948). IT has had a tremendous impact in the design of *efficient and reliable* communication systems (Cover and Thomas 1991; Fano 1961) because it is able to answer two key questions: What is the best possible (minimal) code for our data? and What is the maximal amount of information that can be transferred through a particular channel? In spite of its practical origins, IT is a deep mathematical theory concerned with the *very essence of the communication process* (Fano 1961). IT has also impacted statistics (Kullback 1968) and statistical mechanics by providing a clearer understanding of the nature of entropy as illustrated by Jaynes (1957). These advances, however, are predicated on the specification of the data distributions, which is not realistic for the design of learning machines. In the design of a self-organizing system, the primary objective is to develop an algorithm that will learn an input–output relationship of interest on the basis of input patterns alone. We submit that a thrust to innovate IT is to develop methods to directly estimate entropy from a set of data. With entropic measures, we will be able to utilize the full probability density function (pdf) for optimization and to lift the present restrictions of linearity and Gaussianity for the application of IT to real-world problems. We will achieve this goal by introducing a nonparametric approach to estimate entropy from a discrete set of data and proposing new methods of manipulating mutual information.

In this chapter we will develop information-theoretic criteria that can directly from the sample's train linear or nonlinear mappers either for entropy or mutual-information maximization or minimization. We will start by a brief review of Renyi's entropy and a description of information-theoretic learning (ITL). The Parzen-window method of pdf estimation is fundamental in all our efforts to create algorithms to manipulate entropy. First, we develop the integrated square error (ISE) criterion, which *indirectly* estimates entropy at the output of a mapper by local interactions among its outputs. We present comparisons with principal-component analysis (PCA), and applications to entropy maximization and blind source separation (BSS). The following section covers a more principled approach of designing practical information-theoretic criteria using Renyi's definition of entropy of order two (quadratic entropy). We show that quadratic entropy can be easily integrated with the Parzen-window estimator and leads to an interaction model that is similar to ISE. The pairwise data interactions for the computation of entropy are interpreted as an information potential field and are a powerful analogy between information-theoretical learning and physics. We finally propose the integration of the Cauchy-Schwartz distance and an Euclidean difference with the Parzen window to provide estimators for mutual information. The mutual-information criterion is very general and can be used either in a supervised or unsupervised learning framework. Applications of these algorithms are discussed, first for BSS (as an example of unsupervised learning) and for pose estimation (as an example of supervised learning). Finally we train a multilayer perceptron layer-by-layer as an example of the power of ITL.

7.2   GENERALIZED DEFINITIONS OF ENTROPY

7.2.1   Renyi's Entropy

Information theory is a mathematical formalization of our intuitive notion of information contained in messages. If a message is perfectly known *a priori*, its information content is zero. However, the less predictable a message is, the larger is its information content. Shannon, using an axiomatic approach (Shannon 1948) defined entropy of a probability massfunction $P = (p_1, p_2, \ldots, p_N)$ as

$$H_S(P) = \sum_{k=1}^{N} p_k \log\left(\frac{1}{p_k}\right) \qquad \sum_{k=1}^{N} p_k = 1 \qquad p_k \geq 0 \qquad (7.1)$$

that is, the average amount of information contained in a single observation of a random variable X that takes values $x_1, x_2, \ldots, x_N$ with probabilities $p_k = P(x = x_k)$, $k = 1, 2, \ldots, N$. Entropy measures the average amount of information conveyed by the event x, or alternatively, the amount of missing information on X when only its *a priori* distribution is given. Information theory has been widely applied to the design of communication systems (Cover and Thomas 1991; Fano 1961; Shannon and Weaver 1949). But the definition of entropy can be derived even in a more abstract form. In the general theory of means (Renyi 1976b), the mean of the real numbers $x_1, \ldots, x_N$ with positive weighting (not necessarily probabilities) $p_1, \ldots, p_N$ has the form:

$$\bar{x} = \varphi^{-1}\left(\sum_{k=1}^{N} p_k \varphi(x_k)\right) \qquad (7.2)$$

where $\varphi(x)$ is a Kolmogorov-Nagumo function, which is an arbitrary continuous and strictly monotonic function defined on the real numbers. In general, an entropy measure H obeys the relation

$$H = \varphi^{-1}\left(\sum_{k=1}^{N} p_k \varphi(I(p_k))\right) \qquad (7.3)$$

where $I(p_k) = -\log(p_k)$ is Hartley's information measure (Hartley 1928). In order to be an information measure, $\varphi(\cdot)$ cannot be arbitrary since information is "additive." To meet the additivity condition, $\varphi(\cdot)$ can be either $\varphi(x) = x$ or $\varphi(x) = 2^{(1-\alpha)x}$. If $\varphi(x) = x$ is selected, Eq. (7.3) will become Shannon's entropy. For $\varphi(x) = 2^{(1-\alpha)x}$ Renyi's entropy of order α is obtained (Renyi 1976a), which we denote by H_{R_α}

$$H_{R_\alpha} = \frac{1}{1 - \alpha} \log \left(\sum_{k=1}^{N} p_k^\alpha \right) \qquad \alpha > 0, \quad \alpha \neq 1 \qquad (7.4)$$

There is a well-known relation between Shannon's and Renyi's entropy:

$$H_{R_\alpha} \geq H_S \geq H_{R_\beta}, \qquad \text{if} \quad 1 > \alpha > 0 \quad \text{and} \quad \beta > 1$$

$$\lim_{\alpha \to 1} H_{R_\alpha} = H_S$$

It is important to further relate Renyi's and Shannon's entropies. Let us consider the probability massfunction $P = (p_1, p_2, \ldots, p_N)$ as a point in an N-dimensional space. Due to the conditions on the probability measure ($p_k \geq 0$, $\sum_{k=1}^{N} p_k = 1$), P always lies in the first quadrant of a hyperplane in N dimensions intersecting each axis at the coordinate 1 (Fig. 7.1). The distance of P to the origin is the α root of

$$V_\alpha = \sum_{k=1}^{N} p_k^\alpha = \|P\|^\alpha$$

and the α root of V_α is called the α-norm of the probability massfunction (Golub and Van Loan 1989). Renyi's entropy, Eq. (7.4) can be written as a function of V_α

$$H_{R_\alpha} = \frac{1}{1 - \alpha} \log V_\alpha \qquad (7.5)$$

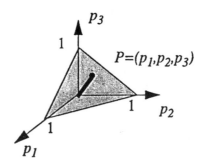

Figure 7.1 Geometric interpretation of entropy for $N = 3$. The distance of P to the origin is related to the α-norm.

When different values of α are selected in Renyi's family, the end result is to select different α-norms. Shannon entropy can be considered as the limiting case $\alpha \to 1$ of the probability distribution norm. Notice that the limit provides an indeterminacy [zero over zero in Eq. (7.5)], but the result exists and is given by Shannon's entropy. With this view, Renyi's entropy is a monotonic function of the α-norm of the massfunction and is essentially a monotonic function of the distance of the probability distribution to the origin. We have considerable freedom in choosing the α-norm (Xu 1998). When $\alpha = 2$, $H_{R_2} = -\log \sum_{k=1}^{N} p_k^2$ is called *quadratic entropy* due to the quadratic form on the probability, and it corresponds to the 2-norm of the probability massfunction.

For the continuous random variable Y with pdf $f_Y(y)$, we can obtain the differential version of Renyi's entropy following a similar route to the Shannon differential entropy (Renyi 1976a):

$$\begin{cases} H_{R_\alpha}(Y) = \dfrac{1}{1-\alpha} \log \left(\displaystyle\int_{-\infty}^{+\infty} f_Y(z)^\alpha \, dz \right) \\ H_{R_2}(Y) = -\log \left(\displaystyle\int_{-\infty}^{+\infty} f_Y(z)^2 \, dz \right) \end{cases} \qquad (7.6)$$

Note that Renyi's quadratic entropy involves the use of the square of the pdf.

Renyi's entropy is just one example of a large class of alternate entropy definitions that have been called *generalized entropy measures* (Kapur and Kesavan 1992). One might wonder why the interest in measures more complex than Shannon's entropy or the Kullback-Leibler direct divergence (also called cross-entropy). Here we will only provide a brief overview of this important question. The reader is referred to the entropy-optimization literature for further study (Kapur and Kesavan 1992; Kapur 1994; Renyi 1976a). The reason for using generalized measures of entropy stems from practical aspects when modeling real-world phenomena through entropy optimization algorithms. It has been found that when we apply the two basic optimization principles based on Shannon's entropy definition (which are Jayne's maximum-entropy principle (MaxEnt) and Kullback's minimum-cross-entropy principle (MinXEnt)), either just one solution from a spectrum of solutions is found or not even "natural" solutions are found. To improve on this situation, researchers have proposed alternative definitions of entropy. An example of a generalized entropy measure in the digital signal-processing arena is Burg's entropy estimator (Burg 1967), which has been successfully applied in spectral analysis (Marple 1988).

In our study of learning from examples, the interest in generalized entropy measures comes from a practical difficulty. We wish to directly estimate entropy from the data samples, without imposing assumptions about the pdf. Shannon's definition of entropy (the sum of terms that are weighted logarithms of probability) is not amenable to simple estimation algorithms, while Renyi's logarithm of the sum of the power of probability is much easier to estimate, and has been utilized in physics (Grassberger and Procaccia 1983). In Section 7.3 we show how a very effective algorithm can be derived. Renyi's entropy has been utilized successfully in nonlinear dynamics to estimate the correlation dimension of attractors. One important question stemming from the use of generalized entropy measures is the justification for the selected measure. We have not yet addressed this question in our research. At this point we can only state that the experimental results obtained with the use of Renyi's entropy estimator and its extension to mutual information have produced practical solutions to difficult problems in signal processing and pattern recognition. Since learning from examples is an inverse problem, we believe that the choice of an appropriate generalized entropy measure will play an important role in the quality of the final solution.

7.2.2 Information Optimization Principles

The most common entropy optimization principles are Jayne's MaxEnt and Kullback's MinXEnt (Kapur and Kesavan 1992). MaxEnt finds the distribution that maximizes Shannon's entropy subject to some explicit constraints. Hence, MaxEnt guarantees that we make no assumptions about possible missing information. MinXEnt finds a distribution, from all possible distributions satisfying the constraints, that minimizes the distance in probability space to the given distribution. The most widely used measure for MinXEnt is the Kullback-Leibler (KL) cross-entropy. Effectively, KL is a measure of directed divergence between the given and the unknown distribution (a directed divergence is a relaxed concept of distance, since it does not need to be symmetric nor obey the triangular inequality). It turns out that MinXEnt (using the KL divergence), with respect to the uniform target distribution, is equivalent to the MaxEnt principle under the same constraints. They are intrinsically different, however, since one maximizes uncertainty while the other minimizes directed divergence between pdf's. Moreover, MinXEnt is invariant to coordinate transformations, which is an advantage for learning, while MaxEnt does not hold this characteristic in the continuous case.

7.2.3 Information-Theoretic Learning

Consider the parametric mapping $g : \Re^K \rightarrow \Re^M$, of a random vector $X \in \Re^K$ (normally $M < K$), which is described by the following equation

$$\mathbf{Y} = g(\mathbf{X}, \mathbf{W}) \tag{7.7}$$

where $\mathbf{Y}$ is also a random vector $\mathbf{Y} \in \Re^M$, and $\mathbf{W}$ is a set of parameters. For each observation $\mathbf{x}_i$ of the random vector $\mathbf{X}$, the parametric system (mapper)

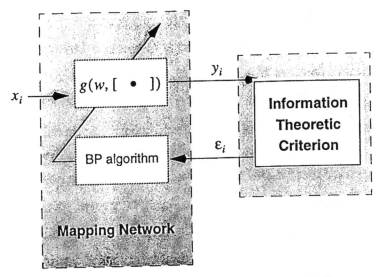

Figure 7.2 Training a mapper (linear or nonlinear) with ITL.

responds with $\mathbf{y}_i = g(\mathbf{x}_i, \mathbf{W})$. Our goal is to choose the parameters $\mathbf{W}$ of the mapping $g(\cdot)$ such that a figure of merit based on information theory is optimized at the output space of the mapper (Fig. 7.2). This is what we call information-theoretic learning. Notice that we are only requiring the availability of observations $\mathbf{x}_i$ and $\mathbf{y}_i$ or random vectors without assuming any *a priori* model for their pdf. Notice also that the mapper can either be linear or nonlinear, and that the criterion may or may not exploit an external input normally called the desired response, that is, ITL includes as special cases both the unsupervised and supervised frameworks. We also want the learning criterion to be external and independent of the mapper. Let us briefly review work done in this area.

By analogy to optimization in Euclidean space, we can adapt the parameters $\mathbf{W}$ of the mapper by manipulating the output distribution $p(\mathbf{Y})$: maximizing output entropy (MaxEnt) or minimizing the cross-entropy among the outputs or among the output and other signals (MinXEnt). The work of Bell and Sejnowski (1995) on BSS is an example of the application of the MaxEnt principle. In the neural network literature, the work of Barlow (Barlow et al. 1989) and Atick (1992) also utilized entropy concepts for learning.

Optimization based on the MinXEnt principle is the one that is potentially wellsuited in solving engineering and in particular learning problems (Deco and Obradovic 1996). Comon (1994), Deco and Obradovic (1996), Cardoso (1997), and Amari (Yang and Amari 1997), among others, utilized the MinXEnt principle to formulate and solve the BSS problem, as demonstrated in previous chapters. One solution to BSS is obtained by minimizing the mutual information (redundancy) among the outputs of a mapper $\mathbf{Y}$, which can be formulated as the KL divergence between the joint pdf of $\mathbf{Y}$ and its factorized marginals as $I(y_1, \ldots, y_n) = \sum_{i=1}^{n} H(y_i) - H(\mathbf{Y})$. The problem arises in estimating the joint output density $H(\mathbf{Y})$. These researchers utilize the well-known result of using a *linear network* to directly compute the output entropy from the input entropy as $H(\mathbf{Y}) = H(\mathbf{X}) + \log|\det(\mathbf{W})|$ where $\mathbf{Y} = \mathbf{WX}$ (Papoulis, 1965). Note that *a full rank k-to-k linear mapping* $\mathbf{W}$ is required in this approach, which is a severe constraint for learning applications (for instance, in subspace mappings as required in classification). The next step is the estimation of the marginal entropy of each output $H(y_i)$ (a scalar problem). Comon (1994) proposed the use of the Edgeworth expansion of the pdf and Amari (Yang and Amari 1997) the Gram-Charlier expansion, which are both well known and equivalent methods (in the limit) of estimating pdf's by the moment expansion method. In practice, the expansions must be truncated (a source of error) and higher-order moments of the pdf estimated from the data, which becomes computationally expensive and requires large amounts of data for robust results. However, after the marginal pdf's are estimated, then a gradient-based algorithm can be formulated to solve the BSS problem (Yang and Amari 1997). Although this method is very appealing from the point of view of a learning criterion, notice that it is not general because the criterion is not totally independent of the topology of the mapper. Recently, Amari and Cardoso proposed a semiparametric model for BSS (Amari and Cardoso 1997).

In the neural-network literature there is still another information-optimization principle, Linsker's principle of maximum information preservation (Infomax), which is a special case of the information loss minimization principle of Plumbey (Plumbey and Fallside 1989). Optimization with mutual information has not been extensively addressed in the optimization literature. Linsker (1989) was interested in finding a principle that self-organizes biological systems. These systems are adaptable, so the issue is to find a criterion to adapt the parameters of the mapper $g(\mathbf{X}, \mathbf{W})$. The goal is to determine the parameters $\mathbf{W}$ such that the output variable $\mathbf{Y}$ conveys as much information as possible about $\mathbf{X}$; that is, a principle for self-organization should maximize the average mutual information between $\mathbf{X}$ and $\mathbf{Y}$ in the presence of noise. For a linear network and under Gaussianity assumptions, the mutual information is maximized by maximizing the output variance (Linsker 1989). Recall that maximization of output variance is basically PCA, for which there are known on-line and local algorithms (Diamantaras and Kung 1996; Oja 1982). Hence, foreseeably, a biological network could self-organize with such a principle. We can see that this method leads to interesting solutions, but it depends on very restrictive assumptions about the pdf's and linearity of the mapping. In fact, Plumbey states (Plumbey and Fallside 1989) that *the big challenge is to extend Linsker's work to arbitrary distributions and nonlinear networks*. This is exactly what we propose to accomplish in this work.

From a theoretical perspective, Infomax is a different principle from Max-

Ent and MinXEnt since it *maximizes* a divergence measure (mutual information). Linsker applied Infomax between the input and the output of deterministic mappers, so the principle reduces to applying MaxEnt at the output of the mapper (Bell and Sejnowski 1995). But Infomax can be applied to any pairs of random variables, such as the outputs of the mapper and any other external random variable. This new application is called here *information filtering*, since it designs a mapper to maximally preserve information about a source while attenuating other information available in the input data. Information filtering for supervised learning applications is exploited later in the chapter.

One of the difficulties of these information-theoretic criteria is that analytic solutions are known only for very restricted cases, for example, Gaussianity and linear volume-preserving mappings [see also Deco and Obradovic (1996)]. Otherwise mathematical approximations and computationally complex algorithms result. A useful neurocomputing algorithm should be applied to any topology, utilize the data directly (on a sample-by-sample basis or in batch), and be a simple learning rule distilled from the mathematics, so we submit that none of the preceding algorithms to train a nonlinear mapper with MinXEnt criterion is "neural." In this respect Bell and Sejnowski's algorithm for maximization of output entropy is paradigmatic. It utilizes a nonlinear mapper (although restricted to be a perceptron), it adapts the weights with a simple batch rule that is not specific to the input data model [as the solution in Yang and Amari (1997)], and globally leads to a solution that maximizes an entropic measure. Recently, Linsker showed that there is a local rule for MaxEnt, which only requires extending the perceptron with lateral connections (Linsker 1997). This is the spirit of a neural computation (Haykin 1994) that we have been seeking all along.

In our opinion, the two fundamental issues in the application of information-theoretic criteria to neurocomputing or adaptive filtering are the choice of the criterion for the quantitative measure of information, and the estimation of the pdf from data samples.

7.3 INFORMATION-THEORETIC LEARNING: UNSUPERVISED LEARNING WITH INTEGRATED SQUARE ERROR

One obstacle of using information-theoretic criteria (entropy or mutual information) is that entropy for discrete random variables based on Shannon definition is a weighted sum of the logarithm of the pdf [Eq. (7.1)] (or an integral function of the logarithm of the probability density for continuous random variables). Since we cannot work directly with the pdf (unless assumptions are made about its form), we rely on nonparametric estimators. Viola proposed such an estimator in Viola et al. (1995), but had to use the sample mean to approximate entropy, which is an approximation that is not an efficient estimator. Moreover, he was not interested in adapting a mapper, but simply in evaluating a measure of similarity based on entropy. Density estimation is an ill-posed problem (Vapnik 1995), and in particular, nonparametric density estimation is very unreliable in high-dimensional spaces (Parzen 1962). The approach described here, however, relies on such estimates in the output space of a nonlinear mapper, where the dimensionality is under the control of the designer, and is generally manageable. Moreover, the ITL algorithms are a function of the integral of the pdf, which is simpler than density estimation. We have to remember that a posteriori probabilities are also a function of the pdf, but we found ways to estimate them reliably.

The Parzen window method (Parzen 1962) will be used in this work. The Parzen estimator is a kernel-based estimator, which estimates the pdf, $f_Y(\mathbf{z})$, of a random vector $\mathbf{Y} \in \mathfrak{R}^M$ as

$$\hat{f}_Y(\mathbf{z}, \mathbf{y}) = \left(\frac{1}{N}\right) \sum_{i=1}^{N} \kappa(\mathbf{z} - \mathbf{y}_i) \tag{7.8}$$

The vectors $\mathbf{y}_i \in \mathfrak{R}^M$ are observations of the random vector, and $\kappa(\cdot)$ is a kernel function that itself satisfies the properties of a pdf. The Parzen window can be viewed as a convolution of the estimator kernel with the observations. We choose the symmetric Gaussian kernel

$$\kappa(\mathbf{z}) = G(\mathbf{z}, \sigma^2\mathbf{I}) = \frac{1}{(2\pi)^{M/2}\sigma^M} \exp\left(-\frac{\mathbf{z}^T\mathbf{z}}{2\sigma^2}\right) \tag{7.9}$$

with covariance matrix $\sigma^2\mathbf{I}$, since we require that $\kappa(\cdot)$ be differentiable everywhere. But other choices besides the Gaussian exist.

7.3.1 Entropy Manipulation with the Integrated-Squared-Error Criterion

Let us consider that the nonlinear mapper $g(\cdot)$ is a multilayer perceptron (MLP). Since the MLP outputs are bounded, for example, within $[-1, 1]$ if a tanh nonlinear function is used in the output layer, *the output entropy will be maximized when the outputs are uniformly distributed.* So, the problem of entropy maximization can be solved by comparing the desired uniform distribution $u(\mathbf{z})$ with an estimate of the output pdf $\hat{f}_Y(\mathbf{z}, \mathbf{y})$ at a point $\mathbf{z}$ over the set of N observations $\{\mathbf{y}\} = \mathbf{y}_1, \mathbf{y}_2, \ldots, \mathbf{y}_N$, with an appropriate metric. Preserving the traditional framework of learning, we will utilize the integrated squares error (ISE) between the pdf of the actual output and the uniform pdf as the criterion for entropy maximization, that is

$$J = \frac{1}{2}\int_D (u(\mathbf{z}) - \hat{f}_Y(\mathbf{z}, \mathbf{y}))^2 \, d\mathbf{z} \approx \sum_{j=1}^{L} \frac{1}{2}(u(\mathbf{z}_j) - \hat{f}_Y(\mathbf{z}_j, \mathbf{y}))^2 \, \Delta\mathbf{z} \tag{7.10}$$

where $\hat{f}_Y(\mathbf{z}, \mathbf{y})$ is the estimated pdf by the Parzen-window method, Eq. (7.8); $u(\mathbf{z})$ is the uniform pdf in the region $D = \{\mathbf{z}|(y^{min} \leq z^r \leq y^{max}, r = 1, \ldots, M)\}$; $y^1, \ldots, y^M$ are the components of $\mathbf{y}$; and $\mathbf{y}_i, \mathbf{i} = 1, \ldots, N$, are output samples that correspond to the inputs $\mathbf{x}_i$. Basically this is an application of the MinXEnt principle discussed in Section 7.2, but where the ISE is used as a proxy of the divergence between the output pdf and the uniform distribution. One practical way to evaluate the integral is to sample a square domain (which contains D) at $L \gg N$ uniformly distributed spatial locations $\mathbf{z}_j \in \mathfrak{R}^M$, $j = 1, \ldots, L$, transforming the integral in Eq. (7.10) into a sum.

Once an external criterion is selected, the gradient method and the chain rule can be used for the minimization of J. As an example, the partial derivative of J with respect to $\mathbf{w}$, a general weight in the mapping, becomes (Fisher 1997)

$$\frac{\partial J}{\partial w} = \sum_{i=1}^{N} \frac{\partial J}{\partial \mathbf{y}_i} \frac{\partial \mathbf{y}_i}{\partial w} = \sum_i \Delta \mathbf{z} \sum_j [u(\mathbf{z}_j) - \hat{f}_Y(\mathbf{z}_j, \mathbf{y})] \frac{\partial}{\partial y_i} \hat{f}_Y(\mathbf{z}_j, \mathbf{y}) \frac{\partial}{\partial w} g(\mathbf{w}, \mathbf{x}_i)$$

$$= -\sum_i \left(\varepsilon_i \frac{\partial}{\partial w} g(\mathbf{w}, \mathbf{x}_i) \right) \tag{7.11}$$

where N is the number of training exemplars, $\varepsilon_i = \sum_j (u(\mathbf{z}_j) - \hat{f}_Y(\mathbf{z}_j, \mathbf{y})) \cdot \partial \frac{\hat{f}_Y(\mathbf{z}_j, \mathbf{y})}{\partial a_i} \Delta \mathbf{z}$ is the computed error distribution between the estimated output distribution and the desired uniform distribution over the observations, and $\partial g(\mathbf{w}, \mathbf{x}_i)/\partial w$ are the mapping sensitivities. Excluding the mapping sensitivities, the remaining terms can be combined to compute *an error term*, ε_i, associated with each training sample. Note that we are adapting an MLP in an *unsupervised mode*, since there is no external desired response. The error is derived from a global property of the output samples. The chain rule was used in Eq. (7.11), and the mapping sensitivities can be computed using the backpropagation algorithm (Principe and Fisher 1997), that is, this method integrates seamlessly with the main stream of neurocomputing algorithms. A more general discussion of this approach is presented in Fisher (1997).

The straightforward approach of numerically approximating the computation of an error between two pdf's as in Eq. (7.11) has one significant drawback: it requires the evaluation of the error term at a sufficient number of points in the output space. Consequently the computational complexity of this algorithm is proportional to $O(N^{N_d+2})$, where N_d is the dimension of the output space and N the number of samples in the training set. So, Eq. (7.10) imposes a fundamental computational limitation to the dimensionality of the subspace mapping. Moreover, one can argue that if a numerical method is chosen to compute J, then Shannon entropy definition could also be directly used and numerically approximated as discussed earlier. The advantage of the ISE is that it leads to a major simplification of the computation, as we show next.

7.3.2 Computation of the Criterion with a Local Interaction Model

Further examination of the gradient of the ISE criterion results in significant reduction in the computational complexity, leading to a local algorithm that bypasses the necessity to approximate the integral numerically. Basically the idea is to search for a method for computing an error for each data sample, instead of performing the computation in a large density of points in the output space. The analytic form of the local kernels will extend the sample interactions over the output space. First, let us note that for any analytic, even kernel function $\kappa(z)$,

$$\frac{\partial}{\partial \mathbf{y}_i} \kappa(\mathbf{z} - \mathbf{y}_i) = \frac{\partial}{\partial \mathbf{y}_i} \kappa(\mathbf{y}_i - \mathbf{z}) = \kappa'(\mathbf{y}_i - \mathbf{z})$$

which leads to

$$\frac{\partial}{\partial \mathbf{y}_i} \hat{f}_Y(\mathbf{z}, \mathbf{y}) = \frac{1}{N} \kappa'(\mathbf{y}_i - \mathbf{z})$$

Therefore we can operate directly with the integral definition of J in Eq. (7.10). The derivative of J with respect to a general weight of the mapper becomes

$$\frac{\partial J}{\partial w} = \sum_i \frac{\partial J \partial \mathbf{y}_i}{\partial \mathbf{y}_i \partial w} = -\sum_i \left\{ \int [u(\mathbf{z}) - \hat{f}_Y(\mathbf{z}, \mathbf{y})] \frac{\partial}{\partial \mathbf{y}_i} \hat{f}_Y(\mathbf{z}, \mathbf{y}) \, d\mathbf{y} \right\} \frac{\partial}{\partial w} g(\mathbf{w}, \mathbf{x}_i)$$

$$= -\frac{1}{N} \sum_i \left\{ \int [u(\mathbf{z}) - \hat{f}_Y(\mathbf{z}, \mathbf{y})] \kappa'(\mathbf{y}_i - \mathbf{z}) \, d\mathbf{z} \right\} \frac{\partial}{\partial w} g(\mathbf{w}, \mathbf{x}_i)$$

$$= -\frac{1}{N} \sum_i \varepsilon_i \frac{\partial}{\partial w} g(\mathbf{w}, \mathbf{x}_i) \tag{7.12}$$

Expanding the error term, ε_i, yields

$$\varepsilon_i = \varepsilon_Y(\mathbf{z}, \mathbf{y}) * \kappa'(\mathbf{z})\big|_{z=y_i}$$

$$= (f_Y(\mathbf{z}) - \hat{f}_Y(\mathbf{z}, \mathbf{y})) * \kappa'(\mathbf{z})\big|_{z=y_i}$$

$$= \left(f_Y(\mathbf{z}) - \frac{1}{N} y(\mathbf{z}) * \kappa(\mathbf{z}) \right) * \kappa'(\mathbf{z})\big|_{z=y_i}$$

$$= (f_Y(\mathbf{z}) * \kappa'(\mathbf{z})) - \frac{1}{N} y(\mathbf{z}) * \kappa(\mathbf{z}) * \kappa'(\mathbf{z})\big|_{z=y_i}$$

$$= f_r(\mathbf{z}) - \frac{1}{N} y(\mathbf{z}) * \kappa_a(\mathbf{z})\big|_{z=y_i}$$

$$= f_r(\mathbf{y}_i) - \frac{1}{N} \sum_{j \neq i} \kappa_a(\mathbf{y}_i - \mathbf{y}_j) \tag{7.13}$$

where $*$ means the convolution operation and $y(\mathbf{z})$ represents the location of the training samples in the output space and is evaluated as

$$y(\mathbf{z}) = \sum_{i=1}^{N} \delta(\mathbf{z} - \mathbf{y}_i)$$

The terms $\kappa_a(\mathbf{z})$ and $f_r(\mathbf{z})$ are termed the *attractor kernel* and the *topology regulating kernel*, respectively. Equation (7.13) overcomes the fundamental limitation implied by Eq. (7.10). Both terms in Eq. (7.13) can be computed analytically (for the Gaussian kernel and the uniform distribution). So we can effectively compute an estimate of the pdf over the full space by performing calculations only at the sample locations using the analytic forms of the local kernels. More importantly, the computational complexity of Eq. (7.12) is only of order N for each y_i, which is quadratic in the number of exemplars N and reduces the complexity to

$$O(N^{N_d+2}) \rightarrow O(N^2) \tag{7.14}$$

7.3.3   Interpretation of the ISE

Of particular interest are the forms of $\kappa_a(\mathbf{z})$ and $f_r(\mathbf{z})$ that compute the error. The analytic forms for the Gaussian kernel (with diagonal covariance) and the uniform distribution are, respectively (Fisher 1997),

$$\kappa_a(\mathbf{z}) = \kappa(\mathbf{z}) * \kappa'(\mathbf{z}) = -\left(\frac{1}{2^{M+1}\pi^{M/2}\sigma^{M+2}}\right) \exp\left(-\frac{1}{4\sigma^2}(\mathbf{z}^T\mathbf{z})\right)\mathbf{z}$$

$$= -\left(\frac{1}{2^{3M/4+1}\pi^{M/4}\sigma^{(M/2)+2}}\right)\kappa(\mathbf{z})^{1/2}\mathbf{z} \tag{7.15}$$

$$f_r(\mathbf{z}) = \frac{1}{d^M}$$

$$\left[\begin{array}{c} \prod\limits_{i \neq 1}\frac{1}{2}\left(\mathrm{erf}\left(\frac{z_i+\frac{d}{2}}{\sqrt{2}\sigma}\right) - \mathrm{erf}\left(\frac{z_i-\frac{d}{2}}{\sqrt{2}\sigma}\right)\right)\left(G_1\left(z_1+\frac{d}{2},\sigma^2\right) - G_1\left(z_1-\frac{d}{2},\sigma^2\right)\right) \\ \cdots \\ \cdots \\ \prod\limits_{i \neq M}\frac{1}{2}\left(\mathrm{erf}\left(\frac{z_i+\frac{d}{2}}{\sqrt{2}\sigma}\right) - \mathrm{erf}\left(\frac{z_i-\frac{d}{2}}{\sqrt{2}\sigma}\right)\right)\left(G_1\left(z_N+\frac{d}{2},\sigma^2\right) - G_1\left(z_N-\frac{d}{2},\sigma^2\right)\right) \end{array} \right]$$

$$\tag{7.16}$$

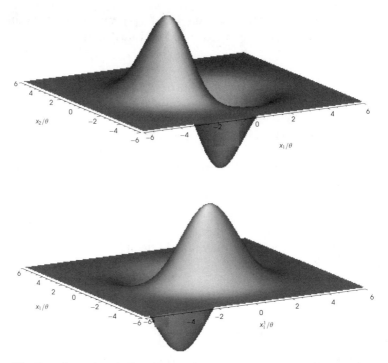

Figure 7.3 Two-dimensional attractor functions. The x_1-component is shown at the top, while the x_2-component is shown at the bottom. The function represents the local influence of each data point in the output space.

where M is the dimension of the kernel, σ is the width of the kernel, d is the extent in all dimensions of the uniform distribution, and $G_1(z, \sigma^2)$ indicates the one-dimensional Gaussian kernel evaluated at z with standard deviation σ, and $\text{erf}(z) = \int_{-z}^{z}(1/\sqrt{2\pi}) \exp(-x^2/2)\, dx$ is the error function. For the multidimensional case we will use circularly symmetry Gaussian kernels, that is,

$$G(z, \sigma^2 I) = \prod_{i=1}^{M} G_i(z_i, \sigma^2) = \prod_{i=1}^{M} \frac{1}{\sqrt{2\pi}\sigma} \exp\left(-\frac{z_i^2}{2\sigma^2}\right)$$

These functions are shown for the two-dimensional case in Figs. 7.3 and 7.4. From these figures we can see that $\kappa_a(\cdot)$ represents the influence that each observation has on its local surrounding, while $f_r(\cdot)$ represents the influence on each sample near the boundary of the region of support.

From this perspective, we see that our estimator of entropy can be modeled as local interactions between samples in the output space, as well as interactions with the boundary. The functional form of the sample-to-sample interaction is

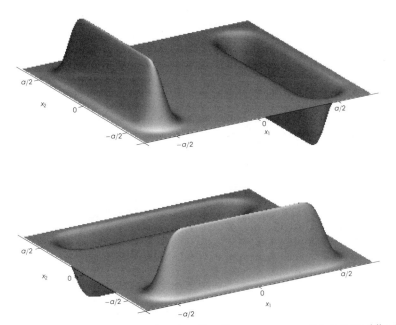

Figure 7.4 Two-dimensional regulating function. The x_1-component is shown at the top, while the x_2-component is shown at the bottom.

dictated by Eq. (7.15), which shows that when maximizing entropy, the distribution error through the kernel acts as a local attractor when the computed pdf error is positive, and as a local repellor when the pdf error is negative. When minimizing entropy the sign of the interactions is the opposite. Let us observe Fig. 7.5a to understand this behavior better in one dimension. In this figure, we show a multimodal pdf of the input data and a uniform distribution between $[-1, 1]$. Assume that we want to maximize entropy. Figure 7.5b shows the error computed by the attractor kernel through the domain. For the regions where the original pdf is concave, the error direction will diffuse the samples, while for the regions where the input pdf is convex, the attractor kernel will concentrate the samples, trying to achieve a uniform distribution after convergence. Figure 7.5c shows that for entropy minimization the behavior is the opposite, that is, the algorithm tries to create a concentration of samples (delta function) near the two peaks of the input pdf.

A similar interpretation can be presented for the topology regulating kernel. Notice, however, that this kernel always forces the samples to be within the range of the sigmoid. Furthermore, the erf evaluations can be omitted without changing appreciably the overall form of the interaction (Fisher 1997). We conclude by saying that the function of the attractor kernel is to model the interaction of data points with each other, while the function of the topology

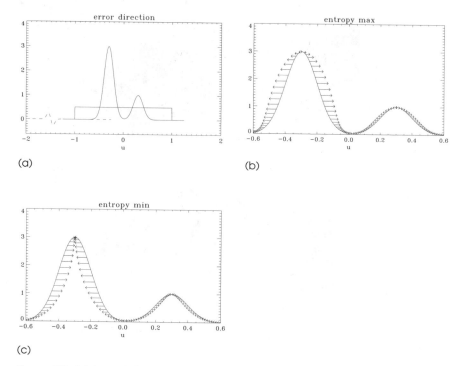

Figure 7.5 Mixture of Gaussians example. The estimated distribution is a mixture of Gaussians, while the desired distribution is uniform between $[-1, 1]$. The kernel gradient, which will be convolved with the difference between the two distributions is shown in the dotted line of (a). In (b) the error direction is shown for every point in the domain for the case of entropy maximization, while (c) depicts the case of minimization of entropy.

regulating term is to model the interaction between the data and the constraints of the desired output distribution.

7.3.4 Creating a Learning Algorithm

The aspects that need to be addressed to create a practical information-theoretic algorithm are the convergence of the algorithm and a stopping criterion. Both are intimately related to the kernel size.

Algorithm Convergence The main implication of the previous discussion is that our proposed cost function leads to the manipulation of a global property of the output data by a process of local attraction and repulsion among samples. Each kernel provides an error that will influence the surrounding samples. But unlike supervised learning where the desired response provides a normalization (the error normally decreases when the system response approaches the desired response), here it is the shape of the influence function and the size of the interactions among samples that will control the error size. We experimentally verified that, in the case of large learning rates (close to divergence), the algorithm tended to oscillate. This difficulty is conquered by imposing a local normalization in each kernel. Let us start with the case of minimization of entropy which, in the limit, becomes attraction to a single point (delta function) within our framework. The degree of the attraction is a function of the size, σ, of the kernel Eq. (7.15), and it is inversely proportional to σ^{M+2}. The error will always point to the center of the kernel, but if we do not scale appropriately oscillations around the center of the kernel may occur. The influence of the samples in our model is additive. So in order to guarantee that, in fact, the net attraction converges to a point, we normalize the gradient at the center of the attractor kernel to unity (i.e., the slope around the origin is unitary). The normalization factor that guarantees this for a Gaussian kernel in any dimensional space is (assuming a radially symmetric kernel) (Fisher 1997)

$$\left\| \frac{\partial}{\partial \mathbf{z}} \kappa_a(\mathbf{z}) \right\|_{z=0} = \frac{1}{2^{M+1} \pi^{M/2} \sigma^{M+2}} \tag{7.17}$$

As we see, stable convergence is a function of the dimensionality of the output space and the kernel size, which leads to the question of how to set the kernel size. The kernel size parameterizes a family of densities. In the context of entropy maximization a principled approach is to choose the value of σ that parameterizes the density with minimum entropy, followed by finding the maximum entropy direction and then iterate. The kernel size that achieves minimum entropy is the one that maximizes the likelihood of the observations.

The log-likelihood function (estimated with the Parzen estimator), L, and associated score function, $S = \partial L / \partial \sigma$, are

$$L = \sum_{y_j \in U} \log(\hat{f}(y_j))$$

$$= \sum_{y_j \in U} \log \left(\frac{1}{|U_j|} \sum_{y_i \in U_j} \kappa(y_j - y_i, \sigma^2) \right) \tag{7.18}$$

$$S = \sum_{y_j \in U} \frac{1}{\hat{f}(y_j)} \frac{1}{|U_j|} \sum_{i \in U_j} \kappa'(y_j - y_i, \sigma^2)$$

$$= \frac{1}{\sigma^3} \sum_{y_j \in U} \frac{1}{\hat{f}(y_j)} \frac{1}{|U_j|} \sum_{i \in U_j} \kappa(y_j - y_i, \sigma^2)((y_i - y_j)^2 - \sigma^2) \tag{7.19}$$

where $U = \{y_1, \ldots, y_N\}$ is the set of data points, $U_j \subset U$ is the subset of points used to estimate $\hat{f}(y_j)$, and the second line of Eq. (7.19) follows from the fact that for the radially symmetric Gaussian kernel $\partial k(y, \sigma)/\partial\sigma = (1/\sigma^3)\kappa(y, \sigma^2) \cdot (y^2 - \sigma^2)$. The optimal value for σ is one of the roots of Eq. (7.19). By inspection we see that if the set U_j contains the data point y_j, the optimal value of σ is zero. This condition never occurs in our model due to the pairwise interaction where $i \neq j$. In general, Eq. (7.19) requires a numerical solution, and any number of fast techniques can be used to find σ to some precision (e.g., line search). This implies a minimax approach to maximizing entropy. Finally, as a consequence of this approach the kernel size can be used directly as the stopping criterion. That is, when the kernel size has reached its maximum, the maximum entropy has been reached. In practice one must set a threshold.

7.3.5 Applications of the ISE Algorithm

We show here several applications of the ISE algorithm to toy problems and also to some real-world data. The goal of this section is to show that the algorithm is practical and provides meaningful results.

Maximum Entropy/PCA Comparison   We wish to illustrate the differences between the well-known PCA approach to feature extraction and to an entropy-driven approach implemented with the ISE algorithm. We begin with the simple case of a two-dimensional Gaussian distribution. The distribution we use is zero mean with a covariance matrix of

$$\Sigma = \begin{bmatrix} 0.1 & 0 \\ 0 & 1 \end{bmatrix}$$

The contour plot of this distribution is shown in Fig. 7.6, along with the image of the first principal-component vector (left panel). We see from the figure that the first principal component lies along the x_0-axis. We draw a set of observations (50 in this case) from this distribution and compute a mapping using an MLP and the entropy maximizing criterion described in the previous section. The architecture of the MLP is 2-4-1, indicating 2 input nodes, 4 hidden nodes, and 1 output node. The nonlinearity used is the hyperbolic tangent function. We are therefore nonlinearly mapping the two-dimensional input space onto a one-dimensional output space. The plot at the right of Fig. 7.6 shows the image of the maximum entropy mapping onto the input space. From the contours of this mapping we see that the maximum entropy mapping lies essentially in the same direction as the first principal components. This result is expected. It illustrates that when the Gaussian assumption is supported by the data, maximum entropy and PCA are equivalent. This result has been reported by many researchers.

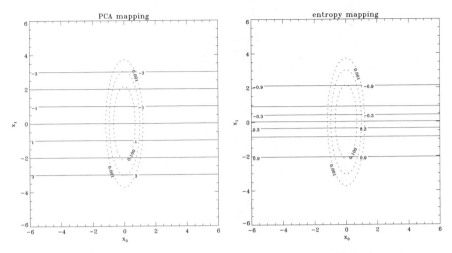

Figure 7.6 PCA vs. entropy–Gaussian case. (Left) Image of PCA features shown as contours. (Right) Entropy mapping shown as contours.

We conduct another experiment to illustrate the difference when we draw observations from a random source whose underlying distribution is not Gaussian. Specifically the pdf is a mixture of Gaussian modes with the following form:

$$f(x) = 1/2(G(\mathbf{x}, \mathbf{m}_1, \mathbf{\Sigma}_1) + G(\mathbf{x}, \mathbf{m}_2, \mathbf{\Sigma}_2))$$

where $G(\mathbf{x}, \mathbf{m}, \mathbf{\Sigma})$ is a Gaussian distribution with mean m and covariance Σ. In this specific case

$$m_1 = \begin{bmatrix} -1.0 \\ -1.0 \end{bmatrix} \quad \Sigma_1 = \begin{bmatrix} 1 & 0 \\ 0 & 0.1 \end{bmatrix}$$

$$m_2 = \begin{bmatrix} 1.0 \\ 1.0 \end{bmatrix} \quad \Sigma_2 = \begin{bmatrix} 0.1 & 0 \\ 0 & 1 \end{bmatrix}$$

It can be shown that the principal components of this distribution are the eigenvectors of the matrix

$$\mathbf{R} = \frac{1}{2}(\mathbf{\Sigma}_1 + \mathbf{m}_1\mathbf{m}_1^T + \mathbf{\Sigma}_2 + \mathbf{m}_2\mathbf{m}_2^T) = \begin{bmatrix} 0.62 & 1 \\ 1 & 0.62 \end{bmatrix}$$

This distribution is shown in the left panel of Fig. 7.7 along with its first

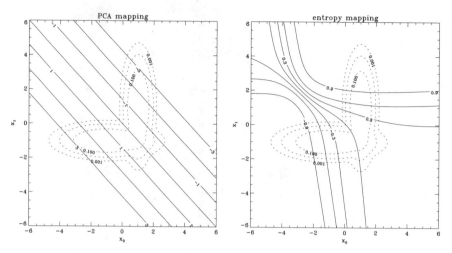

Figure 7.7 PCA vs. entropy–non-Gaussian case. (Left) Image of PCA features shown as contours. (Right) Entropy mapping shown as contours.

principal-component vector mapping. The right panel of Fig. 7.7 shows the image of the maximum entropy mapping for the same network as just discussed. As we can see there are two distinct differences between this mapping and the PCA result. The first observation is that the mapping is nonlinear. The second observation is that the maximum entropy mapping is more tuned to the structure of the data in the input space. This experiment helps to illustrate the differences between PCA and entropy mappings. PCA is primarily concerned with direction finding and is only sensitive to second-order statistics of the underlying data, while entropy explores the structure of the data class better. In a few limited cases, second-order statistics are sufficient (e.g., Gaussian) to describe such a structure, but in general they are not. Note also that we are training an MLP with an unsupervised algorithm.

Maximum-Entropy Mappings   We now present some experimental results using a maximum-entropy feature extractor for inverse synthetic aperture radar (ISAR) data. The mapping structure used in our experiment is a multilayer perceptron with a single hidden layer (4096 input nodes, 4 hidden nodes, 2 output nodes). The network is used to extract two-dimensional features from a sequence of 50 ISAR images obtained from one vehicle when the vehicle was slowly rotated from 0 to 180 degrees of aspect in increments of 3.6 degrees. Examples of the imagery are shown in Fig. 7.8. The network when trained with maximum entropy finds the two-dimensional projection (from a 4096-dimensional space) that produces the largest entropy at the output. The projection that produces this result in general has no intuitive meaning, but notice that in this case the images in the training set are all from the same vehicle. The

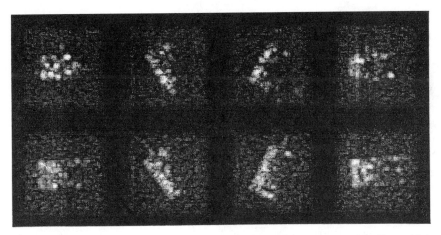

Figure 7.8 Example ISAR images from the vehicles used for experiments. The vehicles were rotated through an aspect range of 0 to 180 degrees. The top and bottom rows are from the training and testing sets (different vehicles).

variability in the training set is created by varying the aspect angle of the vehicle. So we conjecture that a projection that maximizes output entropy should be related to the aspect angle that is the source of uncertainty across the training images. We trained the maximum-entropy feature extractor on a single vehicle (upper row in Fig. 7.8) and tested the mapping in a different vehicle (bottom row of the figure).

We show the two MLP outputs (y_1 and y_2) as axes of a two-dimensional space in Figs. 7.9, 7.10, and 7.11 after 100, 200, and 300 training iterations, respectively. The input images are sequentially presented to the MLP, and for clarity, each output is connected with a line to the next output. In the beginning of training the mapping is concentrated in a limited region of the output space. In the later two plots it is clear that the outputs of the MLP (representing the extracted features) have begun to fill the output space as we can expect from the maximum-entropy criterion. However, *they have also maintained the aspect dependency of the original images*, since adjacent inputs map onto adjacent outputs. This is evidence that while the method increases the statistical independence of the two output features, it is still tuned to the underlying geometric structure of the input space as represented by rotation of the vehicle through aspect. More importantly, this seems to be independent of the vehicle, since the test set produces very similar outputs. In a trained system, the aspect angle of the input vehicle is mapped to specific regions in the two-dimensional output space, which indicates that this maximum-entropy mapping does in fact extract the feature that is most variable among the training set images (the vehicle's aspect angle).

We believe that this is evidence that the mapping has maintained topological neighborhoods in a similar fashion to the Kohonen self-organizing map (SOM)

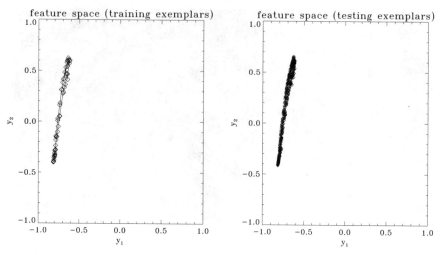

Figure 7.9 Single-vehicle experiment, 100 iterations. Projection of training (top left) and testing (top right) images onto feature space.

(Kohonen 1995). The difference between this approach and the SOM approach is that in this case the mapping is continuous, whereas in the SOM the input space is mapped onto a discrete lattice. The relationship of this maximum-entropy mapping approach to the SOM is a topic that will be left for later research.

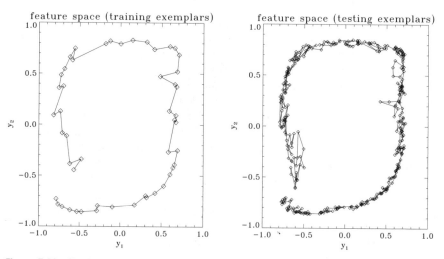

Figure 7.10 Single-vehicle experiment, 200 iterations. Projection of training (top left) and testing (top right) images onto feature space. Adjacent aspect angles are connected by a line.

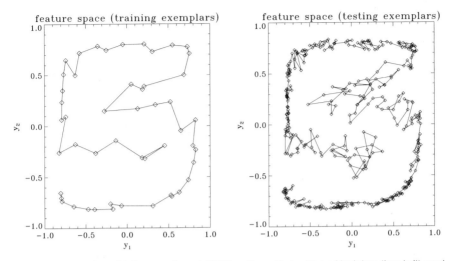

Figure 7.11 Single-vehicle experiment, 300 iterations. Projection of training (top left) and testing (top right) images onto feature space.

Blind Source Separation Blind source separation can be formulated in the following way. The observed data $\mathbf{X} = \mathbf{AS}$ is a linear mixture ($\mathbf{A} \in \mathbf{R}^{m \times m}$ is nonsingular) of independent source signals $\mathbf{S} = (\mathbf{S}_1, \ldots, \mathbf{S}_m)^T$. There is no further information about the sources and the mixing matrix, hence the term "blind." The problem is to find a projection $\mathbf{W} \in R^{m \times m}$, so that $\mathbf{Y} = \mathbf{WX}$ will become $\mathbf{Y} = \mathbf{S}$ up to a permutation and scaling.

In our experiments we compare the ISE algorithm with Bell and Sejnowski's (BELS) method (1995) (described in the previous chapter) for separating instantaneous linear mixtures of sub-Gaussian and super-Gaussian sources. This set of experiments illustrates the comparative sensitivity of both methods to the kurtosis of the source distributions and small sample size. There are two experiments:

Separation of two sources with the same distribution,
Separation of three sources with three different distributions.

In each experiment we conduct 20 Monte Carlo runs using the same mixing matrices. The nominally well-conditioned mixing matrices are either

$$\mathbf{A} = \begin{bmatrix} 2 & 1 \\ 1 & 2 \end{bmatrix} \quad \text{or} \quad \mathbf{A} = \begin{bmatrix} 2 & 1 & 1 \\ 1 & 2 & 1 \\ 1 & 1 & 2 \end{bmatrix}$$

for two and three independent sources, respectively. In each trial the initial estimate of the demixing matrix is the identity matrix. As all sources have unit

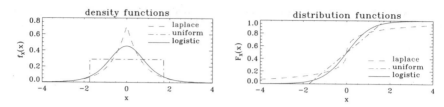

Figure 7.12   Density and CDF for the three cases utilized: Laplace, uniform, and logistic.

variance the initial signal-to-noise ratios (SNRs) are 6 and 3 dB in the two and three source experiments, respectively. The sources are various combinations of three distributions; the Laplace, logistic, and uniform densities with relative kurtosis of 3, 1.2, and −1.2, respectively. The logistic density has a logistic cumulative distribution function (cdf), which is the nonlinearity used for these experiments (i.e., maximum entropy can be achieved for these sources). Figure 7.12 shows the densities and their associated distributions. The ISE updates the weights of the neural network as Eq. (7.12).

A window of 1000 data samples is utilized to update the weights. Samples as far apart as the delay of the first zero of the autocorrelation function are taken from this window to minimize correlations among the samples. Further details of this experiment can be found in Fisher and Principe (1998). In every trial the ISE and BELS methods use the same sample draw, allowing direct trial comparisons. Each method was iterated (in batch) until the coefficients of the separating matrix estimate had converged (in the sense that its Frobenius norm did not change significantly). The number of samples (per source) in each trial is 100.

In the first set of experiments samples were drawn from two sources using the same distributions. This was repeated for each of the distribution types. Table 7.1 summarizes the results, showing when the ISE approach performs better and worse than BELS. "ISE better in 2 sources" is the number of trials (out of 20) in which the ISE method recovered both sources with a higher SNR than BELS. Likewise, "BELS better in 2 sources" is the number of trials in

Table 7.1   Performance for Separation of Two Independent Sources with Like Distributions

	Laplace	Logistic	Uniform
ISE better in 2 sources	5	16	20
BELS better in 2 sources	7	1	0
min SNR (dB)	10 (11)	17 (3)	9 (NM)
ISE best difference (dB)	22	52	VL
BELS best difference (dB)	−28	−24	NM

which BELS similarly performed better than the local interaction approach. "min SNR (dB)" is the lowest SNR for either source for the ISE method over all trials [with the same measure for BELS in parentheses—(NM) indicates that the BELS did not recover both sources in every trial with greater than 0.25 dB SNR]. The "ISE best difference" entry is the best SNR difference for a single recovered source over all trials for the ISE approach versus BELS (VL indicates a trial in which BELS failed to recover either source with greater than 0.25 dB SNR). The "BELS best difference" entry is the opposite; the degree by which SNR of the BELS recovered source exceeded the SNR of the ISE method.

In the Laplace case (higher kurtosis), the two methods are comparable with the BELS approach performing slightly better (e.g., in seven trials BELS recovered both sources with higher SNR versus five for the ISE method). In the other cases, however, the ISE outperformed BELS in nearly all measures. Notably, the ISE exhibited some degree of source separation for all three distribution types. It is, of course, not surprising that BELS failed on the uniform source densities, as this was pointed out in the Bell and Sejnowski paper (1995), but it is encouraging that the ISE did not suffer the same shortcoming.

The last experiment is of more interest; in this case the sources are of both sub- and super-Gaussian distributions. In this set of trials, we use three sources (Table 7.2), drawing from each of the three distributions. Because one of the sources is uniform, it is not surprising that the ISE outperforms BELS. It is surprising, however, that in 14 of the 20 trials the ISE did a superior job in separating the Laplace source (in light of the first two experiments in which the performance was at best comparable). One might conclude that the addition of a single sub-Gaussian source has significantly hindered BELS, although it may also be attributed to the small number of samples. Further experimentation is warranted.

Although the selection of the ISE of Eq. (7.10) seems theoretically arbitrary at this point, the fact is that the method yields one practical method for entropy

Table 7.2   Performance for Separation of Three Independent Sources with Mixed Distributions

	Mixed
ISE better in 3 sources	12
BELS better in 3 sources	0
ISE better in 2 sources	20
BELS better in 2 sources	0
Laplace better	14
Logistic better	18
Uniform better	20
min SNR (dB)	10 (NM)
ISE best difference (dB)	$\approx \infty$
BELS best difference (dB)	-8

maximization or minimization. This work by Principe and Fisher (1997) is the first account in the literature to optimize a mapper with entropy estimated directly from examples. Direct estimation of entropy and mutual information from examples is more desired, and we will do that below by using alternate definitions of entropy.

7.4 INFORMATION-THEORETIC LEARNING: UNSUPERVISED LEARNING WITH RENYI'S QUADRATIC ENTROPY

ITL algorithms are based on a combination of a nonparametric pdf estimator and a procedure to compute entropy. Both the ISE method and Viola's work approximate the definition of Shannon's entropy. The ISE utilizes the mean-square error (MSE) as the criterion to decide how different the output pdf is from the uniform. Viola's method approximates the computation of entropy by the sample mean, which is inefficient in high-dimensional spaces. In this section we overcome the difficulty in approximating Shannon's entropy by utilizing Renyi's generalized entropy. Before we start the derivation of the algorithm, let us state a property of Gaussian functions that will be very useful in the method. Let

$$G(\mathbf{z}, \mathbf{\Sigma}) = \frac{1}{(2\pi)^{M/2}|\mathbf{\Sigma}|^{1/2}} \exp\left(-\frac{1}{2}\mathbf{z}^T\mathbf{\Sigma}^{-1}\mathbf{z}\right)$$

be the Gaussian kernel in M-dimensional space, where $\mathbf{\Sigma}$ is the covariance matrix, $\mathbf{z} \in R^M$. Let $\mathbf{y}_i \in R^M$ and $\mathbf{y}_j \in R^M$ be two data samples in the space, and $\mathbf{\Sigma}_1$ and $\mathbf{\Sigma}_2$ be two covariance matrices for two Gaussian kernels in the space. Then it can be shown that the following relation holds:

$$\int_{-\infty}^{+\infty} G(\mathbf{z} - \mathbf{y}_i, \mathbf{\Sigma}_1) G(\mathbf{z} - \mathbf{y}_j, \mathbf{\Sigma}_2) \, d\mathbf{z} = G((\mathbf{y}_i - \mathbf{y}_j), (\mathbf{\Sigma}_1 + \mathbf{\Sigma}_2)) \tag{7.20}$$

Similarly, the integration of the product of three Gaussian kernels can also be obtained, and so on. Equation (7.20) can also be interpreted as a convolution between two Gaussian kernels centered at $\mathbf{y}_i$ and $\mathbf{y}_j$ and it is easy to see that the result should be a Gaussian function with a covariance equal to the sum of the individual covariances and centered at $\mathbf{d}_{ij} = (\mathbf{y}_i - \mathbf{y}_j)$.

7.4.1 Quadratic Entropy Cost Function for Discrete Samples

Let $\mathbf{y}_i \in R^M$, $i = 1, \ldots, N$, be a set of samples from a random variable $Y \in R^M$ in M-dimensional space. An interesting question is what will be the entropy associated with this set of data samples, without prespecifying the form of the pdf. Part of the answer lies in the methodology presented in Section 7.3 of estimating the data pdf by the Parzen-window method using a Gaussian kernel:

$$\hat{f}_Y(\mathbf{z}, \{\mathbf{y}\}) = \frac{1}{N} \sum_{i=1}^{N} G(\mathbf{z} - \mathbf{y}_i, \sigma^2 \mathbf{I}) \qquad (7.21)$$

where $G(\cdot, \cdot)$ is the Gaussian kernel as above and $\sigma^2 \mathbf{I}$ is the covariance matrix. When Shannon's entropy, Eq. (7.1), is used along with this pdf estimation, an algorithm to estimate entropy becomes unrealistically complex as Viola (1995) also realized. So, we conclude that Shannon's definition of information does not yield a practical measure for ITL. Fortunately, Renyi's quadratic entropy leads to a much simpler form. Using Eq. (7.21) in Eq. (7.6), we obtain an entropy estimator for a set of discrete data points $\{\mathbf{y}\}$ as

$$\begin{cases} H(\{\mathbf{y}\}) = H_{R_2}(Y|\{\mathbf{y}\}) = -\log\left(\int_{-\infty}^{+\infty} f_Y(\mathbf{z})^2 \, d\mathbf{z}\right) = -\log V(\{\mathbf{y}\}) \\ V(\{\mathbf{y}\}) = \frac{1}{N^2} \sum_{i=1}^{N} \sum_{j=1}^{N} \int_{-\infty}^{+\infty} G(\mathbf{z} - \mathbf{y}_i, \sigma^2 \mathbf{I}) G(\mathbf{z} - \mathbf{y}_j, \sigma^2 \mathbf{I}) \, d\mathbf{z} \\ \qquad = \frac{1}{N^2} \sum_{i=1}^{N} \sum_{j=1}^{N} G(\mathbf{y}_i - \mathbf{y}_j, 2\sigma^2 \mathbf{I}) \end{cases} \qquad (7.22)$$

We will simplify the notation by representing $\mathbf{y} = \{\mathbf{y}\}$ whenever possible. The combination of Renyi's quadratic entropy with the Parzen window leads to an estimation of entropy by computing interactions among pairs of samples, which is a practical cost function for ITL. There is no approximation in this evaluation (apart from the pdf estimation).

7.4.2 Quadratic Entropy and Information Potential

We wrote Eq. (7.22) in this way because there is a very interesting physical interpretation for this estimator of entropy. Let us assume that we place physical particles in the locations prescribed by the data samples $\mathbf{y}_i$ and $\mathbf{y}_j$. For this reason we will call them information particles (IPCs). Since $G(\mathbf{y}_i - \mathbf{y}_j, 2\sigma^2 \mathbf{I})$ is always positive and is inversely proportional to the distance between the IPCs, we can consider that a potential field was created in the space of interactions with a local field strength dictated by the Gaussian kernel (an exponential decay with the distance square) $V_{ij} = G(\mathbf{y}_i - \mathbf{y}_j, 2\sigma^2 \mathbf{I}) = G(\mathbf{d}_{ij}, 2\sigma^2 \mathbf{I})$. Physical particles interact with an inverse of the distance rule, but Renyi's quadratic entropy with the Gaussian kernel imposes a different interaction law. Control of the interaction law is possible by choosing different windows in the Parzen estimator. The sum of interactions on the ith IPC is $V_i = \sum_j V_{ij} = \sum_j G(\mathbf{d}_{ij}, 2\sigma^2 \mathbf{I})$.

Now $V(\mathbf{y}) = (1/N^2) \sum_i \sum_j V_{ij}$, which is the sum of all pairs of interactions, can be regarded as an overall potential energy of the data set. We call this potential energy an *information potential (IP)*. So maximizing entropy becomes equivalent to minimizing the IP. Our estimator for quadratic entropy is the negative logarithm of the IP. It was a pleasant surprise to verify that our quest for ITL algorithms ended up with a procedure that resembles the world of interacting physical particles that originated the concept of entropy.

We can also expect from Eq. (7.6) that this methodology can be applied to Renyi's entropy of higher order ($\alpha > 2$). In fact, Renyi's entropy of order α will compute interactions among α-tuples of samples, providing even more information about the complex structure of the data set. These interactions can be estimated with an extension of Eq. (7.20) when the Parzen-window method implemented with the Gaussian kernel is utilized in the estimation. However, the complexity of the algorithm becomes increasingly prohibitive ($O(N^\alpha)$).

7.4.3   Information Forces

Just as in mechanics, the derivative of the potential energy is a force, in this case an information-driven force that moves the data samples in the space of the interactions. Therefore,

$$\frac{\partial}{\partial \mathbf{y}_i} G(\mathbf{y}_i - \mathbf{y}_j, 2\sigma^2 \mathbf{I}) = -G(\mathbf{y}_i - \mathbf{y}_j, 2\sigma^2 \mathbf{I})(\mathbf{y}_i - \mathbf{y}_j)/(2\sigma^2) \qquad (7.23)$$

can be regarded as the force $\mathbf{F}_{ij}$ that IPC $\mathbf{y}_j$ impinges upon $\mathbf{y}_i$, and will be called an *information force (IF)*. If all the data samples are free to move in a certain region of the space, then the information forces between each pair of IPCs will drive all the samples to a state with minimum IP. If we add all the contributions of the IF from the ensemble of samples on $\mathbf{y}_i$, we have the net effect of the information potential on sample $\mathbf{y}_i$, that is,

$$\mathbf{F}_i = \frac{\partial}{\partial \mathbf{y}_i} V(\mathbf{y}) = -\frac{1}{N^2 \sigma^2} \sum_{j=1}^{N} G(\mathbf{y}_i - \mathbf{y}_j, 2\sigma^2 \mathbf{I})(\mathbf{y}_i - \mathbf{y}_j) = \frac{-1}{N^2 \sigma^2} \sum_{j=1}^{N} V_{ij} \mathbf{d}_{ij} \quad (7.24)$$

7.4.4   "Force" Backpropagation

The concept of IP creates a criterion for ITL, which is external to the mapper of Fig. 7.2. The only missing step is to integrate the criterion with the adaptation of a parametric mapper as the MLP. Suppose the IPCs $\mathbf{y}$ are the outputs of our parametric mapper of Eq. (7.7). If we want to adapt the MLP such that the mapping maximizes the entropy at the output $H(\mathbf{y})$, the problem is to find the MLP parameters w_{ij} so that the IP $V(\mathbf{y})$ is minimized. In this case, the IPCs are not free but are a function of the MLP parameters. So, the information forces

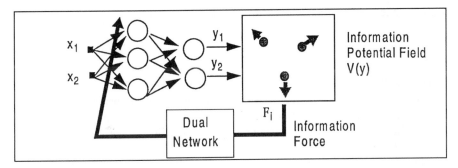

Figure 7.13   Training an MLP with the information potential.

applied to each IPC by the information potential can be backpropagated to the parameters using the chain rule (Rumelhart et al. 1986), that is,

$$\frac{\partial}{\partial w} V(\mathbf{y}) = \sum_{i=1}^{N} \left[\frac{\partial}{\partial \mathbf{y}_i} V(\mathbf{y}) \right]^T \frac{\partial \mathbf{y}_i}{\partial w} = \sum_{i=1}^{N} F_i^T \frac{\partial}{\partial w} g(\mathbf{w}, \mathbf{x}_i) \qquad (7.25)$$

where $\mathbf{y}_i = (y_{i1}, \ldots, y_{iM})^T$ is the M-dimensional MLP output. Notice that from Eq. (7.25) the sensitivity of the output with respect to a MLP parameter $(\partial \mathbf{y}_i / \partial w)$ is the *transmission mechanism* through which information forces are backpropagated to the parameter (Fig. 7.13). From the analogy of Eq. (7.25) with the backpropagation formalism [see Eq. (7.11)] we conclude that $F_i = \varepsilon_i$, that is, *information forces take the place of the injected error in the back-propagation algorithm.* Hence, we obtain a general, nonparametric, and sample-based methodology to adapt arbitrary nonlinear (smooth and differentiable) mappings for entropy maximization (Fig. 7.13). Notice that we are adapting an MLP without a desired response, hence this is an unsupervised criterion. We have established an ITL criterion that *adapts* the MLP with a *global property* of its output sample distribution. It is very useful to analyze this expression in detail and compare it with the well-known MSE. Note that MSE is computed with a single data sample/desired response combination. However, the entropy is estimated with *pairs of data samples*, that is, more information about the data set is being extracted here than with the MSE criterion (in an N-sample data set there are $\binom{N}{2}$ different pairs). As a consequence, we can also expect that the algorithm will be computationally more expensive $(O(N^2))$.

This criterion can be utilized to directly implement Jayne's MaxEnt optimization principle, but instead of requiring analytic manipulations, it solves the problem using the iterative approach so common in adaptive filtering and neurocomputing. The constraints in MaxEnt are here specified by the topology of the mapper. The weights of any MLP Processing element (PE) will be adapted with the backpropagation algorithm (Haykin 1994) as

$$\Delta w_{ij} = \pm \eta \delta_j x_i$$

where η is the stepsize, x_i is the input to the PE, and δ_j is the local error at the PE [see (Haykin 1994)]. If the goal is to maximize output entropy (as required by MaxEnt), the $+$ sign is used, and if the purpose is to minimize output entropy, the $-$ sign is required. Notice that this will change the interactions among IPCs in the output space from repulsion to attraction.

The reader may have noticed the parallel of this derivation with the ISE method presented in Section 7.3. Here we will show that the information force of Eq. (7.24) is the same as the attractor kernel of Eq. (7.15). First note that $\kappa(\mathbf{z}) = G(\mathbf{z}, \sigma^2 \mathbf{I})$, so

$$\kappa'(\mathbf{z}) = \frac{1}{(2\pi)^{M/2} \sigma^M} \exp\left(-\frac{\mathbf{z}^T \mathbf{z}}{2\sigma^2}\right)\left(-\frac{\mathbf{z}}{\sigma^2}\right) = G(\mathbf{z}, \sigma^2 \mathbf{I})\left(-\frac{\mathbf{z}}{\sigma^2}\right)$$

Now according to Eq. (7.15), if we convolve these two kernels,

$$\kappa_a(\mathbf{z}) = \kappa(\mathbf{z})^* \kappa'(\mathbf{z}) = G(\mathbf{z}, 2\sigma^2 \mathbf{I})\left(-\frac{\mathbf{z}}{\sigma^2}\right)$$

since the derivative and the convolution commute. When we apply this equation to all of the output samples we obtain the IF of Eq. (7.24), which becomes the injected error ε_i. *So the attractor kernel of ISE is equivalent to the information force derived in this section.* Notice that when Renyi's quadratic entropy is utilized as the ITL, the regulating kernel term in Eq. (7.13) disappears. Practically, since the MLP provides limited output dynamic range due to the saturating nonlinearities, data samples will fill the space but also be placed around the limits of the output space, mimicking the boundary and the regulation term of Eq. (7.13). This derivation shows that ISE is in fact utilizing Renyi's quadratic entropy to compute the interactions among pairs of data samples in the output space, which provides a more principled justification than the ISE criterion.

We conclude this section by stating that the methodology presented here lays down the framework to construct an *entropy machine*, that is a learning machine that is capable of estimating entropy directly from samples in its output space, and can modify its weights through backpropagation to manipulate output entropy. An electronic implementation using the laws of physics to speed up the calculations is an intriguing possibility. The algorithm has complexity $O(N^2)$ since the criterion needs to examine the interactions among all pairs of output samples. Note that we are extending BELS approach to ICA. Bell's approach is conceptually very elegant, but it cannot be easily extended to MLPs with arbitrary topologies nor to data distributions that are multimodal in nature. On the other hand, Renyi's quadratic entropy becomes essentially a general-purpose criterion for entropy manipulation. Moreover, since the MLP is an arbitrary mapper, we should be able to use this approach to attack the problem of BSS with nonlinear (static) mixtures. Recently Almeida (Marques and Almeida 1999) utilized a very similar idea to separate nonlinear mixtures.

7.5 INFORMATION-THEORETIC CRITERIA: UNSUPERVISED LEARNING WITH QUADRATIC MUTUAL INFORMATION

In Section 7.4 we implemented a nonparametric method to solve MaxEnt. Here we will develop an ITL criterion to estimate the mutual information among random variables that enables the implementation of MinXEnt and Infomax. Mutual information is capable of quantifying the entropy between pairs of random variables so it is a more general measure than entropy and can be applied more flexibly to engineering problems. Mutual information at the output of a mapper can be computed as a difference of Shannon entropies $I(x, y) = H(y) - H(y|x)$. But we have to remember that Shannon entropy is not easily estimated from exemplars. Therefore this expression for $I(x, y)$ can only be utilized in an approximate sense to estimate mutual information. An alternative to estimate mutual information is the KL divergence (Kullback 1968). The KL divergence between two pdf's $f(x)$ and $g(x)$ is

$$K(f, g) = \int f(x) \log \frac{f(x)}{g(x)} \, dx \tag{7.26}$$

where implicitly Shannon's entropy is utilized. Likewise, based on Renyi's entropy, Renyi's divergence measure (Renyi 1976a) with order α for two pdf's $f(x)$ and $g(x)$ is

$$H_{R_\alpha}(f, g) = \frac{1}{(\alpha - 1)} \log \int \frac{f(x)^\alpha}{g(x)^{\alpha - 1}} \, dx \tag{7.27}$$

The relation between the two divergence measures is

$$\lim_{\alpha \to 1} H_{R_\alpha}(f, g) = K(f, g)$$

that is, they are equivalent in the limit $\alpha = 1$. The KL between two random variables Y_1 and Y_2 essentially estimates the divergence between the joint pdf and the factorized marginal pdf's, that is,

$$I_S(Y_1, Y_2) = KL(f_{Y_1 Y_2}(z_1, z_2), f_{Y_1}(z_1) f_{Y_2}(z_2))$$
$$= \int \int f_{Y_1 Y_2}(z_1, z_2) \log \frac{f_{Y_1 Y_2}(z_1, z_2)}{f_{Y_1}(z_1) f_{Y_2}(z_2)} \, dz_1 \, dz_2 \tag{7.28}$$

where $f_{Y_1 Y_2}(z_1, z_2)$ is the joint pdf, and $f_{Y_1}(z_1)$ and $f_{Y_2}(z_2)$ are marginal pdf's. From these formulas, we can also observe that unfortunately none of them is quadratic in the pdf, so they cannot be easily integrated with the information potential described in Section 7.4. Therefore, we propose below new distance measures between two pdf's that contain only quadratic terms to utilize the tools of IP and IF developed in Section 7.3. There are basically four different ways to write a distance measure using L_2 norms, but here we will concentrate on two:

1. Based on the Euclidean difference of vectors inequality we can write

$$\|\mathbf{x}\|^2 + \|\mathbf{y}\|^2 - 2\mathbf{x}^T\mathbf{y} \geq 0 \qquad (7.29)$$

2. Based on the Cauchy-Schwartz inequality we can write

$$\log \frac{\|\mathbf{x}\|^2 \|\mathbf{y}\|^2}{(\mathbf{x}^T\mathbf{y})^2} \geq 0 \qquad (7.30)$$

Notice that both expressions utilize the same quadratic quantities, namely the length of each vector and their dot product. We utilize these distance measures to approximate the KL directed divergence between pdf's, with the added advantage that each term can be estimated with the IP formalism developed in the previous section.

For instance, based on the Cauchy-Schwartz inequality (7.30), we propose to measure the divergence of two pdf's $f(x)$ and $g(x)$ as

$$I_{CS}(f, g) = \log \frac{(\int f(x)^2 \, dx)(\int g(x)^2 \, dx)}{(\int f(x)g(x) \, dx)^2} \qquad (7.31)$$

It is easy to show that $I_{CS}(f, g) \geq 0$ (nonnegativity) and the equality holds true if and only if $f(x) = g(x)$ (identity) if $f(x)$ and $g(x)$ are pdf's. Eq. (7.31) is also a divergence and estimates the distance between the joint quadratic entropy and the product of the quadratic entropy marginals. Eq. (7.31) is intrinsically normalized, but it does not preserve all the properties of the KL divergence. Likewise we also propose to estimate the divergence between two pdf's $f(x)$ and $g(x)$ based on the Euclidean distance (ED) as

$$I_{ED}(f, g) = \int f(x)^2 \, dx + \int g(x)^2 \, dx - 2 \int f(x)g(x) \, dx \qquad (7.32)$$

7.5.1 Mutual Information and the Proposed Measures

One important question is to find out how similar to mutual information are the two newly introduced measures. We will first demonstrate in a simple

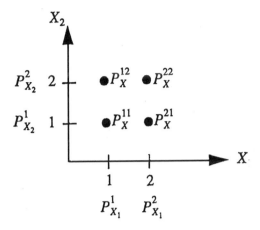

Figure 7.14 A simple 4-sample two-dimensional example.

problem that the two new measures of mutual information (I_{CS} and I_{ED}) produce the same minimization results as the mutual information using the KL divergence. Later in the chapter we also show experimentally that estimators of I_{CS} and I_{ED} based on the information potential can be applied to solve practical problems with a high degree of success.

Let us take a simple case with two discrete random variables X_1 and X_2 (Xu 1998). As shown in Fig. 7.14, X_1 has two values and its probability distribution is $P_{X_1} = (P^1_{X_1}, P^2_{X_1})$, that is, $P(X_1 = 1) = P^1_{X_1}$ and $P(X_1 = 2) = P^2_{X_1}$; similarly, X_2 has the probability distribution $P_{X_2} = (P^1_{X_2}, P^2_{X_2})(P(X_2 = 1) = P^1_{X_2}, P(X_2 = 2) = P^2_{X_2})$. The joint probability is $P_X = (P^{11}_X, P^{12}_X, P^{21}_X, P^{22}_X)$, where $P((X_1, X_2) = (1, 1)) = P^{11}_X$, $P((X_1, X_2) = (1, 2)) = P^{12}_X$, $P((X_1, X_2) = (2, 1)) = P^{21}_X$ and $P((X_1, X_2) = (2, 2)) = P^{22}_X$. Obviously, $P^1_{X_1} = P^{11}_X + P^{12}_X$, $P^2_{X_1} = P^{21}_X + P^{22}_X$, $P^1_{X_2} = P^{11}_X + P^{21}_X$, and $P^2_{X_2} = P^{12}_X + P^{22}_X$.

First, let us treat the case of a fixed distribution X_1, $P_{X_1} = (0.6, 0.4)$. Then the free parameters are P^{11}_X from 0 to 0.6 and P^{21}_X from 0 to 0.4. When P^{11}_X and P^{21}_X change in these ranges, the values of I_s, I_{ED}, and I_{CS} can be calculated. Figure 7.15 shows how these three quantities change with P^{11}_X and P^{21}_X, where the left graphs are the I_s, I_{ED}, and I_{CS} surfaces versus P^{11}_X and P^{21}_X; the right graphs are the corresponding contour plots.

These graphs show that although the surfaces of the three measures are different, they reach the minimum value of zero at the same line, $P^{11}_X = 1.5P^{21}_X$, where the joint probabilities equal the corresponding factorized marginal probabilities. Moreover, the maximum values, although different, are also reached at the same points, $(P^{11}_X, P^{21}_X) = (0.6, 0)$ and $(0, 0.4)$, where the joint probabilities are, respectively,

$$\begin{bmatrix} P^{12}_X & P^{22}_X \\ P^{11}_X & P^{21}_X \end{bmatrix} = \begin{bmatrix} 0 & 0.4 \\ 0.6 & 0 \end{bmatrix} \quad \text{and} \quad \begin{bmatrix} P^{12}_X & P^{22}_X \\ P^{11}_X & P^{21}_X \end{bmatrix} = \begin{bmatrix} 0.6 & 0 \\ 0 & 0.4 \end{bmatrix}$$

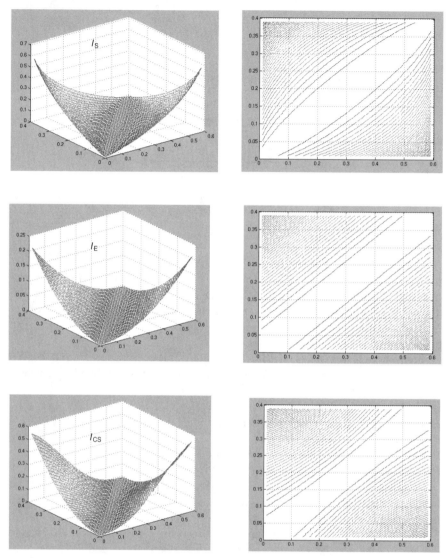

Figure 7.15 Surfaces and contours of I_S, I_{ED}, I_{CS} vs. P_X^{11} and P_X^{21}.

This example illustrates the case where X_1 and X_2 have a one-to-one relation, that is, X_1 can determine X_2 exactly, and vice versa.

If the marginal probability of X_2 is further fixed, for example, $P_{X_2} = (0.3, 0.7)$, then the free parameter becomes P_X^{11} from 0 to 0.3. In this case, both marginal probabilities of X_1 and X_2 are fixed, and the factorized marginal probability distribution is thus fixed and only the joint probability distribution will change. This case can also be regarded as a further constraint of the previous case specified by $P_X^{11} + P_X^{21} = 0.3$. Figure 7.16 shows how the three mea-

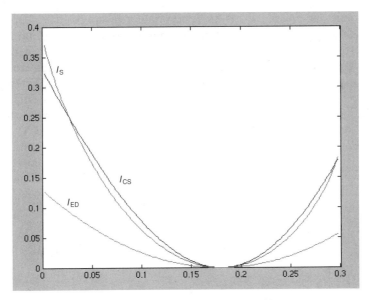

Figure 7.16 Plot of I_S, I_{ED}, I_{CS} vs. P_X^{11}.

sures change with P_X^{11}, from which we can see that the minima are reached at the same point $P_X^{11} = 0.18$, and the maxima are also reached at the same point $P_X^{11} = 0$, that is,

$$\begin{bmatrix} P_X^{12} & P_X^{22} \\ P_X^{11} & P_X^{21} \end{bmatrix} = \begin{bmatrix} 0.6 & 0.1 \\ 0 & 0.3 \end{bmatrix}$$

From this simple example, we can see that although the three measures are different, they have the same minimum and maximum points. It is known that both Shannon's mutual information I_s and I_{ED} are convex functions (Kapur 1994). From the preceding graphs, we can confirm this fact and also come up to the conclusion that I_{CS} is not a convex function in the full space, although in this case it is well behaved near the minimum. This analysis makes I_{ED} preferable for maximization.

7.5.2 Estimators for Quadratic Mutual Information

For two random variables Y_1 and Y_2 [with marginal pdf's $f_{Y_1}(z_1)$, $f_{Y_2}(z_2)$, and joint pdf $f_{Y_1 Y_2}(z_1, z_2)$], the "quadratic mutual information" based on the distance measure, Eq. (7.31), becomes

$$I_{CS}(Y_1, Y_2) = \log \frac{\left(\iint f_{Y_1 Y_2}(z_1, z_2)^2 \, dz_1 \, dz_2 \right) \left(\iint f_{Y_1}(z_1)^2 f_{Y_2}(z_2)^2 \, dz_1 \, dz_2 \right)}{\left(\iint f_{Y_1 Y_2}(z_1, z_2) f_{Y_1}(z_1) f_{Y_2}(z_2) \, dz_1 \, dz_2 \right)^2} \quad (7.33)$$

It is obvious that $I_{CS}(Y_1, Y_2)$ is an appropriate measure for the independence of two variables (minimization of mutual information). We also have experimental evidence that $I_{CS}(Y_1, Y_2)$ is an appropriate measure for the dependence of two variables (maximization of mutual information). Although we are still unable to provide a strict justification that $I_{CS}(Y_1, Y_2)$ is appropriate to measure dependence, we call Eq. (7.33) "Chauchy-Schwartz quadratic mutual information," or CS-QMI for convenience. Now, suppose that we observe a set of data samples $\{y_1\} = \{y_{i1}, i = 1, \ldots, N\}$ for the variable Y_1, $\{y_2\} = \{y_{i2}, i = 1, \ldots, N\}$ for the variable Y_2. Let $y_i' = (y_{i1}, y_{i2})^T$. Then $y = \{y_i', i = 1, \ldots, N\}$ are data samples for the joint variable $(Y_1, Y_2)^T$. Based on the Parzen-window method, Eq. (7.8), the joint pdf and marginal pdf can be estimated as

$$
\begin{cases}
\hat{f}_{Y_1 Y_2}(z_1, z_2) = \dfrac{1}{N} \sum_{i=1}^{N} G(z_1 - y_{i1}, \sigma^2 I) G(z_2 - y_{i2}, \sigma^2 I) \\[2mm]
\hat{f}_{Y_1}(z_1) = \dfrac{1}{N} \sum_{i=1}^{N} G(z_1 - y_{i1}, \sigma^2 I) \\[2mm]
\hat{f}_{Y_2}(z_2) = \dfrac{1}{N} \sum_{i=1}^{N} G(z_2 - y_{i2}, \sigma^2 I)
\end{cases}
\tag{7.34}
$$

Combining Eqs. (7.33), (7.34), and using Eq. (7.22), we obtain the following expressions to estimate $I_{CS}(Y_1, Y_2)$ based on a set of data samples:

$$
I_{CS}((Y_1, Y_2)|y) = \log \frac{V(y) V^1(\{y_1\}) V^2(\{y_2\})}{V_{nc}(y)^2}
$$

where

$$
\begin{cases}
V(y) = \dfrac{1}{N^2} \sum_{i=1}^{N} \sum_{j=1}^{N} \left(\prod_{l=1}^{2} G(y_{il} - y_{jl}, 2\sigma^2 I) \right) \\[3mm]
V^l(y_j, \{y_l\}) = \dfrac{1}{N} \sum_{i=1}^{N} G(y_{jl} - y_{il}, 2\sigma^2 I), \qquad l = 1, 2 \\[3mm]
V^l(\{y_l\}) = \dfrac{1}{N} \sum_{j=1}^{N} V^l(y_j, \{y_l\}) \qquad\qquad l = 1, 2 \\[3mm]
V_{nc}(y) = \dfrac{1}{N} \sum_{j=1}^{N} \left(\prod_{l=1}^{2} V^l(y_j, \{y_l\}) \right)
\end{cases}
\tag{7.35}
$$

In order to interpret these expressions in terms of information potentials we have to introduce some further definitions: We will use the term *marginal* when the IP is calculated in the subspace of each of the variables y_1 or y_2, and *partial* when only some of the IPCs are used. With this in mind, $V(y)$ is the joint information potential (JIP) in the joint space, and $V_i^l = V^l(y_i, \{y_l\})$ is the partial marginal information potential (PMIP), because it is the potential of the sample y_j in its corresponding marginal information potential (MIP) field (indexed by l). $V^l = V^l(\{y_l\})$ is the lth MIP because it averages all the PMIPs for one index l, and $V_{nc}(y)$ is the unnormalized *cross-information potential (UCIP)*, because it measures the interactions between the PMIPs (Xu 1997). We utilize the simplified notation herein. All these potentials can be computed from sums of pairs of interactions among the IPCs in each of the marginal fields. (V_{ij}^l) V_{ij}^l, PMIP (V_i^l), and MIP (V^l) have the same definitions as for Renyi's entropy, but now qualified by the superscript l to describe which field we are referring to.

The argument of the logarithm in the first equation of Eqs. (7.35) can be regarded as a normalization for the UCIP, that is, $V_{nc}(y)$ normalized by the JIP and the MIPs. *The cross-information potential (CIP)* can then be defined as

$$V_c(y) = \frac{V_{nc}(y)}{V(y)} \frac{V_{nc}(y)}{V^1(y) V^2(y)} \qquad (7.36)$$

so quadratic mutual information is measured by the CIP. With the CIP concept and the comparison between Eqs. (7.22) and (7.36), we obtain consistent definitions from entropy to cross-entropy as shown by

$$\begin{cases} H(Y|y) = -\log V(y) \\ I_{CS}((Y_1, Y_2)|y) = -\log V_c(y) \end{cases} \qquad (7.37)$$

which relates the quadratic entropy calculated with the IP and the QMI calculated with the CIP (Xu 1997). Therefore, maximizing the QMI is equivalent to minimizing the CIP, while, minimizing the QMI is equivalent to maximizing the CIP. If we write the CIP as a function of the individual fields, we obtain

$$V_c(y) = \frac{\left(\dfrac{1}{N} \sum_{i=1}^{N} V_i^1 V_i^2\right)^2}{\left(\dfrac{1}{N^2} \sum_{i=1}^{N} \sum_{j=1}^{N} V_{ij}^1 V_{ij}^2\right)(V^1 V^2)}$$

and conclude that $V_c(y)$ is a *generalized measure of cross-correlation* between the MIPs at different levels (at the individual IPC interactions; at the PMIP and MIP levels).

The quadratic mutual information described in 7.35 can easily be extended to the case with multiple variables $\mathbf{Y}_1, \ldots, \mathbf{Y}_k$, as

$$
\begin{cases}
V(\mathbf{y}) = \dfrac{1}{N^2} \displaystyle\sum_{i=1}^{N} \sum_{j=1}^{N} G(\mathbf{y}_i - \mathbf{y}_j, 2\sigma^2 \mathbf{I}) = \dfrac{1}{N^2} \displaystyle\sum_{i=1}^{N} \sum_{j=1}^{N} \left(\prod_{l=1}^{k} G(\mathbf{y}_{il} - \mathbf{y}_{jl}, 2\sigma^2 \mathbf{I}) \right) \\[3mm]
V^l(\mathbf{y}_j, \{\mathbf{y}_l\}) = \dfrac{1}{N} \displaystyle\sum_{i=1}^{N} G(\mathbf{y}_{jl} - \mathbf{y}_{il}, 2\sigma^2 \mathbf{I}), \qquad l = 1, \ldots, k \\[3mm]
V^l(\{\mathbf{y}_l\}) = \dfrac{1}{N} \displaystyle\sum_{j=1}^{N} V^l(\mathbf{y}_j, \{\mathbf{y}_l\}) \qquad\qquad l = 1, \ldots, k \\[3mm]
V_{nc}(\mathbf{y}) = \dfrac{1}{N} \displaystyle\sum_{j=1}^{N} \left(\prod_{l=1}^{k} V^l(\mathbf{y}_j, \{\mathbf{y}_l\}) \right) \\[3mm]
V_c(\mathbf{y}) = \dfrac{V_{nc}(\mathbf{y})^2}{V(\mathbf{y}) \displaystyle\prod_{l=1}^{k} V_l(\{\mathbf{y}_l\})}
\end{cases}
$$

7.5.3 "Forces" in the Cross-Information Potential

The cross-information potential is more complex than the information potential (Xu 1997). Three different information potentials contribute to the cross-information potential of Eq. (7.35), namely, the JIP (V^l), the PMIP (V_l^l), and the MIP (V^l). So, the force applied to each IPC $\mathbf{y}_i$, comes from three independent sources, which we call the *marginal information forces* (*MIF*). The overall marginal force from $k = 1, 2$ that the IPC $\mathbf{y}_i$ receives is, according to Eq. (7.35),

$$
\frac{\partial}{\partial \mathbf{y}_{ik}} I_{CS}((\mathbf{Y}_1, \mathbf{Y}_2) | \mathbf{y}) = \frac{1}{V(\mathbf{y})} \frac{\partial}{\partial \mathbf{y}_{ik}} V(\mathbf{y}) + \frac{1}{V^k(\{\mathbf{y}_k\})} \frac{\partial}{\partial \mathbf{y}_{ik}} V^k(\{\mathbf{y}_k\})
$$

$$
-2 \left(\frac{1}{V_{nc}(\mathbf{y})} \frac{\partial}{\partial \mathbf{y}_{ik}} V_{nc}(\mathbf{y}) \right) \tag{7.38}
$$

Notice that the forces from each source are normalized by their corresponding information potentials to balance them out. This is a consequence of the logarithm in the definition of $I_{CS}(\mathbf{Y}_1, \mathbf{Y}_2)$. Each marginal force k that operates on the data sample $\mathbf{y}_i$ can be calculated according to the following formulas obtained by differentiating the corresponding fields (Xu 1997)

$$\begin{cases} \dfrac{\partial}{\partial \mathbf{y}_{ik}} V(\mathbf{y}) = -\dfrac{1}{N^2} \sum_{j=1}^{N} \left(\prod_{k=1}^{2} G(\mathbf{y}_{ik} - \mathbf{y}_{jk}, 2\sigma^2 \mathbf{I}) \right) \dfrac{\mathbf{y}_{ik} - \mathbf{y}_{jk}}{\sigma^2} \\[2mm] \dfrac{\partial}{\partial \mathbf{y}_{ik}} V^k(\{\mathbf{y}_k\}) = -\dfrac{1}{N^2} \sum_{j=1}^{N} G(\mathbf{y}_{ik} - \mathbf{y}_{jk}, 2\sigma^2 \mathbf{I}) \dfrac{\mathbf{y}_{ik} - \mathbf{y}_{jk}}{\sigma^2} \qquad (7.39) \\[2mm] \dfrac{\partial}{\partial \mathbf{y}_{ik}} V_{nc}(\mathbf{y}) = -\dfrac{1}{N^2} \sum_{j=1}^{N} \left(\prod_{l \neq k} V^l(\mathbf{y}_j, \{\mathbf{y}_l\}) \right) G(\mathbf{y}_{ik} - \mathbf{y}_{jk}, 2\sigma^2 \mathbf{I}) \dfrac{\mathbf{y}_{ik} - \mathbf{y}_{jk}}{\sigma^2} \end{cases}$$

Once the forces that each IPC receives are calculated by Eqs. (7.39), they represent the injected error that can again be backpropagated to all the parameters of the mapper with backpropagation so that the adaptation with QMI takes place. The marginal force for the two variable cases is finally given by

$$F_i^k = -\frac{1}{N^2 \sigma^2} \left[\frac{\sum_j V_{ij}^1 V_{ij}^2 d_{ij}}{\sum_i \sum_j V_{ij}^1 V_{ij}^2} + \frac{\sum_j V_{ij}^k d_{ij}}{\sum_i \sum_j V_{ij}^k} - \frac{2 \sum_j V_j^1 V_{ij}^2 d_{ij}}{\sum_j V_j^1 V_j^2} \right] \qquad k = 1, 2 \quad (7.40)$$

7.5.4 Quadratic Mutual Information with the Euclidean Difference Measure

We can also utilize Eq. (7.32) to express QMI using the Euclidean difference (ED-QMI) of vectors inequality

$$I_{ED}(Y_1, Y_2) = \left(\iint f_{Y_1 Y_2}(z_1, z_2)^2 \, dz_1 \, dz_2 \right) + \left(\iint f_{Y_1}(z_1)^2 f_{Y_2}(z_2)^2 \, dz_1 \, dz_2 \right)$$
$$- 2 \left(\iint f_{Y_1 Y_2}(z_1, z_2) f_{Y_1}(z_1) f_{Y_2}(z_2) \, dz_1 \, dz_2 \right) \qquad (7.41)$$

Obviously, $I_{ED}(Y_1, Y_2) \geq 0$ and equality holds if and only if Y_1 and Y_2 are statistically independent, so it is also a divergence. Basically Eq. (7.41) measures the ED between the joint pdf and the factorized marginals. With the previous definitions it is not difficult to obtain

$$I_{ED}((Y_1, Y_2)|\mathbf{y}) = V_{ED}(\mathbf{y})$$

$$V_{ED}(\mathbf{y}) = \frac{1}{N^2} \sum_{i=1}^{N} \sum_{j=1}^{N} V_{ij}^1 V_{ij}^2 - \frac{2}{N} \sum_{i=1}^{N} V_i^1 V_i^2 + V^1 V^2 \qquad (7.42)$$

Although Eqs. (7.35) and (7.42) differ in form, we can see that V_c is still an overall measure of cross-correlation between two marginal IPs. We have found experimentally that although $I_{ED}(Y_1, Y_2)$ is not normalized, it is better behaved than $I_{CS}(Y_1, Y_2)$ for maximization of the QMI, while they both provide similar results for the minimization of QMI.

It is also not difficult to obtain the formula for the calculation of the information force produced by the CIP field in the case of the ED-QMI of Eq. (7.41)

$$c_{ij}^k = V_{ij}^k - V_i^k - V_j^k + V^k, \qquad k = 1, 2$$

$$F_i^l = \frac{\partial V_{ED}}{\partial y_i^l} = \frac{-1}{N^2 \sigma^2} \sum_{j=1}^{N} c_{ij}^k V_{ij}^l d_{ij}^l \qquad (7.43)$$

$$i = 1, \ldots, N, \qquad l \neq k, \qquad l = 1, 2$$

where c_{ij}^k are cross-matrices that serve as force modifiers.

7.5.5 Interpretation of the CIP

Another way to look at the CIP comes from the expression of the factorized marginal pdf's. From Eq. (7.34), we have

$$f_{Y_1}(\mathbf{z}) f_{Y_2}(\mathbf{z}) = \frac{1}{N^2} \sum_{i=1}^{N} \sum_{j=1}^{N} G(\mathbf{z} - \mathbf{y}_{i1}, \sigma^2 \mathbf{I}) G(\mathbf{z} - \mathbf{y}_{j2}, \sigma^2 \mathbf{I}) \qquad (7.44)$$

This suggests that in the joint space, there are N^2 "virtual IPCs" $\{(\mathbf{y}_{i1}, \mathbf{y}_{j2})^T, i, j = 1, \ldots, N\}$ whose coordinates are given by each of the coordinates of the IPCs, that is, for every real IPC location $(\mathbf{y}_{i1}, \mathbf{y}_{j2})$, N virtual IPCs are placed at points given by the coordinate $\mathbf{y}_{i1}$ of the real IPC and $\mathbf{y}_{j2}, j = 1, \ldots, N$, of all the other real IPCs. The pdf of Eq. (7.44) is exactly the factorized marginal pdf's of both types (real and virtual) of the IPCs. The relation between both real types of IPCs is illustrated in Fig. 7.17 for two extreme cases. In the left panel, the real IPCs are distributed along a diagonal line. In this case, the virtual IPCs are maximally scattered in the joint field, and the difference between the distribution of the real IPCs and virtual IPCs is maximized. In the right panel of Fig. 7.17, the real IPCs are in a more compact distribution in the joint field. In this case, the virtual IPCs occupy the same locations as the real IPCs. In this case, the two fields are the same and the CIP is zero, which corresponds to the case of statistical independence of the two marginal variables Y_1 and Y_2. All the other distributions of IPCs will provide intermediate conditions between these two extremes.

From the preceding description, we can reinterpret the CIP as the square of the Euclidean distance between the IP (formed by real IPCs) and the virtual IP fields (formed by virtual IPCs). CIP is a general measure for the statistical relation between two variables (based merely on the given data). It may also be noted that both Y_1 and Y_2 can be multidimensional variables, and their dimensions even can be different.

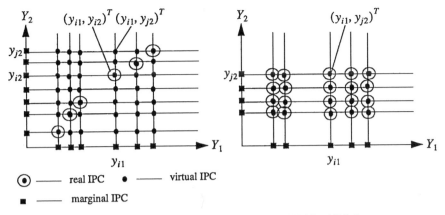

Figure 7.17 Illustration of "real IPC" and "virtual IPC."

7.5.6 An Example of Blind Source Separation

We present below the results of a linear demixing system trained with the CS-QMI criterion (Xu et al. 1998b). From this point of view, the problem can be restated as finding a projection $\mathbf{W} \in R^{m \times m}$, $\mathbf{Y} = \mathbf{WX}$ so that the CS-QMI among all the components of $\mathbf{Y}$ is minimized, that is, all the output signals are independent of each other. For ease of illustration, only a 2-source–2-sensor problem is tested. There are two experiments presented here: Experiment 1 tests the performance of the method on a very sparse data set that was instantaneously mixed in the computer with a mixing matrix [2, 0.5; 1, 0.6]. Two, two-dimensional, different colored Gaussian noise segments are used as sources, with 30 data points for each segment (sparse-data case). The two segments were concatenated and shuffled. Figure 7.18 (left panel) shows the source density in the joint space (each axis is one source signal). As Fig. 7.18 shows, the mixing produces a mixture with both long and short "tails," which is difficult to separate (middle panel). Whitening is first performed on the mixtures to facilitate demixing. The data distribution for the recovered signals are plotted in Fig. 7.18 (right panel). As we can observe, the original source density is obtained with high fidelity. Figure 7.18 also contains the evolution of the SNR of demixing-mixing product matrix ($\mathbf{WA}$) during training as a function of iterations (SNR approaches 36.73 dB). Both figures show that the method works well.

Experiment 2 uses two speech signals from the Texas Instrument (TIMIT) database as source signals (Figure 7.19). The mixing matrix is [1, 3.5; 0.8, 2.6] where the two mixing direction [1, 3.5] and [0.8, 2.6] are similar. An on-line implementation is tried in this experiment, in which a short-time window (200 samples) slides over the speech data (e.g., 10 samples/step). In each window position, the speech data within the window is used to calculate the information

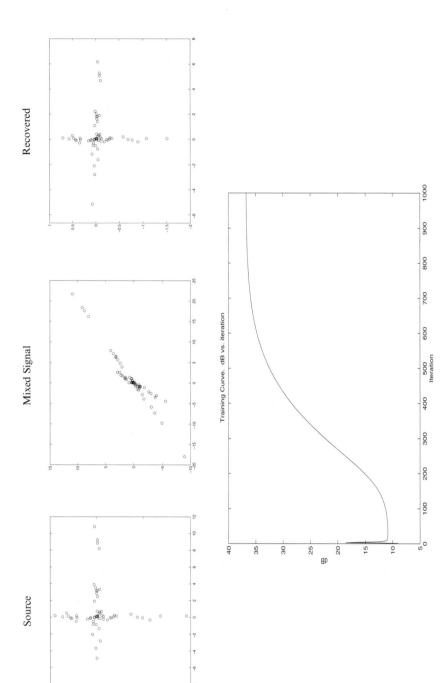

Figure 7.18 Data distributions in the joint space for the sources (left), mixed (middle), and demixed with the proposed method (right). Learning curve on bottom. Notice the 36 dB of SNR

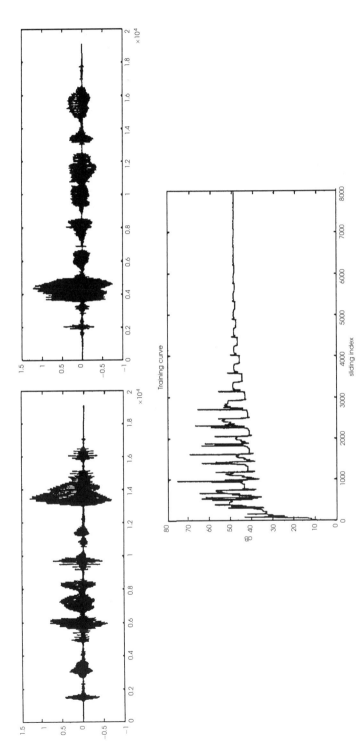

Figure 7.19 Two speech signals from TIMIT that were instantaneously mixed, and resulting training curve. SNR is also around 50 dB.

307

potentials, related forces, and over the speech signal backpropagated forces to adjust the demixing matrix. As the window slides, the demixing matrix keeps being updated. The training curve (SNR as defined above versus sliding index) is shown in Fig. 7.19, which tells us that the algorithm converges fast and works very well with the SNR approaching 49.15 dB. The large spikes in the training curve show the occasional almost perfect demixing matrix estimation while the algorithm is still adapting (notice that during adaptation the algorithm can estimate one of the directions very well, although it is still far away from the optimal solution). In order to obtain a stable result the learning rate is progressively reduced and the result stabilizes approximately at 50 dB. Although whitening is done before the "potential energy" method, we believe that the whitening process can also be incorporated into the ITL algorithm.

These results show that Eq. (7.35) is effective for minimizing mutual information. As far as we know, our work at the Computational NeuroEngineering Laboratory (CNEL) is the first to propose a nonparametric formulation to solve BSS with a general information-theoretic criterion (Xu et al. 1998b). This ITL is an improvement over Amari's work (Yang and Amari 1997) where the information criterion and the learning equations are interrelated to solve the BSS problem. Moreover, since in our approach we have decoupled the mapper from the criterion, the quadratic mutual information becomes essentially a general-purpose criterion for MinXEnt optimization.

7.6   INFORMATION-THEORETIC LEARNING: SUPERVISED LEARNING WITH QUADRATIC MUTUAL INFORMATION

In order to apply quadratic mutual information to train MLPs in a supervised framework, we just need to reassign variables. In a supervised framework we would like to maximize the mutual information between the desired response and the output of the MLP. So the joint space $(\mathbf{Y}_1, \mathbf{Y}_2)$ becomes the space $(\mathbf{d}, \mathbf{y})$ of the desired response function and of the MLP outputs, characterized by a joint pdf $f_{DY}(\mathbf{d}, \mathbf{y})$. The two random variables D, Y are characterized by marginal pdf's $f_D(\mathbf{d})$ and $f_Y(\mathbf{y})$. The desired response can either be a function of the input as in regression $(\mathbf{d} = f(\mathbf{x}))$ or an indicator function $\mathbf{d} = \{d_1, \ldots, d_k\}$ with $d = \{1, 0\}$, as in classification. This framework extends supervised learning since it does not require D and Y to be of the same dimension, because we are describing their relation statistically. We start by illustrating the method with pose estimation, which is a functional mapping problem.

7.6.1   Pose Estimation with Mutual Information

Our problem is to find the aspect angle (also called the pose) of vehicles in synthetic-aperture radar (SAR) imagery. The poor resolution of SAR combined with speckle and the variability of scattering centers makes the determination

Degree = 56.19

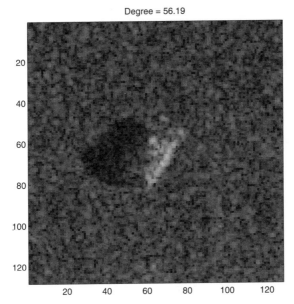

Figure 7.20 SAR image of a personnel carrier.

of pose a nontrivial problem (Fig. 7.20). Knowing the pose of a vehicle can tremendously simplify the task of vehicle recognition.

Generally, the pose estimation can be formulated in terms of maximum *a posteriori* probability (MAP) (Xu et al. 1998a):

$$\hat{\mathbf{a}} = \arg \max_{a} f_{A|X}(\mathbf{a}|\mathbf{x}) = \arg \max_{a} f_{AX}(\mathbf{a}, \mathbf{x}) \qquad (7.45)$$

where $\hat{\mathbf{a}}$ is the estimation of pose $\mathbf{a}$; $f_{A|X}(\mathbf{a}|\mathbf{x})$ is the *a posteriori* probability density function given the image x; and $f_{AX}(\mathbf{a}, \mathbf{x})$ is the joint pdf of the pose and the image. Hence, the key issue here is to estimate the joint pdf. The very high dimensionality of the image (size: 80×80), however, makes it very difficult to obtain a reliable estimation. Dimensionality reduction (or feature extraction) becomes necessary. The output of the mapper $\mathbf{y} = \text{MLP}(\mathbf{w}, \mathbf{x})$ will serve as the feature space for pose estimation. The parameter vector of the mapper is $\mathbf{w}$, in this case a single-layer network of 6400 inputs and two outputs, since a multi-layer perceptron did not provide any performance advantage. So, instead of working directly on the image itself, our pose estimator becomes

$$\hat{\mathbf{a}} = \arg \max_{a} f_{A|Y}(\mathbf{a}|\mathbf{y}) = \arg \max_{a} f_{AY}(\mathbf{a}, \mathbf{y}) \qquad (7.46)$$

The crucial point for this pose estimation scheme is how well y, which can be

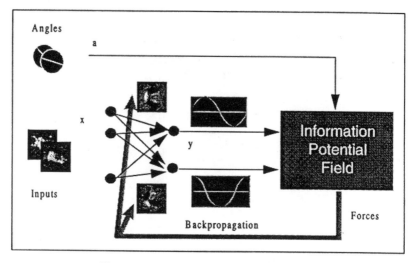

Figure 7.21 Supervised learning using QMI.

interpreted as a feature vector (Duda and Hart 1973), conveys information about the pose. To obtain an effective feature vector, the mutual information between the feature vector and the pose, $I(\mathbf{y}, \mathbf{a})$ is used as the criterion to train the network, that is

$$\mathbf{w}_{\text{optimal}} = \arg \max_{w} I(\mathbf{y} = \text{MLP}(\mathbf{w}, \mathbf{x}), \mathbf{a}) \qquad (7.47)$$

so that the feature conveys the most information about the pose. To implement the idea, the ED-QMI of Eq. (7.41) is used to estimate and maximize $I(\mathbf{y}, \mathbf{a})$.

The system diagram is shown in Fig. 7.21. The true pose angles for each vehicle in the training set were coded in rectangular coordinates in a two-dimensional space. In this case, the maximum of the joint pdf estimation used for network training produces directly the pose estimation in two dimensions. In Xu et al. (1998a) a more involved pose estimator (discrete angles) and in a higher dimension (i.e., with 3 outputs) was proposed, but the results are identical to the ones obtained with this simpler estimator.

The MSTAR public release data set (Veda Inc. 1997) is used here to test our pose-estimation algorithm. Each image chip contains one vehicle roughly centered. Data are scarce, so we train the linear mapper with only 53 images of the vehicle BMP2_c21 (a personnel carrier), taken approximately 3.5° apart (0° to 180°). The algorithm converges in less than 100 batch iterations to a stable configuration where the outputs for the training set approximate a circle in the output space (Fig. 7.22). This solution makes perfect sense from an L_2 norm perspective using our knowledge about the problem.

In the test images of the same vehicle, the algorithm provides outputs that may be far from the circle of the training exemplars, but the response tends to

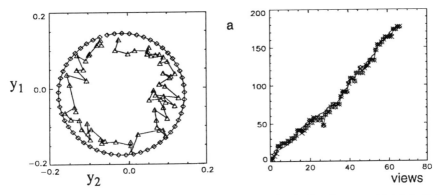

Figure 7.22 System outputs: diamonds training, triangles testing (same vehicle). In the right panel, true (∗) and estimated poses for 65 consecutive views of the testing vehicle from 0 to 180 degrees.

move radially, preserving the accuracy of the pose estimation. Table 7.3 shows the performance of the algorithm on other vehicle types (a tank and other personnel carriers). As we can see, the estimation of the pose for two different BMP types (9563 and 9566) and the BTR (a different personnel carrier) are very close to the resolution of the training-set data (3.5°). So we conclude that our algorithm generalizes the information contained in the training images. The vehicle T72 is a totally different vehicle class (tank), so we consider the mean error of 6.65° better than our expectations, which shows that the pose estimator has a smooth roll-off. Another very interesting characteristic is that the pose estimator implemented with an MLP did not show any overtraining, even though it is trained with 53 exemplars and has more than 12,000 weights (Xu et al. 1998a).

We like to think of the supervised learning method with quadratic mutual information as *information filtering*, since we are projecting the input signal into the space of the system weights just like in linear filtering. However, here we do not use concepts of frequency to choose the projection, but rather information theory to select the projection that best preserves the input information as specified by the desired response (pose in this case). These results and many

Table 7.3 Testing Results in Several Vehicles

Vehicle Class_type	Mean Abs Error (std dev) (degrees)
BMP2_c21	0.54 (0.40) -training set
BMP2_9563	4.25 (3.62)
BMP2_9566	3.81 (3.16)
BTR70_c71	3.18 (2.84)
T72_s7	6.65 (5.04)

others not reported here show that maximizing mutual information using the criterion of Eq. (7.41) is effective. The next section presents yet another example of the power of quadratic mutual information in learning.

7.6.2 Training MLPs Layer-by-Layer

This last example challenges the accepted training methodology for the multi-layer perceptron. During the first neural-network era that ended in the 1970s no algorithm was known to train MLPs. In the late 1980s the backpropagation algorithm was popularized to train MLPs, contributing to the revival of neural computation. Ever since this time backpropagation has been almost exclusively utilized to train MLPs, to a point that several researchers confuse the topology with the training (calling MLPs backpropagation networks). It has been accepted that training the intermediate layers requires backpropagation of errors from the top layers.

We will utilize a modification of Linsker's Infomax concept to train MLPs in a *strictly feedforward fashion* (Xu 1998). We model the MLP network as a communication channel, and train it layer-by-layer such that at each layer the mutual information between the output of the layer and the desired signal is maximized (Fig. 7.23), that is, each layer transfers as much information as possible about the desired response. Notice that we are not using back-propagation of errors across layers. The network is incrementally trained in a strictly feedforward fashion, from the input layer to the output layer. This may seem impossible since we are not using the information of the top layer to train the input layer. The training is simply guaranteeing that the maximum possible information is transferred from the input to each layer from the point of view of the desired response. ITL training makes the desired response explicit to each network layer.

As an example, if the input layer is $R^n \rightarrow R^m : Y = q(X, \theta)$, where θ is the weight set of the layer, and the training set is $\{(X_i, D_i)|i = 1, \ldots, N\}$, where X_i are input patterns and D_i are corresponding desired output signals, then there are actual outputs Y_i corresponding to each input pattern X_i. The mutual information between the output Y and the desired signal D can be estimated by

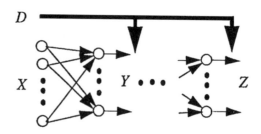

Figure 7.23 MLP as a communication channel. *X*: input signal; *Y*: hidden-layer signal; *Z*: output signal; *D*: desired signal.

the cross-information potential $V_c(b)$ [here we use the ED measure of Eq. (7.41)], where $b = \{(Y_i, D_i)|i = 1, \ldots, N\}$ is the data set to compute the CIP. To maximize the CIP, the marginal information forces at each layer are computed according to Eq. (7.43) and the gradient estimated as

$$\frac{\partial}{\partial \theta} V_c(b) = \sum_{i=1}^{N} (F_i)^T \frac{\partial Y_i}{\partial \theta}, \qquad F_i = \frac{\partial}{\partial Y_i} V_c(b) \qquad (7.48)$$

where $\partial Y_i / \partial \theta$ are sensitivities, and F_i are the information forces in the CIP field for each IPC Y_i. The delta rule is used to adapt the weights.

To test the method we select the "frequency doubler" problem, which is representative of a nonlinear temporal mapping. The input signal is a sine wave (period of 40 samples) and the desired output signal is still a sine wave, but with twice the frequency (as shown in Fig. 7.24). The topology is a focused time-delay neural network (TDNN) with five delays, one hidden layer two PEs with tanh nonlinearity) and one linear output PE (as shown in Fig. 7.24). The training scheme just described is used. The hidden layer is trained first, followed by the output layer. The training curves are shown in Fig. 7.25. The output of the hidden PEs and output PE after training are shown in Fig. 7.25, which tells us that the frequency of the final output is doubled. This problem is not easy with the parameters given because the input delay line "sees" less than one-quarter of the period of the input sine wave. This partially explains the oscillations in the learning curve. Fine-tuning the learning parameters (kernel size and learning rate) makes learning smoother.

This example can also be solved with backpropagation, possibly with a smaller error, so our point is not to propose ITL as a substitute to backpropagation for MLP training. We want to simply illustrate that the backpropagation of errors from layer to layer is not the only possible way to train systems with hidden PEs. The idea of transferring as much information as possible across layers toward a goal, which is achieved by maximizing the mutual information between the output of the layer and the desired response is sufficient to discover complex mappings. Notice also that the proverbial problem of guessing what the hidden PEs should represent is solved with this training method.

7.7 CONCLUSIONS

This chapter describes our efforts to develop an information-theoretic criterion that can be utilized in adaptive filtering and neurocomputing. The optimization criteria should be external to the mapper, and should work directly with the information contained in the samples, without any further assumptions. We found the answer in a combination of a nonparametric density estimator (Parzen windows) and easily computable definitions of entropy and mutual

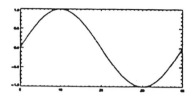

desired response

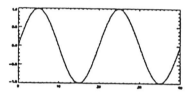

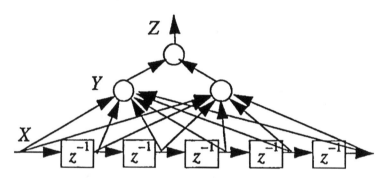

Figure 7.24 Focused TDNN trained as a frequency doubler.

Training Curve CIP vs. Iterations: Left hidden layer, right output layer

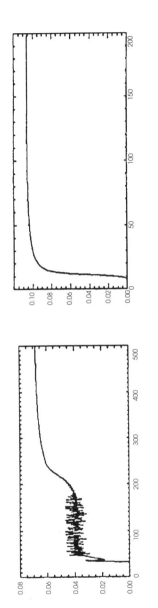

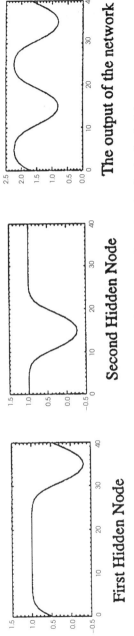

First Hidden Node

Second Hidden Node

The output of the network

Figure 7.25 Training curves and the output of the PEs and net output after training.

315

information. Although Shannon's entropy definition is the only one that obeys all the properties of information, alternate definitions have been shown to be of practical value. We have explored the parametric definition of entropy proposed by Renyi, and settled on the member of the family with order $\alpha = 2$, or quadratic entropy. Renyi's quadratic entropy can be readily integrated with the Parzen-window estimator, yielding without any approximation (besides the pdf estimation step) an optimization criterion that is appropriate for our concept of "neural processing."

We explained how the idea of pdf estimation with Parzen windows leads to the ISE method, which was the first reported practical nonparametric method for ITL. ISE is a criterion external to the mapper, so it can be used with any mapper, linear or nonlinear. An analysis of the computations showed that the pdf estimation can be bypassed and the criterion computed with local interactions among samples with an influence function. The algorithm has a computational complexity of $O(N^2)$, where N is the number of training patterns. We showed that the method is practical and works well, extending the work of Bell and Sejnowski for BSS. But the criterion seems to have other very interesting properties (such as neighborhood preservation) that have not been explored.

With Renyi's quadratic entropy we have a more principled approach to directly manipulate entropy. We provided an interpretation of the local interactions among pairs of samples as an information potential field. The injected error for the mapper can also be interpreted as an information force. This physical analogy raises hope of building an "entropy machine" based on this approach. We also showed the relationship between the information potential and the influence function obtained in ISE.

An important conclusion of this work is the form of the Renyi's entropy estimator. Although entropy is a function of the pdf, we do not need to estimate the pdf to estimate Renyi's entropy. This is due to the fact that Renyi's entropy is a function of the norm of the pdf, which can be estimated directly by interactions among α-plets of data samples. A similar simplification happens in the design of classifiers, where the *a posteriori* probability is estimated without the need to directly estimate the pdf. This is a saving grace that makes ITL practical.

Mutual information was estimated using the CS-QMI and ED-QMI inequalities as measures of divergence between the joint density and the factorized marginals. This is a proxy (approximation) for the KL divergence, but has the advantage of being easily integrated with Renyi's entropy to implement sample estimators. We showed that the minimization of this cost function efficiently separates instantaneously mixed speech sources. The idea is to minimize the mutual information among the outputs of the demixing filter, as described in previous chapters. But with the information-potential method we are using solely the information from samples, so we can potentially separate nonlinearly mixed sources.

But quadratic mutual information transcends the ICA application. It can also be used for supervised learning by interpreting the variables as the desired response and the output of the mapper. We showed that the maximization of the QMI works as an *information filtering criterion* to estimate pose from vehicle images in SAR. We also showed how to adapt an MLP layer-by-layer without error backpropagation. Each layer of the MLP is interpreted as an information filter with the explicit goal of maximizing mutual information between the desired response and the output of the layer. No backpropagation of errors is necessary to discover complex mappings.

Many challenging steps lie ahead in this area of research, but we hope to have shown that information-theoretic learning criteria are flexible, usable, and provide more information about the data than the MSE criterion, which is still the workhorse of neurocomputing.

ACKNOWLEDGMENT

This work was partially supported by DARPA Grant F33615-97-1-1019 and NSF Grant ECS-9900394.

REFERENCES

Amari, S., and J. Cardoso, 1997, "Blind source separation—semiparametric statistical approach," *IEEE Trans. Signal Processing*, vol. 45, no. 11, pp. 2692–2700.

Atick, J., 1992, "Could information theory provide an ecological theory of sensory processing?," *Network* vol. 3, pp. 213–251.

Barlow, H., T. Kaushal, G. Mitchison, 1989, "Finding minimum entropy codes," *Neural Computation*, vol. 1, no. 3, pp. 412–423.

Bell, A. J., and T. J. Sejnowski, 1995, "An information-maximization approach to blind separation and blind deconvolution," *Neural Computation*, vol. 7, no. 6, pp. 1129–1159.

Burg, J., 1967, "Maximum entropy spectral analysis," *Proc 37th Meeting Society of Exploration Geophysics*.

Cardoso, J.-F., April 1997, "Infomax and maximum likelihood for blind source separation," *IEEE Signal Processing Lett.*, vol. 4, no. 4, pp. 112–114.

Comon, P., 1994, "Independent component analysis: a new concept?," *Signal Processing*, vol. 36, no. 3, pp. 287–314.

Cover, T., and J. Thomas, 1991, *Elements of Information Theory* (New York: Wiley).

Deco, G., and D. Obradovic, 1996, *An Information-Theoretic Approach to Neural Computing* (New York: Springer).

Diamantaras, K., and S. Kung, 1996, *Principal Component Neural Networks, Theory and Applications* (New York: John Wiley).

Duda, R. O., and P. E. Hart, 1973, *Pattern Classification and Scene Analysis* (New York: John Wiley).

Fano, R., 1961, *Transmission of Information* (Cambridge, MA: MIT Press).

Fisher, J. W., III, 1997, Nonlinear Extensions to the Minimum Average Correlation Energy Filter, Ph.D. dissertation, Dept. of ECE, University of Florida.

Fisher, J., and J. Principe, 1998, "Blind source separation by interactions of output signals," *IEEE Workshop on Signal Processing*, DSP98, Utah.

Grassberger, P., and I. Procaccia, 1983, "Characterization of strange attractors," *Phys. Rev. Lett.* vol. 50, no. 5, pp. 346–349.

Golub, G., and F. Van Loan, 1989, *Matrix Computations* (Baltimore, MD: Johns Hopkins Press).

Hartley, R. V., 1928, "Transmission of information," *Bell System Tech. J.*, vol. 7.

Haykin, S., 1994, *Neural Networks, A Comprehensive Foundation*, (New York: Macmillan Publishing Company).

Jaynes, E., 1957, "Information theory and statistical mechanics," *Phys. Rev.*, vol. 106, pp. 620–630.

Kapur, J. N., 1994, *Measures of Information and Their Applications* (New York: John Wiley).

Kapur, J., and H. Kesavan, 1992, *Entropy Optimization Principles and Applications*, (San Diego, Associated Press).

Kohonen, T., 1995, *The Self-Organizing Map*, New York Springer Verlag.

Kullback, S., 1968, *Information Theory and Statistics* (New York: Dover Publications).

Linsker, R., 1989, "An application of the principle of maximum information preservation to linear systems," in D. S. Touretzky, *Advances in Neural Information Processing Systems 1*, ed., Morgan-Kauffman. 186–194, San Mateo, CA.

Linsker, R., 1997, "A local learning rule that enables information maximization for arbitrary input distributions," *Neural Computation*, vol. 9, 1661–1665.

Marple, S., 1988, *Modern Spectral Estimation*, Prentice-Hall.

Marques, G., and L. Almeida, 1999, "Separation of nonlinear mixtures using pattern repulsion," *Proc. ICA'99*, Ausois, France, pp. 277–282.

Oja, E., 1982, "A simplified neuron model as a principal component analyzer," *J. of Mathematical Biology*, vol. 15, pp. 267–273.

Papoulis, A., 1965, *Probability, Random Variables and Stochastic Processes*, McGraw-Hill.

Parzen, E., 1962, "On the estimation of a probability density function and the mode," *Ann. Math. Stat.* vol. 33, pp. 1065.

Plumbey, M., and F. Fallside, 1989, "Sensory adaptation: an information theoretic viewpoint," *Int. Conf. on Neural Nets*, vol. 2, pp. 598.

Principe, J., and J. Fisher, 1997, "Entropy manipulation of arbitrary nonlinear mappings," *Proc. IEEE Workshop Neural Nets for Signal Processing*, pp. 14–23, Amelia Island.

Renyi, A., 1976a, "Some Fundamental Questions of Information Theory," Selected Papers of Alfred Renyi, vol. 2, (Budapest: Akademia Kiado) pp. 526–552.

Renyi, A., 1976b, "On Measures of Entropy and Information," Selected Papers of Alfred Renyi, vol. 2. (Budapest: Akademia Kiado) pp. 565–580.

Rumelhart, D. E., G. E. Hinton, and J. R. Williams, 1986, "Learning representations by back-propagating errors," *Nature* (London), vol. 323, pp. 533–536.

Shannon, C., and W. Weaver, 1949, *The Mathematical Theory of Communication*, University of Illinois Press.

Shannon, C. E., 1948, "A mathematical theory of communication," *Bell Sys. Tech. J.*, vol. 27, pp. 379–423, 623–653.

Vapnik, V. N., 1995, *The Nature of Statistical Learning Theory* (New York: Springer).

Veda Incorporated, 1997, MSTAR data set.

Viola, P., N. Schraudolph, and T. Sejnowski, 1995, "Empirical entropy manipulation for real-world problems," *Proc. Neural Info. Processing Sys. (NIPS 8) Conf.*, pp. 851–857.

Xu, D., May 1997, From decorrelation to statistical independence, Ph.D. proposal, Univ. of Florida.

Xu, D., 1998, Energy, entropy and information potential in neurocomputing, Ph.D. Dissertation, Univ. of Florida.

Xu, D., J. Fisher, and J. Principe, 1998a, "A mutual information approach to pose estimation," *Algorithms for Synthetic Aperture Radar Imagery V*, vol. 3370, 218229, SPIE 98.

Xu, D., J. Principe, J. Fisher, and Wu, H.-C., 1998b, "A novel measure for independent component analysis (ICA)" in *Proc. ICASSP'98* vol. II, pp. 1161–1164.

Yang, H. H., and S. Amari, 1997, "Adaptive on-line learning algorithms for blind separation—maximum entropy and minimum mutual information," *Neural Computation*, vol. 9, no. 7.

8

BLIND SEPARATION OF DELAYED AND CONVOLVED SOURCES

Kari Torkkola

ABSTRACT

Blind separation of independent sources from their convolutive mixtures is a problem encountered in many real-world multisensor applications. We present an overview of the problem, a summary of current methods to solve it, and some foreseeable application areas. We exhibit a slight bias toward one of the methods, the information maximization principle, due to its conceptual simplicity and intuitivity. Because the solution requires a network of filters, we discuss several architectures capable of coping with delayed or convolved mixtures, and we show how to derive the adaptation equations for the filters in the networks by maximizing the information transferred through the network. Another bias in this chapter is toward engineering rather than science: We give a slight emphasis to practical applications.

8.1 INTRODUCTION

As already mentioned in previous chapters, blind source separation (BSS) denotes observing mixtures of *independent* source signals, and by making use of these mixture signals only, recovering the original signals. "Blindness" refers

Unsupervised Adaptive Filtering, Volume I, Edited by Simon Haykin.
ISBN 0-471-29412-8   © 2000 John Wiley & Sons, Inc.

to weak assumptions made about the signals and about the mixing. Due to a number of interesting applications in communications and in speech and medical signal processing, among others, BSS has received a great deal of attention in the literature. Much of the work addresses the case of instantaneous mixtures, where the following linear mixture model is assumed:

$$\mathbf{x}(n) = \mathbf{A}\mathbf{s}(n) \tag{8.1}$$

where $\mathbf{s}(n)$ is a vector of sources at discrete time instant n, $\mathbf{A}$ is the mixing matrix, and $\mathbf{x}(n)$ is the observed vector of mixtures (ignoring additive noise). The sources can be separated by finding a matrix $\mathbf{W}$ with the condition $\mathbf{W}\mathbf{A} = \mathbf{P}\mathbf{D}$, where $\mathbf{P}$ is a permutation matrix and $\mathbf{D}$ is a diagonal matrix. Thus the recovered sources

$$\mathbf{u}(n) = \mathbf{W}\mathbf{x}(n) = \mathbf{P}\mathbf{D}\mathbf{s}(n) \tag{8.2}$$

are the original ones up to permutation and scaling. This instantaneous-mixture case is also called independent components analysis (ICA) (Comon 1994).

If $\mathbf{A}$ is a matrix of filters instead of scalars, we are talking about convolutive mixtures. This case occurs in many practical situations, for example, in multi-microphone audio recordings done in enclosures that are not anechoic. This is the subject of the chapter at hand.

The structure of this chapter is as follows. Notational convention is presented in Table 8.1. We begin by briefly reviewing the instantaneous-mixture problem, concentrating on one particular approach, the information-maximization approach. We proceed by pointing out the assumptions that need to be made in order to address the problem of convolutive mixtures.

As an intermediate step toward convolutive mixtures, we first study the cases where the sources may have been delayed with respect to each other. We present network architectures capable of coping with such sources, and derive the adaptation equations for the delays and the weights in the network on the basis of the information-maximization principle.

We discuss mixtures of sources convolved through some kind of a filter (such as impulse response of an acoustic enclosure) and conditions for their separation. We present network architectures for separation, and we derive adaptation rules for the networks based on information maximization. Examples using wide-band sources such as speech are presented to illustrate the method.

Limitations of this approach are discussed next, and two possible solutions are presented, taking the form of an acausal filter network to which the information maximization can also be applied.

We then compare these methods to others, we discuss possible application areas of BSS of convolutive mixtures, and finally, touch on the issue of the usefulness of *blind deconvolution* in the audio realm. We end the chapter by summarizing what the technology yet might lack to be able to produce successful applications.

Table 8.1   Notational Convention

n	Discrete time index
s	Vector of sources
s_i	ith component of a vector
A	Mixing matrix
x	Vector of mixtures
W	Separating matrix
u	Vector of separated sources
$y_i = g_i(u_i)$	Separated source passed through a function approximating the cdf of the source
ŷ	Componentwise nonlinearity applied to separated sources
$f_s(s)$	Probability density function (pdf)of s
$D(f, g)$	Kullback-Leibler divergence between two pdf's
$\det(\mathbf{A})$	Determinant of a matrix
$\mathbf{A}^T$	Transpose
$\mathbf{A}^\dagger$	Complex-conjugate transpose
$\mathbf{A}^*$	Complex conjugate
d_{ij}	Adaptive delay from signal j to signal i
W_{ij}	Adaptive filter from signal j to signal i
w_{ikj}	Weight from signal j to signal i with delay k
$\dot{u}(n)$	Time derivative of u at discrete time n
0	A vector of zeros
$\underline{\mathbf{W}}$	Matrix of FIR polynomials

8.2   INSTANTANEOUS-MIXTURE PROBLEM

8.2.1   Setup

In the simplest case, unobservable sources $s_i(n)$ are mixed using unobservable scalar coefficients a_{ij} (matrix **A**) to produce observable mixtures $x_i(n)$. The problem is to learn coefficients w_{ij} (matrix **W**) that produce the original sources, or scaled and permuted versions of them, as outputs $u_i(n)$. For two sources the process is illustrated in Fig. 8.1.

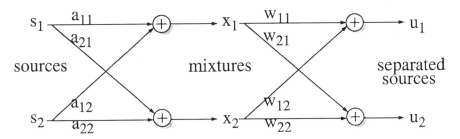

Figure 8.1   Linear mixing and separation of two sources.

On what basis can coefficients w_{ij} be learned? How could such learning be at all possible, as *both* the sources and the mixing are unknown? Let us begin by looking at Fig. 8.2, which depicts how separated sources can be produced from two mixtures if the separating coefficients w_{ij} are known. Note that the mixtures have to be at least slightly different, and that, in general, there needs to be at least as many mixture signals as there are source signals we wish to separate. In order to learn **W**, what could be used as a criterion C that conveys when the separated sources appear at the outputs, and when the outputs are yet mixtures, even without assuming much about the sources?

8.2.2 Criteria for Separation

Obviously the signals u_i will be correlated if they still remain mixtures, and they will be uncorrelated if the separation is successful, assuming that the original sources are also uncorrelated. This is a necessary condition, but is it sufficient? The answer turns out to be no. The point is illustrated in Figure 8.3, in which two speech signals are mixed as follows:

$$x_1(n) = 1.0s_1(n) + 0.7s_2(n)$$
$$x_2(n) = 0.8s_1(n) + 1.0s_2(n) \tag{8.3}$$

The left side of Fig. 8.3 depicts the joint distribution of the mixtures, that is, it plots $x_2(n)$ (vertical axis) against $x_1(n)$ (horizontal axis). The right side shows the distribution of the original sources, $s_2(n)$ against $s_1(n)$.

Principal components analysis (PCA) is one way to produce uncorrelated outputs. PCA finds the direction of largest variance in the data, the next largest direction orthogonal to the previous, and so on. Thus **W** would be a matrix of the eigenvectors of the covariance matrix of **x**. Applying PCA to the mixture distribution results in directions that are plotted as gray arrows on the left side of the Fig. 8.3. Projecting data onto these directions results in outputs that are uncorrelated (since the directions are orthogonal), but alas, yet mixed.

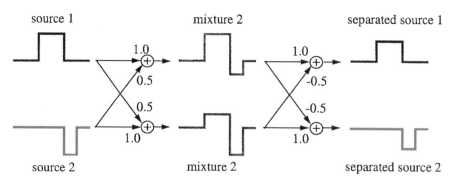

Figure 8.2 Two simple sources mixed and separated.

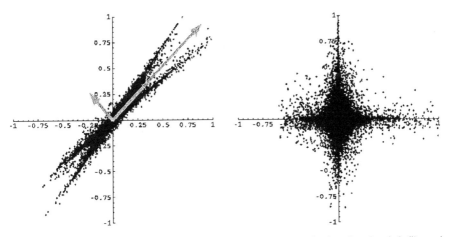

Figure 8.3 A scatterplot, or a joint distribution of the speech signals mixed (left) and unmixed (right). Principal component directions are drawn on top of the mixed signals.

The unmixing directions are not orthogonal in general. They can be seen as the two major components in the left side of the figure. Projecting the data onto these directions results in separation, and in the scatterplot of the right side of Fig. 8.3.

Further assumptions about s_i are needed. A basic assumption is that s_i are independent. This leads to measures of independence as the criterion C. Let us denote a vector of components u_i, by $\mathbf{u}$. If the components of $\mathbf{u}$ are independent, then the probability density function (pdf) should factorize:

$$f_u(\mathbf{u}) = \prod_i f_{u_i}(u_i) \tag{8.4}$$

Divergence between the two sides of Eq. (8.4) thus reflects the independence of the u_i. Kullback-Leibler (KL) divergence (or relative entropy) as a distance measure between two pdf's, f and g, is written

$$D(f, g) = \int f(\mathbf{u}) \log \frac{f(\mathbf{u})}{g(\mathbf{u})} \, d\mathbf{u} \tag{8.5}$$

For the two sides of Eq. (8.4), the divergence is

$$D\left(f_u(\mathbf{u}), \prod_i f_{u_i}(u_i)\right) = \int f_u(\mathbf{u}) \log \frac{f_u(\mathbf{u})}{\prod_i f_{u_i}(u_i)} \, d\mathbf{u} \tag{8.6}$$

which is also the mutual information of $\mathbf{u}$ (Deco and Obradovic 1996). This is the basis of a number of BSS algorithms, for example, Comon (1994) and

Amari et al. (1996), as discussed in previous chapters. To be able to make use of the criterion, the individual pdf's need to be parametrized and estimated, usually in terms of a truncated series of cumulants, which are estimated from the outputs. See Chapter 4 of this volume or Amari et al. (1996), Haykin (1998), and Comon (1994) for more detailed discussion about truncated series of cumulants, and Hyvärinen (1997) for a more general approach on direct entropy estimation based on the maximum-entropy principle.

Another related additional assumption is that at most one of the sources s_i can be Gaussian, which leads to many intuitively simple BSS algorithms. This assumption enables us to use the Central Limit theorem: The sum of sufficiently many signals of any distribution has a Gaussian distribution. Also, the sum of a "small" number of non-Gaussian signals has a distribution that is "closer" to a Gaussian than the pdf's of the sources. Thus all we need to do is to write the KL divergence of the pdf of the output to a Gaussian:

$$
\begin{aligned}
D(f_u, f_G) &= \int f_u(\mathbf{u}) \log \frac{f_u(\mathbf{u})}{f_G(\mathbf{u})} \, d\mathbf{u} \\
&= \int f_u(\mathbf{u}) \log f_u(\mathbf{u}) \, d\mathbf{u} - \int f_u(\mathbf{u}) \log f_G(\mathbf{u}) \, d\mathbf{u} \\
&= -H(f_u) + H(f_G)
\end{aligned}
\tag{8.7}
$$

which is equivalent to the differential entropy of a Gaussian (which has the same mean and covariance as f_u) minus the differential entropy of f_u. This measure is also called *negentropy*. From this measure it is straightforward to derive a stochastic-gradient-ascent adaptation rule to find a $\mathbf{W}$ that maximizes the divergence and thus separates the sources, as shown in (Girolami and Fyfe 1997).

Deviation from Gaussianity can also be measured by higher-order statistics since these are zero for Gaussian pdf's. For example, kurtosis, which is written as $E\{u^4\} - 3(E\{u^2\})^2$ for a zero-mean variable, could be used. Both kurtosis and negentropy require us to know to which direction we are driving the output pdf's from a Gaussian. For example, pdf's that have negative kurtosis (flat pdf's like uniform distribution) require the kurtosis of the output signal to be minimized, whereas for positively kurtotic (sharply peaked) signals kurtosis needs to be maximized. Either we need to know the signs of kurtoses of our sources, or we need to estimate them while learning the $\mathbf{W}$, as discussed in (Girolami and Fyfe 1997).

It might also be possible that the pdf's of the sources are known, or that the shapes are known less perhaps for some parameters that need to be estimated while $\mathbf{W}$ is learned. In this case, we can again use the KL divergence between the output pdf and the known source pdf's as the criterion to be minimized, that is, find $\mathbf{W}$ that drives the outputs toward known distributions. This approach lends itself naturally to the maximum-likelihood estimation (Belouchrani and Cardoso 1994; Cardoso and Amari 1997; Pham et al. 1992). Even if the pdf's are unknown we can estimate them as sums of some simple, for example, Gaussian or logistic distributions, while the separation matrix is being estimated. Either gradient ascent (Bell and Sejnowski 1995; Pearlmutter and Parra 1996) or expectation maximization in some cases (Belouchrani and Cardoso 1995; Moulines et al. 1997) can then be used. This situation lends itself to the following interpretation. The sources can be interpreted as latent signals whose mixture is the only one observable. The general literature on latent models and their estimation is then applicable (Bishop et al. 1997). See especially Roweis and Ghahramani (1999) and Girolami et al. (1997) for the application of the framework to ICA.

All assumptions we have mentioned so far are related to the pdf's of the sources. Other information may be available, and in general, if such information is available about the sources or about the mixing, it should be used in designing the algorithm to learn $\mathbf{W}$. Jutten and Cardoso (1995) give an excellent overview of possible sources of information that can be exploited in "less blind" BSS. Such information is available with man-made signals, especially in communications. We will discuss this application area is Section 8.8.2. For separation of convolutively mixed signals, some other criteria are also available, which are discussed in Section 8.7.

8.2.3 The Information-Maximization Approach

In the rest of this chapter we concentrate on a particular approach, the information maximization, due to its conceptual simplicity.

According to the discussion in the previous section, this method can be categorized as "density matching," and it was presented by Bell and Sejnowski (1995). Their approach has its roots in neural information processing, and their starting point was to maximize the information that can be transferred through a single neuron. Maximizing the information transferred through a layer of neurons results in signal separation as follows.

The separating matrix $\mathbf{W}$ is learned by minimizing the mutual information between components of $\mathbf{y}(n) = g(\mathbf{u}(n))$, where g is a nonlinear function approximating the cumulative density function (cdf) of the sources (see Fig. 8.4 for a two-source network (Bell and Sejnowski 1995). For positively kurtotic signals, like speech, minimizing the mutual information between components of $\mathbf{y}$ was shown to be equal to maximizing the entropy of $\mathbf{y}$, which can be written as $H(\mathbf{y}) = -E[\log(f_y(\mathbf{y}))]$, where $f_y(\mathbf{y})$ denotes the pdf of $\mathbf{y}$.

Intuitively, this can be understood as follows. When a signal is passed through a nonlinear function that approximates its cdf, the pdf of the resulting signal will be close to the uniform density. Uniform density is the pdf that has the maximum entropy of all bounded pdf's. Thus finding a $\mathbf{W}$ that maximizes entropy at the outputs passed through the cdf's produces the true pdf's of the sources at the outputs and thus separates the sources.

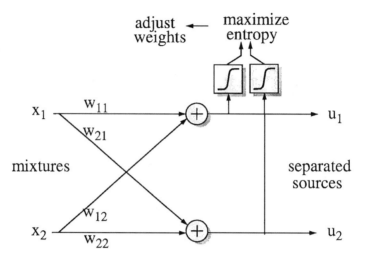

Figure 8.4 The separation network of Bell and Sejnowski for instantaneous mixtures of two sources (ignoring the bias weights). Notice that the separation process is linear. The nonlinearities are only used in learning the weight matrix **W**.

A stochastic-gradient-ascent adaptation was derived as follows. Denoting the determinant of the Jacobian of the whole net by $\det(\mathbf{J})$, $f_y(\mathbf{y})$ can be written as $f_x(\mathbf{x})/\det(\mathbf{J})$ (Jacobian is a matrix with entries of $\partial y_i/\partial x_j$.) Now, maximizing the entropy of the output leads to maximizing $E[\log \det(\mathbf{J})]$, which in turn can be developed into the following stochastic-gradient-ascent rule using instances of $\mathbf{x}(n)$ and $\mathbf{y}(n)$, instead of using the expectation

$$\Delta \mathbf{W} \propto [\mathbf{W}^T]^{-1} + \hat{\mathbf{y}}(n)\mathbf{x}(n)^T \tag{8.8}$$

where $\hat{\mathbf{y}}$ is a vector with the following components $\hat{y}_i$:

$$\hat{y}_i = \frac{\partial}{\partial y_i} \frac{\partial y_i}{\partial u_i} \tag{8.9}$$

This method calls for a nonlinear function that roughly approximates the cdf of the signals to be separated. For speech signals, $y = g(u) = \tanh(u)$ can be used to approximate the cdf. In this case, the $\hat{y}_i$ in Eq. (8.9) has a particularly simple form:

$$\hat{y}_i = \frac{\partial}{\partial y_i} \frac{\partial \tanh(u_i)}{\partial u_i} = \frac{\partial}{\partial y_i}(1 - \tanh^2(u_i)) = \frac{\partial}{\partial y_i}(1 - y_i^2) = -2y_i \tag{8.10}$$

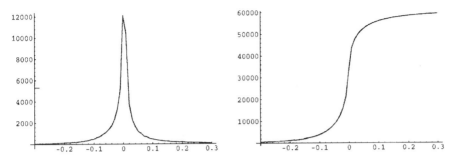

Figure 8.5 A histogram and a cumulative histogram of 2 seconds of a speech signal scaled between minus one and one. Note the sharp peak of a positively kurtotic distribution.

Logistic function $y = g(u) = 1/(1 + e^{-u})$ results in a simple form, too:

$$\hat{y}_i = 1 - 2y_i \qquad (8.11)$$

For ease of comparing these functions to real cdf's, Figure 8.5 contains a histogram and a cumulative histogram of a real speech signal. Both hyperbolic tangent and logistic functions are reasonable approximations, though the actual histogram is even sharper and closer to a Laplacian.

The algorithm can be made more efficient by avoiding the matrix inversion, and independent of the conditioning of the mixing matrix by using the so-called natural or relative gradient instead of the absolute gradient (Amari et al. 1996; Cardoso and Laheld 1996). This amounts to multiplying Eq. (8.8) by $\mathbf{W}^T\mathbf{W}$, giving

$$\Delta\mathbf{W} \propto (\mathbf{I} + \hat{\mathbf{y}}(n)\mathbf{u}(n)^T)\mathbf{W} \qquad (8.12)$$

The idea of a natural gradient was discussed in detail in Chapters 2 and 3 of this volume.

Pearlmutter and Parra (1996) have shown that exactly the same adaptation equation can be derived starting from an entirely different viewpoint, the maximum-likelihood approach to density estimation. The same fact has also been pointed out by others, for example, by Cardoso (1997) and MacKay (1996).

Given (or assuming) the parametric forms of the densities of the underlying source signals, the idea is to find what mixing matrix, when applied to source densities, is most likely to produce the observed mixture density. Alternatively, what unmixing matrix when applied to the observed mixture density, is most likely to produce the assumed source densities. Source densities are assumed independent; thus they factorize.

The KL divergence between these two densities is written as

$$D(f, \hat{f}) = \int f(\mathbf{x}) \log \frac{f(\mathbf{x})}{\hat{f}(\mathbf{x}; \mathbf{w})} \, d\mathbf{x} = H(f) - \int f(\mathbf{x}) \log \hat{f}(\mathbf{x}; \mathbf{w}) \, d\mathbf{x} \qquad (8.13)$$

Here $f(\mathbf{x})$ is a fixed pdf, and $\hat{f}(\mathbf{x}; \mathbf{w})$ is a parametric pdf that can be varied and shaped by varying the parameter vector $\mathbf{w}$.

In source separation $f(\mathbf{x})$ is the pdf of the observable mixture $\mathbf{x}$, and $\hat{f}(\mathbf{x}; \mathbf{w})$ acts as the pdf constructed from single-source pdf's using a current estimate of the unobservable mixing matrix $\mathbf{W}^{-1}$. The purpose is now to find the mixing matrix, or actually its inverse, the separating matrix, and the parameters of the individual source pdf's, such that f and $\hat{f}$ would be as much alike as possible, as measured by the divergence.

The next step is to derive a gradient-ascent rule to update the current estimate of $\mathbf{w}$ that maximizes the likelihood. Since D is not directly available, Pearlmutter and Parra then proceed by taking the gradient of a *sample* of D, which is

$$\hat{D} = H(f) - \log \hat{f}(\mathbf{x}; \mathbf{w}) \tag{8.14}$$

Writing out the natural gradient of $\hat{D}$, the final adaptation equation is as follows:

$$\Delta \mathbf{W} \propto -\frac{d\hat{D}}{d\mathbf{W}} \mathbf{W}^T \mathbf{W} = \left(\left(\frac{f'_j(u_j; w_j)}{f_j(u_j; w_j)} \right)_j \mathbf{u}^T + \mathbf{I} \right) \mathbf{W} \tag{8.15}$$

where $\mathbf{u} = \mathbf{W}\mathbf{x}$ are the sources separated from mixtures $\mathbf{x}$, $f_j(u_j; w_j)$ is the pdf of source j parametrized by w_j, and $(\cdot)_j$ denotes a vector with components j. This appears to be exactly the form of Bell and Sejnowski when f_j is taken to be the derivative of the necessary nonlinearity g_j, which was assumed to be "close" to the true cdf of the source. Thus the information maximization makes implicit assumptions about the cdf's of the sources in the form of the nonlinear squashing function, and does implicit density estimation, whereas in the maximum-likelihood approach, the density assumptions are made explicit. This fact makes it more lucid to derive the adaptation for other forms of densities, and also to extend it to complex-valued variables, which is necessary in digital communications.

8.3 TOWARD MORE REALISTIC ASSUMPTIONS

8.3.1 Why Can Instantaneous BBS Not Be Applied to Real-World Problems?

There are several obstacles to applying the BSS algorithms to real-world situations (Bell and Sejnowski 1995; Pope and Bogner 1994). These include the effect of noise on successful learning of the separating solution, possibly the unknown number of sources (especially noise sources), and the assumption that the source signals are stationary. Most notable is the assumption of the simultaneous mixing of the sources. For example, in any real-world recording, where the propagation of the signals through the medium is not instantaneous (like sound through air or water), there will be differences in the time of arrival between the sources in the mixtures. Instantaneous BSS algorithms cannot tolerate these delays.

A related problem is the multipath arrival of the signals. In general, the sensor is not observing just a single clean copy of the source, but a sum of multipath copies distorted by the environment. This is equivalent to mixing signals with *filtering*. The acoustic environment (such as a room) imposes a different impulse response between each source and microphone pair. Moreover, the microphones may have different characteristics, or at least their frequency responses may differ for sources in different directions. This kind of situations can be modeled as convolved mixtures.

8.3.2 Delays

We first look at the case of sources mixed with delays only. Let us assume for simplicity two sources and their two mixtures (this can easily be generalized to any number of sources):

$$x_1(n) = a_{11}s_1(n) + a_{12}s_2(n)$$
$$x_2(n) = a_{22}s_2(n) + a_{21}s_1(n)$$
(8.16)

Figure 8.2 illustrated how instantaneous BSS can separate the sources from the mixtures (only if the separation weights are correct!). Let us now consider the case in which the signal components in the mixtures are delayed with respect to one another:

$$x_1(n) = a_{11}s_1(n) + a_{12}s_2(n - D_{12})$$
$$x_2(n) = a_{22}s_2(n) + a_{21}s_1(n - D_{21})$$
(8.17)

Figure 8.6 illustrates how the mixing is now changed. It is easy to see that weighted summing of the mixtures will not be able to produce a separated result anymore, because same-source components are misaligned in the two mixtures.

Now a question arises: Can we separate the mixtures with a feedforward network by adding adaptive delays to the branches of the network? This possible structure is illustrated in Fig. 8.7.

By studying Fig. 8.6, we can arrive at a partly positive answer. If the delays were correct, the network could align—for example, source-component 2 in mixture 2 with source-component 2 in mixture 1—apply correct weights to mixtures, and sum them to eliminate source-component 2 altogether. The same applies to source-component 1. In fact, only two adaptive delays are enough, for example, delays d_{12} and d_{21}.

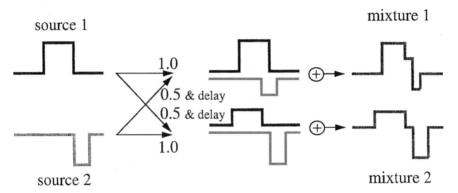

Figure 8.6 Two sources mixed with weights and delays.

But by so doing we create two misaligned copies of source signals in the outputs and the result is thus *filtered* sources instead of pure originals. This "filtering" is simple, though there is only a single echo component summed to the original. We would need to remove those by adding some kind of echo-removal filters for separated outputs. Those filters could be based on blind deconvolution, for example. The whole network can then be trained by a single-objective function that provides separation and deconvolution. This structure only works with two source signals, though.

But there is a simpler way: Let us modify the network structure to a feedback one, and add adaptive delays to feedback branches (Figure 8.8).

Now it is easy to see that this network is able to separate delayed sources without introducing any undesirable side effects to the signals. Let us assume that the weights and delays have converged to their correct values. Then the pure sources can be observed at the output. Now the feedback branches multiply these by the correct weight, and delay them by a correct amount, so that when added to a mixture, the result is a clean source signal at each of the out-

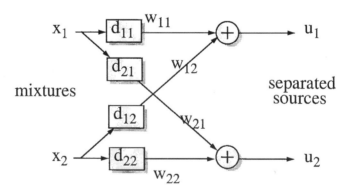

Figure 8.7 Is this structure able to separate delayed sources?

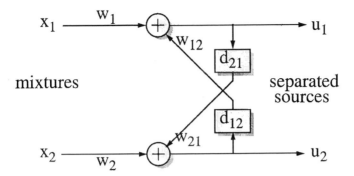

Figure 8.8 Feedback structure for the separation of delayed sources.

puts. This network can be trained using the same separation objective function. We will derive the adaptation in Section 8.4.1.

8.3.3 Convolved Mixtures

We now look at sources mixed with a matrix of filters instead of a matrix of scalars. Assuming we have two speech sources or a speech and a noise source, the aim would be merely to separate the signals without distorting them in any other way. This would be equivalent to observing the signal just with the other signal being absent if all other conditions are kept unchanged.

Let us begin again with the basic feedforward structure by replacing the scalar coefficients with filters. We first determine if a separating solution exists (it is another matter whether and how it can be learned!).

We look at two sources in the z-transform domain for simplicity. The approach can be generalized to any number of sources. For the present, we ignore the noise:

$$X_1(z) = A_{11}(z)S_1(z) + A_{12}(z)S_2(z)$$
$$X_2(z) = A_{21}(z)S_1(z) + A_{22}(z)S_2(z) \tag{8.18}$$

where A_{ij} are the z-transforms of any kind of filters. We can solve the sources S in terms of the mixtures X:

$$S_1(z) = (A_{22}(z)X_1(z) - A_{12}(z)X_2(z))/G(z)$$
$$S_2(z) = (-A_{21}(z)X_1(z) + A_{11}(z)X_2(z))/G(z) \tag{8.19}$$

where $G(z) = A_{11}(z)A_{22}(z) - A_{12}(z)A_{21}(z)$. This suggests a feedforward architecture for separation (which is illustrated later in Fig. 8.12), where $W_{11}(z) = A_{22}(z)/G(z)$, $W_{12}(z) = -A_{12}(z)/G(z)$, and so on, at the ideal separating solution. The existence of a separating solution requires that $G(z) \neq 0$ and that a stable inverse filter of $G(z)$ exists. Note that this implies that it is possible to construct mixtures that cannot be separated with certain filter structures. We return to this subject in Section 8.5.

In addition to the feedforward structure the feedback network of Fig. 8.8 can be employed. We just have to replace the scalar weights in the feedback branches by filters as illustrated later in Fig. 8.13. An FIR-filter actually consists of an individual coefficient for each possible delay.

In this architecture the outputs are defined as

$$U_1(z) = W_{11}(z)X_1(z) + W_{12}(z)U_2(z)$$
$$U_2(z) = W_{22}(z)X_2(z) + W_{21}(z)U_1(z) \tag{8.20}$$

The ideal separating solution in terms of A's and X's, Eq. (8.19), will of course be exactly the same as in the feedforward case, but this time with different filters W_{ij} because of the different network architecture. Using Eqs. (8.18) and (8.20), we can easily derive a solution for perfect separation and deconvolution, that is, $U_1 = S_1$ and $U_2 = S_2$:

$$W_{11}(z) = A_{11}(z)^{-1}, \qquad W_{12}(z) = -A_{12}(z)A_{11}(z)^{-1}$$
$$W_{22}(z) = A_{22}(z)^{-1}, \qquad W_{21}(z) = -A_{21}(z)A_{22}(z)^{-1} \tag{8.21}$$

For this solution to work in practice, A_{11} and A_{22} need to have stable realizable inverses in addition to $G(z) \neq 0$. Besides this solution, we can multiply W_{ii} by any filter resulting in an infinite number of separating solutions in which U_i appear still separated but as filtered versions of the S_i.

8.4 DELAYED AND CONVOLVED MIXTURES: INFORMATION MAXIMIZATION IN THE TIME DOMAIN

8.4.1 A Network Architecture for Delays

We now return to the case of delayed sources. We derive time-domain adaptation rules for the weights and the delays of the separation network of Fig. 8.8.

The feedback architecture for instantaneous BSS was already proposed by Jutten and Herault in (1986, 1991). Platt and Faggin (1992) later added adaptive delays (also filters) that were learned using the minimum output power principle. A learning rule for just a single weight with a delay was suggested in Bell and Sejnowski (1995) based on information maximization, but a general solution to the separation of delayed sources was not presented. We present how mixtures of delayed sources can indeed be separated using the same information-theoretic approach as outlined in Torkkola (1996b).

After adding the nonlinearities approximating the cdf's of the signals, as discussed in Section 8.2.3, the resulting network is depicted in Figure 8.9.

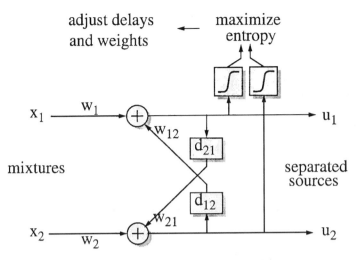

Figure 8.9 The separation network for delays.

For simplicity, let us assume two sources and their two mixtures mixed according to Eq. (8.17) (this can easily be generalized to any number of sources). The network thus computes the following, where u_i are the outputs before the nonlinearities:

$$u_1(n) = w_1 x_1(n) + w_{12} u_2(n - d_{12})$$
$$u_2(n) = w_2 x_2(n) + w_{21} u_1(n - d_{21})$$
$$y_1(n) = g(u_1(n))$$
$$y_2(n) = g(u_2(n))$$

$$(8.22)$$

As with the instantaneous mixture case, we can maximize the entropy at the output by maximizing $H = E[\log \det(\mathbf{J})]$ with respect to the parameters of the network, that is, the weights and the *delays*. The first step in deriving a stochastic-gradient-ascent rule is to take a sample of H, which is $\log \det(\mathbf{J})$.

The determinant of the Jacobian of the network is now

$$\det(\mathbf{J}) = \frac{\partial y_1}{\partial x_1} \frac{\partial y_2}{\partial x_2} - \frac{\partial y_1}{\partial x_2} \frac{\partial y_2}{\partial x_1} = y_1' y_2' D$$

and

$$\log \det(\mathbf{J}) = \log(y_1') + \log(y_2') + \log(D) \qquad (8.23)$$

where

$$D = \left(\frac{\partial u_1}{\partial x_1} \frac{\partial u_2}{\partial x_2} - \frac{\partial u_1}{\partial x_2} \frac{\partial u_2}{\partial x_1} \right) = w_1 w_2, \qquad y_1' = \frac{\partial y_1}{\partial u_1}$$

and

$$y_2' = \frac{\partial y_2}{\partial u_2}$$

The adaptation rule for each parameter of the network can now be derived by computing the gradient of $\log \det(\mathbf{J})$ with respect to that parameter. For w_1 we get

$$\Delta w_1 \propto \frac{\partial \log \det(\mathbf{J})}{\partial w_1} = \frac{1}{y_1'}\frac{\partial y_1'}{\partial w_1} + \frac{1}{y_2'}\frac{\partial y_2'}{\partial w_1} + \frac{1}{D}\frac{\partial D}{\partial w_1} \qquad (8.24)$$

We denote $\hat{y}_i = (\partial y_i'/\partial y_i)$, which depends on the nonlinearity used. Equations (8.10) and (8.11) are examples of $\hat{y}_i$ for the hyperbolic tangent and logistic functions.

Thus we can write for the partial derivatives:

$$\frac{\partial y_1'}{\partial w_1} = \frac{\partial y_1'}{\partial y_1}\frac{\partial y_1}{\partial u_1}\frac{\partial u_1}{\partial w_1} = \hat{y}_1 y_1' x_1$$

$$\frac{\partial y_2'}{\partial w_1} = \frac{\partial y_2'}{\partial y_2}\frac{\partial y_2}{\partial u_2}\frac{\partial u_2}{\partial w_1} = \hat{y}_2 y_2' 0 = 0 \qquad (8.25)$$

$$\frac{\partial D}{\partial w_1} = \frac{\partial (w_1 w_2)}{\partial w_1} = w_2$$

The adaptation rule for w_1 becomes the following from Eq. (8.24) (similarly for w_2)

$$\Delta w_1 \propto \hat{y}_1 x_1 + 1/w_1$$
$$\Delta w_2 \propto \hat{y}_2 x_2 + 1/w_2 \qquad (8.26)$$

For w_{12} the partial derivatives are as follows

$$\frac{\partial y_1'}{\partial w_{12}} = \frac{\partial y_1'}{\partial y_1}\frac{\partial y_1}{\partial u_1}\frac{\partial u_1}{\partial w_{12}} = \hat{y}_1 y_1' u_2(n - d_{12})$$

$$\frac{\partial y_2'}{\partial w_{12}} = \frac{\partial y_2'}{\partial y_2}\frac{\partial y_2}{\partial u_2}\frac{\partial u_2}{\partial w_{12}} = \hat{y}_2 y_2' 0 = 0 \qquad (8.27)$$

$$\frac{\partial D}{\partial w_{12}} = \frac{\partial (w_1 w_2)}{\partial w_{12}} = 0$$

Thus the adaptation for w_{12} will be the following (similarly for w_{21}):

$$\Delta w_{12} \propto \hat{y}_1 u_2(n - d_{12})$$
$$\Delta w_{21} \propto \hat{y}_2 u_1(n - d_{21}) \qquad (8.28)$$

Using either Eq. (8.10) or Eq. (8.11) as $\hat{y}_i$, these rules decorrelate the present squashed output y_i from the other source u_j at delay d_{ij}, which is equivalent to separation. Note that in Eqs. (8.27) and (8.28) the discrete-time indices of u_1 and u_2 are given in parentheses, whereas for all other variables the time is implicitly assumed to be n. All the partial derivatives starting from Eq. (8.23) are also taken at time instance n. This is why we do not expand the cross partial derivatives recursively backward in time. Doing this would be fairly straightforward by introducing recursive intermediate quantities in the fashion of least-mean-square-trained (LMS) recursive adaptive filters (Widrow and Stearns 1985), as done in Torkkola (1997a) for deconvolution. In our experiments the simplified adaptation converged better, though.

The partial derivatives for the delay d_{12} will be

$$\frac{\partial y_1'}{\partial d_{12}} = \frac{\partial y_1'}{\partial y_1}\frac{\partial y_1}{\partial u_1}\frac{\partial u_1}{\partial d_{12}} = \hat{y}_1 y_1' w_{12}(-\dot{u}_2(n-d_{12}))$$

$$\frac{\partial y_2'}{\partial d_{12}} = \frac{\partial y_2'}{\partial y_2}\frac{\partial y_2}{\partial u_2}\frac{\partial u_2}{\partial d_{12}} = \hat{y}_2 y_2' 0 = 0 \qquad (8.29)$$

$$\frac{\partial D}{\partial d_{12}} = \frac{\partial(w_1 w_2)}{\partial d_{12}} = 0$$

where we have taken advantage of the fact that $(\partial/\partial d_{12})u_2(n-d_{12}) = (d/dn)(-u_2(n-d_{12})) = -\dot{u}_2(n-d_{12})$ (Bell and Sejnowski 1995).

The adaptation rules for the delays become the following (again, we have explicitly written the time indices for only $\dot{u}_i$):

$$\Delta d_{12} \propto -\hat{y}_1 w_{12}\dot{u}_2(n-d_{12})$$
$$\Delta d_{21} \propto -\hat{y}_2 w_{21}\dot{u}_1(n-d_{21}) \qquad (8.30)$$

It is notable that the adaptation rules are *local*, that is, to adapt a weight or a delay in a branch of the network, only the data coming in or going out of the branch are needed. For example, adaptation of d_{12} and w_{12} will only depend on u_2 and y_1. Generalization to N mixtures can thus be done simply by substituting other indices to 1 and 2 in Eqs. (8.28) and (8.30), and by summing such terms.

It must also be noted that the values of the delays need to be positive. The zero-delay case would be equal to the instantaneous mixture case with the solution of Eq. (8.8).

8.4.2 Experiments with Delayed Speech Signals

We illustrate the behavior of the resulting adaptation rule using speech examples. In the first example, two sources are artificially mixed according to

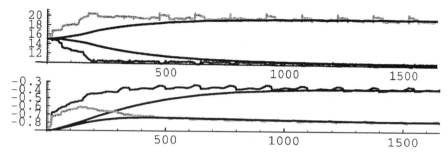

Figure 8.10 Convergence of the delays (top) and feedback weights (bottom). The horizontal axis is the number of iterations, where each iteration consists of accumulating the gradients from 250 samples of the mixtures and updating the weights and the delays. The vertical axis of the top graph is the amount of delay in samples. In the bottom graph the vertical axis is the value of the weight w_{ij}; d_{12} and w_{12} are plotted with a darker shade, d_{21} and w_{21} with a lighter shade. A (very) slow-moving time average of the parameters is also plotted in black. This is the same concept as the familiar "momentum" in backpropagation.

$$x_1(n) = s_1(n) + 0.4s_2(n - 10)$$
$$x_2(n) = s_2(n) + 0.8s_1(n - 20)$$

$$(8.31)$$

Figure 8.10 displays the convergence of the delays and the feedback weights. We can see that the delays converge fast, in less than 200 iterations. The weights start to converge toward the separating solution $(-0.4, -0.8)$ only after the delays have converged. About 400 iterations are needed to get them to converge. The learning rates were kept fixed in this experiment.

We do not need to assume that the sources are stationary, and thus we present the input samples consecutively to the adaptation algorithm. This is also computationally lighter than using random time instances of the input signals because delayed values of the outputs are needed in the adaptation. They can be just buffered as consecutive samples are processed.

A problem is caused by the local nature of the delay adaptation rule. Because speech is very periodic in nature, the rule tends to get trapped in aligning those parts of the waveforms that exhibit strong correlation due to the periodicity of the signals, and not necessarily due to the propagation delay. Figure 8.11 presents a study of the convergence paths of different initial delay values, in the case of two speech signals. The initial values were delays in a grid of 60×60 samples. Those initial values converging to the correct delays $(40, 40)$ are plotted in a darker color. We can clearly see "attractors," which are caused by the periodicity of the speech signal. An interesting thing to note is that the source signal s_1 is a male voice and s_2 female. The attractor points for d_{21} are thus more densely spaced than those for d_{12} because of the shorter pitch period of the female voice. The sampling frequency in this example was 16 kHz.

If the delays converge to a wrong attractor, then the weights obviously are

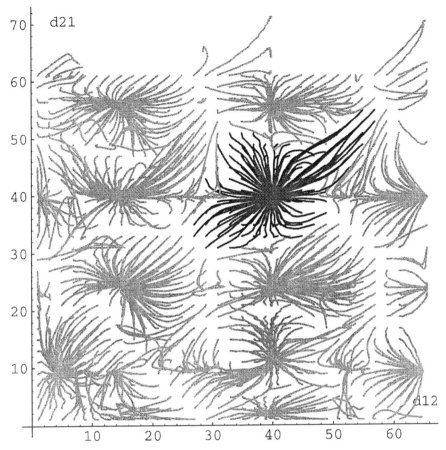

Figure 8.11 An illustration of the effect of the initial condition to delay convergence (see text). The correct delay values were (40, 40).

unable to converge to a separating solution. It is thus crucial that the initial delay values are approximately in the correct range. Here, we can make use of possible knowledge about the recording situation. Even without this knowledge, cross-correlations between the mixtures give good estimates for the initial values of the delays. Another possibility is to remove the disturbing time dependencies within each mixture separately (whitening), learn the delays and weights from the whitened mixtures, and then apply the delays and weights to the original mixtures.

In reality the delays will have a low probability of being exact multiplies of the sampling period. The proposed structure cannot tolerate fractional delays as such. However, it would not be too difficult to replace the integer delays d_{12} and d_{21} in Eq. (8.22) and in Fig. 8.9 by, for example, finite-duration impulse-response (FIR) filters approximating fractional delays. Since the resulting

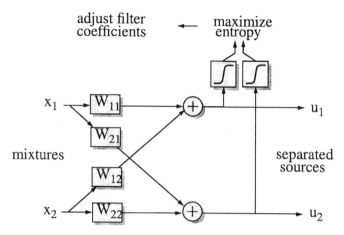

Figure 8.12 A feedforward network with adaptive filters for the separation of convolved mixtures.

structure would be close to the network for separating convolved mixtures, we defer the discussion of fractional delays till later in the section.

If the objective is just to identify the delays as accurately as possible, it is better to resort to global procedures instead of the local rules derived here. For example, Emile and Comon present effective methods based on high-order spectra (Emile and Comon 1998b), or on casting the problem as multichannel moving-average (MA) identification (Emile and Comon 1998a).

8.4.3 Mixtures of Convolved Sources

In this subsection we consider convolved mixtures. We first look at the feed-forward filter network depicted in Fig. 8.12 that is the combined deconvolution and separation network implied in Bell and Sejnowski (1995). The adaptation rules for the filter coefficients are easy to derive directly in the time domain, again by maximizing the entropy at the output.

Assuming causal FIR-filters for W_{ij}, the network carries out the following:

$$u_1(n) = \sum_{k=0}^{L_{11}} w_{1k1}x_1(n-k) + \sum_{k=0}^{L_{12}} w_{1k2}x_2(n-k)$$

$$u_2(n) = \sum_{k=0}^{L_{22}} w_{2k2}x_2(n-k) + \sum_{k=0}^{L_{21}} w_{2k1}x_1(n-k)$$

(8.32)

where w_{ikj} denotes the weight associated with delay k in the filter from mixture j to mixture i. With the exception of $D = w_{101}w_{202} - w_{102}w_{201}$, $\det(\mathbf{J})$ of the network is the same as in Eq. (8.23). Weights with zero delays will thus have different adaptation equations than other weights.

For w_{101} the partial derivatives are as follows:

$$\frac{\partial y_1'}{\partial w_{101}} = \frac{\partial y_1'}{\partial y_1} \frac{\partial y_1}{\partial u_1} \frac{\partial u_1}{\partial w_{101}} = \hat{y}_1 y_1' x_1$$

$$\frac{\partial y_2'}{\partial w_{101}} = \frac{\partial y_2'}{\partial y_2} \frac{\partial y_2}{\partial u_2} \frac{\partial u_2}{\partial w_{101}} = 0 \qquad (8.33)$$

$$\frac{\partial D}{\partial w_{101}} = w_{202}$$

Thus the adaptation rules for zero-delay weights become the following:

$$\Delta w_{101} \propto \hat{y}_1 x_1 + w_{202}/D$$

$$\Delta w_{102} \propto \hat{y}_1 x_2 - w_{201}/D$$

$$\Delta w_{201} \propto \hat{y}_2 x_1 - w_{102}/D \qquad (8.34)$$

$$\Delta w_{202} \propto \hat{y}_2 x_2 + w_{101}/D$$

In matrix form this is exactly the same as Eq. (8.8). The zero-delay weights thus follow the solution in Bell and Sejnowski (1995), as would be expected. For each of the other weights, only one term will be nonzero in the partial derivatives, and we arrive at the rule

$$\Delta w_{ikj} \propto \hat{y}_i x_j(n-k) \qquad (8.35)$$

which is a decorrelation rule when hyperbolic tangent or logistic function is used as the nonlinearity. Again, where time indices have been left out, n is assumed.

However, entropy maximization with this network does not result in the solution of Eq. (8.19). In the instantaneous mixture case the recovered sources can be determined up to permutation and scaling. With convolutive mixtures the indeterminacy is up to permutation and arbitrary filtering. Entropy maximization has the side effect of temporally whitening the outputs, which may not be desirable in some applications, such as in those involving speech. The effect is caused by $W_{11}(z)$ and $W_{22}(z)$, which tend to remove all time redundancies within each signal. A separating solution without the whitening effect exists, though. By rearranging Eq. (8.19) so that the direct filters $W_{11}(z)$ and $W_{22}(z)$ become unity, we get

$$G(z)A_{22}(z)^{-1}S_1(z) = X_1(z) - A_{12}(z)A_{22}(z)^{-1}X_2(z)$$

$$G(z)A_{11}(z)^{-1}S_2(z) = X_2(z) - A_{21}(z)A_{11}(z)^{-1}X_1(z) \qquad (8.36)$$

Unfortunately, after limiting W_{ii} to scalars, the cross-filters $W_{12}(z)$ and $W_{21}(z)$ will not converge to these solutions when the adaptation just discussed is used.

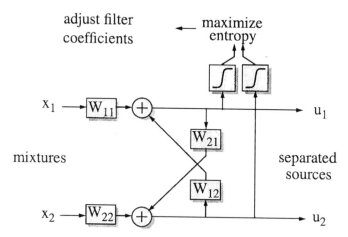

Figure 8.13 A feedback network with adaptive filters for the separation of convolved mixtures.

Rather, they will become mere temporal whitening filters. Obviously, in this case, temporal whitening will contribute more to the total entropy increase than mere separation.

The whitening effect can be avoided by using a different network architecture, namely the feedback structure depicted in Fig. 8.13. Again, maximizing the entropy at the output will result in W_{11} and W_{22} not only inverting A_{11} and A_{22}, but also whitening the sources. This can be avoided by forcing $W_{11}(z)$ and $W_{22}(z)$ to mere scaling coefficients. In the ideal case, W_{12} and W_{21} have the following solutions:

$$W_{11}(z) = 1 \qquad W_{12}(z) = -A_{12}(z)A_{22}(z)^{-1}$$
$$W_{22}(z) = 1 \qquad W_{21}(z) = -A_{21}(z)A_{11}(z)^{-1} \tag{8.37}$$

Why is whitening avoided in this configuration? Since the output of a cross-filter is summed to a *different* branch than where its input comes from, the filter has no chance of removing redundancies from its input signal. Instead, it will learn to remove the redundancies *between* the signals. Thus no deconvolution is performed at all, only separation. We must then be content with $U_1(z) = A_{11}(z)S_1(z)$ and $U_2(z) = A_{22}(z)S_2(z)$, which is what each sensor would observe in the absence of the interfering sources. The existence of this solution requires that $A_{11}(z)$ and $A_{22}(z)$ have stable and realizable inverses, in addition to $G(z) \neq 0$ in Eq. (8.19). If these are not fulfilled, the network is unable to achieve separation. We discuss this situation in Section 8.5. Derivation of the adaptation rules for this architecture[2] follows.

---

[2] It is also possible to switch the order of W_{ii} and W_{ij} filters, that is, place the W_{ii} *after* the summation points.

Assuming causal FIR-filters for W_{ij}, in the time domain the network carries out the following:

$$u_1(n) = \sum_{k=0}^{L_{11}} w_{1k1} x_1(n-k) + \sum_{k=1}^{L_{12}} w_{1k2} u_2(n-k)$$

$$u_2(n) = \sum_{k=0}^{L_{22}} w_{2k2} x_2(n-k) + \sum_{k=1}^{L_{21}} w_{2k1} u_1(n-k)$$

(8.38)

For the Jacobian, we get

$$\log \det(\mathbf{J}) = \log(y_1') + \log(y_2') + \log(D)$$
$$= \log(y_1') + \log(y_2') + \log(w_{101} w_{202})$$

(8.39)

There will now be three different cases: zero-delay weights in direct filters, other weights in direct filters, and weights in feedback cross-filters. Following the steps in previous derivations for all these cases, we get:

$$\Delta w_{i0i} \propto \hat{y}_i x_i(n) + 1/w_{i0i}$$

$$\Delta w_{iki} \propto \hat{y}_i x_i(n-k)$$

$$\Delta w_{ikj} \propto \hat{y}_i u_j(n-k)$$

(8.40)

The zero-delay weights again scale the data to maximize the information passed through the nonlinearity, other weights in the direct branches of the network decorrelate[3] each output from the corresponding input mixture (whitening), and the weights of the feedback branches decorrelate each output y_i from all the other sources u_j at every time instant within the scope of the filters $n-k$ (separation).

The derivation will be equally straightforward for infinite-duration impulse-response (IIR) filters, or filters of any structure. Figure 8.14 depicts a fragment of a feedback separation network that utilizes simple FIR–IIR cascades as the filters. In the time domain, the network is computing the following, where we have introduced intermediate variables x_i^{fb} and u_{ij}^{fb}:

$$x_i^{fb}(n) = \sum_{k=0} w_{ik} x_i(n-k) + \sum_{k=1} w_{ik}^{fb} x_i^{fb}(n-k)$$

(8.41)

$$u_{ij}^{fb}(n) = \sum_{k=0} w_{ikj} u_j(n-k) + \sum_{k=1} w_{ikj}^{fb} u_{ij}^{fb}(n-k)$$

(8.42)

$$u_i(n) = x_i^{fb}(n) + \sum_j u_{ij}^{fb}(n)$$

(8.43)

---

[3] When hyperbolic tangent or logistic functions are used as nonlinearities.

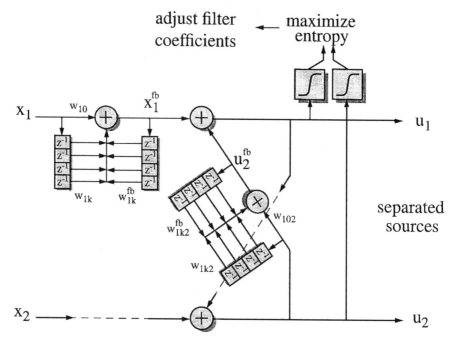

Figure 8.14 A feedback separation network using FIR–IIR cascades as the filters. Dotted lines represent similar filter structures not depicted for clarity. Superscript "fb" refers to feedback quantities.

Using similar steps as before, and by applying the chain rule to the partial derivatives, we arrive at the following adaptation equations:

$$\Delta w_{i0} \propto \hat{y}_i x_i(n) + 1/w_{i0}$$

$$\Delta w_{ik} \propto \hat{y}_i x_i(n - k)$$

$$\Delta w_{ik}^{fb} \propto \hat{y}_i x_i^{fb}(n - k) \tag{8.44}$$

$$\Delta w_{ikj} \propto \hat{y}_i u_{ij}(n - k)$$

$$\Delta w_{ikj}^{fb} \propto \hat{y}_i u_{ij}^{fb}(n - k)$$

If the IIR-filter part is omitted, this is equal to (8.40). Similarly, we can remove the FIR part and only use the IIR-filters. In the latter case, it is necessary to leave the w_{i0} and w_{i0j} as scaling coefficients.

8.4.4 Experiments with Mixtures of Convolved Speech Signals

In this subsection we present an example of separation of two speech signals mixed using the filters in Eq. (8.45). The filters were chosen so that the cor-

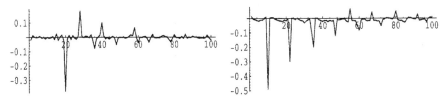

Figure 8.15 Coefficients of separating FIR-filters W_{12} (left) and W_{21} (right).

rectness of the solution would be easy to verify by eye:

$$A_{11}(z) = 1 - 0.4z^{-25} + 0.2z^{-45}$$

$$A_{12}(z) = 0.4z^{-20} - 0.2z^{-28} + 0.1z^{-36}$$

$$A_{21}(z) = 0.5z^{-10} + 0.3z^{-22} + 0.1z^{-34} \tag{8.45}$$

$$A_{22}(z) = 1 - 0.3z^{-20} + 0.2z^{-38}$$

The feedback architecture with FIR-filters was used for separation, and W_{11} and W_{22} were reduced to mere adaptive scaling coefficients to avoid the temporal whitening effect. Both W_{12} and W_{21} had 100 taps each. Figure 8.15 depicts the learned coefficients of each filter.

In this case, W_{12} and W_{21} should follow Eq. (8.37). For example, looking at W_{21}, it is easy to locate the three peaks corresponding to $-A_{21}$. The rest of the peaks correspond to the convolution of $-A_{21}$ with the $1/A_{11}$ (which has an infinite length as an FIR-filter, ideally). We should see the first of these peaks at delays 35 $(10 + 25)$, 47 $(22 + 25)$, 55 $(10 + 45)$, 59 $(34 + 25)$, and 67 $(22 + 45)$. This is the case indeed. The audible separation was nearly perfect.

In some cases the FIR-filters need to be very long to be able to model the inverses of the A_{ii}. Shorter IIR-filters suffice for the same effect. Figure 8.16 presents the coefficients of two IIR-filters of the direct form of length 60. The delay to the first nonzero tap was also learned by the delay adaptation rule, Eq. (8.30). This way we can avoid wasting resources to taps with zero coefficients in the beginning of the filter. The learning rate for the delays needs to be larger than the learning rate for the weights to make the delays "win" and converge first.

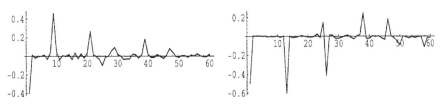

Figure 8.16 Coefficients of separating IIR-filters W_{12} (left) and W_{21} (right).

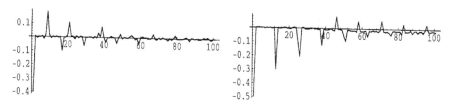

Figure 8.17 Impulse responses of the IIR-filters W_{12} (left) and W_{21} (right).

The impulse responses of the filters are depicted in Fig. 8.17 for up to 100 samples. These are identical to the FIR-filters in Fig. 8.14, which should be the case. Notice that in Fig. 8.17 the impulse responses begin at the first peaks of Fig. 8.15, because the initial delays were learned in this case, too.

In addition to artificial convolutive mixtures, we have also experimented with real-world recordings. For example, with two simultaneous speakers in a small hard-wall conference room using two omnidirectional microphones on a table in the middle of the room, we were able to achieve 12-dB attenuation of the unwanted source components. In this recording situation the microphones clearly had different sources as the strongest components of the mixtures.

No special treatment is required to accommodate fractional delays. The filters are free to also learn those dependencies that are related to filters implementing fractional delays. Figure 8.18 presents an example of a learned separating filter in the case where one source was delayed exactly 14 samples with respect to the other. The filter should have just one nonzero coefficient, which is the case. In Fig. 8.19 the delay was a fractional one, $14\frac{1}{2}$ samples. The learned filter is now exactly an FIR-filter approximating half a sample delay (Laakso et al. 1996).

The reader may have noticed that, for example, if the delays are short (say, 1 and $1\frac{1}{2}$ samples), the fractional delay approximation filters cannot be realized using the presented architecture, since the filters should be noncausal. This issue is examined in the next section.

Also, if there is an instantaneous component in the mixtures, that is, there is no delay between the source components in the mixtures, the presented structures can be cascaded with an instantaneous-mixture separation network or this

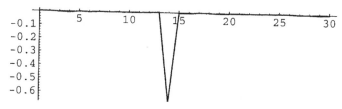

Figure 8.18 Learned filter in the case where a source was delayed 14 samples with respect to the other.

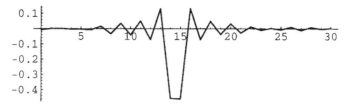

Figure 8.19 Learned filter in the case where a source was delayed 14.5 samples with respect to the other.

can be taken into account from the beginning as described in Cichocki et al. (1996) and Lee et al. (1997a).

8.5 NONMINIMUM-PHASE MIXING CONDITIONS

8.5.1 Restrictions of Causal Filters

In previous sections we have made the assumption that the filters W_{ij} in the networks are causal. This assumption was made due to the simplicity of the derivation of the adaptation equations. However, the assumption introduces restrictions on what kind of recording situations can be separated using this filter construct. For example, in the ideal solution of the feedback network as presented in Eq. (8.37), filter W_{ij} should be able to realize a convolution of a causal filter $(-A_{ij})$ with the inverse of another causal filter (A_{jj}^{-1}). As W_{ij} is a causal filter itself, it is clear that unless A_{jj}^{-1} can be realized as a causal FIR filter, the separation will fail. This restriction also applies to the feedforward network using causal filters. In this case the filters in Eq. (8.19) should be realizable as causal filter structures.

In general, only so-called minimum-phase filters have stable causal inverses.[4] All zeros of a minimum phase filter $F_m(z)$ lie inside the unit circle. Thus all poles of the inverse of such a filter $1/F_m(z)$ are inside the unit circle, too. This is the condition for a stable filter.

8.5.2 Acausal Filters

A filter that has no stable causal inverse can have a stable acausal inverse. A true-phase (or nonminimum phase) filter can be expressed as a product of a minimum-phase filter with an all-pass filter $F(z) = F_m(z)F_a(z)$, where $F_a(z)$ has a unity frequency response, but represents the delay of the nonminimum-phase filter. Thus its inverse. $(1/F_m(z))(1/F_a(z))$ is a product of the inverse of a

---

[4] The term "minimum phase" comes from the following property: Among filters having the same frequency response, the minimum-phase filter has the smallest phase lag.

minimum-phase filter, which is a stable causal filter, with the inverse of an all-pass filter representing time lag, which is time advance. The overall result is an acausal stable filter. See basic digital signal-processing textbooks, for example the one by Oppenheim and Schafer (1975), for details.

This case is illustrated in Fig. 8.20. A minimum-phase filter with transfer function $F(z) = -1 + z^{-1} - 0.5z^{-2} + 0.3z^{-3} - 0.1z^{-4}$ and its inverse are plotted in the top half of the illustration. Zeros of this filter are inside the unit circle. By introducing a one-tap delay after the first tap, we introduce a zero outside the unit circle resulting in a nonminimum-phase filter. This filter $F'(z) = -1 + z^{-1} - 0.5z^{-3} + 0.3z^{-4} - 0.1z^{-5}$ now has an acausal inverse, which is illustrated in the bottom half of the picture.

8.5.3 Implications for Blind Separation

In practice it is not possible to assume that recordings and rooms are so "clean" that the impulse responses that need to be inverted for successful blind separation are minimum phase. We must thus allow acausal filters in the networks to overcome the restrictions.

Acausal filters can be realized by introducing an appropriate delay to the adaptation, which needs to be at least the length of the acausal part of the filters. It now would be possible to derive the adaptation in the time domain following the approach and the steps in Section 8.4.3 (Guo et al. 1999). Instead, we describe briefly two more efficient methods.

8.5.4 BSS with Filters as Polynomials of Matrix Coefficients

Amari et al. (1997a,b) studied the feedforward filter network. They derived an efficient time-domain algorithm by representing the separating filters as a sequence of coefficient matrices $\mathbf{W}_p(n)$ at discrete time n and lag p. The separated output with this notation and causal filters is

$$\mathbf{u}(n) = \sum_{p=0}^{L} \mathbf{W}_p(n)\mathbf{x}(n - p) \tag{8.46}$$

This matrix notation allows the derivation of an equivariant algorithm using the natural-gradient approach. The resulting weight matrix update algorithm, which takes into account the causal approximation of a doubly infinite filter by delaying the output by L samples, is as follows:

$$\Delta \mathbf{W}_p(n) \propto \mathbf{W}_p(n) - g(\mathbf{u}(n - L))\mathbf{v}^\dagger(n - p), \qquad p = 0, \ldots, L \tag{8.47}$$

where $(\cdot)^\dagger$ denotes the complex-conjugate transpose. The nonlinear function g operates componentwise. It is derived from the pdf of the sources and optimally it equals $g_i(u_i) = -(d/du_i)\log f_{u_i}(u_i)$.

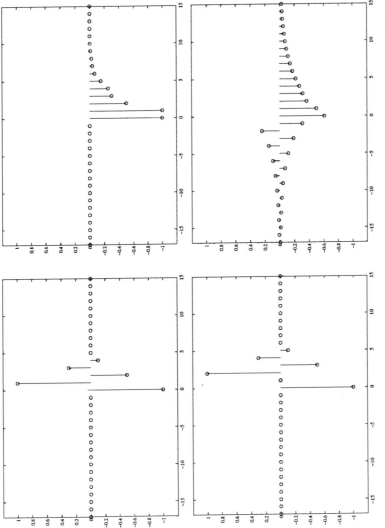

Figure 8.20 A causal FIR-filter (top left), which has a causal inverse (top right), thus being a minimum-phase filter. By a slight modification to the filter (bottom left), a zero is introduced outside the unit circle, which results in the inverse of the filter becoming acausal (bottom right). Horizontal axis to the right of zero represents time lag, and horizontal axis to the left represents time advance.

The $\mathbf{v}(n)$ is reverse-filtered output computed using the L latest samples backwards from the current sample:

$$\mathbf{v}(n) = \sum_{q=0}^{L} \mathbf{W}_{L-q}^{H}(n)\mathbf{u}(n-q) \tag{8.48}$$

The $\mathbf{v}$ needs to be stored for the latest L samples to compute $\Delta\mathbf{W}_p(n)$ for all lags $p = 0, \ldots, L$. The algorithm has thus quite modest computational and memory requirements. Note the similarity to Eq. (8.12) where the role of $\mathbf{u}(n)^T\mathbf{W}$ has now been taken by $\mathbf{v}(n)$. Note also that if $L = 0$, Eqs. (8.47) and (8.48) reduce to Eq. (8.12). Authors present examples of a multichannel equalization task with nonminimum phase channels (Amari et al. 1997a), and of a speech separation task, in which real impulse responses were measured in an anechoic chamber and the signals were then artificially mixed using these impulse responses as mixing filters (Amari et al. 1997b). To achieve 10-dB–15-dB enhancement, about 50 seconds of the mixtures were needed. In Section 8.10 we discuss what might be necessary in real-world applications.

8.5.5 BSS with Matrices of Filters

In the approach of Amari et al. described in the preceding subsection, filters in the z-transform domain are treated as polynomials with matrix coefficients for each time lag. Lambert proposed another approach that makes use of matrices of FIR filters instead (Lambert 1995, 1996). Lambert extended scalar matrix algebra to both matrices of FIR filters (time domain) and matrices of FIR polynomials (frequency domain), as discussed in the next chapter.

This extension is based on the following connection: The eigenvalues of a circulant matrix constructed from the coefficients of a filter h are the discrete Fourier transform coefficients, and the eigenvectors are the Fourier basis functions of a transform with a length of the side of this circulant matrix. Thus we can operate on FIR filters in the frequency domain, for example, applying a function g to a filter h as follows:

$$g(h) = \text{IFFT}(g(\text{FFT}([\mathbf{0}\ h\ \mathbf{0}]))), \tag{8.49}$$

where $\mathbf{0}$ denotes a vector of zeros; h is pre- and postpended with zeros to allow a sufficient accuracy in the double-sided series expansion in computing the effect of the function g on the filter; and FFT denotes fast Fourier transform. Prepending especially allows for expansion of noncausal filters. In fact, we used this method to compute the inverses of filters in Fig. 8.20. The inverse is simply

$$h^{-1} = \text{IFFT}(\mathbf{1}./(\text{FFT}([\mathbf{0}\ h\ \mathbf{0}]))), \tag{8.50}$$

where the "MATLAB"-like operator "./" denotes elementwise division. Convolutions become elementwise multiplications, deconvolutions become divisions. Lambert then goes further and presents other operations for FIR polynomial matrices, such as eigenvalue and eigenvector computation, and whitening (Lambert 1996; Lambert and Nikias 1995).

This methodology leads naturally to the frequency-domain formulation of the information maximization method for BSS, which inherently handles acausal filters in the separating network. In analogy to Eq. (8.12) the adaptation using FIR polynomial matrices can be written as follows (Lambert and Bell 1997; Lee et al. 1997a,b).

$$\Delta \underline{\mathbf{W}} \propto (\underline{\mathbf{I}} + \text{FFT}(\hat{\mathbf{y}})\underline{\mathbf{U}}^{\dagger})\underline{\mathbf{W}} \tag{8.51}$$

where the underline represents frequency-domain quantities having either matrices with FIR polynomials as elements ($\underline{\mathbf{W}}$), or transformed blocks of data as elements ($\underline{\mathbf{U}}$); the nonlinearity $\hat{y}$ operates in the time domain and FFT is applied to a block of such time-domain data; and $(\cdot)^{\dagger}$ denotes complex-conjugate transpose. Written out for a two-by-two feedforward separation network (see Fig. 8.12), the update is

$$\Delta \begin{bmatrix} W_{11} & W_{22} \\ W_{12} & W_{22} \end{bmatrix} \propto \left(\begin{bmatrix} 1 & 0 \\ 0 & 1 \end{bmatrix} - \begin{bmatrix} \text{FFT}(\hat{\mathbf{y}}_1) \\ \text{FFT}(\hat{\mathbf{y}}_2) \end{bmatrix} [\text{FFT}(\mathbf{u}_1) \ \text{FFT}(\mathbf{u}_2)]^* \right)$$
$$\times \begin{bmatrix} W_{11} & W_{22} \\ W_{12} & W_{22} \end{bmatrix} \tag{8.52}$$

where $(\cdot)^*$ denotes the complex conjugate. This kind of a block-frequency domain algorithm can be implemented using standard approaches in the signal-processing literature, for example, using the overlap-save method (Ferrara 1980; Oppenheim and Schafer 1975). Lambert and Bell (1997) and Lee et al. (1997a,b) report successful experiments with as long as 2048 tap filters using real recordings in nonminimum-phase mixing situations. Again, 10–15-dB improvement in separation is reported, which seems to be a standard in this kind of experiment.

8.6 OTHER APPROACHES: TAXONOMY BY PARAMETRIZATION

Any method for separation of convolutive mixtures can be roughly divided into three essential components: (1) parametrization of the separation system; (2) the separation criterion; and (3) the method to optimize the chosen criterion. In the following two sections we attempt to categorize other approaches to separation of convolutive mixtures (also including the presented ones), in light of two possible taxonomies based on the first two components. We concentrate on approaches applicable to acoustic signals.

We discuss only briefly the optimization methods. They can be roughly divided into adaptive and algebraic approaches. The former category can further be subdivided into stochastic-gradient-type algorithms (with or without second-order information, i.e., Newton's method), and into function zero search algorithms or fixed-point methods. The latter category mainly consists of methods to jointly and/or approximately diagonalize a number of matrices.

In this section we discuss what alternatives exist for the parametrization of the separating system, and possibly for the parametrization of the source signals if this is required by the method at hand.

8.6.1   Feedforward, Feedback?

Most real convolutive mixing scenarios with audio signals can be modeled as a feedforward mixing network having FIR filters in its branches. A room with multiple simultaneous sound sources and multiple microphones is an example, where the mixing filters are room impulse responses between each source and each microphone.

The separation system, too, which also ideally inverts the effect of the mixing system, can be modeled as a feedforward network of FIR-filters that approximate the required inverse system filters as in Eq. (8.19) and in Fig. 8.12. Depending on the learning algorithm, the sources will be distorted by filtering, since the convolutive separation problem has the undeterminacy up to arbitrary filtering of the sources. Some separation methods tend to produce temporally whitened outputs, as they aim at redundancy reduction, like the information maximization discussed in this chapter. An advantageous configuration from the viewpoint of parsimony would be to have the separated sources in Eq. (8.19) filtered/distorted by $G(z)$. The filters to be learned are the mixing filters without the $G(z)^{-1}$ factor, not the inverses of the system (the adjunct of the mixing, that is). Because inverses tend to have a larger number of parameters than the actual real impulse responses, this would be a simpler learning task. In fact, the inverses of real room impulse responses might not be realizable at all as stable and "sufficiently short" filters. However, there is no effective way of enforcing this solution in a blind fashion.

Another possible arrangement is the feedback architecture of Fig. 8.13. Note that in Eq. (8.37) the inverses of the *direct* mixing filters are required. If these are minimum-phase filters, they have stable causal inverses. That is, if the direct paths are "good," a feedback network with causal filters is able to invert the mixing. Otherwise, one must use the feedforward network with acausal filters or implement acausal filters within the feedback network, as described in (Guo et al. 1999).

8.6.2   Frequency Domain

The parameters to be learned above are the time-domain coefficients of the filters. However, in the audio case the filters may need to be thousands of taps long to properly invert the mixing. Computationally it may be lighter to move to the frequency domain, as convolutions with long filters in the time domain become efficient multiplications in the frequency domain under certain conditions (Murata et al. 1998; Parra and Spence 1998a). Now there are two avenues to take. In the first one everything including the actual separation could be done in the frequency domain. This has the great advantage of decomposing a convolutive mixing problem into multiple instantaneous mixing problems (i.e., ICA), that can be solved using any desired method. The downside is that now the standard ICA indeterminacy of scaling and permutation appears at each output frequency bin! Reconstruction of the time-domain output signal requires all frequency components of the same source. Various methods to overcome the scaling and permutation problem using different continuity criteria are presented in Soon et al. (1992) and Capdevielle et al. (1995), Murata et al. (1998), Parra and Spence (1998a, 1999), Schobben and Sommen (1998a), Serviere (1998), Serviere and Capdevielle (1996), Smaragdis (1998), and Wu and Principe (1999).

The second avenue is that the actual separation is not done in the frequency domain, but only a single aspect or some aspects of the separation algorithm. The rest is done in the time domain. Filters may be easier to learn in the frequency domain, as components are now orthogonal and not dependent on each other like the time-domain coefficients (Back and Tsoi 1994; Schobben and Sommen 1998a). Examples of methods that apply their separation criterion (independence, higher-order statistics, nonlinearities) in the time domain but do the rest in the frequency domain are reported in Back and Tsoi (1994), Lambert (1995), Lambert and Bell (1997), Lee et al. (1997b,c). A frequency-domain representation of the filters is learned, and they are also applied in the frequency domain. The final time-domain result is reconstructed using, for example, the overlap-save technique. Thus, the permutation and scaling problem does not exist. This was also the case discussed in Section 8.5.5. An example of learning a filter of matrices in time domain, when the criterion is in frequency domain is presented in Matsuoka and Kawamoto (1995). Back and Tsoi (1994) also discuss an example of the Herault-Jutten algorithm in the frequency domain, whereas the nonlinear functions are applied in the time domain.

Another type of parametrization is presented in Soon et al. (1993). The source location is parametrized, and frequency bin information after separation is clustered to produce consistent source-location estimates. Alternative discrete-time operators are considered in Back and Cichocki (1997) for fast sampled signals.

8.6.3 Decomposition

Rather than trying to learn these possibly huge filters all at once, it is possible to decompose the problem, as discussed, for example, in Ohno and Inouye (1998) and Yellin and Friedlander (1996). At this point it is useful to elaborate on the relation between ICA and blind separation of convolutive mixtures (BSCM).

ICA (in the separation context) makes use of *spatial* statistics of the mixture signals to learn a *spatial* separation system. In general ICA needs to use higher than second-order spatial statistics to succeed. If the sources can be assumed to be nonwhite or nonstationary, however, second-order spatiotemporal statistics is sufficient, as shown in Tong et al. (1992). In contrast, BSCM needs to make use of *spatiotemporal* statistics of the mixture signals to learn a *spatiotemporal* separation system. Stationarity of the sources is decisive for BSCM. If the sources are not stationary, only second-order spatiotemporal statistics is enough, as briefly discussed in Weinstein et al. (1993) and later, for example, in Murata et al. (1998) and Parra and Spence (1998a). Stationary sources again require higher than second-order statistics, but in the following fashion. Spatiotemporal second-order statistics can be made use of to decorrelate the mixtures. This step returns the problem to that of conventional ICA, which again requires higher-order spatial statistics. Examples of these approaches are given in Delfosse and Loubaton (1996), Gorokhov et al. (1996), Icart and Gautier (1996) with linear prediction based methods, in Choi and Cichocki (1997b) with an adaptive approach, and in Li and Sejnowski (1995) with a beam-forming approach.

Alternatively, for sources that cannot be assumed nonstationary one can resort to higher-order spatiotemporal statistics from the beginning, as done in numerous papers. We return to this subject in the following section.

Another way to decompose the problem is presented in Ngo and Bhadkam-kar (1999). In this case, the microphone arrangement is rather peculiar, a compact microphone array: two omnidirectional microphones 1 cm apart in a reverberant environment. Now, for each source, the transfer functions differ predominantly by a small delay. Mixtures are modeled as $\mathbf{x} = \mathbf{DR}\mathbf{s}$, where $\mathbf{D}$ is a matrix of band-limited approximations of delta functions at some delays, and $\mathbf{R}$ represents all other acoustic effects. The first stage estimates delays $\hat{\mathbf{D}}$

$$\mathbf{u}_1 = \mathrm{adj}(\hat{\mathbf{D}})\mathbf{x} = \mathrm{adj}(\hat{\mathbf{D}})\mathbf{DR}\mathbf{s} \equiv \mathbf{H}\mathbf{s} \qquad (8.53)$$

The second stage cancels off-diagonal elements of $\mathbf{H}$ (small relative to diagonal elements) using a feedback configuration

$$\mathbf{u}_2 = \mathbf{u}_1 - \mathbf{M}\mathbf{u}_2 = (\mathbf{M} + \mathbf{I})^{-1}\mathbf{H}\mathbf{s} \qquad (8.54)$$

Now $\mathbf{H}$ is easier to estimate than $\mathbf{DR}$ directly, since its off-diagonal elements are small: initial guess $\mathbf{M} = \mathbf{0}$ is likely to lie near a global optimum. Simple decorrelation adaptation is then sufficient to learn $\mathbf{M}$.

8.6.4   Other Parametrizations

Some separation methods require assumptions about the sources (see Section 8.7.1) and parametrize some aspects of them, such as the densities, some parameters thereof (Attias and Schreiner 1998; Girolami 1998; Moulines et al.

1997), or some parameters related to temporal statistics of the sources, such as auto-regressive (AR) parameters (Lee et al. 1997c, Parra and Spence 1998b; Parra et al. 1997).

8.7 OTHER APPROACHES: TAXONOMY BY SEPARATION CRITERIA

The actual separation criterion can be quite independent of the chosen parametrization. In this section we go through most of the criteria that have appeared in the literature, mostly in the context of separation of audio signals.

We have already briefly discussed possible criteria for the instantaneous (or spatial) separation case in Section 8.2.2. These criteria, which mostly make use of the marginal densities of the sources, are also applicable as such to the separation of convolutive mixtures. The criteria can also be applied to the joint densities of the sources over time, with minor modifications. We discuss these criteria briefly in the context of convolutive mixtures, and also elaborate on other criteria that make use of the spatiotemporal cross statistics of the signals.

8.7.1 Criteria Based on Using the Source Densities

Minimizing Mutual Information As discussed in Section 8.2.2, minimizing the mutual information of output components, Eq. (8.6), results in their statistical independence. Equation (8.6) can be written as

$$D(f_u(\mathbf{u}), \prod_i f_{u_i}(u_i)) = \int f_u(\mathbf{u}) \log \frac{f_u(\mathbf{u})}{\prod_i f_{u_i}(u_i)} \, du,$$

$$= -H(\mathbf{u}; W) + \sum_i H(u_i; W) \qquad (8.55)$$

where W represents the parametrization of the system. Thus the mutual information between the output components is equal to the joint entropy of output minus the sum of entropies of individual components. Minimizing this criterion requires entropies of individual components, which are not available, but can be estimated by means of statistical expansions. Since this criterion is not simple to use directly, it has only been used in the convolutive mixture context in a few papers, (e.g., Choi and Cichocki 1997a; Cichocki et al. 1996; Matsuoka and Kawamoto 1995).

Infomax, or Entropy Maximization Now consider passing the outputs through bounded nonlinear functions that approximate the cumulative densities of the sources $\mathbf{y} = g(\mathbf{Wx})$. If outputs $\mathbf{u}$ are the separated true sources, the $\mathbf{y}$ have density close to a uniform density, the density that has the largest entropy among bounded distributions. Maximizing the entropy of $\mathbf{y}$ is thus equivalent to producing the true sources. This criterion makes use of the source densities implicitly.

The criterion was applied to delayed and convolutive mixing in Torkkola (1996a,b), as described in this chapter, and further developed in Amari et al. (1997a,b), Cichocki et al. (1996), Douglas et al. (1997), Guo et al. (1999), Lee et al. (1997a), and Xi and Reilly (1997). Fully or partially frequency domain approaches were developed in Kohler et al. (1997), Lee et al. (1997b,c), Smaragdis (1997a,b, 1998).

Latent-Source Models and Maximum-Likelihood Estimation   Source densities can be made use of explicitly, too. If they are known, or if their shapes are known less, perhaps some parameters that need to be estimated while the separating parameter set **W** is learned, the divergence between the marginal output density and the known source densities can be used as the criterion to be minimized, that is, find **W** that drives the outputs toward known distributions. This approach lends itself naturally to the maximum-likelihood estimation (Belouchrani and Cardoso 1994; Cardoso and Amari 1997; Pham et al. 1992). With convolutive mixing this approach has been taken in (Douglas and Haykin 1997; Girolami 1998; Moulines et al. 1997; Parra and Spence 1998b; Parra et al. 1997).

Even if the densities are unknown we can estimate them as sums of some simple, for example, Gaussian or logistic distributions while the separation matrix is being estimated. Either gradient ascent (Pearlmutter and Parra 1996) or expectation maximization, in some cases (Belouchrani and Cardoso 1995; Moulines et al. 1997), can then be used. The interpretation of sources as latent signals whose convolutive mixture is the only observable has been made in (Attias and Schreiner 1997, 1998).

Central Limit Theorem   Direct application of the Central Limit theorem has not been common in the convolutive mixture case, except in the form of the Bussgang approach discussed in the next paragraph. The reason for this is perhaps that rather than maximizing the divergence of the marginal densities to Gaussians, or rather than maximizing/minimizing the kurtosis or other higher-order statistics (HOS) of the marginal densities, it is more effective to make use of the independence criterion by directly minimizing some cross statistics of the signals, as discussed later.

8.7.2 Bussgang Approach

Bussgang methods have been used as tools for blind deconvolution, where the true source s is estimated from a signal corrupted by convolutional noise by a nonlinear function that optimally equals $g(s) = (\partial f_s(s)/\partial s)/f_s(s)$. Equalization algorithms are derived by finding a filter that minimizes the difference between the output and the true source estimated through g. LMS is typically used as the criterion. Extensions to multichannel blind deconvolution and separation were presented in (Lambert 1995, 1996; Lambert and Bell 1997). Coupled with

FIR-matrix algebra (Lambert 1996; Lambert and Nikias 1995), efficient separation methods seem to result. These approaches are discussed in the next chapter. See also Westner (1999) and Westner and Bove (1999) for application of these methods for the overdetermined mixing case. It is notable that the nonlinearity has the same exact form as in the entropy maximization and maximum-likelihood approaches, leading to similar separation algorithms, as pointed out in Lee et al. (1999).

8.7.3 Spatiotemporal Decorrelation

Let us look at spatial covariance in instantaneous separation

$$\mathbf{R}_x(n) = \mathbf{A}\mathbf{R}_s(n)\mathbf{A}^T + \mathbf{R}_v(n), \quad \mathbf{x}(n) = \mathbf{A}\mathbf{s}(n) + \mathbf{v}(n) \tag{8.56}$$

where $\mathbf{v}$ denotes additive independent noise. If $\mathbf{s}$ is nonstationary, that is, $\mathbf{R}_s(n) \neq \mathbf{R}_s(n + \tau)$, multiple conditions for different choices of τ can be written to solve for $\mathbf{A}$, $\mathbf{R}_s(n)$, and $\mathbf{R}_v(n)$, where the covariances are diagonal matrices.

We can also look at cross covariances over time $\mathbf{R}_x(n, n + \tau) = E\{\mathbf{x}(n)\mathbf{x}(n + \tau)^T\}$. This approach was mentioned in the context of convolutive mixtures in Weinstein et al. (1993) and utilized with instantaneous mixtures in Matsuoka et al. (1995) and Molgedey and Schuster (1994). For the convolutive case we can write in frequency domain for sample averages (Parra and Spence 1998a, 1999)

$$\bar{\mathbf{R}}_x(\omega, n) = \mathbf{A}(\omega)\mathbf{R}_s(\omega, n)\mathbf{A}^H(\omega) + \mathbf{R}_v(\omega, n) \tag{8.57}$$

Again, if $\mathbf{s}$ is nonstationary, we can write multiple linearly independent equations for different time lags and solve for unknowns, or find LMS estimates of them by diagonalizing a number of matrices in the frequency domain (Ehlers and Schuster 1997; Mejuto and Principe 1999; Murata et al. 1998; Parra and Spence 1998a, 1999; Wu and Principe 1997, 1999).

For minimum-phase mixing decorrelation only can provide a unique solution without having to make use of the nonstationarity (Broman et al. 1995; Lindgren and Broman 1996b; Lindgren et al. 1996; Simon et al. 1998). There is a multitude of adaptive approaches (Chan et al. 1996; Douglas and Cichocki 1997; Lindgren et al. 1995, 1996; Sahlin and Broman 1997, 1999; Schobben and Sommen 1998a,b; Van Compernolle and Van Gerven 1992; Van Gerven and Van Compernolle 1992, 1995; Weinstein et al. 1993; Yen and Zhao 1996, 1998), a few algebraic (Lindgren and van der Veen 1996), and many that are derived from anti-Hebbian learning rule considerations (Choi and Cichocki 1997b; Girolami 1998; Girolami and Fyfe 1996; Matsuoka et al. 1995; Principe et al. 1996). However, the fact that practical room responses are not often minimum phase might render many of these methods useless.

8.7.4 Minimization of Cross-HOS

As mentioned in Section 8.6.3, it is possible to construct the criteria for separation from high-order statistics, but by doing it spatiotemporally. The actual criterion is of course independence, which warrants that all higher-order cross statistics between sources vanish. An objective function can be constructed, for example, from fourth-order cross-cumulants, and this can be minimized using stochastic-gradient descent on the filters (Cruces and Castedo 1998; Engebretson 1993; Nájar et al. 1994; Tugnait 1998; Yellin and Weinstein 1996). Alternatively contrast functions can be used as in Comon (1996), Comon and Moreau (1997), and Moreau and Thirion (1996). On the other hand, a simple algorithm can be constructed that aims at just canceling the criteria (Thi and Jutten 1995). HOS can also be used implicitly through nonlinear functions. Examples and analysis are given in Back and Tsoi (1994), Deville and Charkani (1997), Dinc and Bar-Ness (1993), Jutten et al. (1992), Thi and Jutten (1995), and Thi et al. (1992). Platt and Faggin (1992) first derived similar rules employing nonlinear functions, but from the minimum-output power principle. Approaches that utilize HOS in the frequency domain are described in Capdevielle et al. (1995), Serviere (1996), Serviere and Capdevielle (1996), and Shamsunder and Giannakis (1997). An algebraic approach is presented in Tong (1996).

In general, estimation of HOS is more sensitive to noise and outliers than that of second-order statistics. Based on this argument, it would thus seem to be more robust to work with second-order statistics as much as possible.

8.8 APPLICATION AREAS OF BSS FOR CONVOLUTIVE MIXTURES

Since BSS is a relatively new discipline, the contents of this section are speculative. Because the applications of BSS for instantaneous mixtures (i.e., ICA) are covered elsewhere in this volume (see, for example, Chapter 6), we concentrate on two application areas: audio and radio communications.

8.8.1 Audio Applications

BSS seemingly has a large number of potential applications in the audio realm. The generic application is, of course, separation of simultaneous audio sources in a reverberating or echoing environment, that is, in a natural environment, for example, inside a room. We enumerate here only a few actual applications.

A very desirable application area would be signal enhancement by removing noise or other unwanted signal components using blind separation methods as in Girolami (1998) and Shields et al. (1997), for example. In this area only one signal is of interest, the rest are considered a nuisance. Enhancement of voice quality in cell phones would be one important application. Especially interesting is to separate speech from car noise. Since voice coders used in cell phones are optimized for coding speech alone, the combination of excessive noise with a speech signal results in poor sound quality. Some initial experiments in this area can be found in Sahlin and Broman (1997). One quickly realizes, however, that there are multiple noise sources in a car, in fact, an infinite number, since the whole car interior is vibrating and acts as a delocalized noise source. A simple two-sensor speech–noise separation system might not work in this case.

Making voice dialing or speech recognition in general more viable in noisy environments would fall into the same category (Lee et al. 1997b; Yen and Zhao 1996). Spying or forensic applications also fall under the same category, wherein the interest might be in picking up one important signal among others (Pan et al. 1998; Prosoniq 1998).

In audio communications transparency refers to reproduced audio being ideally free from reverberation, noise, acoustical echoes, and other mixed speakers (Schobben and Sommen 1998b). Teleconferencing and speakerphones are two areas where speech-signal aquisition with transparency is desirable. Combining existing multichannel acoustical-echo cancellation technology with BSS has been shown to be useful in a teleconferencing setup (Schobben and Sommen 1998b). Speech-signal aquisition with transparency is also very important for hearing aids, which form another lucrative application area for speech enhancement through BSS.

Passive sonar is an application area that closely resembles audio. The differences are that in passive sonar there is a large number of microphones, the distances between the sources and the microphones are much greater than in audio, but so is the speed of sound in water.

Whether some of the preceding can yet be a viable and profitabe application area is an open question. Current limitations in the methods might render some applications impractical, if not impossible. We discuss these issues in Section 8.10.

8.8.2 Radio Frequency Applications

Besides audio, an extremely fruitful application arena is digital communications. While the basic concept remains the same (multiple transmitters at the same frequency, multiple antennas receiving multiple mixtures, for example), there are a few important differences. Signals in this context are man-made, and thus their properties are completely known in advance. This can (and should be) be exploited in devising separation methods. Another difference is that signals could be transmitted in short bursts, which might call for block-based algebraic methods rather than adaptive methods (van der Veen 1998).

Radio Frequency (RF) applications are surprisingly similar to audio applications in that the wavelengths are approximately the same. For example, an RF signal at 1-GHz frequency has a wavelength of about 0.3 meters, which is the same wavelength as a 1-kHz audio signal has propagating in the air!

However, the delay spreads encountered in cellular communications are usually shorter in terms of filter taps. In audio, typically hundreds of taps are needed, whereas an order of magnitude fewer suffice for RF applications.

In cellular communications the operators try to maximize the usage of the slice of the frequency spectrum allocated for this purpose. Methods that aim at packing multiple simultaneous users into a given bandwith are called *multiple access* methods. Frequency-division multiple access (FDMA) allocates a separate (narrow-band) channel to each simultaneous user. In addition, time-division multiple access alternates the users in a single channel by rapid bursts. In code-division multiple access (CDMA) every simultaneous user utilizes the whole bandwidth. The signals can separated by allocating each user a different code for spreading the spectrum of the transmitted signal. Other simultaneous users using orthogonal codes appear as additive noise.

In spatial-division multiple access (SDMA) the purpose is to separate radio signals of interfering users (either intentional or accidental) from each other on the basis of the spatial characteristics of the signals using smart antennas, array processing, and beam forming (Paulraj and Papadias 1997; van der Veen and Paulraj 1996). SDMA can in principle be combined with the other multiple-access methods. Supervised SDMA methods typically use a variant of LMS, either gradient based or algebraic, to adapt the coefficients that describe the channels or their inverses. This is usually a robust way of estimating the channel, but a part of the signal is wasted as predetermined training data, and the methods might not be fast enough for rapidly varying fading channels.

Unsupervised methods either rely on information about the antenna array manifold, or properties of the signals. Earlier approaches might require calibrated antenna arrays or special array geometries, and they may rely on first estimating the directions of the sources. Less restrictive methods use signal properties only, such as constant modulus, finite alphabet, spectral self-coherence, or cyclostationarity. BSS techniques fall in the last category, since they typically rely only on source signal independence and non-Gaussianity assumptions, as discussed in this chapter.

Two straightforward application areas can be envisioned in the cellular communications realm (Torkkola 1997b):

1. BSS could be used as an enabling technology to increase the capacity of a cellular system in the uplink.
2. BSS could be the basis for interference rejection.

In the first one the aim is to separate simultaneous radio signals occupying the same frequency band, possibly, radio signals that carry digital information. Since linear mixtures of antenna signals end up being linear mixtures of (complex) baseband signals due to the linearity of the downconversion process, it is possible to apply BSS at the baseband stage of the receiver.

In digital communications the binary (or M-ary) information is transmitted as discrete combinations of the amplitude and/or the phase of the carrier signal. After downconversion to baseband the instantaneous amplitude of the carrier can be observed as the length of a complex-valued sample of the baseband signal, and the phase of the carrier is discernible as the phase angle of the same sample. Possible combinations that depend on the modulation method employed are called symbol constellations. M-QAM (quadrature amplitude modulation) utilizes both the amplitude and the phase, whereby the baseband signals can only take one of M possible locations on a grid on the complex plane. In M-PSK (phase-shift keying) the amplitude of the baseband signal stays constant, but the phase can take any of M discrete values. In differential phase-shift keying (DPSK) the information is encoded as the difference between phases of two consecutive transmitted symbols. The phase can thus take any value, and since the amplitude remains constant, the baseband signal distribution is a circle on the complex plane.

Of particular interest is the fact that in communications the signals are artificial; thus the properties of the signals are known exactly. Remember that many BSS methods require some knowledge of the pdf's of the signals. Now the pdf's are predetermined, they need not be estimated, and they can be exploited in BSS resulting in algorithms that converge faster and produce more accurate results than algorithms without this knowledge (Belouchrani and Cardoso 1994; Jutten and Cardoso 1995; Torkkola 1998; van der Veen 1998).

One of the basic signal-processing problems in communications is equalization, which can also be called *deconvolution*. This denotes cleaning the effects of the channel—multipath, intersymbol interference—out of the received signal. This can be done with the aid of a known training symbol sequence, or, which is of interest in this volume, blindly. There are multiple methods in the literature, some also described in this volume (see Chapter 2 in Volume II of this work.). A categorization that is of interest from the viewpoint of BSS is the following. In the basic equalization problem a single source is observed through a single sensor (SISO: single input–single output). Better results can be obtained by using multiple sensor outputs (SIMO: single input–multiple output). Now we step into the realm of array processing. When there are multiple sources observed through multiple sensors (MIMO: multiple input–multiple output), we have exactly the setup of BSS with convolutive mixtures. Cross-fertilizing the equalization and separation disciplines would thus seem beneficial. For example, Douglas and Haykin (1997) and Lambert (1996) take the viewpoint that equalization is single-channel separation and separation is multichannel equalization, and try thus to unify both approaches.

If signal bandwidths are sufficiently narrow compared to the inverse of the delay spread, then equalization is not needed. Instantaneous component separation is enough in this case (Torkkola 1997b). A more common case is wideband, which leads to convolved mixtures. However, using orthogonal frequency division multiplexing (OFDM) as the transmission technique transforms a wideband separation problem into multiple narrow-band separation tasks (Feng and Kammeyer 1999), very much like the frequency domain approaches we discussed in Section 8.6.

Some problems in this application area, many yet unsolved, are listed below. BSS under noisy conditions is not yet sufficiently well studied. Since BSS concentrates on independent sources, it minimizes signal to interfering signal ratio

(S/I), not signal to interference plus noise ratio $(S/(I + N))$. Other, more remote users appear as noise, and they also need to be separated out from the signals of interest.

In TDMA-like systems, where the data are transmitted as short bursts, fast adaptation is required not to miss any of the signals from the beginning of the transmission. Nonstationary channels are a problem in cases when the symbol rate is slow compared to channel changes, such as fading. The BSS algorithm needs to be able to track the changes while they are happening so as not to lose any part of the signal (Torkkola 1997b).

In the second application area, instead of separating multiple equally important user signals, there may just be one signal of interest, which needs to be enhanced relative to interfering signals. This case could be called C/I ratio improvement.

The basic setup of BSS is that there are multiple independent signals mixed at some sensors. This is exactly the case in the uplink direction (user to base station). The base station should have multiple antennas or multisensor arrays for receiving the signals. However, selective spatial transmission to a single user still sharing only the frequency with other users must be handled by another means. This reduces the attractiveness of BSS as a cellular-communications solution. Capacity increase in only one direction is not enough in bidirectional communications. BSS in the uplink could be combined with some other means for downlink, such as with directional transmission, by making use of beam forming, or blind prefiltering, as described in Douglas (1998).

Many standards already contain training symbols for channel equalization purposes. Purely blind methods are thus not needed with these standards. Semiblind approaches, which make use of those training symbols that are available, and in addition, some statistical properties of the signals as described in Section 8.7 make more sense (de Carvalho and Slock 1997).

8.9 BLIND DECONVOLUTION IN AUDIO APPLICATIONS

In this section we discuss the possible uses of just blind deconvolution in audio applications, without considering the separation aspect.

The common element in applicable problem areas is that the signal gets modified by the environment. For example, with a speakerphone the outgoing signal is distorted by the acoustic environment the speaker is in. A hard-wall environment creates echoes and reverberations that garble the outgoing signal.

This problem is not to be confused with the standard echo cancellation problem of speakerphones. The signal of the far-end user that the local user hears from the speaker is also distorted by the same acoustic environment and transmitted back to the far-end phone. This component needs to be suppressed from the outgoing signal to prevent the far-end user hearing his or her own voice as reflected from the local user's environment. Fortunately the undistorted signal component is available, namely, the far-end user's signal arriving at the local user's set. Supervised methods, for example, LMS, can be used to adaptively learn a filter that cancels the echo from the outgoing signal (Sondhi 1967; Widrow and Stearns 1985). In stereo teleconferencing the situation is more complicated, as there are at least two speakers and two microphones (Sondhi et al. 1995).

Unfortunately there is no such reference signal available for the outgoing signal. All that is available is the single distorted signal at the microphone in the speakerphone. This signal is the convolution of the clean signal with the impulse response of the environment.

At first sight blind deconvolution seems to be a good match to this problem. However, an assumption usually made with blind deconvolution either implicitly or explicitly is that the samples of the original signal are independent. Why is this assumption made? Blind deconvolution removes all dependencies from the signal if other assumptions are also fulfilled. Thus if the signal of interest (speech) is already a convolution of a source (vocal cords) with another impulse response (vocal tract), it is impossible to differentiate between the two, and as a result the effect of both convolutions will be removed. Blind convolution would thus produce temporally whitened speech as the output. Some postfiltering would be necessary to restore the spectrum of the original signal.

In certain conditions this problem can be sidestepped as described in Torkkola (1997a). In this scheme the short-time dependencies in the speech signals will be first removed by a whitening filter with a short time span (say, 2–3 milliseconds). This whitener only removes the inherent dependencies in the speech signal, leaving echoes with longer delays intact. Now, blind deconvolution can be applied to learn the echo-removal filter from the whitened signal. Finally, the learned filter will be applied to the original speech signal, which contains both the inherent short-time dependencies and the echo-related dependencies with longer delays. The effect is to remove only the echoes, leaving the speech signal otherwise intact, which is the desired result.

For this to work the microphone needs to be placed so that the first echo arrives no earlier than 2–3 milliseconds from the signal's arrival. Moreover, the echo-cancellation filter learned by blind deconvolution, which is the inverse of the echo-generation process, should not overlap the time span of the whitener so that these two processes (vocal tract–environment) can be kept temporally separate. If this is not true, for example, if the echo canceller needs to be acausal, the approach does not work, and the postfiltering approach needs to be taken to restore the signal.

8.10 DISCUSSION AND CONCLUSIONS

We have discussed architectures and algorithms for blind source separation in the context of convolved mixtures that arises, for example, in audio recordings due to the sound reflections in acoustic environments.

We have derived solutions for the blind separation of this kind of mixture based on the information-maximization principle, and discussed other possible approaches that are found in the literature. We now briefly discuss what might still be needed to bring the technology closer to real-world applications.

1. *Are All Assumptions Really True?* Every method presented in the literature makes some explicit or implicit assumptions about the signals, ambient noise, or the environment. With practical problems it is important to verify that each of them holds true. Assumptions made about the environment might not match the structure of filters designed to invert the mixing. For example, even though we believed that we had a minimum-phase mixing system, the assumption might really not be true in practice (each speaker was closest to his or her own microphone), and needs to be measured. On the same note, it is of course implicitly assumed that the whole mixing system is invertible and realizable within the chosen parametrization. With real room responses, this might not be true. Some methods assume stationarity of the sources, some do not. Speech is certainly nonstationary in a larger time scale, but quasi stationary in a short time scale. Derivation of some methods might expect doubly infinite filters, which in reality can only be approximated. This approximation might cause local minima in optimizing the chosen criterion (Simon et al. 1998).

2. *Are There Always (Much) More Sources Than Sensors?* The number of sources must generally be matched to the number of sensors. We need at least as many microphone signals as there are sources to separate for the basic algorithms. But what if there are (and there will be!) more sources than microphones? How do the algorithms perform in this "noise"? This subject definitely needs further studies, as an additional noise source can render the mixing situation noninvertible.

3. *Is the Reality too Dynamic?* In general, the reported performance figures for static sources (loudspeakers) are higher than figures for real people, even in seemingly static positions. The effect of a speaker turning her head 10–20 degrees, or leaning backward a couple of inches, can have a drastic effect to the impulse response between the speaker and the microphones. These effects have been studied in acoustics, and further cross-fertilization between the fields would seem to be necessary to establish bounds to what performance could be expected from BSS in situations involving live speakers. Brandstein (1998) simulates these effects, and concludes that "any system which attempts to estimate the reverberation effects and apply some means of inverse filtering would have to be adaptable on almost a frame-by-frame basis to be effective." This is quite a pessimistic view of applications that involve on-line adaptation to dynamic situations.

4. *Are There Too Many Parameters?* Another open question not much studied in the literature is: What is the amount of data needed by the adaptation algorithm for successful signal separation? This is obviously a function of the number of parameters in the separation system, which is a function of the "hardness" of the separation problem. Filters with thousands of taps in the

time domain need tens (or hundreds) of thousands of samples to converge. That might be too much for the application.

For a minimum-phase mixing problem two seconds of speech for adaptation is reported to achieve a 12-dB signal to interference improvement in a real room recording with real people talking simultaneously (Torkkola 1996a). For a nonminimum-phase mixing using impulse responses measured in an anechoic chamber, 50 seconds of the mixture signals is reported to result in 10–15-dB enhancement in a three-signal case (Amari et al. 1997a). In a realistic case, two real people talking simultaneously in a real room, 30 seconds is reported to result in 12–15-dB enhancement using separation filters as long as 2048 taps (Lambert and Bell 1997). In another experiment, 10 seconds is reported to suffice for 1024-tap separation filters (Lee et al. 1997b).

These numbers suggest that with the current methods it might not be possible to have enough data to adapt to in real dynamical systems involving real people moving around, or even moving their heads while talking. However, it is not clear how dynamical real situations are. Further studies are needed to understand if moving sources can be tracked and adapted to, and whether on–off sources can be adapted to fast enough. The mixing conditions may also change too rapidly, making adaptation intractable.

Even if there are enough data available, another question is whether is it possible to perform necessary computations in real time. This question, though, might be rendered obsolete in a couple years. Even if we are able to compute in real time, does the algorithm adapt well enough to achieve separation? Is the reported 10–15-dB enhancement enough for the application? What factors set the limits for the quality of the result, and what are the limits?

5. *What Else Can be Done?* It would be reasonable to combine different approaches to make use of every bit of available knowledge. For example, we should combine the nonstationarity-exploiting decorrelation approaches with approaches that make use of the source densities. Initial attempts in this direction are reported in Lee et al. (1998). Another promising approach is decomposing the problem into smaller and independent subproblems instead of trying to solve it all at once with a huge number of parameters, as discussed in Section 8.6.3.

The sensitivity to (unavoidable) noise is much examined in communications, but has not been studied enough in the audio context. Some studies or methods taking noise into account are presented in Comon (1996), Moreau and Pesquet (1997), Moulines et al. (1997), and Parra and Spence (1999).

Using more microphones than there are sources to separate should in theory be able to improve the noise tolerance, in addition to separation quality. Westner (1999), though, reports slightly negative results with real recordings in his initial experiments, although synthetic cases improved dramatically.

An avenue that has not been looked at enough in BSS in an audio context is making use of the fact that the target signal is a speech signal. Again, in digital communications it is straightforward to make use of the known signal properties. Brandstein (1998) advocates explicitly incorporating the nature of the speech signal, including nonstationarity, model of production, pitch, voicing, formant structure, and a model of source radiation, into a beam-forming context. He feels that this is essential to realizing the goal: high-quality speech-signal acquisition from an unconstrained talker in a hands-free environment surrounded by interfering sources. This goal is the same as for many BSS applications. Some recent BSS work to these directions is presented in Wu and Principe (1998) and Wu et al. (1998).

6. *What Is the Best Method for Separating Signals Mixed Convolutively?* Given differing assumptions of each method and the lack of common criteria and databases, this question remains unanswerable. It is possible to derive some theoretical bounds for some algorithms (Lindgren and Broman 1996a; Sahlin and Lindgren 1996; Yellin and Friedlander 1996), but not for all of them. For audio, empirical comparisons using agreed upon databases and measurements may be the only way to find partial answers (Lambert 1999; Schobben et al. 1999). For artificially generated signals, as in digital communications, one only needs to agree upon types of signals and goodness measures.

We conclude the chapter with these open questions, mostly related to problems in the audio realm, and state that as coping with artificial signals seems to be more straightforward than with natural signals, it might be that BSS with convolutive mixtures finds its first real applications in the area of communications.

REFERENCES

Amari, S., A. Cichocki, and H. H. Yang, 1996, "A new learning algorithm for blind signal separation," in *Adv. in Neural Information Processing Syst. 8*, (Cambridge, MA: MIT Press) pp. 757–763.

Amari, S., S. Douglas, A. Cichocki, and H. H. Yang, 1997a, "Multichannel blind deconvolution and equalization using the natural gradient," *Proc. 1st IEEE Signal Processing Workshop on Signal Processing Advances in Wireless Communications*, Paris, France, April 16–18, 1997, pp. 101–104.

Amari, S., S. Douglas, A. Cichocki, and H. H. Yang, 1997b, "Novel on-line adaptive learning algorithms for blind deconvolution using the natural gradient algorithm," *Proc. 11th IFAC Symposium on System Identification*, Kitakyushu City, Japan, July 1997, pp. 1057–1062.

Attias, H., and C. Schreiner, 1997, "Blind source separation and deconvolution by dynamic component analysis," *Proc. IEEE Workshop on Neural Networks for Signal Processing*, Amelia Island, FL, USA, September 1997, pp. 456–465.

Attias, H., and C. Schreiner, 1998, "Blind source separation and deconvolution: The dynamic component analysis algorithm," *Neural Computation*, vol. 10, pp. 1373–1424.

Back, A. D., and A. Cichocki, 1997, "Blind source separation and deconvolution of fast sampled signals," *Proc. Int. Conf. on Neural Inform. Processing*, 1997.

Back, A. D., and A. C. Tsoi, 1994, "Blind deconvolution of signals using a complex recurrent network," *Proc. IEEE Workshop on Neural Networks for Signal Processing*, pp. 565–574.

Bell, A., and T. Sejnowski, 1995, "An information-maximisation approach to blind separation and blind deconvolution," *Neural Computation*, vol. 7, no. 6, pp. 1129–1159.

Belouchrani, A., and J.-F. Cardoso, 1994, "Maximum likelihood source separation for discrete sources," in *Signal Processing VII: Theories and Applications (Proc. of the EUSIPCO-94)*, Edinburgh, Scotland, September 13–16, 1994, Elsevier, pp. 768–771.

Belouchrani, A., and J.-F. Cardoso, 1995, "Maximum likelihood source separation by the expectation-maximization technique: Deterministic and stochastic implementation," *Proc. NOLTA*, Las Vegas, Nevada, USA, December 10–14, 1995, pp. 49–53.

Bishop, C. M., M. Svensén, and C. K. I. Williams, 1997, "GTM: A principled alternative to the self-organizing map," in M. C. Mozer, M. I. Jordan, and T. Petsche, ed., *Advances in Neural Information Processing Systems 9*, (Cambridge, MA: MIT Press) pp. 354–360.

Brandstein, M. S., 1998, "On the use of explicit speech modeling in microphone array applications," *Proc. ICASSP*. Seattle, WA, May 12–15, 1998, pp. 3613–3616.

Broman, H., U. Lindgren, H. Sahlin, and P. Stoica, 1995, "Source separation: A TITO system identification approach," *Technical Report CTH-TE-33*, Chalmers University of Technology, September 27, 1995.

Capdevielle, V., C. Serviere, and J. Lacoume, 1995, "Blind separation of wide-band sources in the frequency domain," *Proc. ICASSP*, Detroit, MI, May 9–12, 1995, pp. 2080–2083.

Cardoso, J.-F., April 1997, "Infomax and maximum likelihood for source separation," *IEEE Lett. on Signal Processing*, vol. 4, no. 4, pp. 112–114.

Cardoso, J.-F., and S.-I. Amari, 1997, "Maximum likelihood source separation: equivariance and adaptivity," *Proc. of SYSID'97, 11th IFAC symposium on system identification*, Fukuoka, Japan, pp. 1063–1068.

Cardoso, J.-F., and B. Laheld, Dec. 1996, "Equivariant adaptive source separation," *IEEE Trans. on Signal Processing*, vol. 44, no. 12, pp. 3017–3030.

Chan, D. B., P. J. W. Rayner, and S. J. Godsill, 1996, "Multi-channel signal separation," *Proc. ICASSP*, Atlanta, GA, May 7–10 1996.

Choi, S., and A. Cichocki, 1997a, "Adaptive blind separation of speech signals: Cocktail party problem," *Proc. International Confererence on Speech Processing (ICSP'97)*, Seoul, Korea, August 26–28, 1997, pp. 617–622.

Choi, S., and A. Cichocki, 1997b, "Blind signal deconvolution by spatio-temporal decorrelation and demixing," In *IEEE Workshop on Neural Networks for Signal Processing*, Amelia Island, FL, USA, September 1997.

Cichocki, A., S.-I. Amari, and J. Cao, 1996, "Blind separation of delayed and convolved signals with self-adaptive learning rate," *Proc. of the Int. Symp. on Nonlinear Theory and Applications (NOLTA)*, Katsurahama-so, Kochi, Japan, October 7–9, 1996, pp. 229–232.

Comon, P., 1994, "Independent component analysis—a new concept?," *Signal Processing*, vol. 36, no. 3, pp. 287–314.

Comon, P., July 1996, "Contrasts for multichannel blind deconvolution," *IEEE Signal Processing Lett.*, vol. 3, no. 7, pp. 209–211.

Comon, P., and E. Moreau, April 1997, "Improved contrast dedicated to blind separation in communications," *Proc. ICASSP*, Munich, Germany.

Cruces, S., and L. Castedo, 1998, "A Gauss-Newton method for blind source separation of convolutive mixtures," *Proc. ICASSP*, Seattle, WA, May 12–15 1998, pp. 2093–2096.

de Carvalho, E., and D. T. M. Slock, 1997, "Cramer-rao bounds for blind, semi-blind, and training sequence based channel estimation," *Proc. 1st IEEE Signal Processing Workshop on Signal Processing Advances in Wireless Communications*, Paris, France, April 16–18, 1997, pp. 129–132.

Deco, G., and D. Obradovic, 1996, *An Information Theoretic Approach to Neural Computing*, (New York: Springer-Berlin).

Delfosse, N., and P. Loubaton, 1996, "Adaptive blind separation of convolutive mixtures," *Proc. ICASSP*, Atlanta, GA, May 7–10, 1996, pp. 2940–2943.

Deville, Y., and N. Charkani, 1997, "Analysis of the stability of time-domain source separation algorithms for convolutively mixed signals," *Proc. ICASSP*, Munich, Germany, April 1997.

Dinc, A., and Y. Bar-Ness, 1993 "A forward/backward bootstrapped structure for blind separation of signals in a multi-channel dispersive environment," *Proc. ICASSP*, Minneapolis, MN, USA, April 27–30, 1993, pp. 376–379.

Douglas, S. C., 1998, "Equivariant algorithms for selective transmission," *Proc. ICASSP*, Seattle, WA, USA, 1998, pp. 1133–1136.

Douglas, S. C., and A. Cichocki, November 1997, "Neural networks for blind decorrelation of signals," *IEEE Trans. on Signal Processing*, vol. 45, no. 11, pp. 2829–2842.

Douglas, S. C., and S. Haykin, 1997, "On the relationship between blind deconvolution and blind source separation," *31st Annual Asilomar Conf. on Signals, Systems, and Computers*, Pacific Grove, CA, USA, November 1997.

Douglas, S. C., A. Cichocki, and S. Amari, September 1997, "Multichannel blind separation and deconvolution of sources with arbitrary distributions," *IEEE Workshop on Neural Networks for Signal Processing*, Amelia Island, FL, USA.

Ehlers, F., and H. Schuster, 1997, "Blind separation of convolutive mixtures and an application in automatic speech recognition in noisy environment," *IEEE Trans. on Signal Processing*, vol. 45, no. 10, p. 2608.

Emile, B., and P. Comon, 1998a, "Estimation of time delays between unknown colored signals," *Signal Processing*, vol. 68, no.1, pp. 93–100.

Emile, B., and P. Comon, July 1998b, "Estimation of time delays with fewer sensors than sources," *IEEE Transactions on Signal Processing*, vol. 46, no. 7, pp. 2012–2015.

Engebretson, A. M., 1993, "Acoustic signal separation of statistically independent sources using multiple microphones," *Proc. ICASSP*, vol. II, Minneapolis, MN, USA, April 27–30 1993, pp. 343–346.

Feng, M., and K.-D. Kammeyer, 1999, "Application of source separation algorithms for mobile communication environment," *Proc. ICA and BSS*, Aussois, France, January 11–15, 1999.

Ferrara, E. R., August 1980, "Fast implementation of LMS adaptive filters," *IEEE Trans. on Acoustics, Speech, and Signal Processing*, vol. ASSP-28, no. 4, pp. 474–475.

Girolami, M., 1998, "Noise reduction and speech enhancement via temporal anti-Hebbian learning," *Proc. ICASSP*, Seattle, WA, USA, May 12–15 1998.

Girolami, M., and C. Fyfe, 1996, "A temporal model of linear anti-Hebbian learning," *Neural Processing Lett.*, vol. 4, no. 3, pp. 139–148.

Girolami, M., and C. Fyfe, 1997, "Generalised independent component analysis through unsupervised learning with emergent Bussgang properties," *Proc. ICNN*, Houston, TX, USA.

Girolami, M., A. Cichocki, and S.-I. Amari, 1997, "A common neural network model for unsupervised exploratory data analysis and independent component analysis," *Technical Report BIP-97-001*, RIKEN Frontier Research Program, Brain Information Processing Group, Saitama, Japan.

Gorokhov, A., P. Loubaton, and E. Moulines, 1996, "Second order blind equalization in multiple input multiple output FIR systems: A weighted least squares approach," In *Proc. ICASSP*, Atlanta, GA, May 7–10 1996, pp. 2417–2420.

Guo, Y., F. Sattar, and C. Koh, 1999, "Blind separation temporomandibular joint sound signals," In *Proc. ICASSP*, Phoenix, AZ, April 1999, pp. 1069–1072.

Haykin, S., 1999, *Neural Networks, A Comprehensive Foundation*, 2nd ed. (New York: Prentice-Hall and IEEE Press).

Herault, J., and C. Jutten, 1986, "Space or time adaptive signal processing by neural network models," in *Neural Networks for Computing, AIP Conf. Proc.*, vol. 151, Snowbird, UT, USA, 1986, pp. 206–211.

Hyvärinen, A., 1998, "New approximations of differential entropy for independent component analysis and projection pursuit," in *Advances in Neural Information Processing Systems 10 (Proc. of NIPS'97)*, (Cambridge, MA: MIT Press). Also published as Report A47, Helsinki University of Technology, Laboratory of Computer and Information Science, August 1997.

Icart, S., and R. Gautier, 1996, "Blind separation of convolutive mixtures using second and fourth order moments," in *Proc. ICASSP*, Atlanta, GA, May 7–10 1996, pp. 3018–3021.

Jutten, C., and J.-F. Cardoso, 1995, "Separation of sources: Really blind?" in *Proc. of the Int. Symp. on Nonlinear Theory and Applications (NOLTA)*, Las Vegas, Nevada, USA, December 10–14, 1995, pp. 79–84.

Jutten, C., and J. Herault, 1991, "Blind separation of sources, part I: An adaptive algorithm based on neuromimetic architecture," *Signal Processing*, vol. 24, no. 1, pp. 1–10.

Jutten, C., L. Nguyen Thi, E. Dijkstra, E. Vittoz, and J. Caelen, 1992, "Blind separation of sources: An algorithm for separation of convolutive mixtures," in J. L. Lacoume, ed., *Higher order statistics: Proc. of the Int. Signal Processing Workshop on Higher Order Statistics*, Chamrousse, France, July 10–12, 1992, Elsevier, pp. 275–278.

Kohler, B.-U., T.-W. Lee, and R. Orglmeister, 1997, "Improving the performance of Infomax using statistical techniques," *Proc. ICANN*, Lausanne, Switzerland.

Laakso, T., V. Välimäki, M. Karjalainen, and U. Laine, January 1996, "Splitting the unit delay [FIR/all pass filters design]," *IEEE Signal Processing Mag.*, vol. 13, no. 1, pp. 30–60.

Lambert, R. H., 1995, "A new method for source separation," In *Proc. ICASSP*, Detroit, MI, May 9–12, 1995, pp. 2116–2119.

Lambert, R. H., May 1996, Multichannel blind deconvolution: FIR matrix algebra and separation of multipath mixtures, Ph.D. dissertation, University of Southern California, Department of Electrical Engineering.

Lambert, R. H., 1999, "Difficulty measures and figures of merit for source separation," *Proc. ICA and BSS*, Aussois, France, January 11–15, 1999.

Lambert, R. H., and A. J. Bell, 1997, "Blind separation of multiple speakers in a multipath environment," *Proc. ICASSP*, Munich, Germany, April 21–24, 1997, pp. 423–426.

Lambert, R. H., and C. Nikias, 1995, "Polynomial matrix whitening and application to the multichannel blind deconvolution problem," *Proc. IEEE MILCOM*, San Diego, CA, Nov. 5–8, 1995.

Lee, T.-W., A. Bell, and R. H. Lambert, 1997a, "Blind separation of delayed and convolved sources," *Adv. in Neural Information Processing Systems 9* (Cambridge, MA: MIT Press) pp. 758–764.

Lee, T.-W., A. Bell, and R. Orglmeister, 1997b, "Blind source separation of real world signals," *Proc. ICNN*, Houston, TX, June 9–12, 1997.

Lee, T.-W., A. Bell, and R. Orglmeister, 1997c, "A contextual blind separation of delayed and convolved sources," *Proc. ICASSP*, Munich, Germany, April 21–24, 1997, pp. 1199–1202.

Lee, T.-W., M. Girolami, A. Bell, and T. Sejnowski, 1999, "A unifying information-theoretic framework for independent component analysis," *Computers and Mathematics with Applications*, in press.

Lee, T.-W., A. Ziehe, R. Orglmeister, and T. Sejnowski, 1998, "Combining time-delayed decorrelation and ica: towards solving the cocktail party problem," *Proc. ICASSP*, Seattle, WA, May 1998, pp. 1249–1252.

Li, S., and T. J. Sejnowski, January 1995, "Adaptive separation of mixed broad-band sound sources with delays by a beamforming Hérault-Jutten network," *IEEE J. of Oceanic Engineering*, vol. 20, no. 1, pp. 73–79.

Lindgren, U., and H. Broman, 1996a, "Monitoring the mutual independence of the output of source separation algorithms," *Proc. ISITA'96*, Victoria, B.C., Canada, September 1996.

Lindgren, U., and H. Broman, 1996b, "On the identifiability of a mixing channel based on second order statistics," *Proc. Radiovetenskap och Kommunikation*, Lulea, Sweden, May 1996, pp. 420–424.

Lindgren, U., and A.-J. van der Veen, "Source separation based on second order statistics—An algebraic approach," *Proc. of the IEEE 8th Workshop on Statistical Signal and Array Processing*, Corfu, Greece, June 24–26, 1996, IEEE, pp. 324–327.

Lindgren, U., H. Sahlin, and H. Broman, 1996, "Source separation using second order statistics." *Signal Processing IX: Theories and Applications (Proc. of the EUSIPCO-96)*, Trieste, Italy, September 1996, Elsevier.

Lindgren, U., T. Wigren, and H. Broman, December 1995, "On local convergence of a class of blind separation algorithms," *IEEE Trans. on Signal Processing*, vol. 43, no. 12, pp. 3054–3058.

MacKay, D., 1996, "Maximum likelihood and covariant algorithms for independent component analysis," *Technical Report, Cavendish Laboratory*, University of Cambridge, draft 3.1.

Matsuoka, K., and M. Kawamoto, 1995, "Blind signal separation based on mutual information criterion," *Proc. NOLTA*, Las Vegas, Nevada, USA, December 10–14, 1995, pp. 85–90.

Matsuoka, K., M. Ohya, and M. Kawamoto, 1995, "A neural net for blind separation of nonstationary signals," *Neural Networks*, vol. 8, no. 3, pp. 411–419.

Mejuto, C., and J. C. Principe, 1999, "A second-order method for blind separation of convolutive mixtures," *Proc. ICA and BSS*, Aussois, France, January 11–15, 1999.

Molgedey, L., and H. Schuster, June 6 1994, "Separation of independent signals using time-delayed correlations, *Physical Review Lett.*, vol. 72, no. 23, pp. 3634–3637.

Moreau, E., and J.-C. Pesquet, June 1997, "Generalized contrasts for multichannel blind deconvolution of linear systems," *IEEE Signal Processing Lett.*, vol. 4, no. 6, pp. 182–183.

Moreau, E., and N. Thirion, 1996, "Multichannel blind signal deconvolution using high order statistics," *Proc. of the IEEE 8th Workshop on Statistical Signal and Array Processing*, Corfu, Greece, June 24–26 1996, IEEE, pp. 336–339.

Moulines, E., J.-F. Cardoso, and E. Gassiat, 1997, "Maximum likelihood for blind separation and deconvolution of noisy signals using mixture models," *Proc. ICASSP*, Munich, Germany, April 21–24, 1997, pp. 3617–3620.

Murata, N., S. Ikeda, and A. Ziehe, 1998, "An approach to blind source separation based on temporal structure of speech signals," *Technical Report BSIS Technical Reports No. 98-2*, RIKEN Brain Science Institute, Japan.

Nájar, M., M. A. Lagunas, and I. Bonet, 1994, "Blind wideband source separation," *Proc. ICASSP*, vol. IV, Adelaide, Australia, April 19–22, 1994, pp. 65–68.

Ngo, J. T., and N. A. Bhadkamkar, 1999, "Adaptive blind separation of audio sources by a physically compact device using second-order statistics," *Proc. ICA and BSS*, Aussois, France, January 11–15, 1999.

Ohno, S., and Y. Inouye, 1998, "A least-squares interpretation of the single-stage maximization criterion for multichannel blind deconvolution," *Proc. ICASSP*, Seattle, WA, May 12–15, 1998, pp. 2101–2104.

Oppenheim, A. V., and R. W. Schafer, 1975, *Digital Signal Processing*, (Englewood Cliffs, NJ: Prentice-Hall).

Pan, H., D. Xia, S. Douglas, and K. Smith, 1998, "A scalable vlsi architecture for multichannel blind deconvolution and source separation," *Proc. IEEE Workshop on Signal Processing Systems*, Boston, MA, October 1998.

Parra, L., and C. Spence, 1998a, "Convolutive blind source separation based on multiple decorrelation," *Proc. IEEE Workshop on Neural Networks for Signal Processing*, Cambridge, UK, September 1998.

Parra, L., and C. Spence, 1998b, "Temporal models in blind source separation," *Adaptive Processing of Sequences and Data Structures—International Summer School on Neural Networks, E.R. Caianiello*, Vietri sul Mare, Salerno, Italy, September 6–13, 1997, Tutorial Lectures, Springer.

Parra, L., and C. Spence, 1999, "Convolutive blind source separation based on multiple decorrelation." *IEEE Trans. on Speech and Audio Processing*, in press.

Parra, L., C. Spence, and B. D. Vries, 1997, "Convolutive source separation and signal modeling with ML," *Proc. Int. Symp. on Intelligent Systems (ISIS'97)*, Regio Calabria, Italy.

Paulraj, A., and C. B. Papadias, 1997, "Array processing in mobile communications," in *Handbook of Signal Processing*, CRC Press.

Pearlmutter, B. A., and L. C. Parra, 1996, "A context-sensitive generalization of ICA," in *Int. Conf. on Neural Information Processing*, Hong Kong, Sept. 24–27, 1996, Springer.

Pham, D., P. Garat, and C. Jutten, 1992, "Separation of a mixture of independent sources through a maximum likelihood approach," in J. Vandevalle, R. Boite, M. Moonen, and A. Oosterlink, ed., *Signal Processing VI: Theories and Applications*, Elsevier, pp. 771–774.

Platt, J. C., and F. Faggin, 1992, "Networks for the separation of sources that are superimposed and delayed," in J. Moody, S. Hanson, and R. Lippmann, ed., *Advances in Neural Information Processing Systems 4*, (San Mateo, CA: Morgan-Kaufmann).

Pope, K., and R. Bogner, 1994, "Blind separation of speech signals," *Proc. of the Australian Int. Conf. on Speech Science and Technology*, Perth, Western Australia, December 6–8, 1994.

Principe, J. C., C. Wang, and H.-C. Wu, 1996, "Temporal decorrelation using teacher forcing anti-Hebbian learning and its application in adaptive blind source separation," *Neural Networks for Signal Processing VI (Proc. IEEE Workshop on Neural Networks for Signal Processing)*, Kyoto, Japan, September 4–6 1996, pp. 413–422.

Prosoniq. *PROSONIQ PANDORA Forensic Blind Signal Separation for Restoration and Forensic Applications*. http://www.prosoniq.com/html/pdforens.html, 1998.

Roweis, S., and Z. Ghahramani, 1999, "A unifying review of linear Gaussian models," *Neural Computation*, vol. 11, no. 2.

Sahlin, H., and H. Broman, 1997, "Signal separation applied to real world signals," *Proceedings of 1997 Int. Workshop on Acoustic Echo and Noise Control (IWAENC97)*, London UK, September 11–12, 1997.

Sahlin, H., and H. Broman, 1999, "A decorrelation approach to blind MIMO signal separation," *Proc. ICA and BSS*, Aussois, France, January 11–15, 1999.

Sahlin, H., and U. Lindgren, "The asymptotic cramer-rao lower bound for blind signal separation," *Proc. of the IEEE 8th Workshop on Statistical Signal and Array Processing*, Corfu, Greece, June 24–26, 1996, IEEE, pp. 328–331.

Schobben, D., and P. Sommen, 1998a, "A new blind signal separation algorithm based on second order statistics," *Proc. IASTED Int. Conf. on Signal and Image Processing*, Las Vegas, USA, October 27–31, 1998.

Schobben, D., and P. Sommen, 1998b, "Transparent communication," *Proc. IEEE Benelux Signal Processing Chapter Symp.*, Leuven, Belgium, March 26–27 1998, pp. 171–174.

Schobben, D., K. Torkkola, and P. Smaragdis, 1999, "Evaluation of blind signal separation methods," *Proc. ICA and BSS*, Aussois, France, January 11–15, 1999.

Serviere, C., 1996, "Blind source separation of convolutive mixtures," *Proc. of the IEEE 8th Workshop on Statistical Signal and Array Processing*, Corfu, Greece, June 24–26, 1996, IEEE, pp. 316–319.

Serviere, C., 1998, "Feasibility of source separation in frequency domain," *Proc. ICASSP*, Seattle, WA, May 12–15, 1998, pp. 2085–2088.

Serviere, C., and V. Capdevielle, 1996, "Blind adaptive separation of wide-band sources," *Proc. ICASSP*, Atlanta, GA, May 7–10, 1996.

Shamsunder, S., and G. B. Giannakis, 1997, "Multichannel blind signal separation and reconstruction," *IEEE Trans. on Speech and Audio Processing*, vol. 5, no. 6, pp. 515–528.

Shields, P., M. Girolami, D. Campbell, and C. Fyfe, 1997, "Adaptive processing schemes inspired by binaural unmasking for enhancement of speech corrupted with noise and reverberation," in L. S. Smith and A. Hamilton, ed., *Neuromorphic Systems: Engineering Silicon from Neurobiology*, World Scientific, pp. 61–74.

Simon, C., G. d'Urso, C. Vignat, and P. Loubaton, 1998, "On the convolutive mixture source separation by the decorrelation," *Proc. ICASSP*, Seattle, WA, May 12–15, 1998, pp. 2109–2112.

Smaragdis, P., 1997a, "Efficient blind separation of convolved sound mixtures," *Proc. of IEEE 1997 Workshop on Applications of Signal Processing to Audio and Acoustics*, New Paltz, NY, Oct 19–22, 1997.

Smaragdis, P., 1997b, Information theoretic approaches to source separation. Master's thesis, Massachusetts Institute of Technology.

Smaragdis, P., 1998, "Blind separation of convolved sound mixtures in the frequency domain," *Proc. Int. Workshop on Independence and Artificial Neural Networks*, Tenerife, Spain, February 9–10, 1998.

Sondhi, M. M., March 1967, "An adaptive echo canceller," *Bell Syst. Tech. J.*, vol. 46, no. 3, pp. 497–511.

Sondhi, M. M., D. R. Morgan, and J. L. Hall, August 1995, "Stereophonic acoustic echo cancellation—An overview of the fundamental problem," *IEEE Signal Processing Lett.*, vol. 2, no. 8, pp. 148–151.

Soon, V., L. Tong, Y. Huang, and R. Liu, 1992, "A wideband blind identification approach to speech acquisition using a microphone array," *Proc. ICASSP*, vol. 1, San Francisco, California, USA, March 23–26, 1992, pp. 293–296.

Soon, V., L. Tong, Y. Huang, and R. Liu, 1993, "A robust method for wideband signal separation," *Proc. ISCAS*, pp. 703–760.

Thi, H.-L. N., and C. Jutten, 1995, "Blind source separation for convolutive mixtures," *Signal Processing*, vol. 45, no. 2.

Thi, H.-L. N., C. Jutten, and J. Caelen, 1992, "Speech enhancement: Analysis and comparison of methods on various real situations," In *Signal Processing VI: Theories and Applications (Proc. of the EUSIPCO-92)*, Bruxelles, Belgium, 1992, Elsevier, pp. 303–306.

Tong, L., 1996, "Identification of multivariate FIR systems using higher-order statistics," *Proc. ICASSP*, Atlanta, GA, May 7–10, 1996.

Tong, L., R. Liu, V. Soon, and Y. Huang, May 1992, "Indeterminacy and identifiability of blind identification, *IEEE Trans. on Circuits and Systems*, vol. 38, no. 5, pp. 499–509.

Torkkola, K., 1996a, "Blind separation of convolved sources based on information maximization," in *IEEE Workshop on Neural Networks for Signal Processing*, Kyoto, Japan, September 4–6, 1996, pp. 423–432.

Torkkola, K., 1996b, "Blind separation of delayed sources based on information maximization," *Proc. ICASSP*, Atlanta, GA, May 7–10, 1996, pp. 3510–3513.

Torkkola, K., 1997a, "Blind deconvolution, information maximization, and recursive filters," *Proc. ICASSP*, Munich, Germany, April 21–24, 1997, pp. 3301–3304.

Torkkola, K., 1997b, "Blind separation of radio signals in fading channels," in *Advances in Neural Information Processing Systems 10*, Denver, CO, December 1–6, 1997, MIT Press, pp. 756–762.

Torkkola, K., 1998, "Blind signal separation in communications: Making use of known signal distributions," *Proc. of the 8th IEEE Digital Signal Processing Workshop*, Bryce, UT, August 9–12, 1998, IEEE.

Tugnait, J. K., "Adaptive blind separation of convolutive mixtures of independent linear signals," *Proc. ICASSP*, Seattle, WA, May 12–15, 1998, pp. 2097–2100.

Van Compernolle, D., and S. Van Gerven, "Signal separation in a symmetric adaptive noise canceler by output decorrelation," *Proc. ICASSP*, vol. 4, San Francisco, California, USA, March 23–26, 1992, IEEE, pp. 221–224.

van der Veen, A.-J., October 1998, "Algebraic methods for deterministic blind beamforming," *Proc. IEEE*, vol. 86, no. 10, pp. 1987–2008.

van der Veen, A.-J., and A. Paulraj, May 1996, "An analytical constant modulus algorithm," *IEEE Trans. Signal Processing*, vol. 44, no. 5, pp. 1136–1155.

Van Gerven, S., and D. Van Compernolle, 1992, "Feedforward and feedback in a symmetric adaptive noise canceler: Stability analysis in a simplified case," in J. Vandevalle, R. Boite, M. Moonen, and A. Oosterlink, ed., *Signal Processing VI: Theories and Applications*, Elsevier, pp. 1081–1084.

Van Gerven, S., and D. Van Compernolle, July 1995, "Signal separation by symmetric adaptive decorrelation: Stability, convergence, and uniqueness," *IEEE Trans. Signal Processing*, vol. 43, no. 7, pp. 1602–1612.

Weinstein, E., M. Feder, and A. Oppenheim, 1993, "Multi-channel signal separation by decorrelation," *IEEE Trans. Speech and Audio Processing*, vol. 1, no. 4, pp. 405–413.

Westner, A., 1999, Object-based audio capture: Separating acoustically-mixed sounds, Master's thesis, Massachusetts Institute of Technology.

Westner, A., and J. V. Michael Bove, 1999, "Blind separation of real world audio signals using overdetermined mixtures," *Proc. ICA and BSS*, Aussois, France, January 11–15, 1999.

Widrow, B., and S. Stearns, 1985, *Adaptive Signal Processing*. (New Jersey: Prentice-Hall).

Wu, H.-C., and J. C. Principe, 1997, "A unifying criterion for blind source separation and decorrelation: Simultaneous diagonalization of correlation matrices," *Proc. of NNSP97*, Amelia Island, FL, pp. 496–505.

Wu, H.-C., and J. C. Principle, 1998, "Simultaneous diagonalization algorithm for blind source separation based on subband filtered features," *Proc. of SPIE—Conference of The Int. Soc. for Optical Engineering*, Orlando, Florida, pp. 466–474.

Wu, H.-C., and J. C. Principe, 1999, "Simultaneous diagonalization in the frequency domain (SDIF) for source separation," *Proc. ICA and BSS*, Aussois, France, January 11–15, 1999.

Wu, H.-C., J. C. Principe, and D. Xu, 1998, "Exploring the time-frequency microstructure of speech for blind source separation," *Proc. ICASSP*, Seattle, WA, USA, 1998, pp. 1145–1148.

Xi, J., and J. P. Reilly, 1997, "Blind separation and restoration of signals mixed in convolutive environment," *Proc. ICASSP*, Munich, Germany, April 21–24, 1997, pp. 1327–1330.

Yellin, D., and B. Friedlander, 1996, "Blind multi-channel system identification and deconvolution: Performance bounds," *Proc. of the IEEE 8th Workshop on Statistical Signal and Array Processing*, Corfu, Greece, June 24–26, 1996, IEEE, pp. 582–585.

Yellin, D., and E. Weinstein, January 1996, "Multichannel signal separation: Methods and analysis," *IEEE Trans. on Signal Processing*, vol. 44, no. 1, pp. 106–118.

Yen, K.-C., and Y. Zhao, October 1996, "Robust automatic speech recognition using a multi-channel signal separation front end," *Proc. 4th Int. Conf. on Spoken Language Processing (ICSLP'96)*, Philadelphia, PA.

Yen, K.-C., and Y. Zhao, 1998, "Improvements on co-channel separation using ADF: Low complexity, fast convergence, and generalization," *Proc. ICASSP*, Seattle, WA, USA, May 12–15, 1998, pp. 1025–1028.

9

BLIND DECONVOLUTION OF MULTIPATH MIXTURES

Russell H. Lambert and Chrysostomos L. Nikias

ABSTRACT

We present cost functions and adaptive cost-function minimization methods applicable to blind separation of multichannel matrix-mixture problems. A Bussgang family of costs is given, consisting of three distinct classes. Blind stochastic-gradient and new adaptation methods are related to finite difference approximation. Where possible, the new costs and update methods are related to the standard tools of adaptive filtering. Theoretical analysis of the general Bussgang cost gives insight into optimal algorithm selection and the prediction of relative convergence speed.

A general tool for the inclusion of multipath in multichannel problems is given in finite impulse response (FIR) matrix algebra. An eigenroutine that computes eigenvalues and eigenvectors of multipath matrices and an efficient multipath matrix-whitening tool, is given along with several examples. Using FIR matrix algebra, methods of single-channel adaptive filtering and source separation of multipath mixtures are merged into a general FIR matrix framework. Within the framework, powerful techniques developed for single-channel equalization can be applied to multichannel source separation and vice versa.

Simulations that demonstrate the utility of multichannel blind least mean squares (MBLMS), multichannel blind recursive least squares (MBRLS), Amari-form finite differencing (Amari-FD), Amari-form blind serial update

Unsupervised Adaptive Filtering, Volume 1, Edited by Simon Haykin.
ISBN 0-471-29412-8 © 2000 John Wiley & Sons, Inc.

(Amari-BSU), and prewhitened MBLMS (W-MBLMS) are given, using real and complex data sources and channels. A successful acoustic-speech separation result attests to the power and robustness of the techniques presented.

9.1 INTRODUCTION

As thermodynamics has its second law, the law of entropy, the corresponding law for signal processing is the Central Limit theorem. The Central Limit theorem states that any finite-variance independent sources, regardless of their initial distribution, will sum at the limit to form a Gaussian distribution. The histogram of a clean speech signal is decidedly more peaked than Gaussian. A speech signal recorded with a microphone far across a room from the source does appear to be Gaussian. The multipath reflections of the room have summed at the microphone to produce an almost Gaussian signal. Using deconvolution tools, it is therefore possible to apply work in learning the inverse room channel and filter the microphone signal with it to obtain the original signal. The deconvolution tools are adjusted to drive the measured signal as far from Gaussian as possible using an inverse filter.

In like manner, consider several microphones in a room and several human speakers. Here, the Central Limit theorem "Gaussianization" is due both to the convolution of distributions from the direct-signal speech signals and also the multipath reflections. We learn the inverse matrix and apply it to the vector of microphones to obtain the original speech signals separated and deconvolved.

In signal-processing and neural-network realms, the terms equalization, source separation, entropy minimization, and information maximization have been used, while in geophysical data processing and speech processing, the term deconvolution has been used to describe the reversal of entropy. The processing that we employ involves cost functions, adaptive cost-function minimization techniques, and some matrix theory involved with the addition of multipath delays or memory to matrices.

9.1.2 Notation Conventions Used in this Chapter

Term	Definition
x	Original source
h	Mixture system
y	Sensor date, output of mixture, input to equalization/separation system
w	Equalization/separation system
$\mathbf{x}$	Column vectors will be boldface, lowercase symbols
$\mathbf{H}$	Matrix variables will be in boldface, uppercase

Term	Definition
$\underline{\mathbf{H}}$	FIR matrices will be underlined boldface, uppercase symbols may be time or freq. domain, depending on context
$*$	Convolution (i.e., $y = x * h$ will denote $y(n) = \sum_{\text{all } k} h(k)x(n-k)$)
L	Number of multipath elements in w
$\underline{y}$	Underlined lowercase variables describe vector containing last L samples of variable in time-reversed order
y^*	Superscript $*$ describes conjugation
W^H	Superscript H describes conjugation and transpose
$E\{\cdot\}$	The expectation operator
$FFT\{\cdot\}$	Forward discrete Fourier transform
$IFFT\{\cdot\}$	Inverse discrete Fourier transform
H_{ij}	Here matrix indices always follow standard row, column order
μ	Small real-valued step size
ΔW	The change made to W by the update equation at each iteration
$w = w + \mu \dfrac{\partial J}{\partial w}$	$w(n+1) = w(n) + \mu \dfrac{\partial J}{\partial w(n)}$ (time index implied and computer convention of left assignment is used in order to ease notation)
$f_a(b)$	Known probability density function form for variable a using variable b
$\Phi_a(\omega_b)$	Known characteristic function form for variable a using variable b
k	Lowercase k is multipath lag index
n	Lowercase n discrete time index
$r(k) = E\{x(n+k)x(n)^*\}$	Correlation at lag k sums over time index n
X	Capital X indicates frequency domain $X = FFT(x)$, $Y = FFT(y)$, $\hat{X} = FFT(\hat{x})$
$R = E\{X^*X\}$	$R(\omega) = E\{X^*X\}$ spectrum estimate is in the frequency domain; frequency dependancy is omitted to ease notational burden

9.1.2 Cost Functions and Adaptive Minimization Methods

For our purposes, we are able to draw a distinction between cost, algorithm, and adaptation method in the following "equation":

$$\text{Algorithm} = \text{cost function} + \text{adaptation method}$$

We develop new blind costs and relate them to the more familiar least-mean-square, (LMS) training-based costs. The most widely used adaptation method is perhaps the gradient approach $w(n + 1) = w(n) \pm \mu(\partial J/\partial w)$, to which the chain rule may be applied,

$$w(n + 1) = w(n) \pm \mu \frac{\partial J}{\partial \hat{x}} \frac{\partial \hat{x}}{\partial w}$$

The above equation is kept general to allow for cost minimization (negative step size) or cost maximization (positive step size). Training based algorithms usually try to minimize a cost such as mean squared error and blind algorithms often maximize a cost such as distance from Gaussianity.

Recently we have seen some new adaptation methods emerge, the serial update of Cardoso and Laheld (1996), and the finite difference approach to be shown later.

There may be some *a priori* known aspect about the original source signals that we wish to exploit such as spectral whiteness, independence, or non-Gaussianity. While the well-known minimum mean-squared-error (MMSE) cost function does not require any of these special signal characteristics and is applicable to a wide range of problems, it is limited to training-based learning. We therefore investigate other specialized costs in order to find blind, unsupervised, or self-organizing capability.

Fig. 9.1 depicts the type of multi-input, multi-output systems we consider. Fig. 9.2 depicts a specific two by two system. The two-input, two-output problem can be explicitly formulated:

$$\begin{bmatrix} y_1 \\ y_2 \end{bmatrix} = \begin{bmatrix} h_{11} & h_{12} \\ h_{21} & h_{22} \end{bmatrix} * \begin{bmatrix} x_1 \\ x_2 \end{bmatrix} \tag{9.1}$$

This means that

$$y_1 = x_1 * h_{11} + x_2 * h_{12} \tag{9.2}$$

$$y_2 = x_1 * h_{21} + x_2 * h_{22} \tag{9.3}$$

and

$$\hat{x}_1 = y_1 * w_{11} + y_2 * w_{12} \tag{9.4}$$

$$\hat{x}_2 = y_1 * w_{21} + y_2 * w_{22} \tag{9.5}$$

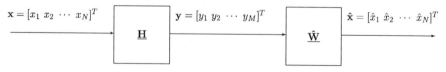

Figure 9.1 General multichannel channel system under consideration. *M* is greater than or equal to *N*.

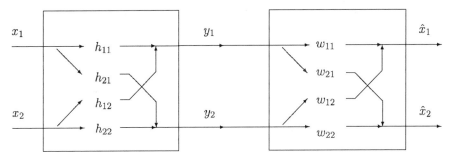

Figure 9.2 Feedforward two-input–two-output blind equalization system.

9.2 SINGLE-CHANNEL/MULTICHANNEL COST FUNCTIONS AND ADAPTATION METHODS

Blind equalization was developed in the field of communications (Bellini 1994; Benveniste and Goursat 1984; Benveniste et al. 1980; Godard 1980; Sato 1975), and somewhat independently by researchers in the area of geophysical data processing under the name "blind deconvolution" (Donoho 1981; Gray 1979; Walden 1985; Wiggins 1978).

In blind equalizers, a cost function J is minimized in much the same way that LMS and recursive-least-squares (RLS) adaptation (with access to the source) minimize the MSE

$$J = E|\hat{x} - x|^2 \qquad \text{MMSE or LMS cost} \qquad (9.6)$$

$$w = w - \mu(\hat{x} - x)y^* \qquad \text{MMSE or LMS update} \qquad (9.7)$$

9.2.1 A More Powerful Cost Available for Non-Gaussian Sources

While MMSE techniques are optimal for use with Gaussian sources, it is possible to make use of prior knowledge of the probability density function (pdf) of sources through the use of the Bussgang nonlinearity

$$g(x) = -E|x|^2 f_x'(x)/f_x(x).$$

This gives rise to an MMSE-like cost function optimized for non-Gaussian data, which we will refer to as MMSE-NG:

$$w = w \pm \mu(\hat{x} - g(x))\underline{y}^* \qquad \text{MMSE-NG update} \qquad (9.8)$$

We will shortly derive the general cost function for this. In the case of non-Gaussian sources, we also show the improved asymptotic performance of this method, as compared to MMSE.

9.2.2 Bussgang Blind Equalization Cost Functions and Methods

For the moment, let us consider real (not complex) data-specific applications. An example of a general blind Bussgang cost function is Gray's variable norm (Gray 1979; Satorius and Mulligan 1993),

$$O_\alpha^2 = \frac{E|\hat{x}|^2}{(E|\hat{x}|^\alpha)^{2/\alpha}}$$

Taking the gradient of O_4^2 gives an update similar to one presented in Godard (1980),

$$w = w + \mu\left(\hat{x} - \frac{E|x|^2}{E|x|^4}|\hat{x}|^2\hat{x}\right)\underline{y}^* \tag{9.9}$$

when the known statistics $E|x|^2$ and $E|x|^4$ are used in place of $E|\hat{x}|^2$ and $E|\hat{x}|^4$. Similarly, O_1^2 yields an algorithm similar to the Sato algorithm (Sato, 1975),

$$w = w + \mu\left(\hat{x} - \frac{E|x|^2}{E|x|}\frac{\hat{x}}{|\hat{x}|}\right)\underline{y}^* \tag{9.10}$$

Other blind or self-organizing cost functions exist that are, perhaps, more directly related to the fundamental idea of information maximization or entropy minimization than Gray's norm (see Walden 1985; Bell and Sejnowski 1995). We propose as a general nonparametric cost,

$$J = E\left\{\log\frac{f_x(\hat{x})}{f_G(\hat{x})}\right\} \tag{9.11}$$

where $f_G(\hat{x})$ is a Gaussian pdf with variance $E|\hat{x}|^2$, which is verified by $(\partial J/\partial w) = (\hat{x} - g(\hat{x}))\underline{y}^*$, the general Bussgang update equation [as seen in Bellini (1994)]. This function $g(\cdot)$ is the same Bussgang nonlinearity mentioned earlier, $(-E|x|^2 f_x'(x)/f_x(x))$ [see Bellini (1994) for details]. The Bellini notation includes the linear term as part of the nonlinearity as in $b(\hat{x}) = \hat{x} - g(\hat{x})$. We will keep the linear and nonlinear terms distinct. This cost was proposed in Godfrey (1978), in which a nonparametric method was employed to estimate the pdfs at each stage of batch iteration. It has since generally been observed [Gray (1979) and Lambert (1996, chap. 5)] that the use of parametric models for the true source pdf's, such as the generalized Gaussian model, even if the models are not exact, greatly facilitates a practical algorithm.[1] Other excellent information theoretic view points on cost functions are given in (Bell and Sejnowski 1995; Burel 1995; Linsker 1989; Walden 1985) and other chapters of this book.

---

[1] The direction of "Gaussianity," more peaked than Gaussian or flatter than Gaussian is of most importance in choosing a model.

Remark 1. $J = E\{\log(f_x(\hat{x})/f_G(\hat{x}))\}$, is a blind cost that shows the relation of the Bussgang property to a measure of mutual information, relative entropy, or Kullback-Leibler (KL) divergence, between the known pdf of the source and a Gaussian pdf of equal variance.

Now the cost function for MMSE-NG, which is training based, is just a slight variant of the blind cost that uses the known data x instead of estimates

$$J = E\left\{\log\frac{f_x(x)}{f_G(\hat{x})}\right\} \tag{9.12}$$

9.2.3 Multichannel Cost Functions

We begin with the groundwork of multichannel Wiener/MMSE filtering (Wiener 1949). In Kailath (1980) and Robinson (1967), the multichannel Wiener solution is found by minimizing the trace of the expectation of the error covariance matrix. Put into our notation, the multichannel MMSE cost is

$$J = \operatorname{tr} E\{(\hat{\mathbf{x}} - \mathbf{x})(\hat{\mathbf{x}} - \mathbf{x})^H\} \tag{9.13}$$

or

$$J = \sum_{\text{all } i} E\{|\hat{x}_i - x_i|^2\} \tag{9.14}$$

The batch solution that minimizes this multichannel cost function uses the traditional least-squares procedure of forming the data matrix and solving for the estimate of the inverse filter:

$$\mathbf{Y}_i = \begin{bmatrix} \mathbf{y_i} & 0 & 0 & 0 & 0 & 0 & 0 & \cdots & 0 \\ 0 & \mathbf{y_i} & 0 & 0 & 0 & 0 & 0 & \cdots & 0 \\ 0 & 0 & \mathbf{y_i} & 0 & 0 & 0 & 0 & \cdots & 0 \\ \cdot & & & \cdot & & & & & \cdot \\ \cdot & & & & \cdot & & & & \cdot \\ \cdot & & & & & \cdot & & & \cdot \\ 0 & 0 & 0 & \cdots & 0 & 0 & 0 & 0 & \mathbf{y_i} \end{bmatrix} \tag{9.15}$$

$$\mathbf{y_i} = \left[y_i\left(\frac{L}{2}+1\right) y_i\left(\frac{L}{2}+2\right) \cdots y_i\left(\frac{L}{2}+T\right)\right]$$

$$\mathbf{Y} = [\mathbf{Y}_1 \quad \mathbf{Y}_2 \quad \mathbf{Y}_3 \quad \cdots \quad \mathbf{Y}_N]^T \tag{9.16}$$

$$\mathbf{x_i} = [x_i(1) \quad x_i(2) \quad \cdots \quad x_i(T)] \tag{9.17}$$

$$\mathbf{X} = [\mathbf{x_1} \quad \mathbf{x_2} \quad \mathbf{x_3} \quad \cdots \quad \mathbf{x_M}]^T \tag{9.18}$$

$$\underline{\mathbf{W}} = (\mathbf{YY}^H)^{-1}\mathbf{YX}^H \tag{9.19}$$

where L is the number of multipath delays, N is the number of sensors, M is the number of inputs, and T is the number of data samples collected at the sensors. An efficient batch solution of this training-based multichannel problem, which takes advantage of the block Toeplitz structure of $(\mathbf{YY}^H)^{-1}$, was given also in Robinson (1967), and adaptive stochastic gradient and exact least squares methods are possible solutions, as well.

The form of $\underline{\mathbf{W}}$ is an (FIR) matrix, which will be discussed shortly.

As we have seen in the Wiener multichannel cost [Eq. (9.14)], it is simply a sum of single-channel costs, we wish to try the same strategy in dealing with blind multichannel problems.

Hypothesis 1. Given sources of known pdfs, and the cost functions J_i needed for single-channel blind equalization of each source, then a multichannel blind cost function can be formed as the sum of the single-channel J_i:

$$J = J_1(\hat{x}_1) + J_2(\hat{x}_2) + \cdots + J_N(\hat{x}_N) \tag{9.20}$$

or

$$J = \sum_{\text{all } i} J_i(\hat{x}_i) \tag{9.21}$$

Remark 1 (Key Result). In view of multichannel MMSE cost function [Eq. (9.13)] and the results of Bussgang blind equalization presented earlier, a plausible multichannel blind cost function is

$$J = \sum_{\text{all } i} E\left\{ \log \frac{f_{x_i}(\hat{x}_i)}{f_G(\hat{x}_i)} \right\} \tag{9.22}$$

9.2.4 Application of Multichannel Cost Using the FIR Matrix Inverse

Definition 1 (FIR Matrix). A matrix with FIR, moving average (MA), or all-zero representation (numerator only) filters as elements is a FIR matrix.

The FIR matrix inverse system is the solution of the MMSE equation [Eq. (9.19)]. We describe the elements of the FIR matrix as

$$\underline{\mathbf{W}} = \begin{bmatrix} w_{11} & w_{12} & w_{13} & \cdots & w_{1N} \\ w_{21} & w_{22} & w_{23} & \cdots & w_{2N} \\ \cdot & \cdot & & & \cdot \\ \cdot & & \cdot & & \cdot \\ \cdot & & & \cdot & \cdot \\ w_{M1} & w_{M2} & w_{M3} & \cdots & w_{MN} \end{bmatrix} \tag{9.23}$$

where each w_{ij} element is an FIR filter.

We first present the stochastic-gradient adaptive implementation of the multichannel Wiener filter (training-based)–multichannel least mean squares (MLMS).

9.2.5 Multichannel Least Mean Squares

Using the cost function $J = \operatorname{tr} E\{(\hat{\mathbf{x}} - \mathbf{x})(\hat{\mathbf{x}} - \mathbf{x})^H\}$, we have the updates in terms of individual components of the inverse system,

$$\frac{\partial J}{\partial w_{ij}} = \frac{\partial J}{\partial \hat{x}_i} \frac{\partial \hat{x}_i}{\partial w_{ij}} = \frac{\partial J}{\partial \hat{x}_i} y_j = (\hat{x}_i - x_i) y_j^* \qquad (9.24)$$

In general notation, we have

$$\frac{\partial J}{\partial \underline{\mathbf{W}}} = \frac{\partial J}{\partial \hat{\mathbf{x}}} \frac{\partial \hat{\mathbf{x}}}{\partial \underline{\mathbf{W}}} = \frac{\partial J}{\partial \hat{\mathbf{x}}} \underline{\mathbf{y}}^H \qquad (9.25)$$

The familiar error $(\partial J / \partial \hat{x}) = (\hat{\mathbf{x}} - \mathbf{x})$ from the LMS algorithm is, here, a column vector instead of a scalar.[2]

The update is simply

$$\underline{\mathbf{W}}(n + 1) = \underline{\mathbf{W}}(n) - \mu(\hat{\mathbf{x}}(n) - \mathbf{x}(n)) \underline{\mathbf{y}}(n)^H \qquad (9.26)$$

where $\hat{\mathbf{x}}(n) = [\hat{x}_1(n)\hat{x}_2(n) \cdots \hat{x}_N(n)]^T$, $\mathbf{x}(n) = [x_1(n)x_2(n) \cdots x_N(n)]^T$, $\underline{\mathbf{y}}(n) = [\underline{y}_1(n)\underline{y}_2(n) \cdots \underline{y}_M(n)]^T$.

Convergence analysis of this multichannel algorithm is similar to the single-channel case (Widrow et al. 1975), with the speed of convergence and allowed step sizes being governed by the eigenvalues of the *block* Toeplitz autocorrelation matrix $\mathbf{R} = \mathbf{Y}\mathbf{Y}^H$.

9.2.6 Multichannel Blind Least Mean Squares

We name the following algorithm multichannel blind least mean squares (MBLMS) because of the similarity of its update equation to the MLMS update equation just given. Using the blind cost function,

$$J = \sum_{\text{all } i} E\left\{ \log \frac{f_{x_i}(\hat{x}_i)}{f_G(\hat{x}_i)} \right\} \qquad (9.27)$$

the updates in terms of individual components of the inverse system are

$$\frac{\partial J}{\partial w_{ij}} = \frac{\partial J}{\partial \hat{x}_i} \frac{\partial \hat{x}_i}{\partial w_{ij}} = (\hat{x}_i - g_i(\hat{x}_i)) y_j^* \qquad (9.28)$$

---

[2] In contrast, several useful multichannel algorithms [see, for example, Widrow et al. (1975), Slock et al. (1992), and Gooch (1983)], use a single input with multiple reference signals, obtaining an error that is still a scalar and a vector of filters instead of a matrix of filters.

In general notation,

$$\frac{\partial J}{\partial \underline{\mathbf{W}}} = \frac{\partial J}{\partial \hat{\mathbf{x}}} \frac{\partial \hat{\mathbf{x}}}{\partial \underline{\mathbf{W}}} = \frac{\partial J}{\partial \hat{\mathbf{x}}} \mathbf{y}^H \tag{9.29}$$

The blind "error" is also a vector:

$$\frac{\partial J}{\partial \hat{\mathbf{x}}} = (\hat{\mathbf{x}} - \mathbf{g}) \tag{9.30}$$

where $\mathbf{g} = [g_1(\hat{x}_1) \ g_2(\hat{x}_2) \cdots g_N(\hat{x}_N)]^T$. The update is

$$\underline{\mathbf{W}}(n+1) = \underline{\mathbf{W}}(n) + \mu(\hat{\mathbf{x}}(n) - \mathbf{g}(n))\underline{\mathbf{y}}(n)^H \tag{9.31}$$

where $\mathbf{g}(n) = [g_1(\hat{x}_1(n))g_2(\hat{x}_2(n)) \cdots g_N(\hat{x}_N(n))]^T$.

For an example using two sources, the blind cost is

$$J = O_4^2(\hat{x}_1) + O_4^2(\hat{x}_2) \tag{9.32}$$

using the assumption that the sources are both of the generalized Gaussian pdf family with shape parameter $= 4$. Performing the derivative calculation, we have updates:

$$w_{11}(n+1) = w_{11}(n) + \mu\left(\hat{x}_1(n) - \frac{E|x_1|^2}{E|x_1|^4}\hat{x}_1(n)^3\right)\underline{y}_1^*(n)$$

$$w_{12}(n+1) = w_{12}(n) + \mu\left(\hat{x}_2(n) - \frac{E|x_2|^2}{E|x_2|^4}\hat{x}_2(n)^3\right)\underline{y}_1^*(n)$$

$$w_{21}(n+1) = w_{21}(n) + \mu\left(\hat{x}_1(n) - \frac{E|x_1|^2}{E|x_1|^4}\hat{x}_1(n)^3\right)\underline{y}_2^*(n) \tag{9.33}$$

$$w_{22}(n+1) = w_{22}(n) + \mu\left(\hat{x}_2(n) - \frac{E|x_2|^2}{E|x_2|^4}\hat{x}_2(n)^3\right)\underline{y}_2^*(n)$$

Here we have replaced the estimated statistic, $(E|\hat{x}_1|^2)/(E|\hat{x}_1|^4)$, with the known statistic.

9.2.7 Multichannel Blind RLS

Using training based cost, Eq. (9.13), we compute the gradient

$$\underline{\mathbf{W}}(n+1) = \underline{\mathbf{W}}(n) - \mathbf{R}^{-1}(\hat{\mathbf{x}}(n) - \mathbf{x}(n))\underline{\mathbf{y}}(n)^H \qquad \text{MRLS} \tag{9.34}$$

Using blind cost, Eq. (9.22),

$$\underline{W}(n+1) = \underline{W}(n) + \mathbf{R}^{-1}(\hat{\mathbf{x}}(n) - \mathbf{g}(n))\underline{y}(n)^H \qquad \text{MBRLS.} \qquad (9.35)$$

9.2.8 Cost Functions Directly Based on the Bussgang Property

In 1952, Bussgang made the simple and powerful discovery (Bussgang 1952), that under certain conditions the autocorrelation of a stochastic noise process would equal the correlation of the sequence and the sequence passed through an amplitude nonlinearity:

$$E\{\hat{x}(n+k)\hat{x}(n)\} = E\{\hat{x}(n+k)g(\hat{x}(n))\} \qquad \text{Bussgang property (Bellini 1994)}$$
$$(9.36)$$

The Bussgang property carries with it the following stipulations: the source must be zero mean, symmetrically distributed, finite variance, and be independently identically distributed (i.i.d.) (the independence stipulation means that no correlation exists in the data—it is *both* magnitude white and without phase correlations.)

The Need for True Phase Equalization/Separation   The autocorrelation $E\{\hat{x}(n+k)\hat{x}(n)\}$ is a single delta function only under the condition of no magnitude or phase correlation. To illustrate this subtle point, consider a filter possessing Fourier transfer function with real coefficients all unity and imaginary coefficients all equal to zero. Multiply these by an all-pass filter made up of complex values on the unit circle e^{-jx}, where x can take any real value. Such a filter will still have a unit magnitude response, but a nonzero phase response.[3] Because the autocorrelation is symmetric, the phase response of the data is lost, similar to the way that the sign of a number is lost when the number has been squared. The frequency domain method of calculating the autocorrelation shows this loss of phase: $X \cdot X^*$, where X is the Fourier transform of x.

The Bussgang method for calculating the autocorrelation provides phase information. As long as the pdf $f(x)$ of the data is not Gaussian, $f'(x)/f(x) \neq x$ and phase will not be canceled out. The further the data are from Gaussian [see (Gray 1979; Hogg 1972) for candidate distance measures], the more powerful will be the blindly converging algorithm that is applied to the data. The Bussgang property is an example of a flexible (not constrained to integer orders) path to true phase information.[4]

---

[3] The matrix problem of source separation also has this phase specification problem. Consider a matrix A. Similarity transform it to diagonal form with its eigenvalues on the diagonal. Now multiply each eigenvalue by e^{-jx} where x can take any real value, similarity transform back, and call the matrix B. We have $AA^H = BB^H$, illustrating the nonuniqueness of second-order matrix correlations and the need for true phase separation.

[4] Higher-order spectra may be used to recover phase information as well (Nikias and Petropulu 1993), and the Bussgang measure $E\{\hat{x}_{i+k}g(\hat{x}_i)\}$ may be viewed as a slice of the bispectrum when $g(\cdot) = x^2$ and the trispectrum when $g(\cdot) = x^3$.

The Bussgang nonlinearity, which is the log derivative of the source pdf, appears when computing the gradient of the information theoretic cost given in Eq. (9.22). The Bussgang property allows us to compute the autocorrelation two ways. Since the Bussgang property only holds at convergence (the point of full equalization or separation), it is proposed that the two different autocorrelation methods be computed and the difference be used as a metric or cost that iteratively guides the algorithm to convergence.

This said, it is more revealing to write the time-domain Bussgang property as

$$\delta(0) = E\{\hat{x}(n+k)g(\hat{x}(n))\}$$

where again the same stipulations used previously apply. The proposed cost is then

$$J = \delta(0) - E\{\hat{x}(n+k)g(\hat{x}(n))\}, \qquad \text{time domain}$$

$$J = 1 - \mathbf{R_g}, \qquad \text{frequency domain}$$

where $\mathbf{R_g} = E\{XX_g^*\}$ and X and X_g are the Fourier transforms of the data x and $g(x)$, respectively.

This form of $1 - \mathbf{R_g}$ was used in the update appearing in Amari et al. (1996). In the adaptive methods used in the simulations, it is the *instantaneous* value of the costs that is used in the update made at each iteration, for example, $J(n) = 1 - \mathbf{R_g}(n)$.

An adaptive matrix-whitening algorithm that utilizes cost $1 - \mathbf{R}$, where R is the matrix correlation $E\{X^*X\}$, was perhaps the first application of the serial update or relative gradient (see Chapter 4 of this volume). Also possible is a phase-only equalizer using $\mathbf{R_g^H} - \mathbf{R_g}$. The combination of the magnitude-whitener cost $1 - \mathbf{R}$ and the phase-equalizer cost $\mathbf{R_g^H} - \mathbf{R_g}$ is one possible way of viewing the EASI approach of Cardoso and Laheld (1996).

9.2.9 Using the Bussgang Property in the Frequency Domain to Form Cost Functions

We start with the standard system of

$$x \rightarrow H \rightarrow y \rightarrow W \rightarrow \hat{x},$$

in the frequency domain,

$$Y = XH \tag{9.37}$$

$$\hat{X} = YW \tag{9.38}$$

$$E\{\hat{X}^*\hat{X}\} = E\{\hat{X}^* \text{fft}\{g(\hat{x})\}\} \quad \text{Bussgang property frequency domain} \tag{9.39}$$

Form 1: Blind LMS. Divide both sides of Eq. (9.39) by W^* (or $\underline{\mathbf{W}}^H$ with conjugate transpose for the multichannel case):

$$E\{Y^*\hat{X}\} = E\{Y^* \text{FFT}\{g(\underline{\hat{x}})\}\} \qquad \text{Bussgang form 1} \qquad (9.40)$$

This gives the update form 1:

$$\Delta W = \mu(\hat{X} - \text{FFT}\{g(\underline{\hat{x}})\})Y^*, \qquad \text{BLMS} \qquad (9.41)$$

or the traditional Bussgang result in the time domain, that is, $(\hat{x} - g(\hat{x}))\underline{y}^*$. The name blind LMS is given because of the updates similarity to the LMS update, is "blind enabled" LMS.

Form 2: INFOMAX/Direct Minimum-Entropy Deconvolution. Using the relations $Y = XH$ and $\hat{X} = YW$, express the left-hand side of Eq. (9.40) in X:

$$WH^*HE\{X^*X\} = E\{Y^* \text{FFT}\{g(\underline{\hat{x}})\}\} \qquad (9.42)$$

Since the source data are independent and $H = W^{-1}$ at convergence, we get

$$\frac{1}{W^*} = E\{Y^* \text{FFT}\{g(\underline{\hat{x}})\}\} \qquad \text{Bussgang form 2} \qquad (9.43)$$

This gives the direct Bussgang update form 2:

$$\Delta W = \mu\left(\frac{1}{W^*} - \text{FFT}\{g(\underline{\hat{x}})\}Y^*\right), \qquad \text{INFOMAX} \qquad (9.44)$$

the INFOMAX (Bell and Sejnowski 1995), or the direct minimum-entropy deconvolution (DMED) (Lambert 1996) result.

Form 3: Direct Bussgang Cost. Starting with the Bussgang property itself (in the frequency domain),

$$E\{\hat{X}^*\hat{X}\} = E\{\hat{X}^* \text{FFT}\{g(\underline{\hat{x}})\}\} \qquad (9.45)$$

defining $R = E\{\hat{X}^*\hat{X}\}$ and $R_g = E\{\hat{X}^* \text{FFT}\{g(\underline{\hat{x}})\}\}$ gives the direct Bussgang update form 3:

$$\Delta W = \mu(R - R_g) \qquad \text{direct Bussgang cost (DBC)} \qquad (9.46)$$

Form 4: Amari Form. Starting with the Bussgang property with the independent data stipulation explicit,

$$1 = E\{\hat{X}^* \text{FFT}\{g(\underline{\hat{x}})\}\} \qquad (9.47)$$

gives the direct Bussgang update form 4:

$$\Delta W = \mu(1 - R_g) \qquad \text{Amari form} \qquad (9.48)$$

9.2.1 Finite-Difference Approximation, the Serial Update, and the Blind Serial Update

The traditional forms of LMS adaptive algorithms are stochastic-gradient methods. The original stochastic-gradient idea was presented in a pioneering paper by Robbins and Monro (1951). One year later, Kiefer and Wolfowitz presented what came to be known as the finite difference approximation method (Kiefer and Wolfowitz 1952). The finite difference approximation to the true gradient of a function is made by evaluating the cost at two points spaced a small distance apart and computing their difference: $J(x) - J(x + \delta)$. A further approximation, possible using stochastic measures on the output variable, which have their minimum at the zero scalar or zero matrix, is

$$\Delta W = \mu J(n) \tag{9.49}$$

where $J(n)$ is the instantaneous value of the cost computed at time point n. We shall use stochastic measures such as functions of the correlation and higher-order correlation.

A serial or relative-gradient update [as introduced in Cardoso and Laheld (1996)] can then be viewed as a finite difference approximation update with a vector step size μW:

$$\Delta W = \mu J(n) W \tag{9.50}$$

The finite difference stochastic measures we refer to as costs, but they are different from traditional scalar costs for stochastic-gradient adaption in that they can be matrix valued. Even the LMS algorithm can be seen to use a type of finite difference approximation with

$$J(n) = R_{xy} - R_{\hat{x}y}$$

where the cross-correlations between x and y, and $\hat{x}$ and y are used. When the original source data x are i.i.d., the cross-correlations at convergence equal H. The vector step-size window of the inverse of H, W has been noted to have a stabilizing effect in the convergence of serial update forms. Although the serial update does not remove eigenvalue spread dependence as does an RLS update, it improves robustness in many experimental cases.

9.2.11 Summary of Algorithms

Now combining costs with adaptation methods, we obtain the algorithms summarized in table form. Cost functions are summarized in Table 9.1, and

Table 9.1 Table of Bussgang-Related Algorithm Cost Functions in the Frequency Domain

Blind Algorithm	Cost
DBC	$R - R_g$
EASI	$(1 - R) + (R_g - R_g^*)$
AMARI	$1 - R_g$
Infomax or DMED	$\left(\dfrac{1}{W^*} - R_{gy}\right)$ or $E\{\log|W|\Phi_x(\omega_{\hat{x}})\}$
BLMS	$R_{\hat{x}y} - R_{gy}$ or $E\left\{\log\dfrac{\Phi_x(\omega_{\hat{x}})}{\Phi_G(\omega_{\hat{x}})}\right\}$

Note: As given, these equations hold for single-channel and all square systems, except BLMS, which also holds for over-determined systems (more sensors then sources). Where possible, the scalar cost for stochastic-gradient adaption is given. This summary holds for single-channel and multichannel problems with or without multipath.

algorithm update forms are summarized in Table 9.2. These algorithms are shown here in the frequency domain and are implementable either via a block adaptive frequency domain overlap and save method or via the time domain with updates occurring at every time point. See Table 9.4 for an example of time-domain serial and finite difference updates.

9.2.12 Convergence

Blind convergence of Bussgang and Bussgang-related costs has been discussed in several papers (Bellini 1994; Benveniste et al. 1980; Donoho 1981; Vembu et al. 1994). In the noiseless case, single-channel Bussgang blind equalizers have been shown to have convergence to a unique solution (the Wiener solution), up to a sign change, an amplitude scale factor, and time-delay shift, under the following conditions:

1. The source signal is non-Gaussian, of finite variance, and independent identically distributed.
2. The forward system we are equalizing is linear and time invariant.
3. The theoretically infinite extent of the FIR inverse-channel estimate is sufficiently well approximated and remains "zero-centered" (Foschini 1985).
4. The system is invertible. No roots of the forward channel can lie on the unit circle.

Table 9.2 Table of FIR Bussgang-Related Algorithm Updates for Blind Equalization and Blind Separation

Blind Algorithm	Update
DBC-FD	$\Delta W = \mu(R - R_g)$
DBC-BSU	$\Delta W = \mu(R - R_g)W$
EASI-FD	$\Delta W = \mu((1 - R) + (R_g - R_g^*))$
EASI-BSU	$\Delta W = \mu((1-R) + (R_g - R_g^*))W$
AMARI-FD	$\Delta W = \mu(1 - R_g)$
AMARI-BSU	$\Delta W = \mu(1 - R_g)W$
Infomax or DMED	$\Delta W = \mu\left(\dfrac{1}{W^*} - \mathrm{FFT}\{g(\hat{x})\}Y^*\right)$
BLMS	$\Delta W = \mu(\hat{X} - \mathrm{FFT}\{g(\hat{x})\})Y^*$

Note: This summary holds in general form for single-channel and multichannel problems with or without multipath. Multichannel forms use underlined boldface notation with conjugation replaced by hermitian transpose.

Remark 2. The authors of this chapter have noted convergence of the Bussgang multichannel blind cost, Eq. (9.22), using several different minimization methods to the Wiener solution up to a permutation, sign change of the sources, a scale factor, and a time-delay shift. Viewing the multichannel Bussgang cost of Eq. (9.22), as a multichannel matrix/vector algebra version of convergent single-channel Bussgang equalizers, its adaptive convergence to the multichannel Wiener solution, Eq. (9.19) has been experimentally obtainable under the following additional assumptions:

1. The source signals are each non-Gaussian, finite variance, uncorrelated (both spectrally white and uncorrelated with other sources), and i.i.d.
2. The theoretically infinite extent of each FIR element of the multichannel inverse system estimate is sufficiently well approximated and remains "zero-centered."
3. The forward system is invertible.

   No roots of the determinant of the forward channel can lie on the unit circle.

   The forward system must be full column rank, having sensors ≥ sources.
4. Care in the choice of adaptive step size is required for multichannel systems. The adaptive step size must be small. In the case of single-channel systems, the step size may cause the system to become unstable if set to a

large value. The same thing will occur with multichannel systems, but in addition to instability, it is also possible that "same source lockon" will occur, that is, more than one row of the estimated inverse channel has an identical source as output.

In the experiment/simulation section we show several examples of convergence to permuted, scaled, and shifted Wiener solutions. When the forward channel is known, a useful measure of convergence, which is insensitive to the permutation, scale factor, and time delay, is intersymbol interference (ISI), computed on the rows and columns of the global system.

9.2.13 Analysis and Performance Measures

Asymptotic performance measures of costs [i.e., Eqs. (9.11) and (9.12)] are given. We measure the general performance of tracking ability of fixed step-size adaptive algorithms. The analysis verifies for sources from the generalized Gaussian family of pdfs' that the Bussgang nonlinearity is optimal to use in equalization/separation problems. This analysis is important, as it gives an idea of how performance will degrade if the source data do not exactly match the parametric model of the data inherent in the nonlinearity used by the blind algorithm. For comparison, the analysis is also carried out for the standard MMSE cost and also the MMSE-NG cost, which exploits non-Gaussianity of the source.

The Generalized Gaussian or Exponential Power Family The generalized Gaussian family of distributions,

$$f_x(x) \propto e^{-(|x|^s/sE|x|^s)} \tag{9.51}$$

is useful because it describes pdf's ranging from impulsive, with $s < 1$, to the Gaussian itself, with $s = 2$, to more bounded pdf's, with $s \gg 1$. The uniform distribution has $s = \infty$. We will use this model in computing performance measures of cost functions set for the generalized Gaussian family

$$J = E\left\{\log \frac{Ke^{-(|\hat{x}|^s/sE|x|^s)}}{f_G(\hat{x})}\right\} \quad \text{and} \quad J = E\left\{\log \frac{Ke^{-(|x|^s/sE|x|^s)}}{f_G(\hat{x})}\right\}$$

which are costs, Eqs. (9.11) and (9.12), specified at the generalized Gaussian family.

9.2.14 Asymptotic-Performance Measures of the Blind Cost and MMSE-NG for the Generalized Gaussian Family

Asymptotic performance is made analytically tractable by considering the system at steady-state convergence; thus, it measures the ability of the cost function to keep the system at convergence when slightly perturbed. At steady-state convergence, sources are separated, and there is no difference between single and multichannel analysis. We make use of the cost-function performance measure given in Benveniste (1987, chap. 4), Donoho (1981), and Martin (1979). It is a measure of tracking ability and may be viewed as the error power expression multiplied by the maximum allowable step size determined by the inverse Hessian [see Lambert and Nikias (1995)]:

$$\mathbf{T} = \frac{E(\partial J/\partial w)^2}{(E(\partial^2 J/\partial^2 w))^2} = \frac{E(\partial J/\partial \hat{x})^2}{(E(\partial^2 J/\partial^2 \hat{x}))^2} \mathbf{R}^{-1} = A(x,s)\mathbf{R}^{-1} \tag{9.52}$$

This measure $\mathbf{T}$ shows the $A(x,s)$ factor depending on the cost function and data type, and the $\mathbf{R}^{-1}$ term depending on the channel (the eigenvalue spread). The factor $A(x,s)$ is what we refer to as the performance measure.

For the Bussgang cost

$$J = E\left\{ \log \frac{Ke^{-|\hat{x}|^s/(sE|x|^s)}}{f_G(\hat{x})} \right\}$$

we calculate $A(x,s)$ (a scalar)

$$A(x,s) = \begin{cases} \dfrac{(E|x|^{2(s-1)})/(E|x|^s)^2 - 1}{(1 - (s-1)(E|x|^{s-2})/(E|x|^s))^2} & \text{if } s \neq 1 \\[4mm] \dfrac{(E|x|^2)/(E|x|)^2 - 1}{(1 - ((E|x|^2)/(E|x_i|))2p_{x_i}(0))^2} & \text{if } s = 1 \end{cases} \tag{9.53}$$

For MMSE-NG

$$J = E\left\{ \log \frac{Ke^{-|x|^s/(sE|x|^s)}}{f_G(\hat{x})} \right\}$$

$$A(x,s) = \begin{cases} \dfrac{E|x|^{2(s-1)}}{(E|x|^s)^2} - 1 & \text{if } s \neq 1 \\[4mm] \dfrac{E|x|^2}{(E|x|)^2} - 1 & \text{if } s = 1. \end{cases} \tag{9.54}$$

By comparison, the MMSE cost $A(x,s) = 1$. (It is not data dependent.)

For the Bussgang cost and MMSE-NG in Fig. 9.3, we plot bounds on $A(x,s)$ by using the optimum cost function for each generalized Gaussian data type. For example, for generalized Gaussian data with $s = 2.5$, we have computed and plotted $A(x, 2.5)$, and likewise $A(x,s)$, for every s in the range

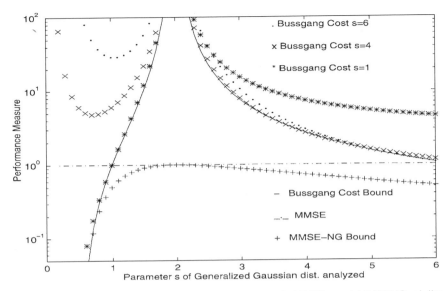

Figure 9.3 Performance measures for Bussgang cost, MMSE, and MMSE-NG at the generalized Gaussian family.

$(0 < s < 6)$. The x-axis here is data type, the parameter s from the generalized Gaussian family. The different curves plotted show the relative figure of merit for the their respective cost functions used on the data type specified on the x-axis (a smaller value means a more powerful cost). Figure 9.3 is useful in determining which algorithm to use for a given source data type. Donoho (1981) has a plot similar to Fig. 9.3; here we have considered more cases (super- and sub-Gaussian data types), and added the MMSE-NG analysis.

In Fig. 9.3 small values on the performance-measure axis denote faster convergence and tracking of the algorithm. Note that the convergence for blind algorithms when the source is Gaussian is impossible. Note that the MMSE and MMSE-NG are equal when $s = 2$, and note the relative flatness of the MMSE-NG curve around the Gaussian data type. Note, for example, that it is not a wise choice to use a Bussgang cost set for $s = 6$, when the data type is Laplace $(s = 1)$.

9.2.15 Conclusions for the Cost-Function Section

We began the section on the foundation of MMSE Wiener filtering cost functions by presenting a way, the Bussgang property, to exploit the non-Gaussianity of source data to obtain blind costs. We have concentrated on the general family of Bussgang-related blind cost functions and have found three major classes:

1. The Blind LMS (stochastic-gradient) types;
2. The Infomax types, which are unique since an inverse separation matrix appears in the update;
3. And what we term the direct Bussgang costs (EASI, Amari, and DBC), which are unique since their updates only use the output of the equalizer/separation matrix.

The serial update technique of applying the current estimate of inverse channel as a vector step size in the update equation was found to be applicable to all classes and can have a stabilizing effect on convergence.

The Bussgang-cost family uses the log derivative of the pdf of the source data to optimally process different data types. The Bussgang cost's global convergence in the single-channel case (Donoho 1981), led us to expect excellent convergence properties in the multichannel case, as the experiment has verified.

9.3 FINITE IMPULSE RESPONSE MATRIX ALGEBRA

9.3.1 Introduction

The authors of this chapter began to see the very general concepts involved with treating the FIR filter the same way that one would a scalar in traditional matrix algebra. What began as a method for whitening multipath mixtures, uncovered a very general tool-FIR matrix algebra. We will first present the algebra of single-channel FIR filters and then show its extension in FIR matrices. Application of FIR-matrix algebra to whitening multipath mixtures is given.

The techniques of blind equalization and blind source separation are overlapping in many instances; in the former case, adaptive filters are employed, and in the latter, adaptive matrices. Existing blind equalization techniques are for single-channel multipath channels. Source-separation algorithms usually involve scalar matrix mixtures and do not allow for multipath. FIR matrices can essentially combine equalization of multipath and separation of matrix mixtures to allow for separation of multipath mixtures. FIR matrix methods employ extensions of standard matrix tools, eigenvalue routines, factorizations, algorithms, and so forth.

9.3.2 Wide Band or Narrow Band?

It is possible to solve an array processing or source-separation problem at a specific frequency for the narrow-band data case using scalar matrices. In many cases, the wide-band solution can be said to be made up of many narrow-band solutions, each at a different frequency. Indeed, if we Fourier transform the elements (each an FIR filter) of an FIR matrix, we obtain many scalar matrices, each applying to a different (independent) frequency.

Yet, many applications require the knowledge of eigenvectors and eigenvalues of the matrix, to be used for diagonalization, inverting the matrix, or performing an arbitrary function on the matrix. The association of eigenvectors and eigenvalues is important, as a particular eigenvalue has a corresponding eigenvector. It is very difficult to keep track of *orders* and *associations* of eigenvalues as a function of frequency and eigenvector coefficients as a function of frequency using many traditional scalar (or multipath-free) matrices at different frequencies (frequency domain), or a filter with scalar matrix coefficients (time domain).

The Need for a Formal FIR Polynomial Matrix Algebra   A problem specific to the source-separation application, is that of permutation of sources. If separation of sources is based on knowledge of statistics only, it is possible that the order of sources permute. Keeping track of these permutations at all frequencies in a wide-band or multipath application could be difficult or impossible. We avoid this problem through the use of FIR polynomial matrices.

9.3.3   A Polynomial Matrix or a Matrix Polynomial? Polynomial Matrix!

Traditional matrix algebra is noncommutative (right and left multiplication are distinct operations, $AB \neq BA$.) It is trivial to verify that source-separation problems of sending vector data or multiple sources through matrix channels also follow this noncommutative property. The order of cascaded channels becomes important when multiple sources are involved. It is trivial to show that adding multipath to the matrix channels does not change this noncommutative requirement. Noncommutivity is a property that again is not easily applied using polynomials with matrix coefficients. We therefore need a polynomial matrix (a matrix of polynomials), a matrix made of polynomial elements with the noncommutative property of matrices left intact. Table 9.3 will summarize these concepts.

9.3.4   Adaptive vs. Batch Processing?

It is possible to solve many multipath matrix problems using block Toeplitz methods, as seen in the rich geophysical data-processing literature (Claerbout 1985; Robinson 1967; Walden 1985). These methods involve batch, off-line processing, where the data are available for the formation of data matrices that can be quite large and contain a great deal of redundancy. We obtain FIR matrices as an isomorphic mapping of a special form of block Toeplitz matrices, block circulant matrices. The FIR matrix representation is more amenable to *adaptive* filtering uses.

Thus, we see the need for FIR matrix algebra where the traditional matrix/vector algebra rules and tools are intact (matrix right/left multiplication, inner product, transpose, etc.) and matrix elements are FIR filters.

The exciting implication of FIR matrix algebra is that many matrix techniques using functions of a matrix (such as source-separation algorithms) can be applied to single-channel or multichannel filter applications, and the area of adaptive filter theory can be viewed as a single-channel case of the source-separation problem. Many of the powerful results of adaptive filtering are easily extended to the multichannel source-separation case and vice versa.

9.3.5 Algebraic Properties and Definitions

By an abstract algebra, we mean a set of elements that satisfy certain properties under an operation such as addition, multiplication. By a group, we mean an algebra that satisfies the following requirements:

1. The product of any two elements is also a member of the set. The set is *closed.*
2. The associative law holds, that is, $A(BC) = (AB)C$.
3. There exists a unit element E such that $EA = AE = A$.
4. There exists in the set an inverse to each element A such that $AA^{-1} = A^{-1}A = E$.

If the group multiplication is commutative, so that $AB = BA$, the group is said to be Abelian.

The group of complex numbers is an example of a commutative (Abelian) group. The general linear group of invertible matrices is an example of a non-commutative group. Just as in the case of the general linear group of invertible matrices, we exclude consideration of the zero element under the division operator. A summary of the algebraic properties of FIR filter and FIR polynomial algebras is given later in Table 9.3.

9.3.6 The Algebra of FIR Filters

Although functions of a matrix such as the matrix exponential and matrix inverse are well documented and understood, very little is said regarding functions of filters or polynomials. The Wiener inverse filter is widely used, and the cepstrum (log of a filter) has received some attention, but a formal "FIR algebra" has not been formally addressed. We first turn our attention to the matter of single-channel FIR algebra and show a unique relation between the matrix functional of block circulant matrices, the polynomial functional, and the discrete Fourier transform (denoted as FFT). When a time-delay shifting of the unit element is permitted, we are able to define a group of FIR polynomials (under elementwise multiplication) or FIR filters (under convolution).

The time delay must be available when nonminimum phase roots (with magnitude greater than 1) are present in the filter. The inverse of a non-minimum phase root must be expanded noncausally for stability.

We adopt the convention of saying FIR polynomial when the frequency domain is meant. The term polynomial is used here, because we use the double-sided z-transform (Laurent series) and write the filters as polynomials in z

(sampling z on the unit circle gives the discrete Fourier domain). FIR filters will refer to the discrete time domain and are mathematically described by a series of scaled and delayed delta functions.

FIR *polynomials* are a group under addition, multiplication, and division, exactly as in the case of complex scalars. In the frequency representation, each coefficient is a complex scalar and is independent from the others. We must stipulate that perfect spectral nulls are not present, and the double-sided Laurent series expansion must be sufficiently long.

FIR filters can be thought of as an algebraic group with the operations of addition, convolution, and deconvolution defined. The nature of FIR filters moves us toward the discrete frequency domain, as even the time-domain methods for computing functions of an FIR filter, such as the inverse, involve the formation of a circulant "data matrix," the eigenvalues of which are simply the discrete Fourier coefficients [see Gonzales and Wintz (1977)]. The eigencolumns of the circulant matrix are nothing other than the discrete Fourier basis functions of the transform of length equal to the number of rows or columns in the circulant matrix.

Linear equalization systems (both blind and training-based) use the algebra of FIR polynomials. We routinely use functions of these filters, the equalization filter is the inverse, the -0.5 power yields a whitening filter, and we can define general functions of a filter h as $g(h) = \text{IFFT}(g(\text{FFT}(h)))$, where the function $g(\cdot)$ is applied to each element in $\text{FFT}(h)$.

9.3.7 FIR Filter Functional Definition

Definition 2 *(FIR Filter Functional Definition)*. The function $g(\cdot)$ acting on an FIR filter h, is logically defined as

$$g(h) = \text{IFFT}\{g(\text{FFT}\{[0 \quad \cdots \quad 0 \ h \ 0 \quad \cdots \quad 0]\})\}$$

where sufficiently many zeros are prepended and postpended to allow the FIR (numerator only) representations to well approximate arbitrary filters. The function $g(\cdot)$ acts elementwise on the Fourier representation of h.

Using this definition, $h * h^{-1} = e$ (the identity element) and $h = \log(\exp(h))$, $h = ((h)^3)^{1/3}$, and so forth. This property can be proven through the use of the matrix functional applied to a circulant data matrix. This is similar to the result, from the area of algebraic coding with Galois finite groups, which states that powers of polynomials are defined by applying the power to each element in the polynomial [see Lin and Costello (1983, 29)].

First, explicitly define

$$f = [0 \quad \cdots \quad 0 \ h \ 0 \quad \cdots \quad 0]$$

as the filter h with prepended and postpended zeros, then construct the square-circulant data matrix[5] X from f:

$$X = \text{circulant}(f)$$

The number of zeros prepended and postpended to h determines the length of series expansion to use in the determination of $g(h)$.

The Wiener inverse filter using a data matrix of this type is

$$f^{-1} = (X^H X)^{-1} X p$$

The vector p serves to select the inverse filter of a certain delay. Here we choose $p = [1 \; 0 \; \cdots \; 0]^T$, the delay used is zero, since zeros have already been prepended in forming f.

In this case of *square* circulant data matrices X, we have

$$f^{-1} = ((X)^{-1} p)^T$$

from which we envision a generalization relating matrix functionals (established result) to polynomial or filter functionals. The matrix functional applied to the circulant matrix X is equivalent to the use of the discrete Fourier representation of the filter.

A filter f (row vector) equals the transpose of the first column of its circulant matrix X:

$$f = (X p)^T \tag{9.55}$$

This follows from the definition of circulant matrices.

We define the mapping function diag$\{\cdot\}$ as the matrix/vector operation that converts the diagonal of a square matrix into a vector or a vector into a diagonal matrix. Thus, diag$\{\cdot\}$ is its own inverse.

Elsewhere in this chapter, a nonlinear functional type is context dependent. In this section, in order to prove a useful theorem, we define specific nonlinear functionals building on the scalar nonlinear function $g_s(\cdot)$. The same function $g_e(\cdot)$ (inverse, log, cube, etc.) can be applied elementwise to a vector which we denote as

$$g_e(a) = [g_s(a_1) g_s(a_2) g_s(a_3) \cdots]$$

where a is a vector. The function can be a matrix functional operating on a

---

[5] The MATLAB command to form a circulant matrix is

$$\text{toeplitz}(f, [f(1) \quad \text{fliplr}(f(2 : \text{length}(f)))])$$

(It is a valuable exercise to check this proof on the computer.) The circulant matrix is a special form of a Toeplitz matrix.

matrix

$$g_M(A) = T \, \text{diag}\{g_e(\text{diag}\{\Lambda\})\} T^{-1}$$

where A is a square matrix and T and Λ are respectively the eigenvector and eigenvalue matrices of A.

Theorem 1. An FIR filter function $g_f\{\cdot\}$ applied to filter f equals the transpose of the first column of the result of applying the same function $g_M\{\cdot\}$ as a matrix functional to the circulant matrix constructed from f. Furthermore, this filter function may be computed by application of the elementwise function to the discrete Fourier coefficients of f, the result of having the inverse Fourier transform applied to it:

$$g_f(f) = (g_M(X)p)^T = \text{IFFT}\{g_e(\text{FFT}\{f\})\} \qquad (9.56)$$

Using the circulant matrix property [Gonzales and Wintz (1997)] that the eigenvalues of a circulant matrix are the discrete Fourier coefficients of the first column of X,

$$\text{FFT}\{f\} = \text{diag}\{T^{-1}XT\} = \text{diag}\{\Lambda\} \quad \text{GW Property 1}$$

where T is the eigenvector matrix of X, we can prove the result.

We find the expression for the FFT operator followed by the IFFT operation (these must go together because of the issue of ordering eigenvalues and eigenvectors)

$$\begin{aligned} f = \text{IFFT}\{\text{FFT}\{f\}\} &= (Xp)^T \\ &= (T\Lambda T^{-1}p)^T \\ &= (T \, \text{diag}\{\text{fft}\{f\}\} T^{-1}p)^T \end{aligned}$$

Now define $a = \text{FFT}\{f\}$, then

$$\text{IFFT}\{a\} = (T \, \text{diag}\{a\} T^{-1}p)^T \quad \text{GW Property 2}$$

Using similarity factorization, $X = T\Lambda T^{-1}$, where Λ is diagonal containing the eigenvalues of X,

$$g_f(f) = (g_M(T\Lambda T^{-1})p)^T$$

the matrix functional can be moved inside to operate only on eigenvalues,

$$g_f(f) = ((Tg_M(\Lambda)T^{-1})p)^T$$

which is the expression of the FFT followed by IFFT. Now $\Lambda =$

$\text{diag}\{\text{FFT}\{f\}\}$, so

$$g_f(f) = \text{IFFT}\{g_e(\text{FFT}\{f\})\}$$

which was the claim.

Corollary 1. There exists an isomorphism between FIR filters and square-circulant matrices with the FIR filter as the first row of the circulant matrix as defined earlier. In the FIR polynomial domain, we have closure under addition, subtraction, elementwise multiplication, and elementwise division. In the square-circulant matrix representation, we have closure under addition, subtraction, matrix multiplication, and matrix division.

FIR Filter Algebra: Example 1. Given a filter h, we can compute its inverse $w = \text{IFFT}\{\text{FFT}\{h\}^{-1}\}$, and obtain the unit element $e = h * w$.
Example:

$$h = \delta(0) \quad +0.5\delta(1) = [\cdots \quad 0 \quad 0 \quad 0 \quad 1 \quad 0.5 \quad 0 \quad 0 \quad 0 \quad \cdots]$$

$$w = \delta(0) \quad -0.5\delta(1) \quad +0.25\delta(2) \quad -0.125\delta(3) \quad +0.0625\delta(4) \quad \cdots$$

$$= [\cdots \quad 0 \quad 0 \quad 0 \quad 1 \quad -0.5 \quad 0.25 \quad -0.125 \quad 0.0625 \quad \cdots]$$

$$h * w = \delta(0) = [\cdots \quad 0 \quad 0 \quad 0 \quad 1 \quad 0 \quad 0 \quad 0 \quad \cdots] \text{ the unit element}$$

FIR Filter Algebra: Example 2. An example from random-variable transformations is illustrative of the algebraic properties of the group of FIR polynomials. We pass a random variable y through a filter w, and wish to determine the pdf of the output x. Starting with a single tap w (scalar gain only), we have the simple transformation:

$$p_x(x) = \frac{1}{|w|} p_y(y) \tag{9.57}$$

Moving to the M-tap FIR filter case, w is now a filter $[w_1 \ w_2 \ \cdots \ w_M]$, shorthand notation for the discrete time filter

$$w = w_1\delta\left(-\frac{M}{2}\right) + w_2\delta\left(-\frac{M-1}{2}\right) + \cdots + w_M\delta\left(\frac{M-1}{2}\right)$$

We have the FIR filter algebra, or time-domain transformation

$$p_x(x) = \frac{1}{|w|} * p_y(y) \tag{9.58}$$

where $*$ denotes the convolution operator. A function $g(\cdot)$ of a time-domain filter is defined by $g(w) = \text{IFFT}\{g(\text{FFT}\{w\})\}$; thus

$$\frac{1}{|w|} = \text{IFFT}\left\{\frac{1}{|\text{FFT}\{w\}|}\right\} \tag{9.59}$$

We propose to sample the pdf's in question and go to the discrete characteristic domain using an M-point discrete Fourier transform, where the general form of the expression will the same as in Eq. (9.57). The FIR polynomial algebra transformation is

$$\text{FFT}\{p_x(x)\} = \frac{1}{|\text{FFT}\{w\}|} \text{FFT}\{p_y(y)\} \qquad \text{or}$$

$$\Phi_x(\omega_x) = \frac{1}{|W|} \Phi_y(\omega_y) \tag{9.60}$$

Starting with a gamma (double-sided chi-square) distributed random variable, we pass it through an FIR channel described by [1 1 −0.75]. Observing the pdf of the output by histogramming the output data, we verified our result. The pdf of the output is given in Eq. (9.60). In Fig. 9.4, the spreading filter $1/|w|$ is plotted. In Fig. 9.5, we plot (9.5a) the true gamma pdf, (9.5b) the spreading filter convolved with the true gamma pdf, (9.5c) the histogram of the unfiltered gamma data, and (9.5d) the histogram of the filtered data. Figure 5b and d are of the same shape. One hundred thousand data samples and a 200-bin histogram were used for these plots.

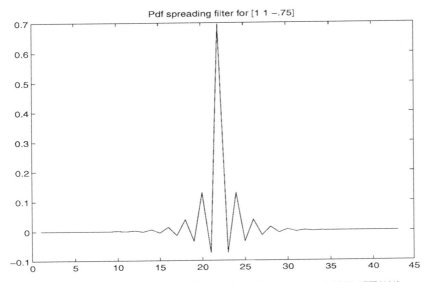

Figure 9.4 The spreading filter of $h = [1\ 1\ -0.75]$, given by IFFT $\{1/|\text{FFT}\{h\}|\}$.

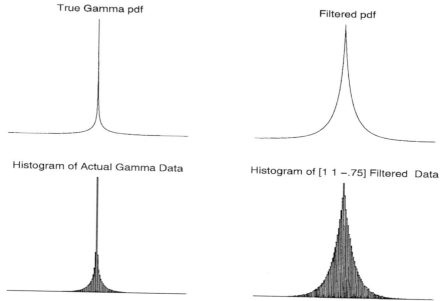

True Gamma pdf

Filtered pdf

Histogram of Actual Gamma Data

Histogram of [1 1 −.75] Filtered Data

Figure 9.5   (a) Top left—Original gamma pdf (theory); (b) top right—pdf of filtered data (theory); (c) bottom left—histogram of unfiltered data (experiment); (d) bottom right—histogram of filtered data (experiment). This shows theory and experiment in exact agreement.

In this random-variable transformation example, a filter appears in Eq. (9.60) exactly as the scalar gain did in Eq. (9.57).

9.3.8   FIR Matrices

History and Introduction   Matrices with time-delayed elements are discussed in the geophysical signal-processing literature (Claerbout 1985; Robinson 1967) and in the controls engineering and linear algebra literature (Kailath 1980; Maciejowski 1989; Rosenbrock 1970; Vardulakis 1991). The geophysical literature describes polynomials with matrix coefficients (as in the Cayley-Hamilton theorem, a matrix satisfies its own characteristic polynomial). The controls literature polynomial matrices are defined using a rational function (or ARMA) model and is an algebraic ring with a division subject to minimum phase-stability conditions. Rational-function tools, which use a denominator in the model, are needed in control problems. However, in system identification problems, where a finite group delay or processing time lag is permitted, one has the luxury of using the FIR representation capable of noncausal expansions and nonminimum phase-inverse system identification.

The elements of the FIR polynomial matrices are taken from the group of FIR polynomials (or filters) described in the previous section. When using FIR filter matrices (time domain), one can think of convolution of filter elements as

Table 9.3 Algebras Involving Polynomials and Matrices, with base Elements $a, b, c, \ldots$ Taken from the Group of Complex Scalars

Algebra	Example	Comments/Properties
Complex scalars	a	Commutative group
Scalar matrices	$\begin{bmatrix} a & b \\ c & d \end{bmatrix}$	Noncommutative group
Rational polynomial (ARMA)	$\dfrac{a + bz^{-1}}{c + dz^{-1}}$	Algebraic ring, group for minimum phase
FIR filter	$w(1)\delta\left(-\dfrac{L}{2}\right) + \cdots + w(L)\delta\left(\dfrac{L-1}{2}\right)$ or $w = [w(1)w(2)\cdots w(L)]$	Commutative group (convolution)
FIR polynomial (MA)	$a + bz^{-1}$	Commutative group (Laurent series)
Polynomial with matrix coefficients Geophysics	$\begin{bmatrix} a & b \\ c & d \end{bmatrix} + \begin{bmatrix} e & f \\ g & h \end{bmatrix}z^{-1}$	Noncommutative property awkward
Polynomial matrix (ARMA) Controls area	$\begin{bmatrix} a+bz^{-1} & e+fz^{-1} \\ c+dz^{-1} & g+hz^{-1} \\ k+lz^{-1} & p+qz^{-1} \\ m+nz^{-1} & r+tz^{-1} \end{bmatrix}$	Algebraic ring, group for minimum phase
FIR filter matrices	$\begin{bmatrix} w_{11}(1)\cdots w_{11}(L) & w_{21}(1)\cdots w_{21}(L) \\ w_{12}(1)\cdots w_{12}(L) & w_{22}(1)\cdots w_{22}(L) \end{bmatrix}$	Noncommutative group (convolution)
FIR polynomial matrices	$\begin{bmatrix} a+bz^{-1} & c+dz^{-1} \\ e+fz^{-1} & g+hz^{-1} \end{bmatrix}$	Noncommutative group (Laurent series)

Note: FIR algebras also require zero padding (both prepending and postpending).

taking the place of multiplication of scalar elements. When using FIR polynomial matrices (frequency domain), one can think of elementwise multiplication of polynomials as taking the place of multiplication of scalar elements.

Properties of these algebraic groups are summarized in Table 9.3. Most important to note for FIR polynomials and filters is the noncommutative property. These groups thus share all the properties of the general linear group of invertible N-by-N matrices. All the powerful results of traditional scalar matrix algebra can be extended to these FIR polynomial matrices, and to FIR filter matrices with convolution replacing multiplication of elements.

Definition 3. The FIR polynomial (frequency domain) matrix unit element is defined as

$$\underline{\mathbf{I}} = \underline{\mathbf{H}}\,\underline{\mathbf{H}}^{-1} = \begin{bmatrix} \bar{1} & \bar{0} \\ \bar{0} & \bar{1} \end{bmatrix} \tag{9.61}$$

where $\bar{1}$ is a sequence of all ones and $\bar{0}$ is a sequence of all zeros.

Definition 4. The FIR filter (time domain) matrix unit element is defined as

$$\underline{\mathbf{I}} = \underline{\mathbf{H}}\,\underline{\mathbf{H}}^{-1} = \begin{bmatrix} 1\,\bar{0} & \bar{0} \\ \bar{0} & 1\,\bar{0} \end{bmatrix} \tag{9.62}$$

Theorem 2. An FIR filter matrix $\underline{\mathbf{H}}$ with filter elements h_{ij}, is uniquely related to a block-circulant matrix $\mathscr{H}$, with square blocks defined by

$$\mathscr{H}_{ij} = \mathrm{Toeplitz}(h_{ij}, [h_{ij}(1) \; h_{ij}(L) \; h_{ij}(L-1) \; \cdots \; h_{ij}(2)]),$$

through the following relation

$$\underline{\mathbf{H}} = (\mathscr{H}\underline{\mathbf{I}}^T)^T$$

where h_{ij} filters have already been prepended and postpended with zeros.

Corollary 2. There exists an isomorphism between FIR polynomial matrices (N by M by L) and scalar matrices (NL by ML) made of special square-circulant blocks as defined earlier.

Theorem 3. An arbitrary function of an FIR filter matrix $\underline{\mathbf{H}}$ with filter elements h_{ij} is uniquely defined by the same function applied to the block circulant matrix $\mathscr{H}$, with square block elements defined by

$$\mathscr{H}_{ij} = \mathrm{Toeplitz}(h_{ij}, [h_{ij}(1) \; h_{ij}(L) \; h_{ij}(L-1) \; \cdots \; h_{ij}(2)]),$$

through the following relation

$$g_{\mathbf{M}}(\underline{\mathbf{H}}) = (g_{\mathbf{M}}(\mathscr{H})\underline{\mathbf{I}}^T)^T$$

where h_{ij} filters have been prepended and postpended with zeros.

The proof of Theorem 2 follows from the application of standard matrix algebra using filters as elements. It is also possible to define functions of an FIR matrix using the standard method of going to the similarity transform factorization $\underline{\mathbf{H}} = \underline{\mathbf{T}}\Lambda\underline{\mathbf{T}}^{-1}$.

Definition 5. A function of an FIR polynomial matrix is defined as

$$g_{\mathbf{M}}(\underline{\mathbf{H}}) = \underline{\mathbf{T}}g_{\mathbf{M}}(\Lambda)\underline{\mathbf{T}}^{-1}.$$

where the columns of $\underline{\mathbf{T}}$ are the eigenvectors of $\underline{\mathbf{H}}$ (with polynomial elements), and $\underline{\mathbf{\Lambda}}$ is a diagonal FIR polynomial matrix with the eigenvalue polynomials of $\underline{\mathbf{H}}$ on the diagonal.

FIR Matrix Algebra: Example 3. Given a two-by-two FIR matrix

$$\underline{\mathbf{H}} = \begin{bmatrix} h_{11} & h_{12} \\ h_{21} & h_{22} \end{bmatrix} \tag{9.63}$$

the inverse of $\underline{\mathbf{H}}$ is

$$\underline{\mathbf{W}} = \frac{1}{(h_{11} * h_{22} - h_{12} * h_{21})} \begin{bmatrix} h_{22} & -h_{12} \\ -h_{21} & h_{11} \end{bmatrix} \tag{9.64}$$

The unit element for 2-by-2 FIR filter matrices is then

$$\underline{\mathbf{I}} = \underline{\mathbf{H}}\underline{\mathbf{W}}$$

$$= \frac{1}{(h_{11} * h_{22} - h_{12} * h_{21})} \begin{bmatrix} h_{11} * h_{22} - h_{12} * h_{21} & h_{11} * h_{12} - h_{11} * h_{12} \\ h_{21} * h_{22} - h_{21} * h_{22} & h_{11} * h_{22} - h_{12} * h_{21} \end{bmatrix} \tag{9.65}$$

$$= \begin{bmatrix} \cdots 0\,0\,1\,0\,0 \cdots & \cdots 0\,0\,0 \cdots \\ \cdots 0\,0\,0 \cdots & \cdots 0\,0\,1\,0\,0 \cdots \end{bmatrix} \tag{9.66}$$

FIR Matrix Algebra: Example 4. Given the 2-by-2 FIR polynomial matrix

$$\underline{\mathbf{H}} = \begin{bmatrix} z - 1 - 3z^{-1} & -1 + 2z^{-1} + 3z^{-2} \\ z - 3 & 3z^{-1} \end{bmatrix} \tag{9.67}$$

compute its eigenvalues and eigenvectors. First, find the eigenvalues by finding the roots of $\|\underline{\mathbf{H}} - \lambda \underline{\mathbf{I}}\|$:

$$\begin{vmatrix} z - 1 - 3z^{-1} - \lambda & -1 + 2z^{-1} + 3z^{-2} \\ z - 3 & 3z^{-1} - \lambda \end{vmatrix} \tag{9.68}$$

$$[(z - 1 - 3z^{-1}) - \lambda][3z^{-1} - \lambda] - (-1 + 2z^{-1} + 3z^{-2})(z - 3)$$

$$\lambda^2 - (z - 1)\lambda + z - 2$$

$$(\lambda - 1)(\lambda - (z - 2))$$

Thus eigenvalues are $\lambda_1 = (z - 2)$ and $\lambda_2 = 1$.

Find eigenvectors via $\underline{\mathbf{H}} \begin{bmatrix} T_{1i} \\ T_{2i} \end{bmatrix} = \lambda_i \begin{bmatrix} T_{1i} \\ T_{2i} \end{bmatrix}$. $\begin{bmatrix} 1 + z^{-1} \\ 1 \end{bmatrix}$ is associated with $\lambda_1 = (z - 2)$, and $\begin{bmatrix} z^{-1} \\ 1 \end{bmatrix}$ is associated with $\lambda_2 = 1$. Hence,

$$\mathbf{\underline{T}} = \begin{bmatrix} 1 + z^{-1} & z^{-1} \\ 1 & 1 \end{bmatrix}$$

$$\mathbf{\underline{T}}^{-1} = \begin{bmatrix} 1 & -z^{-1} \\ -1 & 1 + z^{-1} \end{bmatrix}$$

The reader should verify that $\mathbf{\underline{T}}^{-1}\mathbf{\underline{H}}\mathbf{\underline{T}} = \begin{bmatrix} z - 2 & 0 \\ 0 & 1 \end{bmatrix}$. Note also that the determinant of $\mathbf{\underline{H}}$ equals the product of eigenvalues, and that the trace is the sum of eigenvalues. In general, one must both prepend and postpend zeros (in the time domain) to the FIR matrix elements, in order to allow for the causal and non-causal Z-transform (Laurent series) expansions for the inverse of the determinant of $\mathbf{\underline{T}}$, the matrix of eigenvectors. In the simple example just given, this was unity. Note that for FIR matrices, the eigenvalues and the elements of the eigenvectors have a time delay or multipath dimension.

We present a more general example computed via computer. A numerical eigenvalue routine for the computer is outlined below.

9.3.9 FIR Matrix Eigenroutine

A method for the numerical calculation of FIR polynomial (frequency domain) matrix eigenvalues and eigenvectors has been constructed that can diagonalize an FIR matrix. In diagonal form, whitening, matrix exponential, or any matrix function can be computed.[6] It is recommended that an FIR matrix in the time domain be placed into the frequency domain during the iterations and calculations of the algorithm, and then converted back to the time domain in reporting the final answer.

"Going to the frequency domain" for an FIR matrix simply entails Fourier transforming each matrix element (a filter) in an FIR filter matrix.

The eigenroutine we outline, uses an iterative QR technique described in Golub and van Loan (1989), using Householder transformations such that the Schur (upper triangular) form is obtained. The Schur form has the eigenvalues of the original matrix in question along the diagonal. The eigenvalues thus in hand find the eigenvectors by finding the similarity transform required to go to diagonal form.

This routine numerically finds eigenvalues and vectors of a general FIR polynomial matrix denoted by $\mathbf{\underline{A}}$. In general, we mean that it need not be real or symmetric. For notational ease, we outline the 2-by-2 case. The standard vector algebra concepts of rectangular matrix operations will be important to us (inner products, orthonormalization). This eigenvalue routine outlines some of the general concepts involved with FIR algebra.

---

[6] A similar FIR filter matrix (time domain) routine is possible using convolutions instead of multiplication. The solution is not as elegant in the time domain, as iterative algorithms (QR iterations) using convolutions increase the length of the intermediate result vector at each iteration.

Routine outline:

1. Obtain Schur form (upper triangular). Use a FIR-adapted version of the Householder transformation in order to annihilate the lower triangle of the matrix. At convergence, eigenvalues λ_i of $\underline{A}$ will be the diagonal elements of $\mathbf{Q}^H \underline{A} \mathbf{Q}$:

```
for j=1,2,...
    Z=AQ
    Q=householder(Z)
end
```

2. Form controllability matrix $\underline{C}$, which similarity transforms any matrix into companion matrix form [see Kailath (1980)]:

$$\underline{C} = [\mathbf{b} \quad \mathbf{Ab}] \tag{9.69}$$

where $\mathbf{b}$ is any column vector that is not an eigenvector of $\underline{A}$.

3. Form Vandermonde matrix $\underline{V}$ using the eigenvalues. The Vandermonde matrix similarity transforms a companion matrix into Jordan or diagonal form:

$$\underline{V} = \begin{bmatrix} 1 & \lambda_1 \\ 1 & \lambda_2 \end{bmatrix} \tag{9.70}$$

4. The eigenvectors are now the columns of $\mathbf{T} = \mathbf{CV}^{-1}$ and $\mathbf{J} = \mathbf{T}^{-1}\underline{A}\mathbf{T}$. The columns of $\mathbf{T}$ should be orthonormalized (as shown for t_{11})

$$t_{11} = \text{IFFT} \left(\frac{\text{FFT}(t_{11})}{(|\text{FFT}(t_{11})|^2 + |\text{FFT}(t_{21})|^2)^{0.5}} \right) \tag{9.71}$$

Eigenroutine: Example 1. Using the nonminimum phase-discrete multichannel filter $\mathbf{H}$ shown in Fig. 9.6, the eigenvalue algorithm was applied and found to require about 10 QR power iterations to obtain an accurate Schur form. The nonzero coefficients are $h_{11} = [1 \; 1 \; -0.75]$, $h_{21} = [-0.2 \; 0.4 \; 0.7]$, $h_{12} = [0.5 \; -0.3 \; -0.2]$, $h_{22} = [0.2 \; 1 \; 0]$. The eigenvalue polynomials are shown in Fig. 9.7 and the eigenvector polynomials are shown in Fig. 9.8.

Eigenroutine: Example 2. Eigenvalues and eigenvectors are computed for a 4-by-4 FIR matrix given by

$$\underline{H} = \begin{bmatrix} 1.25 & 0.11z^{-1} + 0.45z^{-2} & -0.48 & -0.22 \\ -0.1z^{-1} & -0.63 + 0.28z^{-1} & 0.83 + 0.23z^{-1} & 0.71 - 0.1z^{-1} \\ 0.52 - 1.57z^{-2} & 0.12 + 0.54z^{-1} & 0.71 - 0.12z^{-1} & -0.57z^{-2} \\ -0.27z^{-1} & -0.38z^{-2} & -0.49z^{-1} & -0.65 + 0.32z^{-1} \end{bmatrix} \tag{9.72}$$

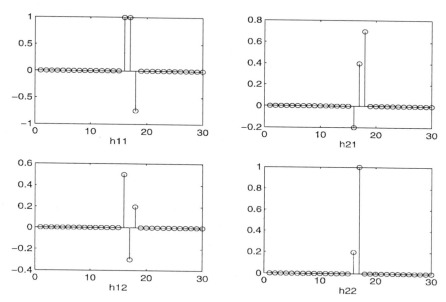

Figure 9.6 Multichannel filter $\underline{\mathbf{H}}$.

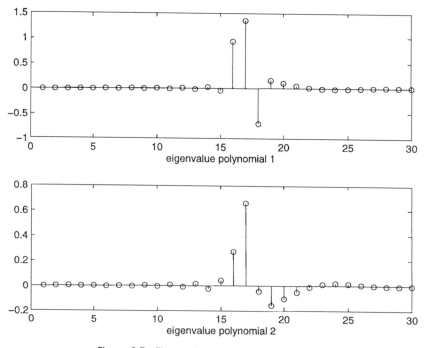

Figure 9.7 Eigenvalue polynomials of $\underline{\mathbf{H}}$, λ_1, λ_2.

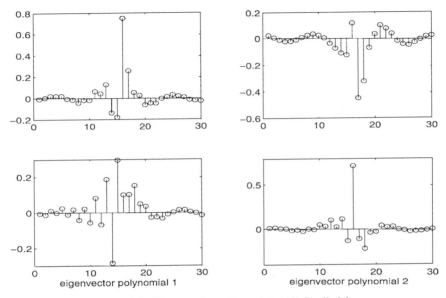

Figure 9.8 Eigenvector polynomials of $\underline{\mathbf{H}}$, $\underline{\mathbf{T}} = [\underline{t}_1 \ \underline{t}_2]$.

Eigenvector matrix $\underline{\mathbf{T}}$ is plotted in Fig. 9.9 and the diagonal matrix $\underline{\mathbf{T}}^{-1}\underline{\mathbf{H}}\underline{\mathbf{T}}$, with eigenvalues on the diagonal is plotted in Fig. 9.10.

9.3.10 FIR Polynomial Matrix Whitening

Consider whitening in a traditional scalar matrix context, for a matrix $\mathbf{A}$,

$$\mathbf{R} = \mathbf{A}\mathbf{A}^H$$

$$\mathbf{C} = \operatorname{diag}[\lambda_1^{-0.5}, \ldots, \lambda_n^{-0.5}]$$

$$\mathbf{Z} = \mathbf{C}\mathbf{T}^H\mathbf{A}$$

such that

$$\mathbf{I} = \mathbf{Z}\mathbf{Z}^H \tag{9.73}$$

We wish to perform prewhitening of multichannel data with multipath by getting an estimate of $\underline{\mathbf{R}} = \underline{\mathbf{A}}\underline{\mathbf{A}}^H$ using $\hat{\underline{\mathbf{R}}} = (1/n)\sum_{i=1}^n \underline{\mathbf{x}}_i\underline{\mathbf{x}}_i^H$ and using it to compute a whitening FIR matrix. The eigenvalues and vectors of $\underline{\mathbf{R}}$ are found using the method described in the previous section. The whitening method is written out for the 2-by-2 case:

$$\underline{\mathbf{A}} = \begin{bmatrix} a_{11} & a_{12} \\ a_{21} & a_{22} \end{bmatrix} \tag{9.74}$$

$$\underline{\mathbf{R}} = \underline{\mathbf{A}}\underline{\mathbf{A}}^H = \begin{bmatrix} r_{11} & r_{12} \\ r_{21} & r_{22} \end{bmatrix} \tag{9.75}$$

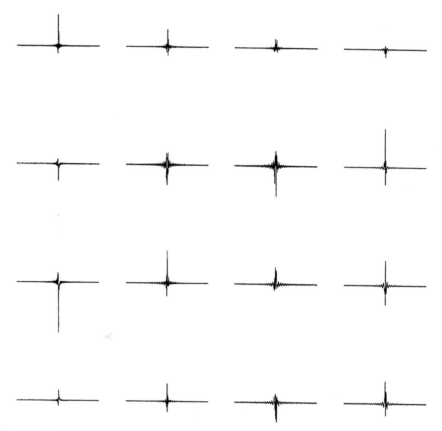

Figure 9.9 Matrix $\underline{T}$ (with eigenvectors as columns) for eigenroutine Example 2. The multipath extent of the matrix is 400 taps.

where in discrete frequency domain notation, we perform matrix correlation

$$
\begin{aligned}
r_{11} &= \text{IFFT}(A_{11}\, A_{11}^* + A_{12}\, A_{12}^*) \\
r_{12} &= \text{IFFT}(A_{21}\, A_{11}^* + A_{22}\, A_{12}^*) \\
r_{21} &= \text{IFFT}(A_{11}\, A_{21}^* + A_{12}\, A_{22}^*) \\
r_{22} &= \text{IFFT}(A_{21}\, A_{21}^* + A_{22}\, A_{22}^*)
\end{aligned}
\tag{9.76}
$$

where $A_{ij} = \text{FFT}(a_{ij})$.

$$
\underline{C} =
\begin{bmatrix}
c_{11} & 0 \\
0 & c_{22}
\end{bmatrix}
=
\begin{bmatrix}
|\lambda_1|^{-0.5} & 0 \\
0 & |\lambda_2|^{-0.5}
\end{bmatrix}
\tag{9.77}
$$

Because a whitening function is nonunique, the choices of the particular

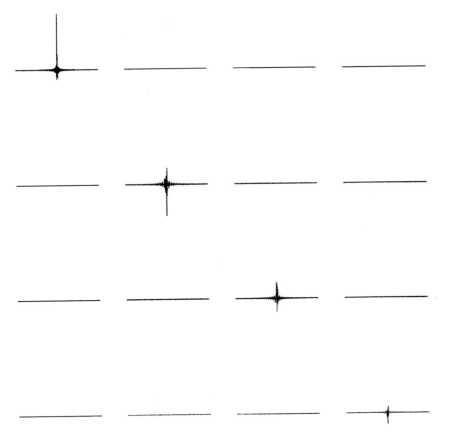

Figure 9.10 Diagonal matrix $\underline{T}^{-1}\underline{HT}$ (with eigenvalues on the diagonal) for eigenroutine Example 2. The multipath extent of the matrix is 400 taps.

whitener to be used is made at this step. Most common realizations of whiteners (or factorizations) are minimum phase (purely causal), maximum phase (purely noncausal), or zero phase symmetric). The zero-phase whitener is obtained using $\mathrm{IFFT}(|\mathrm{FFT}(\lambda_i)|^{-0.5})$. Here, $\underline{T}$ must have orthonormal columns

$$\underline{U} = \underline{CT}^H = \begin{bmatrix} \mathrm{IFFT}(C_{11}\ T_{11}^*) & \mathrm{IFFT}(C_{11}\ T_{21}^*) \\ \mathrm{IFFT}(C_{22}\ T_{12}^*) & \mathrm{IFFT}(C_{22}\ T_{22}^*) \end{bmatrix} \tag{9.78}$$

where again the nonboldface uppercase denotes discrete frequency domain. The data **y**, when multiplied by $\underline{U}$, will be whitened.

If the estimation of $\underline{R}$ were perfect, we would get the factorization of $\underline{R}$,

$$\underline{Z} = \underline{UA} \tag{9.79}$$

where

$$z_{11} = \text{IFFT}(U_{11}\, A_{11} + U_{12}\, A_{21})$$

$$z_{21} = \text{IFFT}(U_{11}\, A_{12} + U_{12}\, A_{22})$$

$$z_{12} = \text{IFFT}(U_{21}\, A_{11} + U_{22}\, A_{21})$$

$$z_{22} = \text{IFFT}(U_{21}\, A_{12} + U_{22}\, A_{22})$$

$$\mathbf{I} = \mathbf{Z}\mathbf{Z}^{H} \tag{9.80}$$

Here, $\mathbf{I}$ is the unit element.

9.4 SIMULATIONS

From a Practical Standpoint In order to avoid the problems associated with estimations of full pdf information, we use a parametric data model, the generalized Gaussian distribution. In this parametric form, the log derivative scaled by the variance is

$$g(\cdot) = \frac{E|\hat{x}|^2}{E|\hat{x}|^s}\,\hat{x}|\hat{x}|^{s-2} \tag{9.81}$$

Most non-Gaussian distributions we have encountered may be well modeled using either $s = 1$ for speechlike super-Gaussian or $s = 4$ for communications datalike sub-Gaussian. In the case of super-Gaussian data, we keep a running average estimate of the statistics $E|\hat{x}|^2$ and $E|\hat{x}|$, and so the nonlinearity is continually changing as the convergence proceeds. In the case of sub-Gaussian data, we fix $E|\hat{x}|^2$ and $E|\hat{x}|^4$ to the known values $E|x|^2$ and $E|x|^4$.

9.4.1 Measures of Difficulty

We preface the simulation section with a discussion of measures of difficulty for general separation/equalization problems.

A given linear blind equalization or source-separation problem has the following difficulty measures associated with it:

1. The distance(s) of the original source distribution(s) from Gaussian. Distance from Gaussian measures may be calculated using the KL divergence or Gray's variable norm, to cite two examples (statistical measure of the source).

2. The dispersion or initial intersymbol interference imposed by the channel. This is a measure of how much the channel will "Gaussianize" the signal (channel difficulty measure).

3. The eigenvalue spread of the channel (channel difficulty measure).

4. The nature of the noise in the channel.

Experience with speech as a source may lead us to list other statistical measures: the degree of stationarity of the source, the whiteness of the sources, and the identically distributed nature of the sources. For the present, we consider zero-mean, symmetric, stationary, i.i.d. sources only, and therefore chiefly focus on the channel-related measures and figures of merit for separation algorithms.

Even in the context of zero-mean, symmetric, stationary and i.i.d. sources, distance from Gaussianity is of fundamental importance. This is what our algorithms use as a cost function/learning rule. The sources become more Gaussian as they traverse the channel the separation matrix is to adapt until its outputs are maximally far from Gaussian.

Distance from Gaussianity   There are many possible Gaussianity measures: KL divergence, normalized entropy, Gray's variable norm, to cite a few [see (Gray 1979; Hogg 1972; Walden 1985; Wiggins 1978)]. A particularly simple and powerful standard measure is normalized kurtosis:

$$\frac{E|x^4|}{\left(E|x^2|\right)^2} \tag{9.82}$$

The pure-Gaussian value is 3.

A super-Gaussian random source will have normalized kurtosis of greater than 3. For example, filter a Laplace source (normalized kurtosis = 6) with a fairly dispersive channel, and then measure normalized kurtosis of the output. It may be something close to 3 (Gaussian), but slightly higher.

A sub-Gaussian random source will have normalized kurtosis of less than 3. A uniform source has normalized kurtosis = 1.8. After filtering these data, the output may be something close to 3 (Gaussian), but slightly less. We may need to know *a priori* whether a data type is sub- or super-Gaussian, so that we can tell the algorithm which way to head.

Intersymbol Interference   ISI is a measure from the area of single-channel blind equalization (communications engineering). It is useful as a dispersion measurement that is insensitive to overall gain and mean group delay, two signal properties that blind equalization is unable to correct. If the original source data were known, then we would be in the area of training-based learning and a measure such as mean square error (which is sensitive to channel gain and group delay) could be used. Single-channel intersymbol interference is defined as follows.

Definition 6. Single-channel intersymbol interference (ISI)

$$\frac{\sum_k |s(k)|^2 - \max_k |s(k)|^2}{\max_k |s(k)|^2} \tag{9.83}$$

where s is the global system $s = w * h$.

We define multichannel ISI, which includes measures of both multipath and matrix mixture, similarly. We have simply added the matrix dimension to the single-channel measure. It is important to measure rowwise and columnwise ISI. For an N-by-N system, there will be N elements for both row ISI and column ISI that all must go to near zero in order to reach convergence.

Definition 7. Multichannel row intersymbol interference (Row ISI)

$$\text{Row ISI}_i = \frac{\sum_i \sum_k |s_{ij}(k)|^2 - \max_{j,k} |s_{ij}(k)|^2}{\max_{j,k} |s_{ij}(k)|^2} \tag{9.84}$$

where s_{ij} are the filter elements of global system $\underline{\mathbf{S}} = \underline{\mathbf{W}}\underline{\mathbf{H}}$.

Definition 8. Multichannel column intersymbol interference (Column ISI)

$$\text{column ISI}_j = \frac{\sum_j \sum_k |s_{ij}(k)|^2 - \max_{j,k} |s_{ij}(k)|^2}{\max_{j,k} |s_{ij}(k)|^2} \tag{9.85}$$

where s_{ij} are the filter elements of global system $\underline{\mathbf{S}} = \underline{\mathbf{W}}\underline{\mathbf{H}}$.

Multichannel ISI is a measure of how close the global system is to a scaled and/or permuted identity FIR matrix. We have found monitoring all the ISI_i terms to be beneficial. Initial ISI terms are defined as earlier using $s = h$, and is a difficulty measure for any algorithm (training-based or blind) that will learn the inverse channel. ISI uses an L_2 (square) norm measure; other measures, such as the L_1 (absolute value) norm, are equally viable. The measure using the L_1 norm is commonly known as *closure*.

Row ISI is probably the most important and its steady-state (at convergence) values are dependent on the algorithm step size and the original source data types. Column ISI will detect that the separation matrix has multiple rows locked onto a single source. At convergence, each column ISI value is the sum of the row ISI values. Examples are in the simulation section.

It may be advantageous to keep the interference due to multipath and that due to other channels separate in the performance measure. Assuming the permutation of sources is known, one may use interchannel interference (ICI) and single channel ISI, as in Joho and Mathis (1999).

Eigenvalue Spread In single-channel equalization problems, eigenvalue spread is one of the measures of difficulty. It is defined as $|\lambda|_{\max}/|\lambda|_{\min}$, where λ's are from the Toeplitz correlation matrix $\mathbf{R}$ of the channel's impulse response. Matrix $\mathbf{R}$ is of size $L \times L$, where L is the length of the equalizer.

In scalar-matrix source-separation problems, the mixing-matrix condition number serves as the measure of "invertibility." In this case, the eigenvalues are taken from the mixing matrix or, alternatively, from $R = HH^H$, where H is the mixing matrix. This correlation-based eigenvalue spread measure is general in that one need not know the mixing matrix itself, only its correlation, something that can be estimated from the received mixed sources (assuming that the sources are white).

For multichannel problems of this type, the block Toeplitz (circulant) correlation matrix **R** of size $NL \times ML$, where N is the number of inputs and M is the number of output channels, is also what determines the eigenvalue spread of the problem.

The eigenvalue spread of the *block* Toeplitz autocorrelation matrix $\mathbf{R} = \mathbf{YY}^H$ using Eqs. (9.15) and (9.16), measures the condition of the channel. The eigenvalues of the block Toeplitz matrix formed from the received mixed data naturally combine the two invertibility issues for a multichannel multipath problem. The matrix condition aspect and the single-channel spectral null aspect are combined here into a single mathematical measure.

Just as in the single-channel case, if RLS is used to adapt the equalizer taps, an estimate of the $\mathbf{R}^{-1}$, which is block Toeplitz (at convergence) and of size $NL \times ML$, is obtained as part of the algorithm. However, if the luxury of off-line computation is available, then data prewhitening is possible. If the data are prewhitened, the use of a less computationally expensive algorithm, such as LMS, is possible, using a FIR polynomial matrix of size $N \times M \times L$. We present an experimental example using the multichannel prewhitening procedure given in the previous section.

When data prewhitening is not possible, RLS adaptive minimization is the most quickly converging method because it is robust to eigenvalue disparity.

9.4.2 Simulation Results for 2-by-2 System with Uniformly Distributed Sources

Here we do a side-by-side comparison of MBLMS, MBRLS, pre-whitened MBLMS (W-MBLMS), Amari blind serial update (BSU), and Amari finite difference (FD). Our tests will use the time-domain versions of these algorithms. The use of the data prewhitening procedure followed by MBLMS (W-MBLMS) is presented in order to show the benefits of the multichannel prewhitening technique. The channel given in Eigenroutine Example 1 was used (see Fig. 9.6), with an eigenvalue spread of 49. The two sources were i.i.d. and uniformly distributed. The explicit updates used for MBLMS, MBRLS, and W-MBLMS were given in Eq. (9.32), and the updates used for MBLMS and W-MBLMS were given in Eq. (9.33).

Results are given in Figs. 9.11, 9.12, 9.13, 9.14, and 9.15, with the ISI corresponding to each row of the global system **WH** being plotted. In each case a steady-state ISI of 0.01 is reached. Note that in comparing Figs. 9.11 and 9.13, prewhitening followed by MBLMS is faster to converge than RLS, since the second-order step of the equalization/separation is already accomplished in the prewhitened case. The prewhitening step produces unit-eigenvalue spread, meaning the block Toeplitz multichannel autocorrelation matrix formed using the two received data channels has all eigenvalues equal to 1.

MBRLS is fastest to converge among MBLMS, BSU, and FD, but is also the most computationally expensive.

The serial update and finite difference approaches appear to be equals in this test, but we have observed in another test with a higher steady-state ISI target (set for faster convergence) that the serial update tends to be more robust.

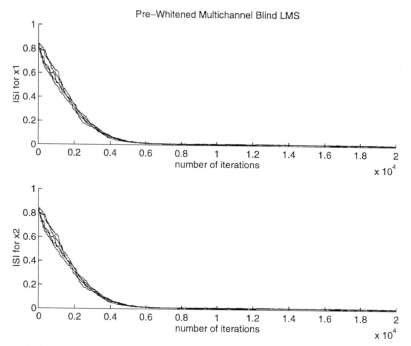

Figure 9.11 Convergence of five realizations of uniformly distributed data using the prewhitening procedure followed by the multichannel blind LMS algorithm (W-MBLMS). Steady-state ISI is 0.01.

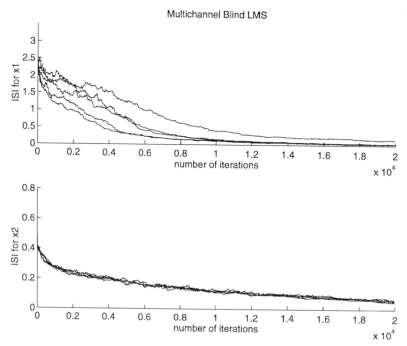

Figure 9.12 Convergence of five realizations of uniformly distributed data using the multichannel blind LMS algorithm with no prewhitening. Steady-state ISI is 0.01.

Multichannel Blind RLS

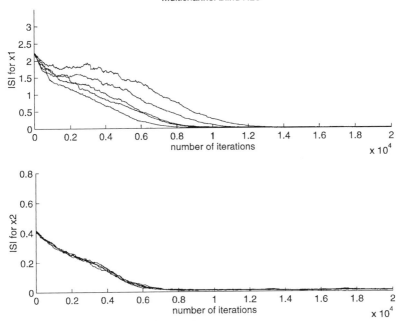

Figure 9.13 Convergence of five realizations of uniformly distributed data using the multichannel blind RLS algorithm. Steady-state ISI is 0.01.

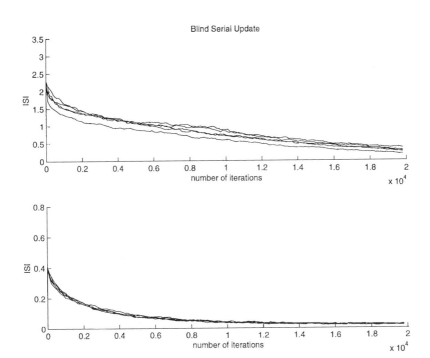

Figure 9.14 Convergence of five realizations of uniformly distributed data using the multichannel blind serial update algorithm using Amari-form. Steady-state ISI is 0.01.

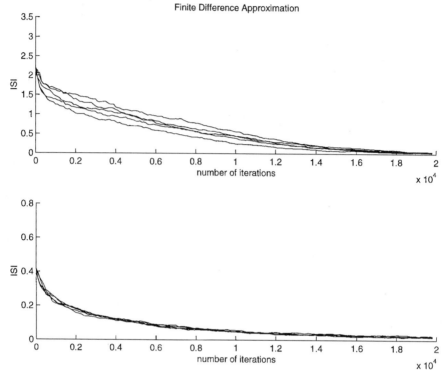

Figure 9.15 Convergence of five realizations of uniformly distributed data using the multichannel Amari-form blind finite difference algorithm Steady-state ISI is 0.01.

9.4.3 Simulation Results for High Eigenvalue Spread 4-by-4 System Using Different Data Types

In this test, the 4-by-4 system used is given by

$$
\underline{\mathbf{H}} =
\begin{bmatrix}
1.25 & 0.11z^{-1} + 0.45z^{-2} & -0.48 & -0.22 \\
-0.1z^{-1} & -0.63 + 0.28z^{-1} & 0.83 + 0.23z^{-1} & 0.71 - 0.1z^{-1} \\
0.52 - 1.57z^{-2} & 0.12 + 0.54z^{-1} & 0.71 - 0.12z^{-1} & -0.57z^{-2} \\
-0.27z^{-1} & -0.38z^{-2} & -0.49z^{-1} & -0.65 + 0.32z^{-1}
\end{bmatrix}
$$

$$(9.86)$$

This channel has an eigenvalue spread of 1625. The inverse of such an extreme channel (the inverses have the same eigenvalue spread of the forward channels) cannot be learned via noneigenvalue spread robust algorithms (like LMS). The blind RLS algorithm was the only algorithm that would solve this extreme type of problem. Source 1 was uniformly distributed, source 2 was a binary ± 1 sequence, sources 3 and 4 were gamma distributed. Delay length of the inverse system was 50 taps. Plotted are row ISI in Fig. 9.16, column ISI in Fig. 9.17, and the converged inverse channel in Fig. 9.18.

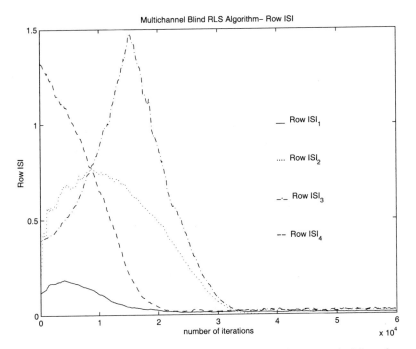

Figure 9.16 Row ISI monitoring of convergence. Eigenvalue spread of the channel is 1625.

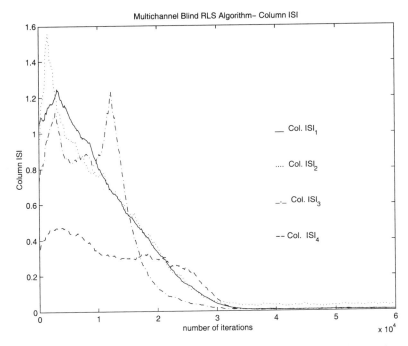

Figure 9.17 Column ISI monitoring of convergence. Steady-state column ISI values are 0.02. Eigenvalue spread of the channel is 1625.

Inverse Channel

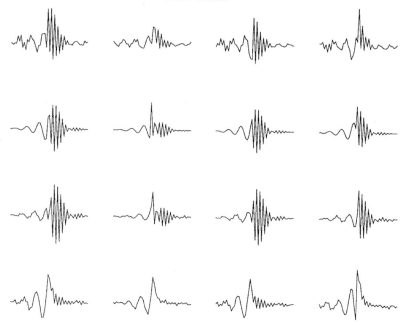

Figure 9.18 Inverse FIR matrix channel learned via the multichannel blind RLS algorithm. Eigenvalue spread of the channel is 1625.

Global system at convergence

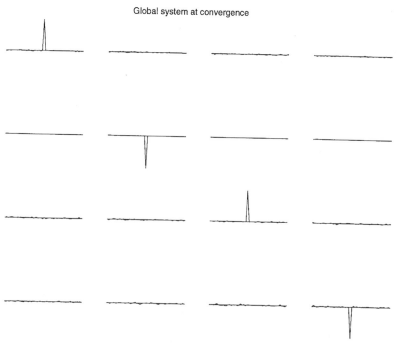

Figure 9.19 Global system at convergence. The fact that it is diagonal (or a permutation thereof) shows convergence has been attained.

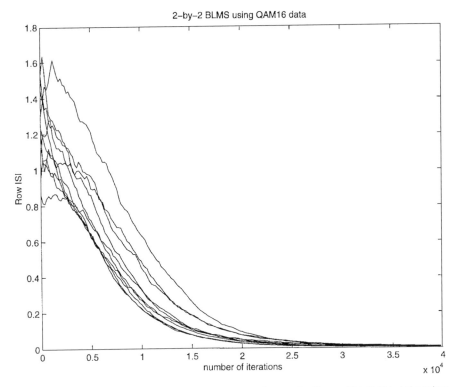

Figure 9.20 Multichannel row ISI convergence of five realizations of 16-QAM data using MBLMS with a complex nonminimum-phase channel. Steady-state ISI is 0.01.

The global system (used in computing ISI measures) is plotted at convergence in Fig. 9.19. Note the sign flips of sources 2 and 4.

9.4.4 Simulation Results of a 2-by-2 Complex Data System

We demonstrate blind separation and equalization of two complex 16-QAM data sources that were mixed by the following nonminimum phase channel

$$H_{11} = 1.2 + 1.i + (1.3 + 1.5i)z^{-1} + (-.84 - .92i)z^{-2}$$
$$H_{12} = .34 + .23i + (-.41 - 1.45i)z^{-1} + (.25 - .12i)z^{-2}$$
$$H_{21} = -.63 + .34i + (-1.47 - .21i)z^{-1} + (.25 - .34i)z^{-2}$$
$$H_{22} = 1.5 - .67i + (-1.65 + .16i)z^{-1} + (-.25 - 1.35i)z^{-2} \qquad (9.87)$$

The MBLMS algorithm was employed. Multichannel row ISI for each of the two sources is plotted as the convergence proceeds for five Monte Carlo experiments in Fig. 9.20 (multichannel column ISI also converged to zero). The

Inverse Channel

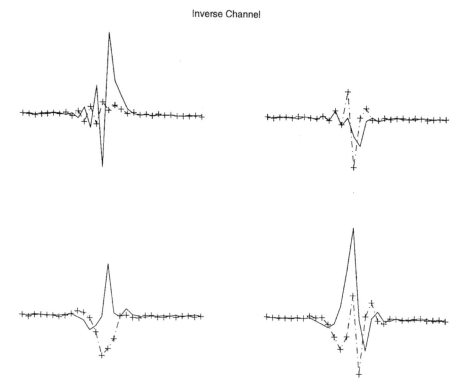

Figure 9.21   Converged inverse of complex nonminimum-phase channel. Real values are shown with a solid line, imaginary values with a +.

converged inverse channel is shown in Fig. 9.21, showing both real and imaginary parts. Scatter plots of the outputs as the convergence proceeds in Fig. 9.22, show the opening of the eye.

Perfect Interference Mitigation Using Fractional-Spaced Sampling

Cochannel interference mitigation in the field of data communications is an important area that is part of source-separation theory. The fractional-spaced approach assumes that an invertible mixture has occurred and that we are only interested in recovering one principal signal. If we wish to cancel $N - 1$ interferers, then we oversample N times obtaining the needed extra sensor data. See, for example, Im and Werner (1995) where, in the separation derivation, matrices of frequency-domain filters are used as in the FIR-matrix approach we have presented. In many cases the separation filters can be adapted using decision-directed or other methods, but our MBLMS algorithm is a viable choice.

Using a blind source-separation algorithm such as MBLMS, which is flexible to allow the updates of only one row of the separation matrix, we have a fractional-spaced interference canceler, a subset of the full separation problem. In

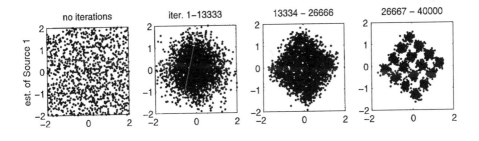

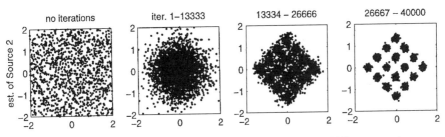

Figure 9.22 Scatter plots of the stages of convergence using MBLMS, a complex non-minimum-phase channel, and 16-QAM data.

the previous 2-by-2 experiment with 16-QAM data, both rows of the separation matrix were updated and both sources were recovered. Updating only one row recovers one principal signal and solves the interference mitigation problem.

9.4.5 Results of Acoustic-Speech Separation Experiment

Speech being neither stationary nor white, it is more difficult to blindly separate than i.i.d. data generated on a computer. Nevertheless, using the block frequency-domain methods, some encouraging results have been obtained. Speech was modeled as Laplace distributed (used cost $O_2^1(\hat{x}_1) + O_2^1(\hat{x}_2)$), and 110 seconds of data were presented to the adaptive algorithm using a frequency-domain overlap-and-save method as in Ferrara (1980). The 110-second data segment was presented four times to the blind algorithm. A two-microphone recording was made in a room approximately 11 feet by 10 feet with two persons talking simultaneously. Distance between the microphones was 1.5 feet. The talkers were located approximately 5 feet from the center of the microphones, and the talkers were stationary and separated by approximately 4 feet. The sample rate was 8000 samples per second. An inverse FIR matrix of dimensions 2 by 2 by 2048 was used with the frequency-domain serial update method. The time domain inverse matrix is plotted in Fig. 9.23. Plots of 10 seconds of the original microphone signals are in Fig. 9.24, and plots of

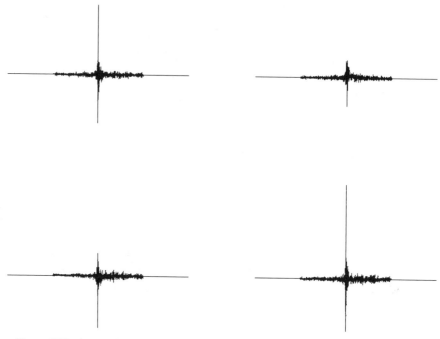

Figure 9.23 Inverse FIR matrix found in true acoustic test with two speech sources.

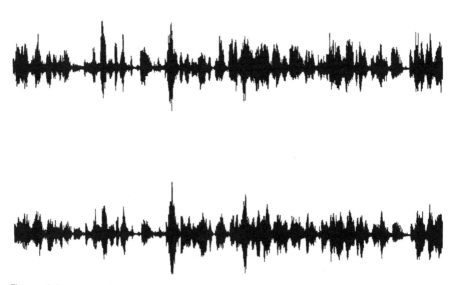

Figure 9.24 Original microphone signals of two mixed-speech sources used in true acoustic speech test.

Figure 9.25 Processed separated speech outputs of true acoustic speech test.

10 seconds of the two outputs of the separation process are in Fig. 9.25. The separation obtained in the output is approximately 18–20 dB. The audio of an early speech-separation result is available in Lambert and Bell (1997) (on CDROM proceedings).

9.5 FEEDBACK METHODS FOR SEPARATION AND EQUALIZATION

Inspired by the Jutten and Herault feedback approach (Jutten and Herault 1991), we present algorithms using feedback architectures that make use of our derived multichannel cost, Eq. (9.22), and variations.

9.5.1 Feedback Inverse System Architectures

Referring to Fig. 9.26, we see that for a 2-by-2 system, there are two possible forms for an inverse system using feedback. This can also be seen in the 2-by-2 equations

$$y_1 = x_1 * h_{11} + x_2 * h_{12}$$
$$y_2 = x_1 * h_{21} + x_2 * h_{22}$$

(9.88)

Depending on how we solve for x_1 and x_2, we obtain different sets of feedback-separation equations.

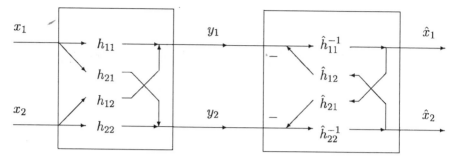

Figure 9.26 Multichannel blind equalization system using a feedback-inverse model.

Inverse System with Feedback: Form 1

$$\hat{x}_1 = (y_1 - \hat{x}_2 * \hat{h}_{12}) * \hat{h}_{11}^{-1} \tag{9.89}$$

$$\hat{x}_2 = (y_2 - \hat{x}_1 * \hat{h}_{21}) * \hat{h}_{22}^{-1} \tag{9.90}$$

Inverse System with Feedback: Form 2

$$\hat{x}_1 = (y_2 - \hat{x}_2 * \hat{h}_{22}) * \hat{h}_{12}^{-1} \tag{9.91}$$

$$\hat{x}_2 = (y_1 - \hat{x}_1 * \hat{h}_{11}) * \hat{h}_{21}^{-1} \tag{9.92}$$

The preceding sets of separation equations should be resampled or repeated at least N times for a general N-channel problem.

The feedback forms are implementations of a multichannel ARMA model. A restriction imposed by our use of any feedback form is that the roots of the determinant be minimum phase or inside the unit circle.

Separation and Equalization Using Linear and Nonlinear Terms with Feedback

Separation and equalization using linear and nonlinear terms with feedback (SELANTF) is an algorithm that solves the combined separation and equalization problem with cost, Eq. (9.22), $J = \sum_{\text{all } i} E\{\log(f_{x_i}(\hat{x}_i))/(f_G(\hat{x}_i))\}$. Using feedback form 1, described by Eqs. (9.89) and (9.90), we obtain estimates of the separated sources at each time step. Instead of a linear feedforward output equation as in Eqs. (9.4) and (9.5), we here use a feedback equation. Our general N-channel separation formulations are

$$\hat{x}_i = \left(y_i - \sum_{j \neq i} \hat{x}_j * \hat{h}_{ij}\right) * \hat{h}_{ii}^{-1} \tag{9.93}$$

with this set of N equations [Eqs. (9.93)] being repeated at least N times.

For the two-source, two-sensor case, the gradients are

$$\frac{\partial J_1}{\partial \hat{h}_{11}^{-1}} = \frac{\partial J_1}{\partial \hat{x}_1} \frac{\partial \hat{x}_1}{\partial \hat{h}_{11}^{-1}} = \frac{\partial J_1}{\partial \hat{x}_1} (y_1 - \hat{x}_2 * \hat{h}_{12})^* \tag{9.94}$$

$$\frac{\partial J_2}{\partial \hat{h}_{21}} = \frac{\partial J_2}{\partial \hat{x}_2} \frac{\partial \hat{x}_2}{\partial \hat{h}_{21}} = \frac{\partial J_2}{\partial \hat{x}_2} (\hat{x}_1 * \hat{h}_{22}^{-1})^* \tag{9.95}$$

$$\frac{\partial J_1}{\partial \hat{h}_{12}} = \frac{\partial J_1}{\partial \hat{x}_1} \frac{\partial \hat{x}_1}{\partial \hat{h}_{12}} = \frac{\partial J_1}{\partial \hat{x}_1} (\hat{x}_2 * \hat{h}_{11}^{-1})^* \tag{9.96}$$

$$\frac{\partial J_2}{\partial \hat{h}_{22}^{-1}} = \frac{\partial J_2}{\partial \hat{x}_2} \frac{\partial \hat{x}_2}{\partial \hat{h}_{22}^{-1}} = \frac{\partial J_2}{\partial \hat{x}_2} (y_2 - \hat{x}_1 * \hat{h}_{21})^* \tag{9.97}$$

The general N-source, N-sensor channel update formulations are

$$\hat{h}_{ij} = \hat{h}_{ij} + \mu \frac{\partial J}{\partial \hat{x}_i} (\hat{x}_j * \hat{h}_{ii}^{-1})^* \tag{9.98}$$

for the forward cross channels, and

$$\hat{h}_{ii}^{-1} = \hat{h}_{ii}^{-1} + \mu \frac{\partial J}{\partial \hat{x}_j} \left(y_i - \sum_{j \neq i} \hat{x}_j * \hat{h}_{ij} \right)^* \tag{9.99}$$

for the inverse direct channels.

9.5.2 Simulation Results for SELANTF

The minimum-phase channel used for this 2-by-2 experiment was

$$H_{11} = 1.13 + 1.23i + (.23 - .57i)z^{-1} + (-.44 - .32i)z^{-2}$$
$$H_{12} = .34 + .23i + (-.41 - .45i)z^{-1} + (.24 + .12i)z^{-2}$$
$$H_{21} = -.63 + .34i + (-.47 - .21i)z^{-1} + (.28 - .34i)z^{-2}$$
$$H_{22} = 1.51 - 1.67i + (-.65 + .16i)z^{-1} + (.22 - .35i)z^{-2} \tag{9.100}$$

The two source distributions were 16-QAM.

Experiments with SELANTF have shown that levels of intersymbol interference should not be much greater than 0.5 in order for convergence to be expected. The multichannel row ISI plots from five Monte Carlo realizations of the experiment are shown in Fig. 9.27. The scatter plot of the recovered sources is shown as the convergence proceeds in Fig. 9.28.

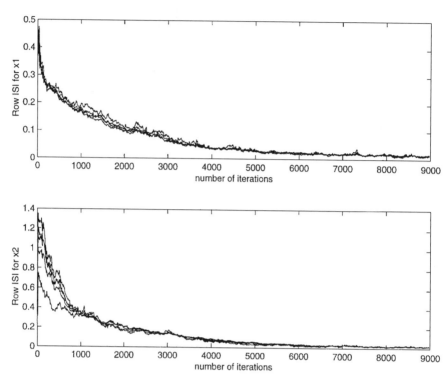

Figure 9.27 Multichannel row ISI convergence of five realizations of 16-QAM data using

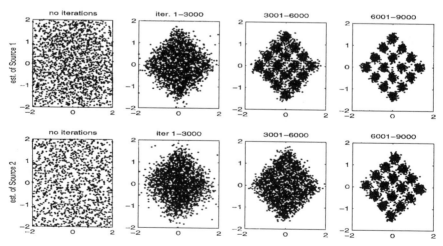

Figure 9.28 Scatter plots of the stages of convergence using SELANTF and QAM16 data.

9.5.3 Feedback Methods for the "Separation Only" Problem

The constraint of unity-direct channels in the "separation only" case allows simpler costs such as total energy minimization to be used. Because of the constraint, these methods [separation using linear terms with feedback (SNTF) and separation using nonlinear terms with feedback (SLTF)] will separate Gaussian sources. The separation-only or unity-direct channel's special case of SELANTF we have termed SLANTF (separation using linear and nonlinear terms with feedback).

Separation Using Linear Terms with Feedback   This is an algorithm with only second-order terms in the cost function

$$J = E|\hat{x}_1|^2 + E|\hat{x}_2|^2 + \cdots + E|\hat{x}_N|^2 \tag{9.101}$$

or

$$J = \operatorname{tr} E\{\hat{\mathbf{x}}\hat{\mathbf{x}}^H\} \tag{9.102}$$

SLTF minimizes the total energy of the output.

Separation Using Nonlinear Terms with Feedback   As before, we can also perform minimization of output norms other than two, obtaining non-linear terms in the update equation using

$$J = \sum_{\text{all } i} E\{-\log f_{x_i}(\hat{x}_i)\} \tag{9.103}$$

Both SLTF and SNTF algorithms are described by the separation equations, Eqs. (9.93), which are repeated at least N times (direct channels are unity), and general cross-channel updates

$$\hat{h}_{ij} = \hat{h}_{ij} + \mu \frac{\partial J}{\partial \hat{x}_i} \hat{x}_j^* \tag{9.104}$$

9.5.4 Conclusion for the Feedback Section

An advantage of using a feedback model is that the system can be described with parsimony, meaning a small number of parameters can be used to describe the system. A feedforward FIR matrix system will usually require more taps then a feedback model. The disadvantage is that we are restricted to stable ARMA systems with minimum-phase determinants.

We have seen that the Bussgang nonlinearity is also useful in feedback systems, providing an algorithm tuned to the specific source-data types.

Table 9.4   MATLAB Example of Amari-form Time-domain Blind Serial and Finite Difference Updates

```
% 2 by 2 Blind Source Separation using Finite Difference and Serial Update
% Russell Lambert 1998
clear
%%%%%%%%%%%%%%%%%%%%%%%%%%%%%%%%%%%%%%%%%%%%%%%%%%%%%% Begin Parameter Set
L=32;                        %length of multipath filters
delay=L/2;                   %delay (allow for noncausal inverse)
%method='serial',mu=.00023   %for serial update
method='finite',mu=.00023    %for finite difference update
%sigtype='gamma'       %signal type more peaked than Gaussian
sigtype='uniform'      %signal type flatter than Gaussian
last=60000                   %number of iterations
istep=100                    %plot filter and compute ISI every istep
%%%%%%%%%%%%%%%%%%%%%%%%%%%%%%%%%%%%%%%%%%%%%%%%%%%%%% End Parameter Set
switch sigtype
case 'gamma'
   s=[1 1], x(:,1)=randn(last,1);x(:,1)=sign(x(:,1)).*x(:,1).^2;
   x(:,2)=randn(last,1);x(:,2)=sign(x(:,2)).*x(;,2).^2;
case 'uniform'
   s=[4 4], x(:,1)=rand(last,1)-.5;x(:,2)=rand(last,1)-.5;
end
x(:,1)=x(:,1)./sqrt(cov(x(:,1)));x(:,2)=x(:,2)./sqrt(cov(x(:,2)));
ea2=[ mean(abs(x(:,1)).^2) mean(abs(x(:,2)).^2)] %use known statistics
eas=[ mean(abs(x(:,1)).^s(1)) mean(abs(x(:,2)).^s(2))]
h(1,:,1)=[1 1 -.75];h(2,:,1)=[.5 -.3 .2];  %FIR       filter matrix
h(1,:,2)=[-.2 .4 .7];h(2,:,2)=[.2 1 0]
H(1,:,1)=fft(h(1,:,1),L); H(2,:,1)=fft(h(2,:,1),L);   %FIR polynomial matrix
H(1,:,2)=fft(h(1,:,2),L); H(2,:,2)=fft(h(2,:,2),L);
y(:,1)=filter(h(1,:,1),[1],x(:,1)+filter(h(1,:,2),[1],x(:,2)));
y(:,2)=filter(h(2,:,1),[1],x(:,1)+filter(h(2,:,2),[1],x(:,2)));
w=zeros(2,L,2);w(1,delay,1)=1.;w(2,delay,2)=1.;
unit=zeros(2,L,2);unit(1,delay,1)=1.;unit(2,delay,2)=1.;
hatx=zeros(last,2);ghatx=zeros(last,2);
m=1;
for n=L:last
   hatx(n,1) = w(1,:,1)*y(n:-1:n-L+1,1)+w(1,:,2)*y(n:-1:n-L+1,2);
   hatx(n,2) = w(2,:,1)*y(n:-1:n-L+1,1)+w(2,:,2)*y(n:-1:n-L+1,2);
   ghatx(n,:)=hatx(n,:).*(abs(hatx(n,:)).^(s-2));
   wghatx(1) = w(1,:,1)*ghatx(n:-1:n-L+1,1)+w(1,:,2)*ghatx(n:-1:n-L+1,2);
   wghatx(2) = w(2,:,1)*ghatx(n:-1:n-L+1,1)+w(2,:,2)*ghatx(n:-1:n-L+1,2);
   hatRg(1,:,1)=(ea2(1)/eas(1))*ghatx(n-L/2+1,1)*conj(hatx(n:-1:n-L+1,1));
   hatRg(1,:,2)=(ea2(1)/eas(1))*ghatx(n-L/2+1,1)*conj(hatx(n:-1:n-L+1,2));
   hatRg(2,:,1)=(ea2(2)/eas(2))*ghatx(n-L/2+1,2)*conj(hatx(n:-1:n-L+1,1));
   hatRg(2,:,2)=(ea2(2)/eas(2))*ghatx(n-L/2+1,2)*conj(hatx(n:-1:n-L+1,2));
   hatRg_w(1,:,1)=(ea2(1)/eas(1))*wghatx(1)*conj(hatx(n:-1:n-L+1,1));
   hatRg_w(1,:,2)=(ea2(1)/eas(1))*wghatx(1)*conj(hatx(n:-1:n-L+1,2));
   hatRg_w(2,:,1)=(ea2(2)/eas(2))*wghatx(2)*conj(hatx(n:-1:n-L+1,1));
   hatRg_w(2,:,2)=(ea2(2)/eas(2))*wghatx(2)*conj(hatx(n:-1:n-L+1,2));
   switch method
   case 'serial'
      w = w + mu * (w-hatRg_w);     %serial update
   case 'finite'
      w = w + mu * (unit-hatRg);  %finite difference
   end
```

Table 9.4 (cont.)

```
    if(rem(n,istep)==0),
        figure(1),subplot(221),plot(w(1,:,1)),subplot(222),plot(w(1,:,2))
        subplot(223),plot(w(2,:,1)),subplot(224),plot(w(2,:,2)),drawnow
        W(1,:,1)=fft(w(1,:,1));  W(2,:,1)=fft(w(2,:,1));
        W(1,:,2)=fft(w(1,:,2));  W(2,:,2)=fft(w(2,:,2));
        S11=W(1,:,1).*H(1,:,1)+W(1,:,2).*H(2,:,1);
        S12=W(1,:,1).*H(1,:,2)+W(1,:,2).*H(2,:,2);
        S21=W(2,:,1).*H(1,:,1)+W(2,:,2).*H(2,:,1);
        S22=W(2,:,1).*H(1,:,2)+W(2,:,2).*H(2,:,2);
        stest1=[abs(ifft(S11)).^2 abs(ifft(S12)).^2];
        stest2=[abs(ifft(S21)).^2 abs(ifft(S22)).^2];
        isi1sc=(sum(stest1)-max(stest1))/max(stest1)
        isi2sc=(sum(stest2)-max(stest2))/max(stest2)
        isi_1(m)=isi1sc;
        isi_2(m)=isi2sc; m=m+1;
    end
end
count=1:istep:(m-1)*istep;
figure(2),clf,subplot(211);plot(count, isi_1)
xlabel('number of iterations'),ylabel('ISI')
title(['Update method' method 'Signal type' sigtype]), hold on
subplot(212);plot(count,isi_2),xlabel('number of iterations')
ylabel('Multichannel Row ISI')
mean(isi_1(fix(length(isi_1)-length(isi_1)/4):length(isi_1)))  %steady state ISI
mean(isi_2(fix(length(isi_2)-length(isi_2)/4):length(isi_2)))
```

9.6 CONCLUSIONS

Source separation is an exciting research area, bringing together concepts from general areas such as signal processing (array processing, adaptive filters), neural networks, information theory, probability, group theory/algebra, and geophysical data processing. In an effort to understand the multipath dimension of the source-separation problem, we have related source-separation concepts to the standard tools (both cost functions and adaptation methods) of single-channel adaptive filtering.

We presented the Bussgang family of cost functions with three general classes, blind LMS, Infomax, and direct Bussgang costs. Most of these costs can be implemented in either the time or frequency domain, or the batch or continuously adaptive modes. The blind LMS and Infomax classes use stochastic-gradient-type adaptation, whereas the direct Bussgang cost class uses a new adaptation method, which we have related to stochastic finite difference approximation.

The MBLMS Bussgang cost was analyzed by computing a measure of asymptotic tracking ability that yields useful guidelines for algorithm selection and how critical parametric data models are.

We introduced FIR matrix algebra as a tool for multichannel and multipath problems and show its application in whitening multipath matrix channels and in source-separation adaptive algorithms. Of the blind adaptive algorithms developed in this work, because of its robustness to eigenvalue spread, the MBRLS algorithm possesses the greatest speed of convergence while being the most computationally expensive. Serial update and finite differencing implementations are also powerful and practical algorithms. Block-oriented frequency-domain implementations, as used in the speech-separation experiment, have been found to be efficient and robust ways to perform separation on systems with hundreds and thousands of taps of multipath delay.

Much exciting research remains to be done in this area. Computational requirements can be extreme for these multichannel problems, and this area has come of age at a time when computer performance can sustain the effort.

REFERENCES

Amari, S., A. Cichocki, and H. H. Yang, 1996, "A new learning algorithm for blind signal separation," *Advances in Neural Information Processing Systems 8*, MIT Press.

Bell, A. J., and T. J. Sejnowski, 1995, "An information-maximization approach to blind separation and blind deconvolution," *Neural Computation*, vol. 7, pp. 1129–1159.

Bellini, S., 1994, "Bussgang techniques for blind deconvolution and equalization," in S. Haykin, ed., *Blind Deconvolution* (Englewood Cliffs, NJ: Prentice Hall) pp. 8–52.

Benveniste, A., and M. Goursat, Aug. 1984, "Blind equalizers," *IEEE Trans. on Communic.*, vol. COM-28, pp. 871–883.

Benveniste, A., M. Goursat, and G. Ruget, June 1980, "Robust identification of a non-minimum phase system: blind adjustment of a linear equalizer in data communications," *IEEE Trans. on Automatic Control*, vol. AC-25, pp. 385–399.

Benveniste, A., M. Metivier, and P. Priouret, 1987, *Adaptive Algorithms and Stochastic Approximations* (Berlin: Springer-Verlag).

Burel, G., 1995, "Blind separation of sources: a non linear neural algorithm," *Neural Networks*, vol. 27, no. 5, pp. 937–947.

Bussgang, J. J., 1952, "Cross correlation functions of amplitude distorted signals," *M.I.T. Research Lab of Electronics Tech.*, No. 216.

Cardoso, J.-F., and B. Laheld, Jan. 1996, "Equivariant adaptive source separation," *IEEE Trans. Signal Processing*.

Claerbout, J. F., 1985, *Fundamentals of Geophysical Data Processing* (Blackwell Scientific Publications).

Donoho, D., 1981, "On minimum entropy deconvolution," in *Applied Time Series Analysis II* (New York: Academic Press) pp. 565–608.

Ferrara, E. R., 1980, "Fast Implementation of LMS Adaptive Filters," *IEEE Trans. Acoust., Speech, Signal Processing*, vol. ASSP-28, no. 4, pp. 474–475.

Foschini, G. J., 1985, "Equalizing without altering or detecting data," *AT&T Tech. J.*, vol. 64, no. 8, pp. 1885–1910.

Godard, D., 1980, "Self-recovering equalization and carrier tracking in two-dimensional data communication systems," *IEEE Trans. on Communic.*, vol. 28, pp. 1867–1875.

Godfrey, R., 1978, "An information theory approach to deconvolution," *Stanford Exploration Project*, vol. 15, pp. 157–182.

Golub, G. H., and C. F. V. Loan, 1989, *Matrix Computations* (Baltimore, MD: Johns Hopkins University Press).

Gonzales, R. C., and P. A. Wintz, 1977, *Digital Image Processing* (Reading, MA: Addison-Wesley).

Gooch, R. P., 1983, Adaptive pole-zero filtering: The equation error approach, Ph.D. dissertation, Stanford.

Gray, W. C., 1979, Variable norm deconvolution, Ph.D. dissertation, Stanford.

Hogg, R. V., 1972, "More light on the kurtosis and related statistics," *J. Am. Stat. Soc.*, vol. 67, no. 338, pp. 422–425.

Im, G.-H., and J.-J. Werner, Dec. 1995, "Bandwidth-efficient digital transmission over unshielded twisted-pair wiring," *IEEE J Sel. Areas in Communic.*, vol. 13, pp. 1643–1655.

Joho, M., H. Mathis, and G. S. Moschytz, 1999, "An FFT-Based Algorithm for Multichannel Blind Deconvolution," in *IEEE International Symposium on Circuits and Systems*, (Orlando, FL), pp. III-203–206.

Jutten, C., and J. Herault, 1991, "Blind separation of sources, part 1: an adaptive algorithm based on neuromimetic architecture," *Signal Processing*, vol. 24, pp. 1–10.

Kailath, T., 1980, *Linear Systems*, Prentice Hall.

Kiefer, J., and J. Wolfowitz, 1952, "Stochastic estimation of the modulus of a regression function," *Ann. Math. Stat.*, vol. 23, pp. 462–466.

Lambert, R. H., 1996, Multichannel blind deconvolution: FIR matrix algebra and separation of multipath mixtures, Ph.D. dissertation, University of Southern California.

Lambert, R. H., and A. J. Bell, 1997, "Blind separation of multiple speakers in a multipath environment," in *ICASSP-Int. Conf. on Acoustics Speech and Signal Processing*, Munich, Germany, IEEE.

Lambert, R. H., and C. L. Nikias, Oct. 1995, "A sliding cost function algorithm for blind equalization," in *29th Asilomar Conf. on Signals Systems and Computers*, Pacific Grove, CA, IEEE.

Lin, S., and D. J. Costello, 1983, *Error Control Coding: Fundamentals and Applications* (Englewood Cliffs, NJ: Prentice-Hall).

Linsker, R., 1989, "An application of the principle of maximum information preservation," in *Advances in Neural Information Processing Systems 1*, Morgan-Kauffman.

Maciejowski, J. M., 1989, *Multivariable Feedback Design* (Reading, MA: Addison-Wesley).

Martin, R. D., 1979, "Robust methods for time series autoregressions," in *Robustness in Statistics* (New York: Academic Press).

Nikias, C. L., and A. P. Petropulu, 1993, *Higher Order Spectral Analysis: A Nonlinear Signal Processing Framework* (Englewood Cliffs, NJ: Prentice-Hall).

Robbins, H., and S. Monro, 1951, "A stochastic approximation method," *Ann. Math. Stat.*, vol. 22, pp. 400–407.

Robinson, E. A., 1967, *Multichannel Time Series Analysis with Digital Computer Programs*, Holden-Day.

Rosenbrock, H. H., 1970, *State-space and Multivariable Theory* (New York: John Wiley).

Sato, Y., 1975, "A method of self-recovering equalization for multilevel amplitude-modulation system," *IEEE Trans. Communic.*, vol. 23, pp. 679–682.

Satorius, E. H., and J. J. Mulligan, 1993, "An alternative methodology for blind equalization," *Digital Signal Processing*, vol. 3, pp. 199–209.

Slock, D. T. M., and L. C. Chisci, 1992, "Modular and numerically stable fast transversal filters for multichannel and multiexperiment RLS," *IEEE Trans. Signal Processing*, vol. 40, pp. 784–802.

Vardulakis, A. I. G., 1991, *Linear Multivariable Control: Algebraic Analysis and Synthesis Methods* (New York: John Wiley).

Vembu, S., S. Verdu, and R. Kennedy, 1994, "Convex cost functions in blind equalization," *IEEE Trans. Signal Processing*, vol. 42, pp. 1952–1960.

Walden, A. T., Dec. 1985, "Non-Gaussian reflectivity, entropy and deconvolution," *Geophysics*, vol. 12, pp. 2862–2888.

Widrow, B., J. R. Glover, J. M. McCool, et al., 1975, "Adaptive noise cancelling: principles and applications," *Proc. IEEE*, vol. 63, pp. 1692–1716.

Wiener, N., 1949, *Extrapolation, Interpolation and Smoothing of Stationary Time Series* (New York: Technology Press and Wiley).

Wiggins, R. A., 1978, "Minimum entropy deconvolution," *Geoexploration*, vol. 16, pp. 21–35.

INDEX